THE ECOLOGICAL BASIS
FOR RIVER MANAGEMENT

FRONTISPIECE   An example of a river managed without ecological input. Photograph taken in 1994 on the campus of an English University by Professor Geoff Petts

THE ECOLOGICAL BASIS
FOR RIVER MANAGEMENT

Edited by

DAVID M. HARPER
University of Leicester, UK

and

ALASTAIR J.D. FERGUSON
National Rivers Authority, UK

JOHN WILEY & SONS
Chichester ● New York ● Brisbane ● Toronto ● Singapore

Copyright © 1995 by John Wiley & Sons Ltd,
Baffins Lane, Chichester,
West Susex PO19 1UD, England

Telephone National (01243) 779777
International +44 1243 779777

Reprinted February 1996

Other Wiley Editorial Offices

John Wiley & Sons, Inc., 605 Third Avenue,
New York, NY 10158-0012, USA

Jacaranda Wiley Ltd, 33 Park Road, Milton,
Queensland 4064, Australia

John Wiley & Sons (Canada) Ltd, 22 Worcester Road,
Rexdale, Ontario M9W 1L1, Canada

John Wiley & Sons (SEA) Pte Ltd, 37 Jalan Pemimpin #05-04,
Block B, Union Industrial Building, Singapore 2057

Library of Congress Cataloging-in-Publication Data
The ecological basis for river management / edited by David Harper and
    Alastair Ferguson.
        p.   cm.
    Includes bibliographical references and index.
    ISBN 0-471-95151-X
    1. Stream ecology. 2. Ecosystem management. 3. Water quality
management. 4. Fish habitat improvement.   I. Harper, David M.
II. Ferguson, Alastair J.D.
QH541.5.S7E26   1994
627′.1—dc20   94-18514
                                                       CIP

British Library Cataloguing in Publication Data
A catalogue record for this book is available from the British Library

ISBN 0-471-95151-X

Typeset in 10/12pt Times by MHL Typesetting Ltd, Coventry
Printed and bound in Great Britain by Bookcraft Ltd, Bath, Avon

Contents

List of Contributors

Dr Patrick D. Armitage Institute of Freshwater Ecology, River Laboratory, East Stoke, Wareham, Dorset BH20 6BB, UK

Professor Angela H. Arthington Centre for Catchment and In-stream Research, Griffith University, Nathan, Queensland 4111, Australia

Dr Maurizio Battegazzore CNR-IRSA (Water Research Institute), Via della Mornera 25, 20047 Brugherio (Milan), Italy

Melanie Bickerton Freshwater Environments Group, Department of Geography, Loughborough University of Technology, Loughborough, Leicestershire LE11 3TU, UK

David R. Blühdorn Centre for Catchment and In-stream Research, Griffith University, Nathan, Queensland 4111, Australia

Dr Philip J. Boon Aquatic Environments Branch, Research and Advisory Services Directorate, Scottish Natural Heritage, 2 Anderson Place, Edinburgh EH6 5NP, UK

Dr John Bowers School of Business and Economic Studies, University of Leeds, Leeds, UK

Dr Florentino Braña Departamento de Biología de Organismos y Sistemas (Zoología), Universidad de Oviedo, 33071 Oviedo, Spain

Dr Andrew Brookes Technical Department, National Rivers Authority, Thames Region, Kings Meadow House, Kings Meadow Road, Reading, Berkshire RG1 8DQ, UK

Dr Paul Carling NERC Institute of Freshwater Ecology, Windermere Laboratory, Ambleside, Cumbria LA22 0LP, UK

Lois Child International Centre of Landscape Ecology, Loughborough University, Leicestershire LE11 3TU, UK

Dr Ian G. Cowx   University of Hull, International Fisheries Institute, Hull HU6 7RX, UK

Dr D.Trevor Crisp   NERC Institute of Freshwater Ecology, c/o Northumbrian Water plc., Lartington Treatment Works, Lartington, Barnard Castle, County Durham DL12 9DW, UK

Dr W.W. Crozier   Aquatic Sciences Research Division, Department of Agriculture for Northern Ireland, Newforge Lane, Belfast BT9 5PX, UK

Dr J.W. Eaton   Department of Environmental and Evolutionary Biology, University of Liverpool, PO Box 147, Liverpool L69 3BX, UK

Professor Peter J. Edwards   Geobotanisches Institute ETH, Zurichbergstrasse 38, CH-8044 Zurich, Switzerland

Professor Ron Edwards   National Rivers Authority, Welsh Region, St Mellons Business Park, Cardiff CF3 0LT, UK

Dr J. Malcolm Elliott   NERC Institute of Freshwater Ecology, Windermere Laboratory, Ambleside, Cumbria LA22 0LP, UK

Dr Alastair J.D. Ferguson   National Rivers Authority, Anglian Region, Kingfisher House, Orton Goldhay, Peterborough, UK

Dr Michael E. Foulkes   Faculty of Science, University of Plymouth, Plymouth, Devon PL4 8AA, UK

Professor Diego García de Jalón   Laboratorio de Hidrobiologia, E.T.S.I. de Montes, Universidad Politecnica de Madrid, Spain

Ramón Garrido   Departamento de Biología de Organismos y Sistemas (Zoología), Universidad de Oviedo, 33071 Oviedo, Spain

Dr Alun S. Gee   National Rivers Authority, St Mellons Business Park, Cardiff CF3 0LT, UK

Dr Anthony M. Gower   Faculty of Science, University of Plymouth, Plymouth, Devon PL4 8AA, UK

Dr Angela M. Gurnell   School of Geography, University of Birmingham, Edgbaston, Birmingham B15 2TT, UK

Dr Alessandra Guzzini CNR-IRSA (Water Research Institute), Via Reno 1, 00198 Rome, Italy

Dr David M. Harper Ecology Unit, Department of Zoology, University of Leicester, Leicestershire LE1 7RH, UK

Dr J.H. Harris Fisheries Research Institute, PO Box 21, Cronulla, NSW 2230, Australia

A. Jan Hendriks Department of Ecotoxicology, Institute For Inland Water Management and Waste Water Treatment (RIZA), Maerlant 16, 8200 AA, Lelystad, The Netherlands

Dr Phil Hickley National Rivers Authority, Severn-Trent Region, Sapphire East, Solihull B91 1QT, UK

Professor Alan G. Hildrew School of Biological Sciences, Queen Mary and Westfield College, University of London, Mile End Road, London E1 4NS, UK

Dr Richard Howell National Rivers Authority, Welsh Region, St Mellons Business Park, Cardiff CF3 OLT, UK

Dr Carles Ibañez Departament d'Ecologia, Universitat de Barcelona, Diagonal 645, 08020 Barcelona, Spain

Frank H. Jones National Rivers Authority, St Mellons Business Park, Cardiff CF3 0LT, UK

G.J.A. Kennedy River Bush Salmon Station, 21 Church Street, Bushmills, County Antrim, Northern Ireland BT57 8QJ, UK

Dr Martin Kent Faculty of Science, University of Plymouth, Plymouth, Devon PL4 8AA, UK

Professor Stanimir Kostadinov Faculty of Forestry, University of Belgrade, 11030 Belgrade, Kneza Viseslava 1, Yugoslavia

ir. J.P.A. Luiten Institute for Inland Water Management and Waste Water Treatment (RIZA), Maerlant 16, 8200 AA, Lelystad, The Netherlands

Dr Ian Maddock Department of Geography, Worcester College of Higher Education, Henwick Grove, Worcester, Worcestershire, UK

Dr Peter S. Maitland Fish Conservation Centre, Easter Cringate, Stirling FK7 9QX, UK

Dr Richard H.K. Mann NERC Institute of Freshwater Ecology, Eastern Rivers Laboratory, c/o Monks Wood, Abbots Ripton, Huntingdon, Cambs PE17 2LS, UK

Professor Roberto Marchetti CNR-IRSA (Water Research Institute), Via della Mornera 25, 20047 Brugherio (Milan), Italy

Chris Marsh National Rivers Authority, Severn-Trent Region, Sapphire East, Solihull B91 1QT, UK

Dr Christopher F. Mason Department of Biology, University of Essex, Colchester, Essex CO4 3SQ, UK

Dr Kevin J. Murphy Centre for Research in Environmental Science and Technology, Botany Building, University of Glasgow, Glasgow G12 8QQ, UK

Dr Graham Myers Faculty of Science, University of Plymouth, Plymouth, Devon PL4 8AA, UK

Alison M. Newall Department of Geography, University of Reading, Reading RG6 2AB, UK

Dr A.G. Nicieza Departamento de Biología de Organismos y Sistemas (Zoología), Universidad de Oviedo, 33071 Oviedo, Spain

Dr Rick North National Rivers Authority, Severn-Trent Region, Upper Severn Area, Shelton, Shrewsbury SY3 8BB, UK

Jay O'Keeffe Institute For Water Research, Rhodes University, PO Box 94, Grahamstown, 6140 South Africa

Professor Timothy O'Riordan School of Environmental Sciences, University of East Anglia, Norwich NR4 7TJ, UK

Dr Steve J. Ormerod Catchment Research Group, Pure and Applied Biology, UWCC, PO Box 915, Cardiff CF1 3TL, UK

Dr Romano Pagnotta CNR-IRSA (Water Research Institute), Via Reno 1, 00198 Rome, Italy

Professor Geoff Petts School of Geography, University of Birmingham, Edgbaston, Birmingham B15 2TT, UK

Judith Pillinger Biology Department, The Open University, Milton Keynes MK7 6AA, UK

Dr Paulo Pinto Dep. Biologia, Universidade De Évora, Largo Dos Colegiais, 7001 Évora Codex, Portugal

Professor Narcís Prat Departament d'Ecologia, Universitat de Barcelona, Diagonal 645, 08020 Barcelona, Spain

J. Rabaça Dep. Biologia, Universidade De Évora, Largo Dos Colegiais 7001, Évora Codex, Portugal

A. Ramos Direcçáo Regional de Ambiente e Recursos Naturais, 7000 Évora, Portugal

Claire Redmond National Rivers Authority, Anglian Region, Kingfisher House, Orton Goldhay, Peterborough PE2 0ZR, UK

M. Revez Dep. Planeamento Biofisico, Universidade de Évora, Largo dos Colegiais, 7001 Évora, Portugal

Felipe G. Reyes-Gavilán Departamento de Biología de Organismos y Sistemas (Zoología), Universidad de Oviedo, 33071 Oviedo, Spain

Dr Colin S. Reynolds NERC Institute of Freshwater Ecology, Windermere Laboratory, Ambleside, Cumbria LA22 0LP, UK

Irene Ridge Biology Department, The Open University, Milton Keynes MK7 6AA, UK

Dr Alberto Rodrigues-Capitulo Instituto de Limnología, CC 712 (1900), La Plata, Argentina.

Maria da Graça Saraiva Dep. Arquitectura Paisagista, Instituto Superior De Agronomia, Tapada Da Ajuda, 1399 Lisboa Codex, Portugal

Dr Anne Schulte-Wülwer-Leidig International Commission for the Protection of the Rhine against Pollution, Postfach 309, Hohenzollernstrasse 18, D-56068 Koblenz, Germany

David Simmons Geographical and Multimedia Solutions, Newlands, Beech Corner, Durley, Southampton S03 2AR, UK

Dr Colin Smith Ecology Unit, Department of Zoology, University of Leicester, Leicester, Leicestershire LE1 7RH, UK

Dr Stephen Swales Fisheries Research Institute, PO Box 21, Cronulla, NSW 2230, Australia

M.M. Toledo Departamento de Biología de Organismos y Sistemas (Zoología), Universidad de Oviedo, 33071 Oviedo, Spain

Dr Louise de Waal Division of Environmental Science, School of Applied Sciences, University of Wolverhampton, Wolverhampton WV1 1SB, UK

Dr Max Wade International Centre of Landscape Ecology, Loughborough University, Loughborough, Leicestershire LE11 3TU, UK

John Walters Biology Department, The Open University, Milton Keynes MK7 6AA, UK

Dr Nigel J. Willby Department of Environmental and Evolutionary Biology, University of Liverpool, PO Box 147, Liverpool L69 3BX, UK

Tom Youdan National Rivers Authority, Anglian Region, Clover Hill, Park View, Kettering, Northamptonshire NN16 9RJ, UK

Preface

This book is the result of an international conference of the same name held at Leicester University in March 1993 to mark the first four years of the National Rivers Authority (NRA) in England and Wales.

The initial objective of the conference was to promote discussion of recent results of research into the ecological basis of river management, between the scientific community and the management community. To this end the core of the conference was made up of invited speakers, some coming from research and some from management fields, addressing topics which have been mainstream issues within the NRA's short lifespan. This was enhanced by representatives of international river management projects (particularly the largest, the Rhine) adding dimensions not present in the British Isles. The editors were delighted that all but two of the original conference presentations were submitted for publication.

The conference was also open to poster papers from the international community, provided they addressed the theme of ecology as the basis for management, and some 45 excellent posters were critically discussed in evening sessions during the conference. A selection of those poster papers submitted for publication has been included in the book, where they provide different perspectives on the major problems, particularly from parts of the world other than England and Wales.

The book follows the same thematic divisions as the conference, addressing the six main functional aspects of river management. The first section, chapters 1–10, deals with issues of water quantity (or water resources). Inevitably, the basis of the management of water quantity is not so much ecological but environmental — a combination of hydrology, geomorphology and ecology, as reflected in the title of chapter 1. Nevertheless, the ultimate goal of management of water quantity is almost always ecological quality, as chapter 3 indicates.

The second section, (chapters 11–17), on water quality, is perhaps the more "traditional" area for the involvement of ecological principles, but as chapter 11 suggests, this is by no means clearly established (the question is addressed more widely in the context of catchment planning in chapter 37). Chapter 12 demonstrates elegantly how the management of one aspect of water quality — acidification — has different ecological ramifications at each level of its hierarchy and is far more complex than appears at first sight. Conventional ecological investigations of water quality also require new intellectual approaches (chapter 13), increasingly sophisticated analytical tools (chapters 14, 15 and 16), but simple ecological tools are still valuable (chapter 17).

Management of the natural river environment (Section 3; chapters 18–25) might be considered the most obvious area for the application of ecological knowledge, but here there are still many areas of uncertainty and difficulty, not least the problem of managing landscapes with information based upon species, a common theme of these chapters.

This comment cannot be made for Section 4, (chapters 26−31), the management of fish stocks, where a knowledge of population-level ecology has been essential in successful management for centuries. Our knowledge base is now good (albeit with gaps, chapters 26−28) but there are still problems in its application (chapters 29−31).

Fish stocks present quite different problems of management than fisheries, which are considered in Section 5 (chapters 32−35) with other human reactional activities. It is perhaps unusual to consider the ecological basis for managing people, but it is clear from these chapters that the goals of management are all ecological and so the basis for management decisions must be also.

Five chapters in the final section (chapters 36−40) address wider issues in catchment management. There is no doubt that all river management is moving towards the catchment scale of planning, with "sustainable management" as the current buzz-word. The possibilities, as well as the difficulties, are well illustrated in the comparison between rivers in Europe and South Africa (chapters 38−39), and the conflicts of integrating ecological with economic goals (chapter 40).

The editors would like to express their thanks to Professor Ron Edwards, Professor Geoff Petts and Dr Roger Sweeting, who with them formed the steering group for the original conference; to Kate Penny and staff of the Professional Development Unit, Leicester, who handled all the organization, and to Steve Ison and postgraduates of the Leicester Zoology Department who provided all the "volunteer" labour during the conference.

We are very grateful to those people who responded to our invitations to guide and nuture the conference deliberations: Lord Cranbrook, Chairman of English Nature, gave a challenging opening address; Lord Crickhowell, Chairman of the National Rivers Authority, a stimulating after-dinner address and Professor Calow a thought-provoking summary. Professor Brian Wilkinson, Dr Jan Pentreath, Lord Cranbrook, Mr David Le Cren, Lord Moran and Professor Brian Moss provided firm but sensitive chairmanship of the six sessions.

We are grateful to those authors of the book who, together with Dr Geoff Phillips and Dr Paul Hart, acted as peer-reviewers for other papers.

We hope that this volume will provide a powerful stimulus to the integration of ecology into river and catchment management, and reserve our final thanks to the authors of the forty chapters who have made this possible.

David M. Harper and Alastair J.D. Ferguson

1

Linking Hydrology and Ecology: The Scientific Basis for River Management

GEOFF PETTS

University of Birmingham, Birmingham, UK

IAN MADDOCK

Worcester College of Higher Education, Worcester, UK

MELANIE BICKERTON

Loughborough University of Technology, Loughborough, UK

and ALASTAIR J.D. FERGUSON

National Rivers Authority, Peterborough, UK

INTRODUCTION

Growing demands for water supply, especially in dry areas for domestic and agricultural consumption, and for hydroelectric power, mean that schemes for river regulation, water abstraction and inter-basin transfer will continue to be advanced (Petts, 1994). Simultaneously, the strengthening of demands for environmental protection will require that improved approaches are developed for assessing the impacts of water resource schemes and especially for allocating water to in-river needs. The setting of priorities requires a decision on Boon's (1992) scale of conservation options from dereliction, through mitigation, to preservation. All river regulation schemes impose an unnatural flow regime and in most cases the concern is to maintain a minimum flow for mitigation or, in *post hoc* assessments, restoration purposes. Yet the needs of birds, fish and invertebrates will rarely be the same; in most cases they are very different from those of canoeists and anglers, and may conflict with the human functions imposed on rivers: effluent dilution and transport, navigation, hydroelectric power production, and abstractions for supply.

The Ecological Basis for River Management. Edited by D.M. Harper and A.J.D. Ferguson. © 1995 John Wiley & Sons Ltd

This chapter examines the link between hydrology and ecology, and focuses on the association between low flows, habitat degradation and biotic response. The application of methods to determine minimum flows for the protection of in-river needs is illustrated for the River Glen, in eastern England. A more expansive treatment of the link between hydrology and ecology, including the important ecological roles of flood flows and intermediate flows, as well as low flows, for maintaining the ecological structure and function of floodplain, riparian, and channel systems is presented by Petts and Maddock (1994). This chapter is also intended to provide a framework for the following detailed chapters on hydraulics (Chapter 2), channel morphology (Chapter 3), invertebrate responses to low flows (Chapter 5), and impacts of flow regulation (Chapter 7).

THE CONTEXTS FOR LINKING HYDROLOGY AND ECOLOGY

The history of scientific development and river management provide two separate, although increasingly interrelated, traditions for the examination of links between flow and biota. Major advances are recent, not least in the context of the history of river regulation (Petts, 1984). Throughout Europe, major flow regulation schemes date from the 19th century, although the peak of dam-building activity was the 1960s. However, in 1970 river management remained an "art" rather than a science, as illustrated by Fraser's statement (Fraser, 1972, p. 277) that: "Discharge recommendations are often based more on a biologist's or engineer's guess than on a quantified evaluation of the relationship between discharge and the ecology of the stream, its aesthetics and other in-place uses."

The management context

Until recently, scientific development remained divorced from practical river management and decision making. In his closing summary to Oglesby, Carlson and McCann's book on *River Ecology and Man*, published in 1972, J. W. Leonard noted (p. 459): "the specialist can no longer responsibly fall back on the Philistine attitude that he will do his thing and if they (decision-makers and other specialists) are not intelligent enough to make use of his golden discoveries, well and good! but if not it's their hard luck."

There can be no doubt that the failure of scientists to communicate their findings effectively across disciplines as well as to managers and decision-makers has contributed to progressive environmental degradation. However, the traditional paradigm "man versus nature", manifest most strongly in the progressive and continuing human ambition to control rivers, has left a dramatic legacy of environmental change. Throughout history societies have sought to regulate rivers. The popular pioneering vision was of Man's struggle to tame natural rivers, large labour forces working long hours in harsh conditions, and entrepreneurs motivated by the desire for economic growth. This story is as much a social history as a technological one (Cosgrove and Petts, 1990); its penultimate chapter is the era of the mega-project, which opened in the second half of the 19th century and reached its zenith in the mid-1970s. The Modernist vision was the control of nature, the extraction of wealth from rural areas and its dispersal through cities.

Today, the traditional paradigm is being replaced by "people within environment". The change from reductionist and isolationist scientific research to truly interdisciplinary and collaborative studies is being paralleled by a new acceptance that for economic development to be sustainable, water and land management must be environmentally sound. The current vision has concern for the quality of life, focuses on investment in the environment, and is expressed by symbols of social advancement: restored landscapes and protected wild areas.

Ecological concerns of river management in the UK

The Water Act 1973 created ten regional water authorities in England and Wales from the 26 river authorities that had been set up by the Water Resources Act 1963. Further reorganization in the Water Act 1989 created the National Rivers Authority (NRA). One of the environmentalists' long-standing criticisms of the 1973 Water Act was that regulatory powers were given to the water authorities who also controlled the sewage treatment works and the abstractions that are responsible for many environmental problems. This criticism was overcome by the 1991 Act: the NRA retained the water authorities' regulatory and river basin functions within the public sector whilst privatizing the utility functions. The NRA has a statutory responsibility to further conservation, improve fisheries and promote recreation through regulatory, operational and advisory activities. In particular, the Authority is required to: "so exercise its functions as to further the conservation and enhancement of natural beauty and the conservation of flora, fauna and geological or physiological features of special interest" (Section 48 of the Wildlife and Countryside Act 1981 and Section 16.1 of the Water Resources Act 1991).

The concept of a "minimum acceptable flow" in river management in the UK has long been established as important for fisheries and for the dilution of sewage and industrial effluents (Sheail, 1984, 1987). The first occasion when considerable attention was paid to environmental requirements in assessing compensation needs was in 1888 when the Halifax Corporation Waterworks Act set the precedent for releasing compensation water as a constant discharge to maintain the quality of a local beauty spot. Specific attention to the needs of fisheries was first raised during the promotion of the Bill by the Corporation of Birmingham (1892) to construct reservoirs in the Elan and Claerwen valleys in Wales. In 1919 an Act to impound Haweswater in the Lake District for the first time stipulated the total quantity of water to be discharged each year, as well as making provisions for the daily compensation flow.

However, the infrastructure necessary for the detailed consideration of flows to maintain in-river needs was not in place until the Water Resources Act 1963. The Act made rights to impound and abstract water obtainable by licence from the river authorities, and Section 48 of the Act put the power to determine compensation flow requirements firmly in the hands of the River Authorities. That power remains with the National Rivers Authority.

The hydrological context

River flows vary with the pattern of precipitation and evapotranspiration. At the seasonal scale, these relationships are analysed as seasonal flow regimes (Vivian, 1989; Gustard, 1992). Marked runoff variations also occur from year to year, and decade to decade, with the natural variations in weather. Thus, Probst (1989) has analysed hydroclimatic fluctuations of some European rivers from 1800 to 1980 and identified four major dry periods (1863 − 75, 1883 − 1910; 1942 − 1956 and the early-mid 1970s) separated by humid phases.

River regulation and inter-basin transfer seek to reduce the natural within-year and between-year flow variability, reducing floods, increasing low flows and in some cases reducing or increasing the total discharge of the river (Petts, 1984). However, major changes of the pattern of river flows can also be caused by other human activities throughout a river catchment, including forestry and agricultural practices, and urban expansion (e.g. Newson, 1992). Hydrological changes may be indicated by reference to a range of statistics, such as mean flow, flood frequency, flow duration percentiles, or the dry-weather flow defined as the annual average, seven-day low flow (Hindley, 1973). However, the identification of anthropogenic change is often problematic because of the natural variability of river flows.

Biota will be more or less adapted to the annual flow regime; to the flow variability and to the magnitude, timing and predictability of high- and low-flow periods. However, the relationship between flow and the distribution of biota is often complicated by other factors, the importance of which differ according to spatial scale. At the regional and catchment scales, hydrological parameters are often secondary to the influence of water quality, especially temperature (a function mainly of climatic setting and altitude) and solute load (a function of the runoff sources within the catchment, i.e. rock types, land uses, urban areas, etc.). Reviews of the water-quality characteristics of rivers are presented by Walling and Webb (1992) and Webb and Walling (1992).

At the scale of a river sector, the influence of flow on the distribution of biota is often affected by changing hydraulic conditions rather than by any hydrological parameter *per se*. Under low to moderate flows, the instream hydraulics expressed by flow velocity and water depth, or more complex variables such as shear stress, Froude number and boundary Reynold's number, reflect channel morphology, especially bed form (e.g. Carling, 1992). Channel morphology is determined by the range of flows experienced by the river but the dominant flow for channel form is generally indexed by the bankfull discharge or the annual high flow having a frequency of 1.5 years (e.g. Church, 1992). However, the shape of the channel is also influenced by the size and load of sediment transported. Changes of channel form will occur consequent upon a change of the bankfull discharge and/or a change of sediment load, or upon any channelization works. These changes will alter the hydraulic conditions within the channel under low flows, impacting the biotic communities within the river.

The historical background

Modern approaches to the scientific study of rivers are founded in two important works that were based on research undertaken during the 1950s and 1960s: *Fluvial Processes in Geomorphology* published in 1964 by L. B. Leopold, M. G. Wolman and J. P. Miller, and *The Ecology of Running Waters* published by H. B. N. Hynes in 1970. Because of the lag between anthropogenic change of river flow regimes and the development of an appropriate science, scientific advance has focused on post-project impact assessment and, most recently, on restoration and rehabilitation measures.

At a Symposium held in 1970, on Conservation and Productivity of Natural Waters (Edwards and Garrod, 1972), organized by the British Ecological Society and the Zoological Society of London, Morgan (p. 143) noted: "In the long term it would be desirable to gain knowledge of the wider implications of the amounts of water let down rivers on the ecosystem as a whole." Importantly, Morgan urged the focusing of efforts on the intricate relationships between animals and plants and their environment if better management is to be achieved. A second important catalyst for linking flows and ecology was also the first major work focusing on human impacts on rivers; *River Ecology and Man* (Oglesby, Carlson and

McCann, 1972) evolved in response to a new, and rapidly growing, insistence that river quality — in terms of nature conservation, aesthetics and recreation — be weighed along with consumptive and ecologically destructive uses.

Significant advances linking hydrology and ecology were slow to be made for two major reasons. First, throughout the 20th century the intensifying scientific tradition was for reductionist and isolationist approaches, with the reinforcement of boundaries around scientific disciplines. Physical sciences and biological sciences were effectively divorced. Secondly, whilst geomorphologists and hydrologists were establishing the river and its drainage basin as their fundamental unit of study, ecologists focused more on lakes and other standing-water bodies. This second reason is explained by Ryder and Pesendorfer (1989) by two facts: rivers represent only 0.004% of the world's freshwaters and in the northern hemisphere, where most research has taken place, there is an inordinately high ratio of lakes to large rivers. Moreover, the clearly circumscribed form of ponds and lakes make them more amenable for research than long, morphologically complex and highly dynamic rivers.

Nevertheless, interaction between disciplines was advanced during the 1970s and two important concepts for modern river ecology were introduced from geomorphology: space and time scales, and energy equilibrium theory. These concepts were fundamental to the establishment of links between hydrology and ecology. Energy equilibrium theory forms the basis of the River Continuum Concept (Vannote et al., 1980) which states that structural and functional characteristics of stream communities are adapted to conform to the most probable state or mean state of the physical system. Thus, producer and consumer communities establish themselves in harmony with the dynamic physical conditions of a given river reach, whilst downstream communities are fashioned to capitalize on the inefficiencies of upstream processing. The important role of space and time scales for studies of river systems, founded in the seminal work of S. A. Schumm and R. W. Lichty (1958), became firmly accepted in the biological sciences. Thus, Minshall (1988) wrote: "The explicit recognition and utilisation of appropriate spatial and temporal scales is essential to the development of an accurate and robust stream theory having a truly global perspective" (p.279).

In the UK, detailed studies were undertaken during the 1970s of the River Tees below Cow Green reservoir (Armitage, 1976, 1978) and the geomorphological basis for studies of flow regulation was established by Petts (1979). Subsequently, influential reviews of the effects of flow regulation on instream biota were published by Brooker (1981), Milner et al. (1981) and Petts (1984). Nevertheless, in a study of compensation flows in the UK, Gustard et al. (1987) concluded that the primary research need remained the development of quantitative relationships between freshwater biota and the physical and chemical variables at a scale appropriate to the river reach. A review of the environmental effects of flow regulation in the UK was edited by Petts and Wood (1988).

Linking science and management

Scientifically based approaches to the practical problem of allocating water for in-river needs began to be developed from the mid-1960s. These developments were initiated in north-west North America where the economic value of salmonid fisheries and conflict with hydro-power development were great. Here, people concerned with protecting rivers had recognized that effects of stream regulation on aquatic habitat needed to be demonstrated not only by scientific research but also through rigorous legal, administrative and technical means (Nestler, Milhous and Layzer, 1989). A review of the status of instream flow methods (Stalnaker and Arnette, 1976), the formation of the Co-operative Instream Flow Service Group of the US Fish and Wildlife Service in 1976, and a major conference on instream flow needs promoted by the American Fisheries Society (Orsborn and Allman, 1976) established the foundation for research on practical approaches to assessing in-river needs.

A simulation model (PHABSIM) was released in 1978. The model was developed from the conceptual framework provided by the Instream Flow Incremental Methodology (IFIM): an approach to problem solving that refers to an institutional policy of slightly modifying procedures or positions from those previously established (Bovee, 1982). At the core of the methodology is a marriage of concepts taken from open-channel hydraulics, hydrology, sedimentology, aquatic ecology and environmental engineering. The computer programs embody an approach to the organization, implementation, execution and documentation of an instream-flow study.

Through refinement and expansion into a library of models that can perform hydraulic simulation and habitat analyses, IFIM became employed widely throughout USA in licensing and flow adjudication decisions. In other countries with different legal systems, acceptance of the approach has been rather cautious, and in some, such as southern Africa (Gore, Layzer and Russel, 1992) and the UK (Bullock and Gustard, 1992), PHABSIM may be used as one of a suit of approaches, each being appropriate to the severity, or perceived cost, of a flow allocation problem.

ASSESSING FLOW NEEDS: THE RIVER GLEN

To illustrate the application of hydrological and hydraulic approaches for assessing in-river needs, data are presented for the River Glen in eastern England (Figure 1.1) which drains low-lying hills formed of permeable Lincolnshire Limestone. Mean annual rainfall (1961–85) is 623 mm and runoff to the main gauging station at Kates Bridge is 107 mm ann^{-1}. Summer flows in the river are maintained by spring flows from the limestone aquifer and reports suggested that by 1976 abstractions had already lowered groundwater levels and spring flows below their natural state (Petts, 1992). These reports also suggested that the declining flows had deleterious effects on the biology and fisheries of the river, although the link between flows and biota is complicated by natural flow variations; a series of dry years (1971–76) following a series of wet years (1965–69).

In 1990 the flow in the Shillingthorpe sector of the middle river (Figure 1.1) declined dramatically after 9 July and the river became dry on 6 August. The following example uses a range of methods to assess the flow required to meet in-river needs within the Shillingthorpe sector, and validates the assessment against the biological recovery in the river after flow augmentation by an inter-basin transfer.

Approaches

Models describing relationships between flow and biota are, as yet, imprecise. Four approaches have been employed to assess the impact of low flows within the Shillingthorpe sector or the River Glen: a hydrological approach; a physical habitat approach; an empirical model based upon between-site distributions of biological quality in relation to habitat attributes, applied to evaluating hydrological change by space–time substitution; and a

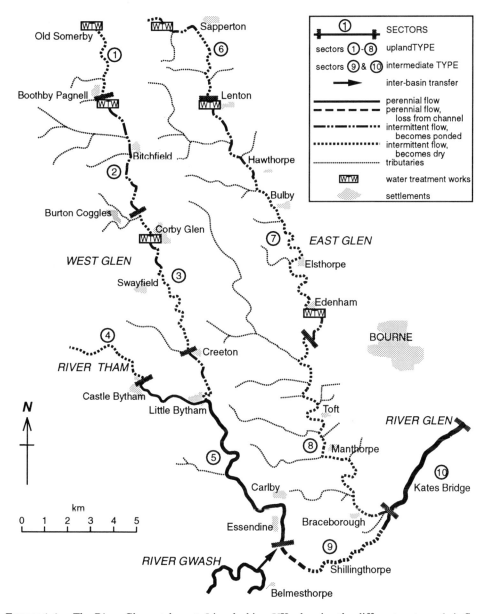

FIGURE 1.1. The River Glen catchment, Lincolnshire, UK, showing the different sectors, their flow characteristics, and the location of the water transfer from the River Gwash

simulation model (PHABSIM) derived from habitat – flow relationships within a representative reach and a general description of habitat suitability for target species.

Definition of river sectors

Based on a detailed field survey of physical and hydraulic characteristics, and analyses of invertebrate records and fish data held by the National Rivers Authority, the river was divided into ten sectors (Figure 1.1). Summary data for sectors 5 and 9, the latter being the Shillingthorpe sector, are presented in Table 1.1 to illustrate the type of information used in sector delineation. This preliminary step in the assessment of in-river needs is important for two reasons. First, because of the influence of water quality and physical habitat on relationships between flow and biota, sectors with different characteristics require different flows to maintain habitat for target species. Secondly, any need for physical habitat restoration, in channelized, dredged or "maintained" sectors, and the potential benefits of restoration works, can also be assessed. In some cases, physical habitat restoration, with regard not only to channel morphology but also riparian conditions (e.g. to provide shade or buffer zones to reduce loads of fine sediments), may greatly increase the benefits of flow augmentation — it may also reduce the flow allocation required!

Hydrological analysis

Flow indices, such as the 95th percentile flow (Q_{95}) or a percentage of Average Daily Flow (ADF), have often been used to asses the flow required to maintain habitat for biota. This approach assumes that natural biota are adapted to the normal range of extreme events. On the River Glen, during the period 1981−87, a period of about average annual rainfall, Q_{95} was 0.076 m^3 s^{-1}. The 10% ADF, 20% ADF and 30% ADF values were also defined, these percentages having been identified as useful guides for indexing respectively: the threshold flow below which major degradation occurs, the flow to protect aquatic habitat, and the flow associated with near-optimum habitat in small streams (Orth and Leonard, 1990). All values are given in Table 1.4.

TABLE 1.1.   Characteristics of Sectors 5 and 9 on the River Glen, based on a 1989 survey

Characteristic	Sector number	
	5	9
Flow	Perennial	Intermittent
Channel scale		
Strahler order	4	4
Shreve magnitude	43	49
Slope	0.0016	0.0015
Average width (m)	5.6	6.5
Channel structure		
Riffle (%)	37	25
Pools (%)	7	6
Shallow run (%)	18	8
Stagnant run (%)	12	15
Deep run (%)	26	42
Dry run (%)	0	4
Fish biomass (g m^{-2})	12.5	8.0
Dominant fish	Dace, brown trout	Dace, chub
Invertebrate quality	A−A+	A−A+

Strahler order and Shreve magnitude are indices of drainage network size and structure

A flow recession approach was also used. An Environmental Assessment of the proposed water resource development of the Glen (Petts, 1990) defined the flow recession for the Shillingthorpe gauging station using data for the 1975 drought year, validated against data series for two other years chosen at random (1979 and 1988). A second-order polynomial regression relationship fitted each data series with adjusted r^2 values of greater than 0.94 (p = 0.0001). A dry-summer flow recession for 1975 was established as a typical extreme-flow year with declining flows of 0.175 m^3 s^{-1}, 0.12 m^3 s^{-1} and 0.08 m^3 s^{-1} in July, August and September respectively: 0.08 m^3 s^{-1} equates to the 93rd percentile flow for the 1981–87 record.

Rapid habitat assessment
This approach establishes relationships between physical habitat and flow. Data for a 90 m long representative reach within sector 9 was used to determine the most effective flow for sustaining the optimum area of wetted bed (Figure 1.2). Wetted bed area is seen to decrease rapidly below a flow of 0.042 m^3 s^{-1} and this provides a useful guide for maintaining instream habitat.

An empirical model
This approach requires relationships between an index of biological quality and environmental variables to be established. In England and Wales, invertebrate assemblages are routinely monitored across a wide network of sites for biological monitoring of water quality. The Biological Monitoring Working Party (BMWP) system is widely used as an indicator of water quality (for a recent review, see Metcalfe-Smith, 1994), but the index is

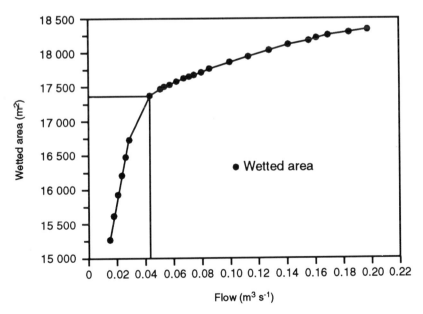

FIGURE 1.2. The relationship for wetted channel-bed area and flow established for the Shillingthorpe sector of the River Glen highlighting the inflexion point which may be used as an indicator for the assessment of minimum flow requirements

also significantly influenced by physical habitat. For 28 sites throughout the Anglian Region, all of a similar size to the River Glen at Shillinghorpe and all routine biological monitoring sites adjacent to flow gauging stations having data available for the period 1980–90, Maddock (1992) established relationships between BMWP (having a range from 25 to 180) and 19 environmental variables. At each site, measurements or observations of a wide range of variables describing flow, channel form, bed sediments, cover, etc., were made at 20 transects spaced approximately at every seven channel widths or at every riffle, whichever was closer. Catchment characteristics, including scale variables such as drainage area, stream order, etc., were also included in the database.

The derived model isolated four primary variables (Table 1.2). Whilst "chemical quality" was, not surprisingly, the most influential variable, flow and physical-habitat descriptors were also significant. The best flow index was found to be the quotient of the 95th percentile flow and channel width. Catchment attributes yielded insignificant correlations at this regional scale of analysis.

The model, given in Table 1.2, was validated against data for sites from six different sectors of the River Glen surveyed in late August 1991, yielding adjusted r^2 values of 0.731 for long-term average BMWP data and 0.959 for BMWP values derived from the autumn 1991 field survey. The latter *emphasizes* the importance of the flow type variable which describes the hydraulic habitat conditions at the time of field survey. This was important here because field data were obtained during a period of extreme low flows.

PHABSIM

A representative reach was located within the Shillingthorpe sector, comprising a riffle–pool–riffle–pool–riffle sequence with a maximum width of 6.6 m. Five transects were established within the reach and the database comprised 156 point measurements of water surface elevation and velocity for each survey at three different calibration flows, i.e. low (0.0267 m^3 s^{-1}), medium (0.183 m^3 s^{-1}, about Q_{65}) and high (2.474 m^3 s^{-1}, about Q_4).

TABLE 1.2.   Relationship between the Biological Monitoring Working Party score (BMWP) and habitat attributes for streams in the Anglian Region, UK. The adjusted r^2 is 0.88

BMWP = ((Chemical score $\times$ 19.79) $-$ 39.39) + ((Q_{95}/w) $\times$ 466.70) + (Cover $\times$ 0.85) + (Flow type $\times$ 52.66)

where:

- the chemical score is derived from the National Water Council classification system based upon dissolved oxygen levels, biological oxygen demand and ammonia concentrations
- Q_{95}/w is the 95th percentile flow duration statistic divided by average channel width
- cover is a measure (%) of the wetted cross-sectional area with cover objects (instream and overhanging vegetation, cobbles and boulders)
- flow type is a composite variable incorporating the average wet width, as a percentage of channel width, and the proportion of the reach with visible flow

---

The first two variables describe the average conditions at the site over the period 1980–90. The cover variable describes macroscale differences between sites, differentiating especially between reaches with natural, vegetated margins, and heavily-maintained often-channelized reaches. Flow type provides an index of bed configuration as well as a flow index to assess habitat quality at a particular point in time.

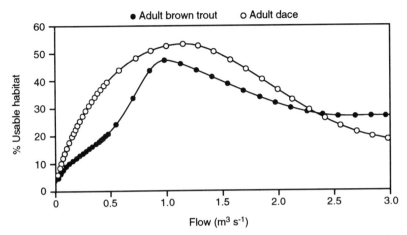

FIGURE 1.3. Relationships between usable habitat and flow for adult brown trout and dace for the Shillingthorpe sector of the River Glen established using PHABSIM

Microhabitat suitability curves utilized in this study were originally developed by Armitage and Ladle (1991) based on experience and local knowledge of UK conditions. Curves have been used for two life stages (juvenile and adult) for the dominant species, dace and chub, and a target species for the sector, brown trout. They were expressed as suitability functions of depth, velocity and substrate. In all cases, usable habitat was found to increase progressively throughout the range of low flows (e.g. Figure 1.3) and to reach an optimum at about 1 m^3 s^{-1} on average.

For all species and life stages, the optimum flow to provide maximum (potential) usable habitat is high and this was shown to reflect the low physical-habitat diversity, especially as indexed by depth variability, within the site (Maddock, 1992). Optimum flows range from 0.368 to 1 m^3 s^{-1}; for the 1981–87 record these flows encompass the range from about Q_{35} to Q_{10}! With the exception of chub for which little habitat is available (and field observations confirm that in summer habitat is confined to the upper parts of two large pools) the Average Daily Flow, about 0.5 m^3 s^{-1}, is estimated to provide 47% and 75% of the maximum usable habitat for adult trout and dace respectively.

In order to consider a minimum flow for the sector, two arbitrary criteria were applied (Table 1.3). First, 50% of the maximum available habitat was used. With the exception of juvenile brown trout, the necessary flows remain high. Secondly, usable habitat was estimated for a discharge of 0.076 m^3 s^{-1}, the Q_{95}.

Application of results

The results of the four approaches outlined above are summarized in Table 1.4. To maintain a minimum acceptable character in the river, manifest by a good invertebrate community but with limited fish habitat, a minimum flow of 0.04 m^3 s^{-1} would be acceptable. The maintenance of good fish habitat, however, requires a much higher minimum flow of at least 0.1 m^3 s^{-1}, providing about 10% of potential usable habitat for adult dace and trout. For the 1981–87 period, 0.04 m^3 s^{-1} and 0.1 m^3 s^{-1} equate to 99th percentile and 87th percentile flows respectively. Best fish habitat is provided when flows are above 0.3 m^3 s^{-1} throughout

TABLE 1.3. Flow-habitat information derived from PHABSIM for the Section 9

Fish species and stage	Optimum discharge (m³ s⁻¹)	Potential usable area (PUA) (% of reach)	Discharge at 0.5 PUA (m³ s⁻¹)	Usable area at 0.076 m³ s⁻¹ (Q_{95}) (% of max)
Brown trout Adult	0.99	48	0.55	19
Juvenile	0.37	20	0.06	55
Dace Adult	1.13	53	0.25	22
Juvenile	0.85	29	0.35	11
Chub Adult and juvenile	1.42	34	0.90	1

the year, with flows of at least $0.5 \text{ m}^3 \text{ s}^{-1}$ in April and May, a situation that occurs naturally only once in every 5−7 years, on average!

Giving due regard to life-stage requirements, especially for spawning, the results suggested that high ecological values could be sustained by the following flow rules:

(i) from November through to the end of May, flows should be maintained above $0.3 \text{ m}^3 \text{ s}^{-1}$

TABLE 1.4. Flow assessment for Sector 9

Discharge (m³ s⁻¹)	Recommendation and derivation
1.000	To provide optimum fish habitat[a]
0.300	To maintain habitat for trout and dace at about 50% of potential[a]
0.155	To provide optimum habitat[b]
0.100	To protect aquatic habitat[c]
0.100	Minimum flow to sustain BMWP of 120[d]
0.080	Minimum flow in driest month during a "normal" dry summer
0.076	95th percentile flow (1981−87)
0.050	Minimum for fish habitat[e]
0.040	Optimum wetted bed width[f]
0.010	Minimum for LQI = A[g]

Discharge (m³ s⁻¹)	Actual 7-day minimum flows during drought (1989−91)
0.026	Minimum flow, summer 1989
0.000	Minimum flow, summer 1990
0.010	Minimum unsupported flow, summer 1991
0.068	Minimum supported flow, summer 1991

[a] PHABSIM for dace and brown trout
[b] 30% of average daily flow (Orth and Leonard, 1987)
[c] 20% of average daily flow (Orth and Leonard, 1987)
[d] Derived from an empirical model of regional data
[e] 10% of average daily flow (Orth and Leonard, 1987)
[f] field data
[g] LQI is a water-quality index based on benthic macroinvertebrates

(ii) in June, July, August and September minimum flows should be maintained above the natural dry-summer recession (respectively 0.2, 0.175, 0.12 and 0.08 m^3 s^{-1}).

The results also indicate that available habitat could be significantly enhanced by maintaining flows in September and October at above 0.10 m^3 s^{-1}. Furthermore, exceptionally, during extreme droughts, minimum flows in summer could be reduced to 0.05 m^3 s^{-1}. Whilst this would severely limit available habitat for fish, the assessment suggested that refuge habitats would be maintained and once flows are restored, the recovery of fauna should be rapid. A detailed discussion of the issue of recovery following disturbance is presented by Milner (1994).

Effects of flow augmentation

Following the construction of an inter-basin transfer, flows during the summer of 1991 were maintained at about 0.13 m^3 s^{-1} on average. The minimum supported daily flow was 0.068 m^3 s^{-1}. The inter-basin transfer from the River Gwash, supplied by releases from the Rutland Water reservoir, was introduced on 21 May when the natural flow, about 0.01 m^3 s^{-1}, was only a fraction of the 1975 dry-year May flow (0.225 m^3 s^{-1}). The actual rate of augmentation depended upon abstraction rates, channel losses and fishery requirements, and a five-year allocation from Rutland Water.

The 1989−91 period was one of extreme low flows (Figure 1.4(a)). A simulation of changes in usable habitat for adult brown trout using PHABSIM for the drought period in comparison to that available during an average year is presented in Figure 1.4(b). Two points are worthy of note: first, the catastrophic loss of habitat in the second half of 1990 (the river ran dry!) and second, the loss of adult trout habitat during the winters of 1989 and 1991. A survey of physical habitat was completed on 14 November 1991 whilst the Gwash−Glen transfer was in operation. In contrast to the reach upstream of the transfer, there was a significant (about 25%) improvement in physical habitat availability as defined by wetted area (Figure 1.2).

From the PHABSIM estimates, the usable area for both adult dace (the dominant fish species in this sector) and adult brown trout was shown to reach 30% of the estimated maximum usable area for each species at the maintained flow of 0.13 m^3 s^{-1}. Although it was not possible to re-survey fish stocks, in order to achieve partial appraisals of the ecological benefits of the transfer and of the results of the instream flow assessment, primary and secondary sources of invertebrate data were examined. A full account is provided by Bickerton (1992).

The empirical model developed for this study predicted that flow augmentation, truncating the flow duration curve at 0.1 m^3 s^{-1}, would maintain the BMWP score at above 120. Data from NRA invertebrate records are shown in Figure 1.5. Within the Shillingthorpe Sector the channel bed was dry for three months during the 1990 summer and BMWP scores fell to zero. However, by July 1991 — after three months of flow augmentation — the BMWP had risen to 142 compared with the long-term average of 141 for this site. Further downstream, at Kates Bridge, BMWP scores of 167 and 162 were recorded during July 1991 and October 1991 respectively. Prior to this, the previous highest score for the site from 13 samples spanning the last 15 years had been 158. At unsupported sites upstream of the transfer and on the neighbouring East Glen which also experienced extreme low flows, BMWP scores remained below 60.

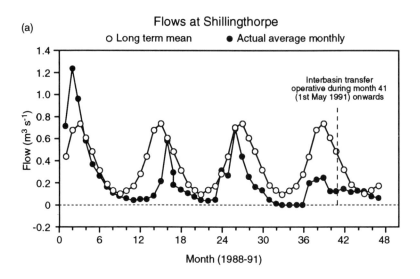

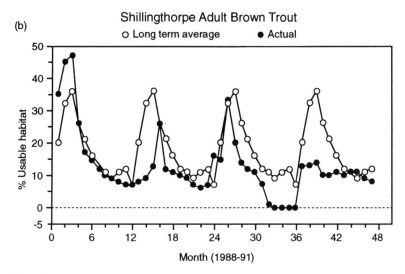

FIGURE 1.4. Flow and habitat changes within the Shillingthrope sector of the River Glen for the period 1988–91. (a) A comparison of the actual monthly flow with the average monthly flow (1981–87). (b) A comparison of the predicted available habitat for adult brown trout under average flows and actual flows. The inter-basin transfer became operative in month 41

Analysis of data for pairs of sites immediately upstream and downstream of the transfer indicate that differences remained between the October 1991 samples and the pre-drought community, probably reflecting incomplete recovery (Bickerton, 1992). By October 1991 three taxa had reappeared both upstream and downstream of the outflow, while three more only reappeared downstream. Of these last, the Piscicolidae (Annelida: Hirudinea) and Molanidae (Insecta: Trichoptera) require perennial flows and their reappearance suggests a major improvement due to the transfer.

Flow augmentation of the magnitude supporting the River Glen has allowed the rapid recovery of the benthic invertebrate community and is likely to be of benefit for fisheries not least by maintaining continuous flow between sectors 5, 9 and 10 (Figure 1.1). It is clear from the above, however, that the specification of precise recommendations from relationships between flow and biota remains problematic for managers who must evaluate ecological needs against other uses. Our ability to predict biological responses, as opposed to habitat changes, remains weak (Armitage, 1994) but the objective assessment of habitat changes in relation to flow contributes valuable information to the decision-making process. Furthermore, flow allocation for in-river needs must be evaluated in relation to opportunities for physical habitat improvement (e.g. Hey, 1994; Wade, 1994), restoration of river margin habitats (e.g. Large and Petts, 1994), and river corridor rehabilitation (e.g. Larsen, 1994).

PROSPECT

Over the past decade there has grown increased awareness that many problems for environmental protection and restoration within and along rivers throughout Europe and North America reflect the historic legacy of modernization (Petts et al., 1989; Cosgrove and Petts, 1990). Competition for water resources remains a major issue for the next century especially in dry lands characterized by accelerating urbanisation and in the developing countries requiring hydroelectric power and water for irrigation and domestic supplies. However, throughout the humid temperate zone, regional water shortages arise from variations in weather patterns and demographic change, with increasing proportions of populations living in relatively low-rainfall regions. Perhaps it is too futuristic to expect national planning systems to assess the carrying capacity of natural resources at the regional

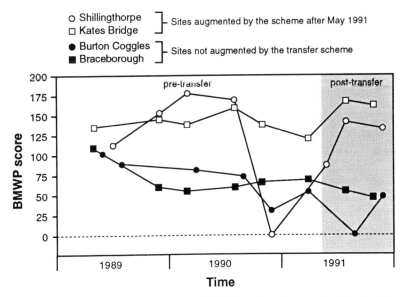

FIGURE 1.5. Variation of an invertebrate score (BMWP) at four sites on the Glen (see Figure 1.1 for locations) during 1989−91 demonstrating ecological recovery following the implementation of the transfer scheme in May 1991

scale before producing development plans for urban and industrial expansion? Water resource management is forced to adopt a reactive strategy, requiring additional storage schemes and the use of inter-basin transfers to provide for water deficits created by the development process. Simultaneously, society is developing a new environmental awareness; there has established a momentum for environmental restoration and a realization that sustainable development can be achieved by working with nature rather than by trying to control nature.

Further interdisciplinary scientific research is required to establish a sound basis for decision-making concerned with the artificial manipulation of river flows, giving due regard to the full range of flows experienced. To date most attention has focused on the development of approaches to setting Minimum Acceptable Flows. Four general approaches have been developed:

(i)   relating regulated flows to the natural flow regime;
(ii) providing a flow that will maintain some index of habitat quality;
(iii) using a paired-river or regional approach to establish flows for restoration or protection of a specific river or sector by reference to one or more rivers or sectors of similar type; and
(iv) a simulation approach integrating habitat preferences for individual taxa with detailed information of instream conditions over a range of flows.

In all cases, the link between hydrology and ecology is based on fundamental assumptions: that biota are adapted to the natural flow regime and that, to a significant extent, their presence/absence and abundance are determined by physical parameters (velocity, depth, shear stress, substrate, etc.) that are influenced by discharge. In most cases, ergodic reasoning (space−time substitution) is implicit in the approaches for predicting ecological changes over time. Relatively little is known about the changes of physiology and behaviour of species to hydrological stress, or of changes in competitive interactions and food-web links at the community level following hydrological change. The need to develop quantitative models linking hydrology and ecology is at the heart of the scientific basis for river and water-resource management; the development of functional ecosystem models remains a long-term ambition.

ACKNOWLEDGEMENTS

The work on the River Glen, described in this paper, was partly funded by the National Rivers Authority. The authors wish to acknowledge the NRA's financial support and the assistance of its staff in providing data.

2

Implications of Sediment Transport for Instream Flow Modelling of Aquatic Habitat

PAUL CARLING

NERC Institute of Freshwater Ecology, Ambleside, Cumbria, UK

INTRODUCTION

Fluvial habitats within the UK are increasingly being modelled for management purposes; for example defining a preferred discharge regime for target species where flow levels will be controlled artificially (Bullock and Gustard, 1992). Often the simulation utilizes the concept of Instream Flow Incremental Methodology, whereby the physical habitat component is addressed by applying general flow modelling techniques, such as PHABSIM (Milhous, Updike and Schneider, 1989). Various criticisms can be levelled at PHABSIM (Scott and Shirwell, 1987) but the question of incorporating sediment dynamics in ecological management models of regulated or disturbed rivers has not been adequately addressed (Parker and Andres, 1976). Existing models commonly assume (a) that a fixed channel morphology and bed roughness exists as discharge varies through time and (b) that discrete "representative" time-slice and sectional surveys suffice to describe the spatial heterogeneity of hydraulic and sedimentological environments and, consequently, the adjustment of ecological patch dynamics to changes in velocity and water depth. Often ignored is the effect of artificially altering the discharge regime on the hydrodynamic processes of coarse and fine sediment transport, deposition and intrusion of fines into the bed sediments; the resultant reach-scale changes in transport relationships as well as long-term trends in sediment supply and channel response. Although such a limited approach to modelling is pragmatic (Gore, Layzer and Russell, 1992), and reflects logistic constraints of time and money, the dynamic response of the river to the altered flow regime cannot always be ignored safely, as sediment has important effects on stream biology (Erman and Lignon, 1988). Consequently, given the lessons learnt from the study of natural rivers as well as the adjustment of rivers regulated by reservoir construction (Carling, 1988) or channelization (Brookes, 1992), the clear-water/fixed-bed perspective may be untenable for certain management applications.

Richardson and Simons (1976) argued that in managing rivers the sediment load is often inexplicably overlooked. Any change in the supply of sediment to the river or a change in the fluid discharge, either in magnitude or distribution through time, will have a potentially basin-wide effect on (a) the spatial distribution of bed sediment and (b) the channel form and

TABLE 2.1.   Degrees of freedom of a river channel (from Hey, 1982)

Adjustment of:	may occur in response to input of:
Width, velocity Depth, hydraulic radius Slope Velocity, wetted perimeter Plan shape: maximum flow depth, sinuosity, meander arc length	Water Sediment Bed and bank material Valley slope

capacity at the reach scale. These adjustments, occurring over greatly different time-scales, appear first at the reach length and are then propagated throughout a greater or lesser part of the channel network. The driving force is the adjustments to the sediment transport processes which, locally at least, may be rapid.

The balance between forcing function, process and response at the reach scale was conceptualized by Lane (1955) as:

$$Q_s D = QS \tag{2.1}$$

A decrease in the fluid discharge (Q) owing to abstraction, for example, or a change in the bed gradient (S) owing to re-sectioning may have repercussions on the nature of sediment load (Q_s). Likewise, a change in the quantity or size-distribution (D) of the sediment load input to a river reach may have dramatic effects on river channel gradient, bed roughness, capacity and hence discharge regime (Table 2.1). Although simple, it is worthwhile keeping Equation (2.1) in mind when first formulating a management strategy for a controlled river.

A simple example of the reflexive nature of these adjustments is provided by Petts' (1979) study of the effect of regulation on the River Rede, wherein channel capacity and sediment size changed over a 50 km reach downstream of the Catcleugh impoundment following regulation (Figure 2.1). Although it is not always possible or necessary to measure or model

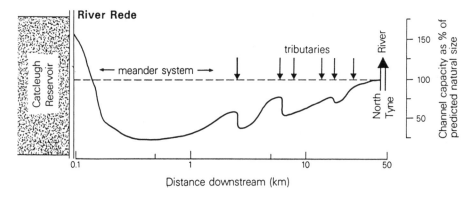

FIGURE 2.1.   Effect of regulation on the channel capacity of the River Rede. Note the effect of unregulated tributaries in injecting coarse sediment which the Rede is incompetent to transport (after Petts, 1979)

the sediment transfer system through a river of ecological interest, the ecologist should be aware of some of the sedimentological adjustments that may ensue from regulation and consider recent techniques designed to shed light on future changes to the system. It is the purpose of this paper to draw attention to some of these issues. Ecologists can draw their own conclusions with respect to specific ecological implications.

SCALES OF INVESTIGATION

A number of paradigms have been proposed within which ecological research can usefully be organized (Cummins, 1992), and consequently it is useful to try to review sediment processes at similar scales. Within the purview of this paper it is possible to separate these on a basis of time and spatial scope (Table 2.2).

Microscale

At the smallest spatial and temporal scale it can be argued that an organism's morphological adaptation and disposition is a response to the boundary layer which constitutes its immediate surroundings (Davis, 1986). This habitat usefully can be characterized in terms of turbulence intensity (root mean square of velocity fluctuations) which is measured over a spatial scale of millimetres or centimetres within a time framework of seconds or minutes (e.g. Statzner and Higler, 1986). At this scale, organisms are affected by such sediment-related processes as abrasion, impact, burial, displacement or modified feeding behaviour. Invertebrates, for example, can be classified by their morphological adaptation of stream lining to reduce drag such that organisms with a high drag coefficient will prefer environments with low turbulence levels, whilst low-drag configurations enable invertebrates to maintain station in regions of high turbulence (Statzner, 1988). Organisms will seek environments characterized by time-average hydraulic characteristics which are stable over time-scales greater than the microscale.

Mesoscale

At this level one might expect spatial scales of tens of metres to hundreds of metres, a time framework measured in hours or days and hydraulic complexity mediated by the flow hydrograph. It is at this scale of resolution that for the present purpose the concept of patch dynamics (Townsend, 1989) usefully can be placed. At this scale, typified by the concept of the "river-reach", environments can be defined by bulk flow parameters such as the Reynold's number which defines turbulence regime (Carling, 1992) and the Froude number,

TABLE 2.2. Conceptual framework for scale of sediment dynamics

	Dimension	River temporal scale	River spatial scale	Biological paradigm
Microscale	>mm<1 m (sec/min)	Turbulence	Individual particle	Boundary layer
Mesoscale	>1 m<1 km (h/day)	Hydrograph	Bedform/reach length	Patch dynamics
Macroscale	>1 km (months/years)	Discharge trends	Multi-reach/catchment	Continuum

which defines sub-critical and super-critical flow (Davis and Barmuta, 1989). A typical ecological division of physical environments is a step−pool system in steep mountain torrents (Chin, 1989) and the riffle−pool sequence in lowland rivers (Logan and Brooker, 1983). In the former case, in streams with gradients >0.05, flow downstream of step crests may be locally super-critical during low flows although this is not usually the case with lowland pool−riffle sequences. In both cases whereas the values of bulk parameters such as mean velocity and mean shear stress tend to converge as stage rises, environments may remain hydraulically and sedimentologically distinctive throughout a range of discharges (Carling, 1991). During low flows often smooth-turbulent flow characterizes the bottom of pools, which are covered with a carapace of fine sand, whilst rough-turbulent flow occurs downstream of the gravel riffle crests. At higher flows the sand is re-suspended, pools roughen and some of the sand may infiltrate the riffle bed sediments which remain stable. At the very highest flows some of the coarser pool sediments may be entrained and moved over the downstream riffle. It is these mesoscale processes active within the channel, as mediated by changes in sediment supply within the catchment, that are readily correlated to changes in channel planform through time. These changes are augmented primarily through bank cutting and the deposition and downstream translocation of a variety of bars of which point bars are the best known examples.

Macroscale

At this scale, spatial changes typically exceed 1000 m in dimension and occur over periods of months to years. An example is the evolution of meander geometry and the eventual development of cut-off channels and sloughs. However, with a change in discharge volume or sediment load the type of channel can also change. For example, a river may change from a single thread to a multi-thread channel or from a gravel-bed to a sand-bed channel. Changes in form and the nature of the bed sediment occur at scales in excess of the reach-length, and often tend to be driven by process changes occurring upstream, including those outwith the channel (e.g. Gurtz and Wallace, 1984). Consequently, reach dynamics are linked, and the appropriate scale is of the order of the catchment dimension. Within this compass the River Continuum concept comes into perspective; in as much as community patterns are related to system-wide changes in the physical complex (Vannote et al., 1980; Cummins, 1988).

EXAMPLES OF SEDIMENT DYNAMICS AT THE THREE SCALES

Microscale

The movement of suspended or bed particles at this scale is of interest to the ecologist in as much as they mediate the supply of nutrients and the removal of metabolic waste as well as having direct effects, such as physical impact and the scouring of invertebrates or algae from the bed (Sorenson et al., 1977; Iwamoto et al., 1978). Limitations of space prevent an examination of all the effects here although as the biomass of some taxa may be reduced by high suspended loads, the mass of others may increase to compensate. However, it is evident that further understanding of the motion of bed particles (e.g. overturning) and their biological relevance (e.g. McAuliffe, 1984; Malmqvist and Otto, 1987) is closely related to an improved understanding of the nature of turbulent flow. Previously such studies have

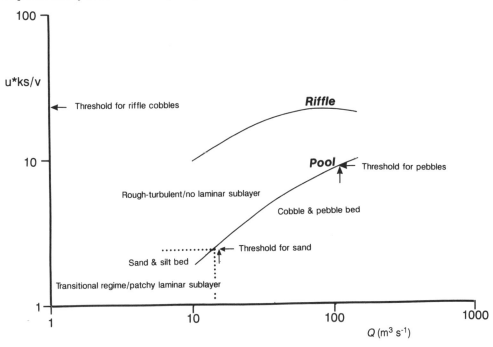

FIGURE 2.2. Variation in the Roughness Reynolds number as a function of discharge over a pool and riffle on the River Severn near Montford Bridge. The discharge regime is near-natural and riffle cobbles are never entrained by in-bank flows. Pebbles move over a stable cobble-bed in the pools during high flows, but are static over a wide-range of lower flows. Very low flows result in the blanketing of pools by sand and silt. Peak flows would need to be less than 14 m³ s⁻¹ for pools to be permanently sand-filled

been restricted to the laboratory but increasingly it is possible to obtain data from natural streams. At the crudest level, a reduction in flow volume may reduce the velocity and depth within a pool so that the Roughness Reynold's number reduces and concomitantly shear stress and the turbulence intensity can be expected to decrease, with repercussions on fine sediment dynamics (Figure 2.2). However, downstream of riffle-crests, the reduction in pool depth may result in an increase in the water surface slope, an increase in the turbulence levels with a concomitant effect on sediment sorting and compaction (Clifford et al., 1994).

Instrumentation, in the form of small electromagnetic current meters (ECMs), is now becoming available and is capable of measuring flow fluctuations with the desired temporal resolution to determine statistical aspects of flow turbulence in sediment-laden waters. Simpler ECMs, although not suitable for study of turbulence can be used to define the spatial resolution in the mean velocity field at intervals of centimetres or millimetres in the vertical (Figure 2.3(a)). By turning the instrument through 45° normal to the apparent downstream flow (U), and measuring U_{45}, the cross-stream flow velocity component (V) can be estimated from simple trigonometry (Figure 2.3(b)). Likewise new instrumentation can provide information on the dynamics of both suspended and bedload motion within a similar time-frame as the turbulent flow fluctuations; whereas previously long-averaging periods of many minutes were the norm. These instruments include a passive acoustic system to obtain high

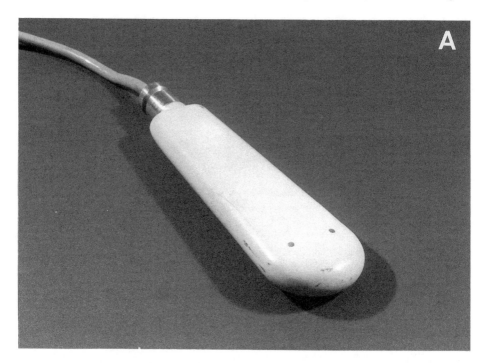

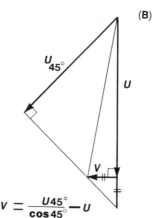

$$V = \frac{U45°}{\cos 45°} - U$$

FIGURE 2.3. (a) New generation of small dorsally-flattened electro-magnetic flow meters. Dimensions: length: 150 mm; height: 17 mm. (b) Calculation of relative contribution of cross-stream velocity vector (V). The maximum downstream velocity vector (U) is found by yawing the flow meter. A second vector (U_{45}) is found by yawing the meter to 45° from the direction of U. Estimate of the component V follows from trigonometry

temporal resolution time series of bedload motion non-intrusively (Williams and Tawn, 1991) as well as an optical backscatter sensor to detect suspended sediment concentration (Lapointe, 1992).

Measuring processes at this scale provides statistical information on the unsteady nature of sediment mobility (Figure 2.4) which may be suitable to calibrate models of river sediment dynamics appropriate to higher levels of time- and spatial-averaging. In particular, occasional large-scale turbulent perturbations in the flow give rise to rapid incursions (sweeps) or excursions (ejections) of fluid fingers towards or away from the bed (bursting). For flows competent to move sediment, this unsteadiness in the fluid results in pulses of bed material being entrained into the bedload or into suspension (Williams, Thorne and Heathershaw, 1990; Lapointe, 1992). It is probable that such unsteadiness is responsible for flushing of deposited fines from the surface interstices of gravel beds or the penetration of fines deep into the sub-surface.

Mesoscale

It is appropriate to pay especial attention to sedimentary processes defined at this time and length scale, not least because "resetting" of benthic community structure occurs over similar scales (Townsend, 1989), but also because the initial response of the sediment load in regulated streams is rapid at the reach scale even if more extensive changes in channel form, in bed sedimentology and in roughness occur over longer time-spans (Gilvear, 1987; Petts, 1987). Siltation of openwork gravel beds disturbed by small trout (*Salmo trutta* L.) cutting redds can occur within a few days during base flows and more rapidly during freshets (Carling and McCahon, 1987). This is especially so during waning hydrographs, although it is not widely recognized that fine sediment will also become entrapped, in deep openwork gravel during flows competent to entrain sand, by turbulent pulses injecting fines into the bed (Carling, 1984). Such a siltation process can smother salmonid fish eggs by occluding intra-gravel convective transport (Moring, 1982; Thibodeaux and Boyle, 1987) and prevent

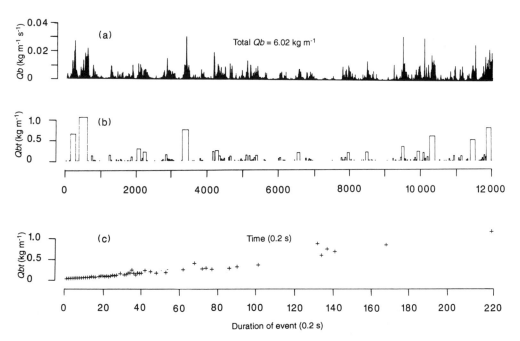

FIGURE 2.4. (a) Time-series of gravel-movement detected using hydrophone. (b) Inferred discrete transport events and (c) distribution of event durations (after Williams and Tawn, 1991)

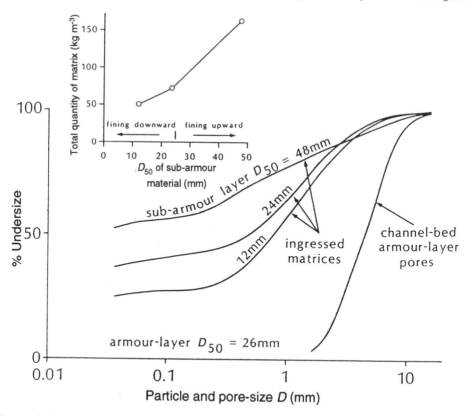

FIGURE 2.5.   Size-distributions of both the armour-layer interstices within a gravel-bed river and the infiltrated fines where the sub-armour gravels are differentiated by size (three examples of the median; D_{50}). Inset is the quantity of fines that infiltrate when either the gravel fines upwards, downwards or there is no vertical gradation (modified from Reid and Frostick, 1985)

alevin emergence by sealing pore-space at the surface (Phillips et al., 1975; Hausle and Coble, 1976). It is known that if the gravel-bed coarsens towards the surface, then generally fines can penetrate deeply within the bed (Figure 2.5). In contrast if the surface gravels are finer than those at depth, then there is a propensity for fines to lodge close to the surface, forming a surface seal with openwork gravels below (Frostick, Lucas and Reid, 1984; Reid and Frostick, 1985). Although the detailed physics of the infiltration process are poorly researched, aspects of the bulk process have been studied both in the laboratory (Diplas and Parker, 1985) and in the field (Lisle, 1982). Prolonged periods of low flow, competent to disturb but not entrain the coarser bed sediment, will result in compaction of the bed as it becomes silted (Reid, Frostick and Layman, 1985) and during these periods (Figure 2.6), pools may fill with fine sediment (Lisle and Hilton, 1992). The siltation process is accentuated by deepening during channelization (Nuttall, 1972; Simon, 1989; Brookes, 1992; Hupp and Bazemore, 1993) with distinct ecological ramifications (Pearson and Jones, 1975; Whitaker, McCuen and Brush, 1979; Iversen et al., 1991).

On the rising limb of the hydrograph, sediment tends to be more prone to entrainment than deposition, but unsteadiness in bedload transport in particular remains characteristic (Kuhnle, 1992). This may be induced by a number of factors, including reach-scale processes such as particle sorting interaction within the bedload (Iseya and Ikeda, 1987) as well as macroscale processes such as the accessing of differing stores of different-sized bed-material as stage rises, and the variation in the transit time of bedload-waves moving down tributaries into the main channel. Often the supply of available sediment can be exhausted by the time the hydrograph peaks so that a dearth results on the waning limb of the hydrograph (Figure 2.7(a)). As a result, bedload may move through a reach as discrete sheets (Dietrich et al., 1989). In the example given (Figure 2.7(a)), small pebbles entrained upstream entered the study reach and in part infilled the surface roughness of a cobble-bed during the early part of the flood event. This resulted in smoothing of the bed and a decrease in the hydraulic roughness. As the sheet passed through the reach, the fine pebbles were re-entrained so the bed coarsened and roughened towards the end of the event (Figure 2.7(b)).

Generally high flows of the order of two-third bankfull at least are required to mobilize the coarse bed sediments in natural rivers. Not all of the bed width is activated simultaneously (Figure 2.8). Instead the area of disturbance increases (Carling, 1987) and insect abundance may decrease as discharge rises (Cobb, Galloway and Flannagan, 1992). Bed mobilization is required to flushout fines deposited in the interstices during low flow (Reiser et al., 1989; Diplas and Parker, 1992). Without this mobilization only the surface layer approximately one to three maximum grain diameters (e.g. $3D_{90\%}$) thick can be flushed by higher flow rates. Commonly when flows are reduced artificially, downstream of a dam for example, the

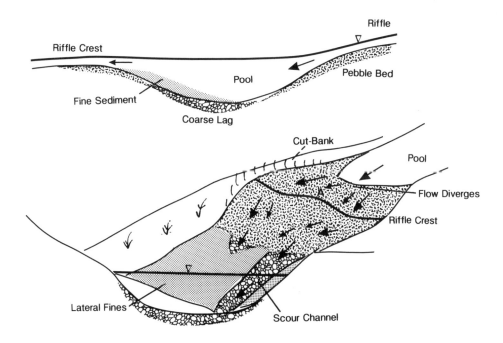

FIGURE 2.6. Schematic indication of the build-up of fine sediment deposits over lag-gravels in the base of pools when regulated and reduced discharges are no longer competent to resuspend fine material. Riffle sediments may cease to be regularly entrained and winnowed so that the gravel void space fills with fines. The siltation process is most evident along the channel margins and on the medial bar (A). Cut-banks previously a source of fines become inactive

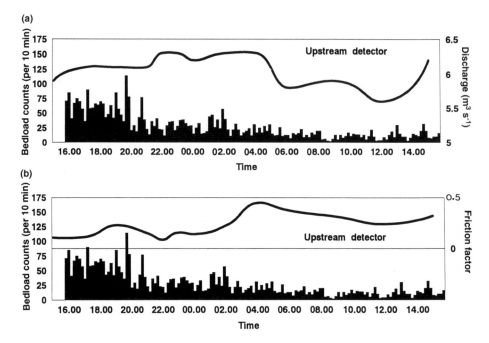

FIGURE 2.7. Passage of a sheet of pebbles over a static cobble bed. (a) Number of particles in motion is at first suddenly high as the front of the sheet arrives and then declines exponentially with time although fluid discharge is variable. (b) The Darcy-Weisbach hydraulic roughness initially declines rapidly as the surface cobble interstices are infilled by pebbles, but then increases as the pebbles are re-entrained in the latter half of the event (after Carling et al., in prep)

larger fractions of the bed material are rarely if ever entrained whilst the smaller fractions continue to be disturbed on occasion. Over time, the bed coarsens and compacts as the finer pebbles are transported downstream. Given this scenario it is even more difficult if not impossible to flushout any fines deposited at depth and consequently a natural bed which was periodically reworked by high flows will both coarsen and become heavily silted under reduced regulated flows (Sear, 1992).

So far we have concentrated on temporal aspects associated with longitudinal changes at the reach scale. However, recent research has demonstrated the importance of near-bank effects in inducing strong lateral gradients in the sedimentation process. It has long been known that flow around a bend, for example, will induce differential sorting across the section, with finer material such as sand migrating towards the top and inner portion of the developing point bar, whilst coarser material such as gravel migrates outwards towards the toe of the point bar, resulting in a gradient in habitat, from low-energy/fine-sediment to high-energy/coarse-sediment. The upper slopes of the point bar are only inundated infrequently, by high flows which characteristically carry the greatest concentration of fine suspended sediment. This material settles to create a silty or muddy environment on the inside of the bend. If the bend is tight enough then a recirculating slow-flow region will develop over the point bar wherein the residence time of the water is substantially greater than that within the main channel flow (Rubin, Schmidt and Moore, 1990). The potential for deposition of fines is then enhanced as velocities and turbulence levels are reduced.

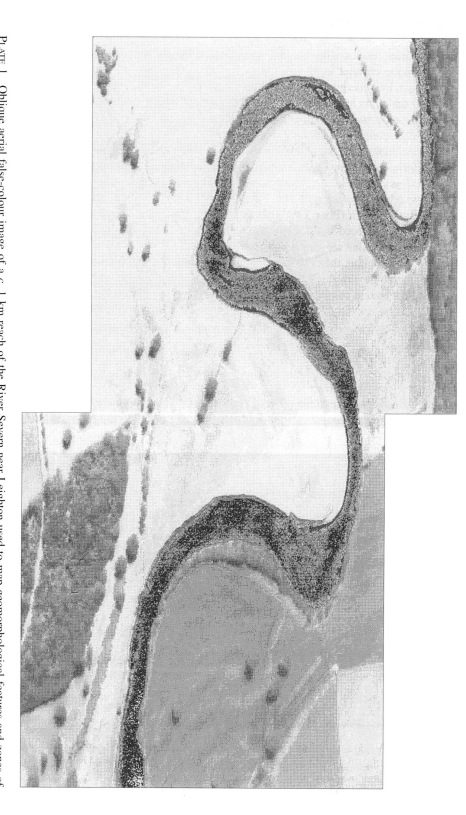

PLATE 1 Oblique aerial false-colour image of a *c.* 1 km reach of the River Severn near Leighton used to map geomorphological features and zones of deposition. Areas shown in red are warmer than 18 °C, have velocities < 0.03 cm s⁻¹ and are typified by fine silt deposition. Cobble-bed riffle areas are cooler (*c.* 17 °C) and are shown in yellow, whilst the coldest waters (*c.* 16 °C) are found in the pebble-bed pools (blue). Flow right to left

(© NERC, reproduced with permission)

Similar effects, as noted in the case of meanders, occur with the presence of sudden expansions in the channel width or in the vicinity of partial cut-offs and inactive tributary junctions. Such low-energy "slack-water" or "dead zone" environments tend to be persistent at the reach scale. As stage changes, the physical dimensions of individual "dead-zones" may change, some may be destroyed whilst others develop, but the overall total volume of the reach-scale component changes little. These areas provide important nurseries and refugia during high flows, for planktonic algae (Reynolds, Carling and Beven, 1991), fishes (Harvey, 1987; Heggenes, 1988; Copp, 1989), benthos and aquatic plants (Gaschignard, Persat and Chessel, 1983; Biggs and Close, 1989). Carling, Orr and Glaister, (1994) used remote-sensed thermal imagery during the summer low-flow period to map dead-zone areas within the River Severn (Plate 1). Field-work demonstrated that where water temperatures on this occasion exceeded 17°C, flow velocities were less than 0.03 m s^{-1}, and the gravel-bed was covered by fine sediment of an average size of 7.5 μm. Outside of the dead zones, or once velocities within the dead zone exceeded 0.03 m s^{-1}, the carapace of fines was re-suspended. In the Severn this occurred in summer at about two-thirds bankfull discharge. The deposition and re-suspension of fine sediment within one large dead zone has been modelled by Tipping, Woof and Clarke (1993); the driving force being exchange of fluid and suspended sediment with the main flow.

Macroscale

Abundant fines on the surface of a regulated or engineered channel may result from a decrease in the competence of the main stream relative to sediment-rich tributaries such that a gravel-bed river can change to a sand-bed river (Petts and Thoms, 1986). Alternatively, a

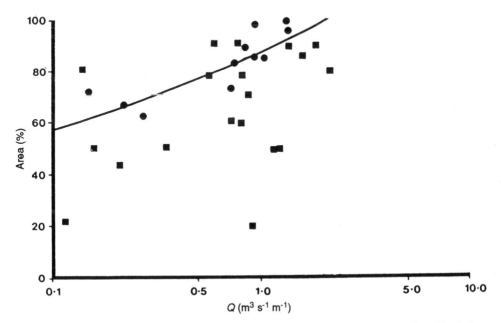

FIGURE 2.8.   The area of gravel-bed disturbed increases as a function of discharge. Great Eggleshope Beck, Carl Beck (re-drawn after Carling, 1987)

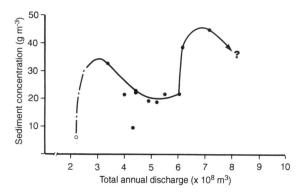

FIGURE 2.9.   Annual suspended sediment discharge for the River Tees at Broken Scar as a function of annual discharge (1971 to 1981). The point at 2.2×10^8 m³ ann⁻¹ represents the suspended sediment discharge expected with regulated base-flow alone and no storm-flow. A rapid increase occurs with a storm-flow regime but then declines exponentially as sediment sources are exhausted. For the highest fluid discharges ($>6 \times 10^8$ m³ ann⁻¹), a threshold is exceeded and new sediment stores are accessed. The outlier represents the drought year 1976 when little sediment reached the river from the land surface

change in the quantity of material supplied to the channel (Platts and Megahan, 1975; Dietrich et al., 1989) can lead to bed sediment changes owing to limitations on stream competence and capacity to transport all material within a given grain-size fraction. Such unsteadiness in sediment supply leads to the development of "waves" of bed sediment passing through a river system such that locally the bed level may aggrade and subsequently degrade as the wave passes. Model studies have shown that channel type may vary between multi-channel (braided) and single thread due to the unsteady changes in the quantity of sand-sized sediment supplied to the channel (Hoey and Sutherland, 1991). Field evidence also indicates that the quantity of gravel supplied to a channel can mediate the change from single to braided form (Harvey, 1991). Basically a progressive shoaling, owing to sediment deposition, in a downstream direction forces the flow to widen out of a single-thread pattern (Carson, 1984).

Annual suspended sediment loads also can change dramatically owing to natural factors or man's intervention. An example of the former can be given for the River Tees (Figure 2.9). If the total system were regulated, suspended sediment concentrations would only be of the order of 6 g m⁻³; sourced largely from biogenic particles and limited bank erosion. Under the present regulated annual discharge regime (up to 3×10^8 m³ ann⁻¹), the suspended sediment load is much higher owing to catchment inputs but supply exhaustion effects are evident. However, in very wet years ($>6 \times 10^8$ m³ ann⁻¹) either additional sediment sources are accessed or sediment delivery processes intensify so that suspended sediment concentrations rise again. Given that deposition rates are directly proportional to the concentration of the suspension, then during base-flow periods the gravel bed might actually be siltier in wet years than during dry years owing to the higher rate of supply to the channel. In a similar vein, changes in channel form may occur owing to a gross change in the suspended sediment load induced by man. Richards (1979) demonstrated that fine-grained kaolin waste produced by hydraulic mining in Cornwall has polluted streams since the 18th century. Deposition of the cohesive material on the banks has resulted in progressive channel narrowing at the catchment scale; again with ecological implications (Herbert et al., 1961).

Instability is inherent in the rate at which the sediment transfer system functions, and, in the case of river meandering, is driven by the rate of channel migration. Sediment is supplied by bank cutting on the outside of bends and deposited downstream on the inside of bends as the point bars referred to above. In the UK, typical natural migration rates depend on the nature of the bank materials and the power of the stream. However, migration rates typically vary between one and two metres per year (Lewin, 1987; Hooke, 1980) although Lewin (1982) recorded channel shifts of up to 50% of the channel width in one year owing to channel adjustment to artificial straightening. The process of channel meandering has always resulted in periodic natural straightening of given reaches as bends are cut off. Through this action, a diversity of sedimentary habitats are continually forming and being replaced; these range from muddy ox-bow lakes, to partial (gravel-bedded) cut-offs and sandy sloughs, point and medial bars and channel thalwegs. This process, although natural, is often affected by man's intervention in cutting through meander-necks and in trying to stabilize straightened channels artificially. Usually, however, diversity in sedimentary environments declines as man back-fills cut-offs, and dredges point bars. Alternatively an artificial reduction in discharge may result in the deposition of in-bank lateral berms (Gregory, 1982) and stabilization of the meander pattern as banks are no longer undercut and entrainment of bed sediment by high flows is rare. A good example of the natural pattern of meander evolution mediated by man's intervention is that of the 1.5 km reach of the River Severn at Dolhafren downstream of Caersws in the Welsh County of Powys (Figure 2.10). Map and aerial photographic evidence exists from 1834 until present. Major meander cut-offs have occurred on three occasions within *c.* 160 year period, following which channel instability and sediment mobility declined, whilst habitat diversity formerly increased with the development

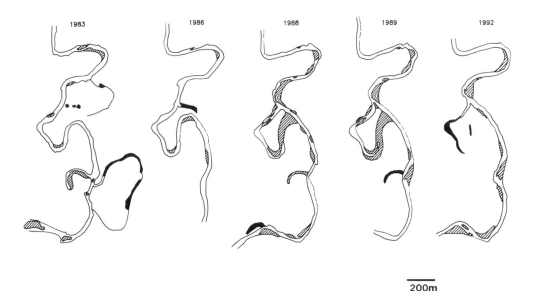

200m

FIGURE 2.10. River Severn near Caersws, showing changes in areas of different sedimentary environments following meander cut-off. Unhatched areas are permanently submerged thalweg; black areas are muddy sloughs; shaded areas are gravel bars exposed during low flow

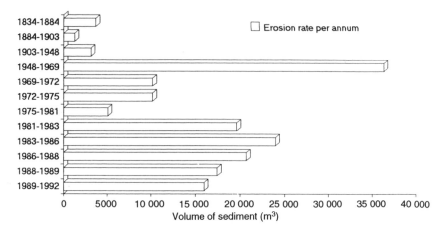

FIGURE 2.11.   Variation in the average annual rate of erosion through time for a 1.5 km reach of the
River Severn near Caersws. Major cut-offs occurred immediately prior to 1834, *c*. 1969 and again in
1988 (from Carling, P. A. and Orr, H. G., unpublished)

of ox-bow lakes and sloughs. In contrast, the rate of bank migration and consequently the
amount of mobile sediment is greatest in the periods of time immediately prior to cut-offs. At
this time, bend extension is most rapid and gravel point bars have their greatest areal extent.
Since 1988, however, ox-bows and sloughs have largely been infilled for agricultural
reclamation and point bars dredged in an attempt to stabilize the channel course. The
evidence is that channel instability has not been reduced significantly by re-sectioning; if
anything, it is now greater than previously occurred following natural cut-offs (Figure 2.11).

CONCLUSIONS

Three scales of adjustment of sediment dynamics to artificial changes in river discharge or
sediment load have been identified. It is argued that these scales are consistent with the
scales at which ecological adjustment is also frequently addressed and so understanding of
the changes to the sediment system are required at all three levels of detail if the effects of
sediment on stream ecology are to be fully determined. However, in modelling ecological
changes in modified natural rivers it is important to consider at the outset the time and spatial
scales of interest. For example, where changes in the river channel are local and induced by
engineering structures at the reach scale with a design life of only one or two decades, it
might be concluded that processes operative within the catchment over time-scales of many
decades are not relevant to immediate management. At this level of explanation, standard
engineering models which incorporate changes in bed level owing to sediment transport,
may provide sufficient detail to predict the effect on the biota. However, with the example of
flow regulation within a catchment-wide perspective, as the pattern of deposition is mediated
by the channel configuration and vice versa, it is evident that not only are the scales of
examination intimately linked, but that it is inevitable that changes in the nature of the
sediments and channel form at the reach-length will be propagated to larger scales over
sufficiently long time-scales (Petts, 1984, 1987). How these changes are to be modelled and
channels managed with respect to these longer time-spans is less clear. Progress is now

being made in extending reach-scale sediment transfer models to multi-reach configurations (Howard, 1992) but much more needs to be known concerning the relaxation times associated with physical perturbations. Coupled with this need is the uncertainty induced by changes in land use which may induce unique changes in point or diffuse sources of sediment. However, although deterministic modelling may not be possible at the macroscale, it is quite possible to make reasonable inferences as to the most probable response of a river's sedimentary system and incorporate these considerations into ecological management plans.

3

The Importance of High Flows for Riverine Environments

ANDREW BROOKES

National Rivers Authority, Reading, UK

INTRODUCTION

Rivers typically experience a wide range of flows causing channel change over a relatively short time period. Understanding and recognising this variability is essential to the development of an improved ecological basis for river management. At one extreme there may be little or no flow in river channels. 1992 saw the fourth successive year of drought in the south-east of England, caused by a prolonged period of below-average rainfall, leading to inadequate replenishment of underground water resources. At the other extreme, some rivers in England and Wales have suffered from catastrophic flooding in the recent past, leading to very significant channel change over a period of only a few hours.

The response of stream channels to the full range of flow events of various magnitudes and frequencies is a fundamental issue in fluvial geomorphology. Whilst studies have focused primarily on the response of channels to, and recovery from, infrequent floods there is a need for sequential observations of channel changes relative to a variety of flows. Changes of channel characteristics over the short term (depth, velocity and shear distribution), substrate and space availability can have significant impacts on benthic invertebrates, fish and aquatic macrophytes.

Drawing on published experiences from a number of countries, this chapter looks at the significance of high flows for riverine environments and presents some interim recommendations applicable to the management of rivers in England and Wales. Key issues which are addressed are:

(i)   a definition of channel-forming flows,
(ii) ecological impacts,
(iii) channel typology,
(iv) the significance of human impact,
(v)   the need for rehabilitation and restoration devices to be stable over the whole range of flows.

The general paucity of published research means that more comprehensive guidance cannot be produced at this stage, and areas for further development are highlighted.

The Ecological Basis for River Management. Edited by D.M. Harper and A.J.D. Ferguson. © 1995 John Wiley & Sons Ltd

CHANNEL CHANGES AT HIGH FLOW

There have been a few studies in the UK of the response to, and recovery from, infrequent or catastrophic floods (Newson, 1975; McEwen and Werritty, 1988). In August 1952 a major flood occurred in Lynmouth, Devon, as a consequence of an extremely severe rain storm over Exmoor at the head of the steep, narrow valleys of the River Lyn. This was one of the three heaviest 24-h rainfalls ever recorded in the UK. The subsequent flood caused the loss of 34 lives, the destruction of 90 houses and 130 cars. However, in geomorphological terms the flood was significant because up to 100 000 tonnes of boulders were transported and deposited near the town (Kidson, 1953). It is estimated that the West Lyn River dissipated over 200 megawatts of energy in flowing through Lynmouth.

Super-critical velocities of between 6 and 9 m s^{-1} were estimated, carrying boulders up to 7.5 t. Sudden surges of flow observed during the flood occurred as a result of the formation and subsequent failure of boulder and tree dams across the channel. The peak flow occurred about 3.5 h after the beginning of the storm and may well have attained a value of 510 m^3 s^{-1}. This is equivalent to the highest daily discharge for the River Thames at Teddington. It has been calculated that the Lynmouth Flood had a return period of about 1 in 50 000 years. The role of a flood can be defined in terms of the "work" done (in moving sediment) and "effectiveness", i.e. the amount of channel change caused (Wolman and Gerson, 1978). The effectiveness of a flood has been shown to vary between upland and lowland areas and is often dependent on a supply of sediment.

Large floods can have very pronounced long-lasting effects on river channels but it is clear from the results of various studies that high flows of less magnitude may have a greater impact because of their greater frequency. Wolman and Miller (1960) drew an analogy with the work of a dwarf, a man and a giant in clearing trees in a forest. The dwarf works for long periods but achieves small results similar to the low flows in river channels which predominate for most of the time. The man works less frequently but more effectively and is equivalent to the flows which occur in a river channel once or several times a year. The giant sleeps for long periods but can cause significant damage over a short time such as catastrophic river floods which occur perhaps only once in several hundred years at a particular site. The conclusion of Wolman and Miller was that the most effective processes are those which occur perhaps once or several times every two or three years. Bankfull discharge has often been referred to as the channel-forming discharge and occurs on average 0.6% of the time, or about 2.2 times per year in England and Wales (Nixon, 1959). At bankfull discharge the surface slope of a river becomes uniform, the whole wetted perimeter is washed, and sediment transport is at a maximum.

A reach of channel or a given cross-section must transmit varying amounts of water passed into the section from the channel upstream. With the rise in discharge accompanying the passage of a flood there is an increase in velocity and shear stress on the bed. As a consequence the bed of a channel may scour during high flow. Pools may scour at high flow and fill at low flows whereas riffles conversely may scour at low flow and aggrade at high flow (Keller, 1971).

Researchers have tended to favour shear stress criteria for stability and bed movement. From a practical river management point of view, however, local shear stresses are difficult to measure and conceptualize, compared to velocities. Neill and Hey (1982) argued that researchers should pay more attention to expressing results in velocity terms for practical application. Table 3.1 gives an indication of the maximum permissible water velocities for

TABLE 3.1. Maximum permissible water velocities for the design of stable channels (flow depth less than 0.9 m) (after Webber, 1971)

Material	Roughness coefficient[a] n (s m$^{-1/3}$)	Clear water velocity, V (m s^{-1})	Water-transporting colloidal silts velocity, V (m s^{-1})
Fine sand, colloidal	0.020	0.45	0.75
Sandy loam, non-colloidal	0.020	0.55	0.75
Alluvial silts, non-colloidal	0.020	0.6	1.0
Fine gravel	0.020	0.75	1.5
Stiff clay, very colloidal	0.025	1.15	1.5
Alluvial silts, colloidal	0.025	1.15	1.5
Coarse gravel, non-colloidal	0.025	1.2	1.8
Cobbles and shingles	0.035	1.5	1.7

[a] Based on a straight channel of uniform section

the design of stable channels with a flow depth of less than 0.9 m.

Clearly a far higher velocity is required for the movement of cobbles and shingles than for fine sand. Hjulstrom's work (1935) predicts that medium sand is the most easily eroded fraction, higher velocities being required to move both finer and coarser grains. However, in this context mean velocity is not the most relevant parameter. One complicating factor is that the threshold for bed erosion may vary by more than an order of magnitude depending on whether the bed is loosely or tightly packed.

High flows can be a major controlling and limiting factor. Movement of large bedload material takes place only at high discharges and there is normally a threshold discharge below which no movement occurs. When the flow drops below the threshold, then sediment is deposited and will generally remain there until a flood of equal or greater magnitude recurs. This may be several days, months, years or even decades later. During the passage of a flood, part of the increase in sediment load is obtained from the channel itself, part from sources beyond the channel. As a consequence, channel morphology may change during a flood: dunes and ripples will tend to move, and depth and width of the channel may change through the processes of scour and deposition. In the short term much of this adaptation of channel form is temporary; scour during a flood is replaced by deposition as the flood passes.

Lewin, Macklin and Newson (1988) concluded from a review of British rivers that many channels are sensitive to relatively small changes in sediment supply and runoff, and adjust their size and shape more frequently, and more rapidly, than generally appreciated. Major floods can cause significant channel change, particularly channel widening, but over months and years after the flood a widened channel usually narrows through deposition and so recovers (Stevens, Simons and Richardson, 1975). Only when high-magnitude floods occur frequently are the natural recovery processes prevented. Rivers in England and Wales appear to be able to recover rapidly because of a plentiful supply of sediment.

ECOLOGICAL IMPLICATIONS OF HIGH FLOWS

Fish and other stream inhabitants require a habitat characterized by a variety of high-flow conditions and shelter areas which provide protection from excessive water velocity. As a result of high flow there is a natural sorting of bed sediments on riffles and point bars, providing good environments for bottom-dwelling organisms in streams that fish and other animals depend on for food supply. There a few published studies which report how floods sweep away many animals. Jones (1951) reported that summer floods on the River Towy in Wales reduced the invertebrate population from $300-1000$ m^{-2} to $40-48$ m^{-2}. Scouring tends to control the nature of the substrate, which in turn controls the density of invertebrates. As early as 1929, Percival and Whitehead showed how animal density falls off sharply as erosion increases. The pattern of siltation following a flood in certain watercourses has been shown to reduce invertebrate numbers (Gilbert, 1989). Aquatic macrophytes, particularly those rooted in silt, may also be washed out during a high flow (Haslam, 1978; Brookes, 1986). Edwards (1969) described the abrasive effect of sediment on plants; the bombardment of small particles was observed to cause small pittings and scratchings on the plant body.

Table 3.2 summarizes some key geomorphological factors of natural alluvial river channels which are important for biological diversity. Typical time-scales over which these

TABLE 3.2. Key geomorphological factors influencing river ecology over different time-scales (after Brookes, 1994)

Factor	Typical time-scale of change change	Methods used to determine response	Knowledge of biological
Cross-sectional shape (e.g. bank collapse)	Hours to years (e.g. during extreme flood events)	Direct observation and measurement	Limited
Cross-sectional size	Hours to years	Direct observation and measurement	Limited
Pool and riffle sequence	Hours to years	Direct observation and measurement	Limited
Point bars	Hours to years	Direct observation and measurement	Limited
Islands	Hours to years	Direct observation and measurement	Some knowledge
Subtrate (including sedimentation)	Hours to years	Direct observation and measurement	Well documented
Pattern such as lateral migration	$100-150$ years	Historical sources such as maps and documents	Limited
Pattern such as braided to meandering	$250-15\ 000+$ years	Sedimentary evidence; dating techniques (floral and faunal evidence; artifactual remains)	Not applicable

TABLE 3.3. Summary of the response of aquatic vegetation to morphological adjustment (after Brookes, 1983)

Morphological adjustment	Vegetation response	Principal cause
Width increase	Increased biomass in the channel; species composition unchanged	Increased bed area for colonization by plants
Depth increase (resulting from degradation of bed)	Reduced biomass in channel; species composition unchanged	Light required for plant growth is attenuated with depth
Coarsening of substrate (resulting from degradation of the bed)	Reduced biomass in channel; changed species composition	Mixed-substrate suitable for rooting of plants is destroyed
Bedrock exposure (resulting from degradation of the bed)	Elimination of plants from exposed areas	Bedrock unsuited to rooting or attachment of plants
Planform adjustment	Relocated channel colonized by plants; biomass and species composition remains the same	Substrate and channel morphology may be similar in relocated channel

factors are observed to change are also shown. Whilst there are numerous studies which have examined the biological response of a changed substrate, for example as a result of scour during a flood, there is little knowledge of the impacts of changes of other factors. In England and Wales, Brookes (1983) attempted to show how channels widened as a result of high flows can become colonized by vegetation. Some results of this study are shown in Table 3.3.

CHANNEL TYPOLOGY

It is important to appreciate which types of river channel are susceptible to change or adjustment at high flow. This is necessary in river management for a number of reasons such as anticipating the effects that particular engineering structures and modifications may have on channel morphology and relating these to the recovery of biological populations. It is also essential for the appraisal of successful rehabilitation and restoration projects which work with nature rather than against it. There have been a number of approaches to river classification throughout the world but few have attempted to integrate physical, biological and human impacts. Working in the former USSR, Popov (1964) identified several "channel processes" based on the type and arrangement of gravel or sand bars in a channel and the extent to which a channel is constrained by valley walls. The types included braided (mid-channel bars), freely meandering and confined meandering. "Channel processes" are a good indication of the level of activity in a channel. The ability of flowing water to form and change a channel and transport sediment depends on gravity, which propels water downslope, and the friction between the water and the bed and banks which tends to resist downslope movement. The variation of channel slope and discharge may form a useful basis for classifying natural river channels. Ferguson (1981) classified British rivers in terms of their degree of activity and sinuosity. A summary of this typology is shown in Table 3.4. This includes three groups, namely active braided channels, active meandering channels and inactive sinuous or straight channels. Active braided channels (group 1) include channels with local division and extensive braiding on large rivers. In these channels erosion and deposition is not concentrated at bends and point bars. By contrast, active meandering channels (group 2) do erode the outside of a bend, with corresponding deposition on the inside, forming a point bar. Group 3 includes inactive sinuous and straight channels. In these channels high flows are insufficiently powerful to overcome the resistance of the banks to erosion and the courses do not migrate across the floodplain. This type of channel includes lowland clay rivers.

Although no attempt has yet been made to map the geographical distribution of these three groups, experience indicates that group 1 is relatively rare, group 2 less so and group 3 is particularly widespread. Groups 1 and 2 tend to be restricted to upland areas of England and Wales, with relatively high channel slopes and discharges.

A further classification which may be applicable to a wide range of areas is that developed by Palmer (1976) based on the earlier work of Bauer. This comprises four geohydraulic zones along a river system which are each defined by a characteristic combination of valley cross-section, channel pattern, gradient and bedload size. These are an upland boulder zone

TABLE 3.4.   Classification of natural British river channels based on sinuosity and activity (adapted from Ferguson, 1981)

Group	Description	Types	Distribution
Active meandering channels	Erosion and deposition not concentrated at bends and point bars; low sinuosity	Local channel division in small streams; Extensive braiding on large rivers	Very limited (e.g. some parts of Wales and the Welsh Borderland)
Active braided channels	Erosion of outside of bend with deposition on the inside, forming a point bar	Freely meandering across floodplain; confined meandering within valley	Limited (e.g. some of Wales and Welsh Borderland, parts of upland Britain including SW England)
Inactive sinuous and straight channels	Sinuous to straight channels which are not migrating across their floodplains; insufficiently powerful to overcome the resistance of the banks to erosion	Rock-cut: tree-lined gravel-bed rivers; clay lowland rivers	Widespread (the majority of British Rivers, especially in lowland areas)

with a steep gradient and mainly coarse sediments; a floodway zone with a coarse (sand to cobble-sized) sediment load and a moderate gradient; a pastoral zone characterized by a fine bedload material (silt and sand), a sinuous channel and low gradient; and an estuarine zone which has an area of periodic river gradient reversal and is typified by braiding channels in tidal washes. The environments for vegetation, fish and wildlife are significantly similar within each zone and different between zones. However, progression from coarse to fine sediment and steep to low gradient is not necessarily continuous and orderly.

It is important to appreciate that stable natural rivers are not necessarily static. Open systems have an ability for self-regulation. Negative feedback mechanisms moderate the impact of external factors such that a system can maintain a state of equilibrium in which there is some degree of stability. A channel may therefore be described as stable even when it is actively meandering. A channel can adjust its boundaries to obtain and maintain a steady state equilibrium over a short time-scale.

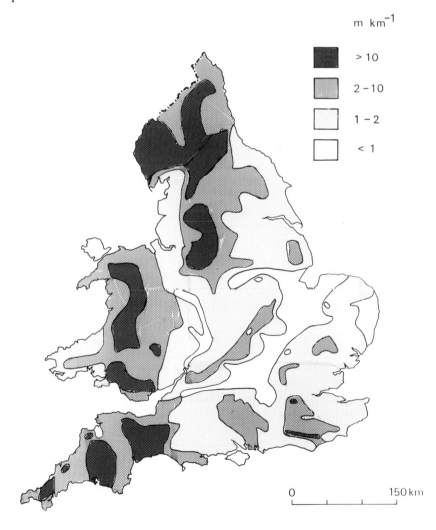

m km^{-1}

> 10

2 – 10

1 – 2

< 1

0   150 km

FIGURE 3.1. Channel slope in England and Wales

As a very tentative first step a map of channel slope may have use in determining the location of sites in England and Wales which are likely to change significantly during a flood (Figure 3.1). Stream power has been recognized by fluvial geomorphologists as an important independent variable determining adjustments of river morphology (Richards, 1982; Brookes, 1987). Stream power is the rate of energy expenditure per unit length of channel (the rate of doing work) and is highly dependent on the channel slope. The mapped values of stream slope (S 1085) originally developed for the Flood Studies Report (NERC, 1975) provide a useful basis. S 1085 is the slope between 10% and 85% of the main stream length in metres per kilometre. Natural alluvial rivers with channel change may be generally coincident with slopes above 2 m km^{-1}, whilst sites which do not naturally erode have slopes below this value. However one of the major problems with this approach, which will need to be extended, is the local variability of channel slope which varies along individual rivers. Neighbouring reaches may have a 10-fold difference in gradient due to varying sizes of river and to geological variation. In particular the balance between discharge and sediment size varies at the local scale (Hack, 1957). For example, Richards (1982) showed that the meandering Afon Elan in Wales (an upland river) no longer has the stream power needed to deform its channel boundaries through active bed scour and bank erosion.

HUMAN IMPACT

Rivers unchanged by human activity are rare, if not non-existent, in England and Wales. Human activity may exacerbate erosion, sedimentation and flooding. In many lowland areas channel modifications are a major part of any classification. Brookes and Long (1992) estimated that for a catchment to the north of London about 96% of the watercourses had been altered directly as a result of straightening, widening or deepening, and this may be typical of much of the River Thames catchment. A channel typology developed for Thames Region has identified 10 major channel types for each catchment, all of which are based on human impact (e.g. semi-natural, channelized, roadside ditches) (National Rivers Authority, 1992). Table 3.5 gives an indication of the types of impacts that may occur, ranging from influences which arise in the upstream catchment to direct channel modification. Channels which under natural circumstances would be inherently stable become subject to channel change, particularly where the bed and bank materials are non-cohesive. Up to 10 000 years ago much of England and Wales was over-run by glaciers and ice sheets, or by river systems

TABLE 3.5. Types of human impact causing channel change

Erosion of bed or banks	Sedimentation
Runoff from urban areas and roads	Construction activitives
River regulation (e.g. water storage by reservoirs)	Mining activities
Gravel extraction from channel	Urban sources
Structures (bridges, weirs)	Agricultural and forestry drainage
Boat wash	
Tree clearance	

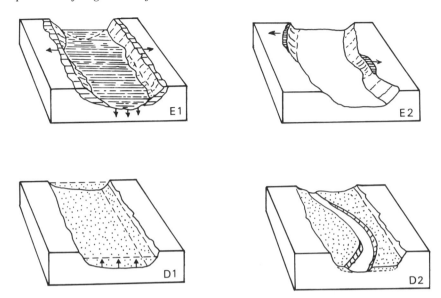

FIGURE 3.2.   Types of channel adjustment (erosion, E1 and E2; deposition, D1 and D2)

which used to drain the former ice margins, and this has given rise to large volumes of
erodible sediment.

Examples of channel change arising from human impact are common throughout England
and Wales. Figure 3.2 shows typical erosion experienced in channels. Erosion (type E1) of
the bed and banks of a watercourse, usually after a high flow, can arise for a number of
reasons. For example, straightening reduces sinuosity and locally increases the slope. The
increase of slope means that the sediment discharge must also be increased by erosion of the
bed and banks. Erosion is a negative feedback mechanism that works towards restored
stability by lowering the channel gradient and increasing the bed material size. Downstream
from an urban area erosion may also take place as a consequence of the frequent and greater
flood discharges. The creation of impervious surfaces and building of more efficient
drainage systems in urban areas increases the volume of runoff for a given rainfall and the
magnitude of flood discharges. For Catterick in North Yorkshire, Gregory and Park (1976)
found that channel capacities were up to 150% larger in urban channels. However from
various studies of the impacts of urbanization no distinction has yet been drawn between the
relative contributions of width and depth to the total change of channel capacity (Knighton,
1984). Erosion (type E2) is characterized by localized bank migration. A typical cause may
be instability in an artificially over-deepened reach, leading to bank collapse. Erosion at high
flow below weirs and other structures may also lead to localized adjustment immediately
downstream.

A variety of human impacts lead to siltation in a channel during low flow which, if not
stabilised by the growth of vegetation, may then be washed out during higher flows (types
D1 and D2). Transient deposits of sediment occur below construction sites, as a result of
agricultural and forestry practices upstream in a catchment, or as a result of direct channel
modification. A cycle of sedimentation and colonization by vegetation typically occurs at
low flows in chalk streams, later to be washed out at discharge rises (Dawson, 1976).

REHABILITATION, RESTORATION AND MITIGATION

Consideration of high flows is essential when designing river rehabilitation or restoration projects or mitigating the impacts of river engineering works. It is necessary to develop solutions to river problems which are sustainable over a range of flows. It is also important to consider the impact of such solutions, which may change their characteristics through time, on the flood flows themselves.

Design for all flows

Flood channels are built with the primary intention of conveying a flood discharge of a particular size. If this involves widening the bed then the velocities at low flow are reduced and so too is the sediment discharge, leading to deposition. Mitigation works for over-wide flood channels include techniques for maintaining a more natural low-flow width, thereby avoiding problems of sedimentation across the whole river cross-section (Brookes, 1991).

In the design of mitigation works for engineered channels, or for the rehabilitation of previously engineered watercourses, it is often desirable to reinstate a suitable substrate or to place rocks for ecological or other reasons. It may be important to ensure stability and this can be achieved by calculating the shear stress exerted on the bed of a channel at the design flow. This will give an indication of the nature and size of material to be placed. Several approaches can be used, including the maximum permissible tractive force approach. Du Boys' equation (1879) for the unit tractive force (τ_0) is:

$$\tau_0 = \omega R\, s$$

where ω is the specific weight of water; R is the hydraulic radius and s is the bed slope.

Maximum permissible values of unit tractive force have been derived from laboratory and field data. A typical value for the movement of coarse gravels in water transporting colloidal silts is 32 N m^{-2} (or a velocity of 1.8 m s^{-1}). For cobbles and shingles the value is 53 N m^{-2}, or a velocity of 1.7 m s^{-1} (Webber, 1971). If the shear stress is not too high then appropriately sized materials can be reinstated. Examples of the application of this method to urban stream channels near London are given by Brookes (1991).

Where the shear stress is too high then alternative methods of placing the substrate can be tried, including rolling angular material into the bed of a channel, and the placement of low weirs across the channel to locally restrain and trap gravels moving along the channel during high flows. Such material will be deposited on the receding limb of a hydrograph. However, this technique relies on a continuous supply of suitably sized materials, moving from upstream.

Where space allows it may be possible to construct a more natural earth channel. Two-stage or multi-stage channels confine the normal range of flows to the original channel, whilst the flood flows are contained within a larger channel constructed above bankfull by widening out the floodplain. Low-flow channels are more effective for conveying bed sediment, because higher velocities are maintained at moderate flows.

The significance of increased resistance for calculations of water level

The hydraulic resistance of a channel can be calculated from the discharge, water surface slope and channel cross-section. Several resistance equations have been developed, including the Chezy equation (1769) and Manning equation (1889). Many authors recommend the Darcy-Weisbach friction equation because of its dimensional correctness and sound theoretical basis. All of the equations assume that the resistance approximates that of a steady, uniform flow but in natural channels the resistance problem is much more complex.

For water level calculations the resistance to flow is incorporated as a number or friction coefficient. The method still commonly used throughout the water industry in England and Wales is Manning's "*n*" value. The larger the value of 'n', the rougher the surface and the more the resistance to flow. The consequence of increased resistance is to give a higher slope to the water surface for a given discharge.

One method of estimating the value of *n* is that developed by the Soil Conservation Service in the USA (SCS, 1963; Urquhart, 1975; French, 1986). This involves the selection of a basic *n* value for a uniform, regular and straight channel and then adding correction factors to modify this value. Each factor must be considered and evaluated independently. Following estimation of a basic value for *n* (Step 1), Step 2 involves estimating modifying values for vegetation, which range from low (0.005) to very high (0.100). The effect of vegetation is to retard flow primarily around stems, tree trunks, branches, etc., and secondarily through the reduction of flow area. The relative importance of vegetation on *n* is a function of the depth of flow and the density, distribution, and type of vegetation. Some aquatic macrophytes may actually wash out during a flood flow, whilst other species such as reed may actually bend with the flow, thereby having less effect on *n* (Watson, 1987; Dawson and Charlton, 1988). Tables for estimating *n* appear in French (1986). However, there needs to be an improved understanding of the effect of different types of vegetation on flow.

Step 3 involves modification for channel irregularity, and must consider both changes in the flow area and changes in cross-sectional shape. In natural channels such irregularities arise from deposition and scour, leading to asymmetrical cross-sections on bends and symmetrical cross-sections in straight reaches. Abrupt changes, such as those found in more natural channels can result in a significantly higher value of *n*. Values range from 0.000 to 0.015 for the changes in channel shape and cross-sectional size and from 0.000 to 0.020 for changes in the irregularity of the surface of the channel.

Step 4 modifies for obstructions. This is determined by the nature and size of individual obstructions which include sediment deposits, boulders, tree roots, fallen trees and logs. Values range from negligible (0.000) to severe (0.060). The effect of obstructions depends on the extent to which they reduce the flow at various depths of flow, the shape of the obstructions and the position and spacing. Step 5 involves modification for channel alignment. Meandering channels will significantly increase the value of *n*. The value is found by summing the values obtained in Steps 1 to 4, to form a subtotal "*n*1". A ratio of a meandering channel (lm) to a straight channel (ls) is then calculated. A modifying value can then be obtained from tables for various values of the ratio lm/ls. The final value of *n* is estimated (Step 6) by summing the results of Steps 1 to 5.

Estimating hydraulic resistance in channels is difficult. Using the Soil Conservation Service method to illustrate the significance of modifying values, it is possible to estimate the potential impact that mitigation devices and river rehabilitation can have on *n*. Three case studies, taken from south-central England, show the range of *n* values that are obtained for different types of channel.

Case study: flood control channel with mitigation
As part of the Lower Colne Flood Alleviation Scheme, intended to reduce flooding and

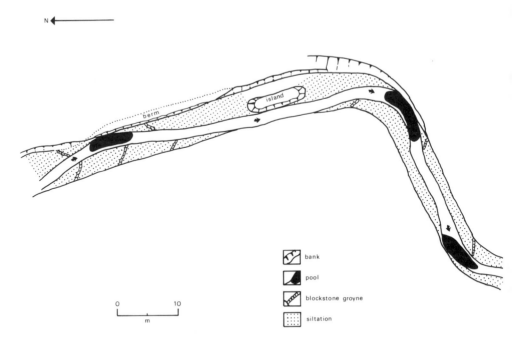

FIGURE 3.3.   Wraysbury River: mitigation works for a flood control channel

provide a 1:100 year level of protection, works were proposed to the Wraysbury River, at Poyle. The channel is adjacent to an industrial park. Although a number of alternative options were carefully evaluated, the chosen option involved widening the existing channel to convey the flood discharge. This option had a significant impact on the existing channel and required an Environmental Statement to be produced (National Rivers Authority, 1991a).

However, the works, which were finished early in 1992, incorporated a number of measures to mitigate the effect of widening the existing channel. A low-flow notch was excavated to a depth of about 200 mm below the design bed level. Low flows are confined to this notch, which approximates the anticipated low-flow width for a natural channel at that location and by a series of carefully sited blockstone groynes which train the flow (Figure 3.3). Water depth of the order of 300 mm is retained at low flow, with velocities of about 0.3 m s^{-1}. A natural pool–riffle sequence was also created (Figure 3.3).

The groynes are also intended to allow for deposition of silt loads carried in the flow, deposited in the wider, shallower areas of the bed. A certain amount of vegetation is anticipated to colonize these areas. Table 3.6 gives an estimation of 0.059 for n, which includes modifying values for the vegetation and for obstructions (blockstone groynes).

The hydraulic analysis undertaken for the design flow conditions made allowance for the groynes, extending over about one-half the width of the channel. The additional flow area provided by the notch and the berm mitigates the effects of the groynes. The low-flow notch itself

TABLE 3.6. Estimation of roughness "n" for the Wraysbury River, a flood channel in south-central England

Step	Comment	Modifying Value
1	Estimation of basic value of "n"	0.018
2	Modification for vegetation (growth of willows, and other bankside vegetation; prolific growth of instream vegetation)	0.015
3	Modification for channel irregularity (designed to allow natural variation of channel shape to develop).	0.010
4	Modification for minor obstructions (appreciable due to instream habitat devices formed from limestone blocks, etc.)	0.015
5	Modification for channel alignment	0.001
6	Estimated value of "n"	0.059

and berm provide an additional flow area. However, there is a need to know how roughness changes over time, particularly if woody vegetation invades, and how to manage the channel in an environmentally acceptable way in the longer term. Clearly for complex projects such as these there is a need for post-project appraisal and the development of a maintenance plan.

Case study: concrete-lined channel proposed for rehabilitation

Table 3.7 estimates the n value for a concrete-lined reach of the River Ravensbourne in south London, proposed to be realigned as part of a development proposal. Given the uniform

TABLE 3.7. Estimation of roughness "n" for a concrete-lined channel in south London

Step	Comment	Modifying value
1	Estimation of basic value of "n" for a concrete channel	0.015
2	Modification for vegetation	0.000
3	Modification for channel irregularity	0.000
4	Modification for obstructions (assumes no urban debris)	0.000
5	Modification for channel alignment (minor)	0.000
	Estimated value of "n"	0.015

FIGURE 3.4.   River Ravensbourne: existing concrete channel

nature of the existing channel, with a concrete finish, a low value for n was estimated (0.015) (Figure 3.4). There are negligible modifying values for vegetation, channel irregularity, obstructions or channel alignment. Hydraulically the existing channel is relatively efficient. The desired environmental solution for the realigned channel was to introduce more sinuosity to the channel, increasing n by about 10%. A low-flow channel with pools and riffles was proposed (modifying value of between 0.025 and 0.050), and varying cross-sections (10% increase of n). The estimate for n was 0.029 for the environmental solution, which represents almost double the original value. Such a high value of *n* would not permit the conveyance of the desired flood flow. Open park land along the left bank could not be used to create a wider two-stage channel, to compensate for loss of conveyance due to an increase of *n*, because of various constraints, including a number of mature trees. Furthermore the need to protect a valuable groundwater resource from contamination meant that floodwaters could not be allowed to inundate the park, and the bed of the channel had to be lined with concrete. In this instance the final solution proposed, although extreme, is a replacement concrete-lined channel but with some bed undulation to form pools and the addition of a limited wetland berm with emergent vegetation.

Case study: rehabilitation of semi-rural watercourse

A 100 m length of the Redhill Brook in Surrey was proposed for realignment in 1991 as part of a development proposal. As a condition of the land drainage consent, the realigned section had to be designed with a natural low-flow width, varying channel cross-sections, and with pools, riffles, point bars and planted with vegetation (Figure 3.5). The low-flow width varies from 2.5 m on each bend to 3.0 m in the straight sections.

There are therefore several modifying values for *n* (Table 3.8), including those for vegetation growth and channel irregularity. The value for *n* is 0.046, slightly higher than the original channel. Given that this is a relatively high value and the capacity of the channel is small, then there is likely to be over-bank flooding. For this particular site this is unlikely to be a problem given that the channel is confined within a relatively steep-sided valley and that a narrow culvert above the site constricts the flow through a railway embankment. However, it is important to realise that in other circumstances recommendations for the addition of natural characteristics to channels can have implications on flood conveyance.

TABLE 3.8. Estimation of roughness "*n*" for the Redhill Brook, a restored channel in Surrey

Step	Comment	Modifying value
1	Estimation of basic value of "*n*" for an earth channel	0.020
2	Modification for vegetation (some growth of bankside vegetation and reeds; limited instream vegetation due to shade)	0.010
3	Modification for channel irregulatiry (natural variation of channel shape, causing main flow to shift from side to side)	0.010
4	Modification for obstructions (negligible: no exposed roots, boulders or fallen trees)	0.000
5	Modification for channel alignment (appreciable degree of meandering)	0.006
6	Estimated value of "*n*"	0.046

FIGURE 3.5. Redhill Brook: river rehabilitation works before flow was diverted to the new course

DISCUSSION AND RECOMMENDATIONS FOR FURTHER WORK

Understanding the significance of high flows in changing channel morphology is a key component of the ecological basis for river management. Whilst conventional ecological surveys may record the location of pools, riffles and other channel features, the value of a geomorphological approach is in identifying which channels are likely to change and the nature of that change. This approach is founded on knowledge and experience acquired in

England and Wales, and elsewhere, over the past 25 years. There are problems in identifying the causative agent of change and in particular disentangling natural change from that induced by human impact. Accurate prediction of river and catchment response to various impacts is often problematic because the sensitivity of geomorphic and hydrologic systems is variable spatially and temporally. However, identifying the causative agent is a fundamental prerequisite to solving present river problems.

Whilst geomorphological studies have focused primarily on the response of channels to, and recovery from, infrequent floods there is a need for sequential observations of channel change relative to a variety of flows. Geomorphological factors of natural alluvial river channels which are important for biological diversity change over a variety of time-scales, ranging from hours to hundreds of years. Whilst numerous studies have examined the biological response of a changed substrate, for example as a result of scour during a flood, there is little knowledge of the impacts of changes of other factors. A valuable area of research would be to relate changes of cross-sectional shape and size, point bars and islands, to biological adjustment. In the longer term, over time periods in excess of 100 years, ecological changes may accompany changes of river channel pattern.

Another way forward is the development of a channel typology for England and Wales, in particular incorporating stability assessment into an ecological classification. If physical environment is combined with human impact then there may be in excess of 50 channel types.

A fundamental area for further research in England and Wales is the development of sustainable solutions to river problems. Mitigation, enhancement, rehabilitation and restoration measures need to be designed to be stable across the whole range from low to high flows. Design calculations for capital, maintenance and enhancement works need to take roughness into account, particularly allowing for the effect of vegetation, channel alignment, obstructions and channel irregularity. There is a need to calculate roughness values for a range of schemes already carried out, in particular improved understanding of the effect of different types of vegetation on flow. From an ecological point of view there is a requirement to understand how roughness changes over time and how to manage the channel in an environmentally acceptable way in the longer term. Given the large extent of environmental improvements recently carried out on watercourses in England and Wales, relatively little work has been carried out on monitoring and appraising changes after construction. If such work was undertaken and documented then more effective maintenance plans could be developed in the future.

ACKNOWLEDGMENTS

The views expressed in this paper are those of the author and not necessarily those of the National Rivers Authority. Permission to publish this paper by the Regional General Manager of Thames Region is gratefully appreciated.

4

The Ecological Basis for Torrent Control in Mountainous Landscapes

STANIMIR KOSTADINOV

University of Belgrade, Belgrade, Yugoslavia

WATER EROSION AND SEDIMENT LOSS

Water erosion is a complex process which starts with the detachment of particles from the surface of soil by rain or from the stream channel bed by friction. The particles are then transported from the upper to the lower parts of the catchment by the energy of surface water running down the slopes or in the channel (Figure 4.1). Erosion processes in the drainage area disturb the hydrological cycle in such a way that is reflected in a drastic decrease in the quantity of water usable by man and an increase (over three orders of magnitude) in destructive floods. When sediment reaches waterways and disturbs the natural runoff regime, torrents are formed. These are two-phase discharge events in which the content of the solid phase (sediment) may be up to 60% of the total volume of the torrent.

Water erosion and torrents cause enormous damage to agriculture, waterpower engineering, traffic, settlements and occasionally to human life. Most of the territory of Serbia is mountainous, and approximately 90% of the area has been affected by erosion processes of different intensities. Sediment yield and transport, related to erosion, is also considerable (Table 4.1). Damage caused by erosion is initially loss of the upper fertile soil horizon, coupled with loss of organic and mineral nutrients from steep and ploughed land, leading to inadequate water and water relations in the soil.

Several substances are transported alongside sediment which may constitute a danger to the environment. Animal manure, fertilizers and pesticides constitute the main threats to water quality, together with the physical blockages of waterways and reservoirs by sediment accumulation. Suspended sediment makes water more difficult to treat for potable supply and to use for agricultural and industrial purposes. Ecological effects include reduced light penetration hence lower photosynthesis in rivers, and sediment accumulation in gravels damaging fish and invertebrate habitat.

The sediment composition and transport have been studied in three catchments in west Serbia which are 60–80% forested and 2–10% tilled. A high concentration of organic matter was found as well as mineral nutrients (Table 4.2) (Kostadinov, Stanojevic and Topalovic, 1992). These data may be compared with an agricultural average of 3 kg t^{-1} N, 1.7 kg t^{-1} P and 20 kg t^{-1} K (Zaslavski 1987).

The Ecological Basis for River Management. Edited by D.M. Harper and A.J.D. Ferguson. © 1995 John Wiley & Sons Ltd

FIGURE 4.1.   Erosion on a hill slope in the southern Morava catchment

TABLE 4.1.   Sediment yield and transport in Yugoslavia

Republic	Annual yield		Annual transport	
	Total (m³ ann⁻¹)	Specific (m³ ann⁻¹ km⁻²)	Total (m³ ann⁻¹)	Specific (m³ ann⁻¹ km⁻²)
Serbia	37×10^6	421.6	9×10^6	105.8
Montenegro	3×10^6	275.1	2×10^6	152.2
Yugoslavia	41×10^6	401.8	11×10^6	112.1

TABLE 4.2. Nutrient content (kg tonne^{-1}) of torrential sediment in three catchments in Yugoslavia

Catchment	Year	Humus	N	P	K	Ca	Mg
Lonjinski	1986	92.2	5.3	0.9	4.9	14.0	8.7
potok	1987	60.7	3.4	0.6	3.6	6.0	5.1
	1988	48.1	2.7	1.0	3.6	7.6	5.1
Djurinovac	1986	93.1	4.2	0.8	5.5	12.0	6.6
potok	1987	94.1	4.3	0.7	6.1	13.2	11.2
	1988	84.8	4.0	—	—	—	—
Dubosnicki	1986	31.4	1.6	0.6	2.0	10.4	7.0
potok	1987	68.9	5.0	0.9	2.8	22.2	11.2
	1988	209.3	11.2	2.3	3.1	10.2	9.8

FIGURE 4.2. Example of a check dam

STRATEGIES FOR TORRENT CONTROL

The main strategy for torrent control is twofold: the control of catchment erosion processes coupled with channel works to check the high kinetic energy of torrents and increase resistance.

Works in the catchment consist of forest and engineering works such as different afforestation practices (pits, batches, bench terraces), the construction of contour furrows, wattles and shelterbelts and coppice reclamation. These supplement well-known agricultural engineering practices such as crop rotation, terracing, contour farming and strip-cropping. Both sets of measures need to be integrated with regional management practices emphasising erosion control and administrative structures (legislation) which control such activities as tree clearance and over-grazing. Wide-ranging, extension educational measures are also necessary.

Channel works consist of two classes of structures: cross-sectional structures such as check dams, sills and groins, the aim of which is to check the high kinetic energy of torrents (Figure 4.2); and longitudinal structures such as embankments and revetments, the aim of which is to increase the scouring resistance of the channel. Small-scale storage structures, such as reservoirs for both checking both water and sediment, may also serve as irrigation or recreation facilities in other seasons.

TORRENT CONTROL IN YUGOSLAVIA

Torrent control measures had commenced in the last century, but most organised works commenced at the beginning of the 20th century, associated with protection of railways. Typically several check dams were constructed upstream of the railroad with protective linings of masonry or concrete upon it. Until the end of the Second World War, there was little or no catchment-based erosion control. The situation changed markedly after the 1960s, with the expansion of highway construction; the influence of ideas from outside the country resulted in a more holistic approach. Table 4.3 summarises total control works over nearly a century in Serbia; all the afforestation was undertaken post-1960, with a rate of 1.5 $\times$ 10^3 ha ann^{-1} by the 1980s.

TABLE 4.3.   Torrent control works in Serbia, 1868–1988

Nature of control technique	Extent
Works in the catchment area:	
Walls, terraces, contour ditches	13 610 ha
Afforestation by different methods	32 720 ha
Forest reclamation	7 802 ha
Pasture and measdow reclamation	35 500 ha
Orchards and tree planting in endangered zones	494 ha
Works in the channel:	
Transverse stone, masonry and concrete structures	474 981 m³
Longitudinal stone and concrete structures: volume	868 360 m³
length	4 448 km

FIGURE 4.3.   Example of non-ecological torrent control

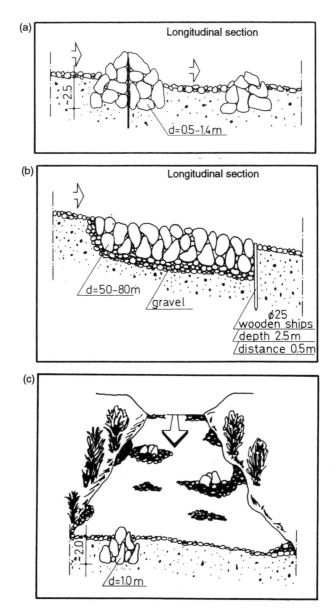

FIGURE 4.4.   Examples of the application of ecologically designed (a) sills, (b) rip-rap and (c) piles of stones

THE ECOLOGICAL BASIS FOR TORRENT CONTROL

Biological and bioengineering works in the catchment, such as forest and agricultural engineering, by their very nature have an ecological basis, re-creating conditions for soil structural improvement and vegetational diversity. Channel works, by contrast, have been more difficult to modify to ecological principles. Transverse structures or masonry revetment and channel lining, for example, can prevent the migration of fish. Many longitudinal channels are lined with masonry which destroys the main streambed habitats, and revetments often destroy the hydraulic connection between stream and groundwater (e.g. Figure 4.3).

Ecological designs have most potential in rural areas, where the need for physical protection is more relaxed. Large stones instead of concrete offer a solution more ecologically and often economically more acceptable for river training (Figure 4.4). Facines made of willow or other wood offer more acceptable protection than concrete or masonry (Figure 4.5).

Ecological ideas are gaining ground in mountainous eastern European countries suffering from high torrent erosion. They are most applicable in lower reaches and floodplains (gradient <10%) and for rivers with permanent flow. They are least applicable in urban areas where a high degree of protection is required or in streams whose only flow is during torrents.

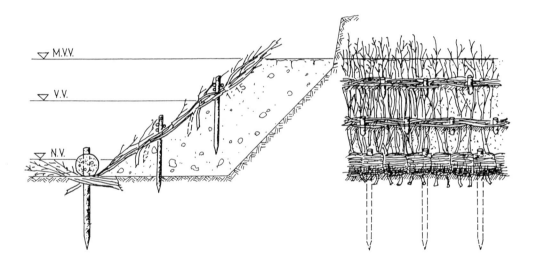

FIGURE 4.5.   Examples of the application of ecologically designed wattles and fascines

5

Faunal Community Change in Response to Flow Manipulation

Patrick D. Armitage

NERC Institute of Freshwater Ecology, Wareham, Dorset, UK

INTRODUCTION

Water flow in stream channels is conveniently described by water resource managers and hydrologists in terms of discharge, i.e. the amount of water passing a given point per unit time. Discharge is however, of little direct interest in studies which are primarily focused on stream biology. In such studies it is important to know the hydraulic conditions where the animals and plants live (Adams et al., 1987; Statzner, Gore and Resh, 1988; Dodds, 1991a).

Current has long been recognized as the most significant feature of running water (Hynes, 1970b). Variations in current or water velocity, through its action on the substratum, result in a diversity of microhabitats to which the benthic fauna is adapted. As discharges are reduced so the pattern of velocity distribution will change across and along a channel. Baxter (1961) noted that most of the beds of Scottish salmon rivers were still covered by water at a discharge as low as one-eighth of the mean annual discharge but small streams began to contract at one-half the mean discharge. At one-eighth, only one-third to one-half of the bed remained wet, dependent on channel shape. In rivers with dense stands of instream macrophytes the impact of low discharge may be reduced by the maintenance of high water levels as a result of macrophyte growth (Hearne and Armitage, 1993).

This chapter examines the responses of benthic faunal communities to both changes in velocity and macrophyte growth in two experimental channels with low discharges. The specific objective is to test the hypothesis that artificial flow manipulation will create a greater range of community types than can exist in unmanaged control sections. Ancillary objectives include examination of the response of communities of macroinvertebrates to low flow and an assessment of the role of such experiments in aiding stream management decisions.

STUDY AREA AND METHODS

The two channels (channels 3 and 4) are situated in the grounds of the Institute of the Freshwater Ecology's River Laboratory at East Stoke in Dorset. They are fed by valved pipes from the Mill Stream which is a side channel of the River Frome (Figure 5.1). Both

---

The Ecological Basis for River Management. Edited by D.M. Harper and A.J.D. Ferguson. © 1995 John Wiley & Sons Ltd

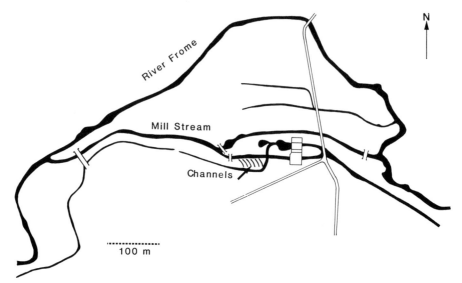

FIGURE 5.1.   Location of the channels in relation to the main River Frome and the Mill Stream

channels are identical in shape with a trapezoidal cross-section and a base substrate dominated by gravel (estimated volumetric proportions of particle sizes, 85% 11−25 mm, 5% 2−11 mm, 5% 0.35−2 mm, 5% <0.35 mm). The intake is larger in channel 4 with the result that discharge is about 3.4 × that of channel 3. The characteristics of the experimental channels are listed in Table 5.1 and discharge in the Mill Stream throughout the study period (April−September) is shown in Figure 5.2. The water chemistry was similar in the parent River Frome, the Mill Stream and in the two channels throughout the study period: a summary of conditions in June is presented in Table 5.1.

At the start of the experiment in April the substratum in both channels was identical, with macrophyte cover less than 1%. Areas were selected in the downstream third of each channel and a grid system marked out (Figure 5.3). Zones for flow manipulation were identified within this grid. Flow velocities were determined 5−10 mm above the substratum in each cell with a Kent miniflow electric current meter (propeller diameter 10 mm) and depth measurements were made. Velocities over macrophytes were measured 5 mm above the plant. All measurements were made from small bridges laid across the channel and at no time did the operator stand in the stream.

Following these physical measurements, biological samples (1−24 in Figure 5.3) were taken. A PVC corer (internal diameter 100 mm) was driven into the channel bed at the sampling point. Any contained macrophyte was removed and preserved and the remaining substrate was disturbed. Animals and detritus in suspension were pumped into a 500 μm sieve, resuspended and preserved in formalin solution, sorted into 70% alcohol and identified to family level.

After this pre-manipulation estimate of the faunal community in each channel was made, the flow-modifying funnel device was placed in the position indicated in Figure 5.3. This resulted in the formation of five zones: A, slow (samples 1, 5, 9); B, fast (2, 6, 10); C, medium slow (3, 7, 11); D, medium fast (4, 8, 12); and E, control (13−24). It was recognized that this latter zone could be affected by slight backing up of water but it represented a better option than using an adjacent untreated channel. The mouth of the

TABLE 5.1. Physical and chemical features of channels 3 and 4, with comparative chemical data for the River Frome and the Mill Stream (R = *Ranunculus*, E = *Elodea*, A = filamentous algae)

Physical feature	Channel 3	Channel 4
Length (m)	10.2	24
Width (m)	1.3	1.25
Intake diameter (cm)	10.8	24.8
Mean velocity, April (cm s^{-1})	10.7	20.5
Mean depth, April (cm)	5.8	10.7
Discharge, April (litre s^{-1})	8.1	27.4
Macrophyte cover (%):		
April	<5 R	<5 R
June	43 A	38 R/A
September	63 A	61 R/E

Chemical feature	Channel 3	Channel 4	Frome	Mill Stream
Date	23 June	23 June	23 June	23 June
pH	7.92	7.92	7.93	7.98
Calcium (Ca)(mg litre^{-1})	101	103.6	101.2	103.2
Conductivity (mhos)	505	510	500	510
Alkalinity (HCO$_3$) (mg litre^{-1})	4.09	4.08	4.06	4.09
Nitrate (NO$_3$) (mg litre^{-1})	3.67	3.7	3.63	3.8
Phosphate (PO$_4$) (μg litre^{-1})	1.09	110	102.5	107.5
Silica (mg litre^{-1})	2.75	2.76	2.81	2.83

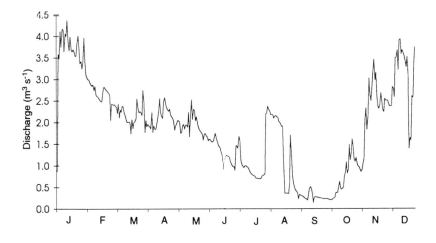

FIGURE 5.2. Discharge in the Mill Stream

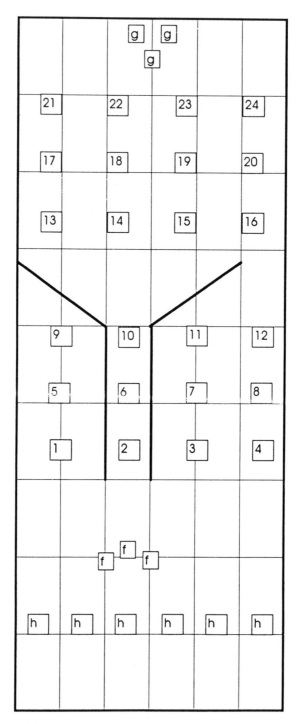

FIGURE 5.3. Sampling grid, showing the position of the flow manipulating device and the location of
sites

funnel was constructed of marine ply but the sides were made of glass to reduce the effects of light exclusion. The funnel measured 95 cm across its mouth and the straight section was 104 cm. The device was left in situ throughout the experiment and both physical and biological observations were made in June and in September in both channels. Samples 3 and 4 from channel 4 in April could not be processed because of decomposition.

A further "habitat zone" was created with the use of artificial substrate samplers. A set of four samplers were placed downstream of the experimental area (f in Figure 5.3) in May and sampled in June. Subsequently two sets of three were placed both downstream and upstream (g in Figure 5.3) of the experimental area and sampled in September. The samplers (Welton et al., 1982) were 100 mm lengths of PVC pipe (internal diameter 80 mm) with 11 mm diameter holes regularly spaced at 4−6 mm intervals. A plastic mesh bag (apertures 20 mm knot to knot) was placed inside, supported by a wire frame. A 100 mm square sheet of black polythene was placed at the base of the bag to retain the contents on removal. Each sampler was filled with gravel from the channel and placed on the substratum, not buried as in Welton et al., (1982). The objective was to create a new habitat not to attempt to simulate natural conditions. Samplers were removed by lifting upstream of a 500 μm net to retain any displaced fauna.

Further sets of samples were taken in June and September across a transect of the channels, downstream of the main experimental area (h in Figure 5.3). These samples were taken by corer as for the main experiment.

RESULTS

Depth and velocity distribution

Changes in these two parameters as a result of flow manipulation and subsequent seasonal effects are shown in Figure 5.4. Variability in depth and especially velocity increased markedly after placement of the funnels. Pre-manipulation velocities ranged from about 9 to 22 cm s^{-1} in both channels. This range increased from 0 to 65 in channel 3 and from 0 to 85 cm s^{-1} in channel 4 after placement. In the control zone E, variation in depth and velocity increased with time as a result of macrophyte growth and the greatest range of values was observed in September (0−39 cm s^{-1} in channel 3, 0−27.5 cm s^{-1} in channel 4).

The changes in velocity were rapidly accompanied by changes in the substratum and by June the base substrate of gravel in the two slow-flow zones (A and C) was overlain by a layer of silty mud. In the two fast-flow zones (B and D) the fine particles were washed away, exposing a clean gravel surface. The control zones were predominantly silty but possessed a range of substrate conditions linked closely with the distribution of macrophytes and algal growths.

The distribution of vegetation

The seasonal changes in the distribution of macrophytes and algae are illustrated in Figure 5.5. There were major differences in the type of cover in each channel. Although there were small patches of macrophytes in June (*Ranunculus* and *Elodea*) in channel 3, by September the area was dominated by filamentous algae (*Cladophora*). In contrast, channel 4 rapidly became dominated by dense growths of *Ranunculus* and *Elodea*. In both channels the fast zones (B) remained free of dense vegetation throughout the study period although

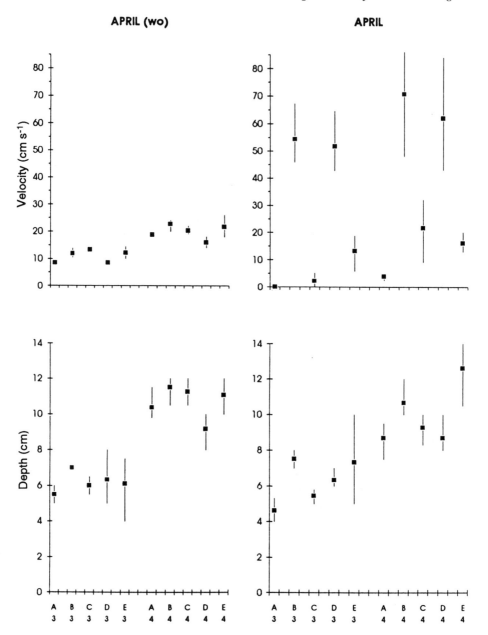

FIGURE 5.4. The range of velocity and depth in the habitat zones in channels 3 and 4 before (April
(wo)) and after (April, June and September) flow manipulation

observations in July showed a severe reduction in fast-flow area. Zone D was more affected
than B in channel 3 because of encroachment of algae growing on the stone lining of the
channel. After a period of high discharge in the Millstream in August (Figure 5.2) both
channels were relatively unaffected but there was a slight increase in exposed gravel in the
fast zones.

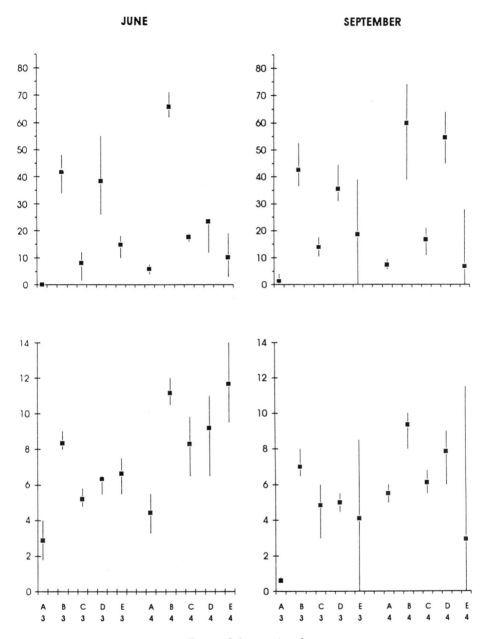

FIGURE 5.4. *continued*

The faunal composition

The major contributor to the fauna of the channels is the Mill Stream. Although no samples were taken from the stream during the experiment, comparison of the qualitative composition of the channels with that of the Mill Stream fauna sampled in April, July and September 1992 show close similarity (Sorensen) of 80% (channel 3) and 79% (channel 4).

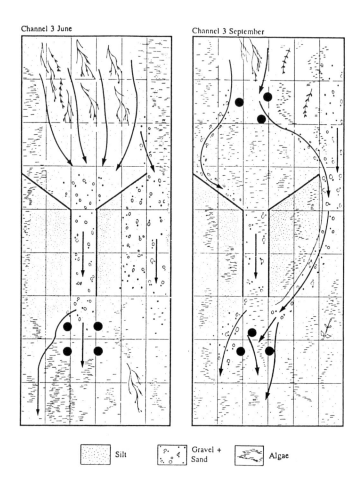

FIGURE 5.5.   The distribution of vegetation in channels 3 and 4 in June and September. The black circles represent the position of the artificial samplers

The similarity did not extend to quantitative data but since the method used in the Mill Stream (i.e. timed net sweeps) differs from that used in the channels this is not surprising. The chief differences are due to the large numbers of oligochaetes and Sphaeriidae in the two channels.

Figure 5.6 shows the faunal composition of the two channels based on total numbers taken on the three sampling occasions in the zones A, B, C, D and E. The main differences were due to very high densities of *Hydra* (included in "others") in channel 4, higher densities of gastropod molluscs (Valvatidae, Hydrobiidae and Lymnaeidae) in channel 3, and higher densities of Crustacea and Trichoptera in channel 4. Seasonal changes were marked by increasing densities between April and September.

Zone A in both channels was dominated by Oligochaeta and Sphaeriidae. Zone B showed some difference between channels, with channel 3 being dominated by gastropods and channel 4 with a slightly more varied fauna with Crustacea dominant. Zone C was similar to B in composition but with more crustaceans and Trichoptera in channel 4. Zones D and B

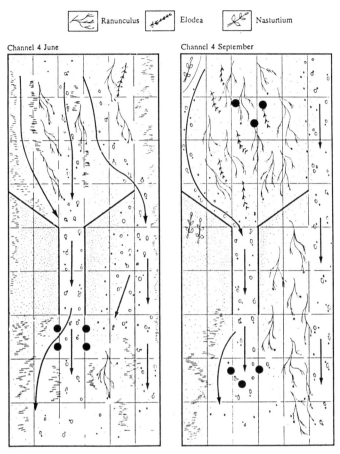

FIGURE 5.5. *continued*

were similar in each of the two channels. In channel 4, zone D supported a greater range of taxa than B. The composition of the artificial samplers (zones F and G) and the extra core samples downstream of the experimental area (zone H) in September was generally similar to that of the control sections but the proportions of the faunal groups were more even.

Temporal and spatial variation

Faunal composition and abundance varied markedly between habitat zones. The mean and confidence limits for total numbers and total taxa per sample are presented in Figure 5.7 for the three sampling occasions. Analysis of variance was used to determine the significance of differences between zones and between channels. The results are presented in Table 5.2. In April, before the flow manipulation device was installed, there were significant differences in the numbers of animals and numbers of taxa in each channel. In June neither numbers of taxa or numbers of invertebrates differed significantly between the channels. By September the numbers of taxa were significantly different in the two channels but the densities were not. The differences between habitat zones (treatments) were not significant in April.

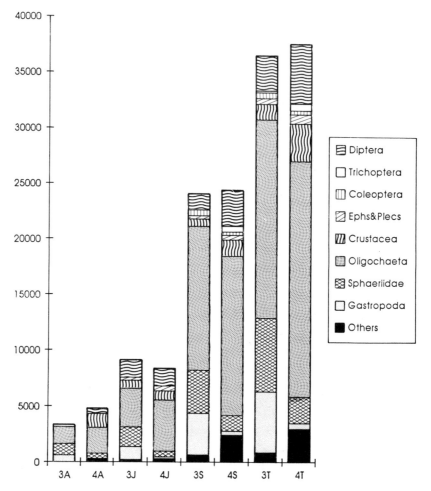

FIGURE 5.6. The variation in faunal composition and density based on the total number of invertebrates taken in habitat zones A, B, C, D and E, in channels 3 and 4 in April, June and September and in all seasons (T)

Subsequent zone differences in both numbers of taxa and faunal abundance were significant. In single factor analyses, taking each channel separately, between-treatment differences in the numbers of taxa were significant in June and September but not in April. Faunal numbers showed more anomalous results with significant differences between treatments in channel 3 in April and June and no significant difference in September. In contrast, in channel 4, differences between habitat zones were significant only in September. Considering the overall variation, differences between channels were not significant for total densities but all other differences (seasons and treatments) were significant for both numbers of taxa and total densities.

Incremental data

The rate at which the number of taxa taken in each zone increased was examined in the two

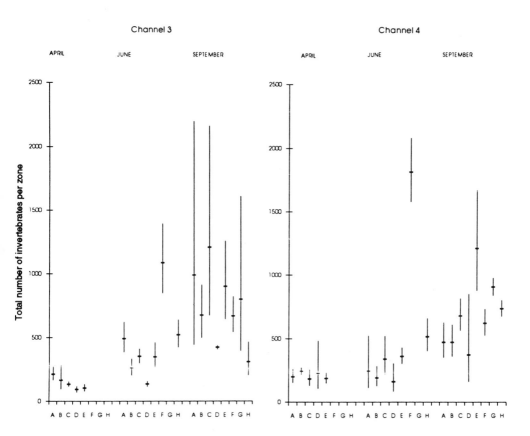

FIGURE 5.7.   The total numbers of invertebrates and total number of taxa per habitat zones (A−H) in April, June and September in channels 3 and 4. (Geometric mean and 95% confidence limits)

channels (Figure 5.8). In the regularly sampled zones with three replicates per season the least number of taxa were recorded in zone A which had negligible flowthrough of water. Zone C, another "slow velocity" habitat also had low numbers of taxa. The highest number of taxa was recorded in the zone with the fastest flow (B).

In the zones sampled only in summer and autumn, or only in autumn, the most taxa were found in the artificial sampler zone (F) below zone B. This was true for both channels. Comparison is made difficult by the varying numbers of samples taken per zone but the top half of Figure 5.11 compares like with like. Here zones A, B, C and D have been combined to compare with the control zone E. By the end of the experiment the combined A, B, C and D zones contained slightly more taxa than the control zone E.

Relationship with environmental variables

The relative importance of the three variables velocity, depth and occurrence of macrophyte, in determining the faunal characteristics of the two channels was examined by means of

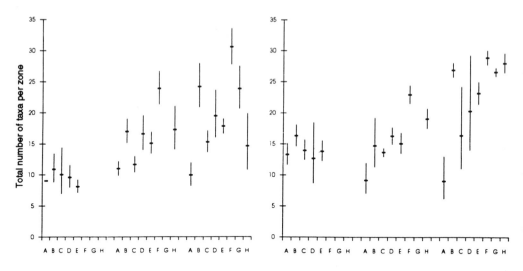

FIGURE 5.7. *continued*

multiple regression analysis (Table 5.3). In the analysis of all samples, relatively small amounts of the overall variance are explained by the regression equations. This is because of the "noise" in the data set. When this is reduced by considering the mean values of the faunal parameters in the habitat zones which were sampled in every season at the same intensity, the $R^2\%$ values are generally higher. An indication of the relative importance of the variables is given by the probabilities in the table. The abundance of Baetidae, Gammaridae, Hydropsychidae and Orthocladiinae is most influenced by velocity (high abundances/high velocities). In the mollusc families Valvatidae, Hydrobiidae and Sphaeriidae, abundance is related to depth (low depth/high abundance). Some organisms such as Ephemeridae are influenced both by velocity and depth (high abundance/high velocity and increased depth). Oligochaeta and total abundance per sample are most dependant on the occurrence of macrophytes and high mean numbers of taxa per sample are related to high velocities. Although these analyses help to explain why certain faunal groups are distributed as they are, the best indicator of changing conditions is the composition of the faunal community considered as a whole.

Community analysis

The entire matrix of 67 taxa and 186 samples was analysed using CANOCO (Ter Braak, 1988). The data were ordinated by canonical correspondence analysis (CCA) which attempts to explain the responses of the taxa by ordination axes that are constrained to be linear combinations of, in this case, the three variables velocity, depth and the occurrence of macrophytes. Figure 5.9 plots ordination sample scores (axis 1 = x axis, axis 2 = y axis) which are linear combinations of the environmental variables. Channel/season components have been separated from the total plot to illustrate more clearly the shifts in community composition throughout the study period. The polygons circumscribe the data points for each

TABLE 5.2. Results of ANOVAs on log 10-transformed values of numbers of taxa and total invertebrate densities, where (A) channel data for each season are combined, (B) channel data are treated separately, and (C) seasonal data from both channels are combined. The degrees of freedom (df) for the F-statistic are given with the associated probability values (p). Significant values of $p < 0.05$ are in italic

A

Month Source of variation	April			June			September		
	df	F	p	df	F	p	sd	F	p
Taxa									
Channel	1	57.59	0	1	0.17	0.682	1	8.01	*0.007*
Treatment	4	1.77	0.153	4	8.92	0	4	28.32	*0*
Error	40			42			42		
Total	45			47			47		
Numbers									
Channel	1	17.65	0	1	0.56	0.457	1	0.23	0.636
Treatment	4	2.05	0.106	4	6.84	*0*	4	4.79	*0.003*
Error	40			42			42		
Total	45			47			47		

B

Source of variation	April			June			September		
	df	F	p	df	F	p	sd	F	p
Taxa									
Channel 3									
Treatment	4	1.76	0.179	4	4.13	*0.014*	4	16.6	*0*
Error	19			19			19		
Total	23			23			23		
Channel 4									
Treatment	4	0.68	0.613	4	3.92	*0.017*	4	11.79	*0*
Error	17			19			19		
Total	21			23			23		
Numbers									
Channel 3									
Treatment	4	3.17	*0.037*	4	3.45	*0.028*	4	1.27	*0.315*
Error	19			19			19		
Total	23			23			23		
Channel 4									
Treatment	4	0.40	803	4	2.39	0.087	4	4.3	*0.014*
Error	17			19			19		
Total	21			23			23		

C

Stacked data Source of variation	Taxa			Numbers		
	df	F	p	df	F	p
Seasons	2	53.24	0	2	133.18	0
Channels	1	22.31	0	1	1.37	0.244
Treatments	4	14.17	0	4	6.56	0
Error	134			134		
Total	141			141		

habitat zone and present a visual representation of the range of community scores within each season and channel.

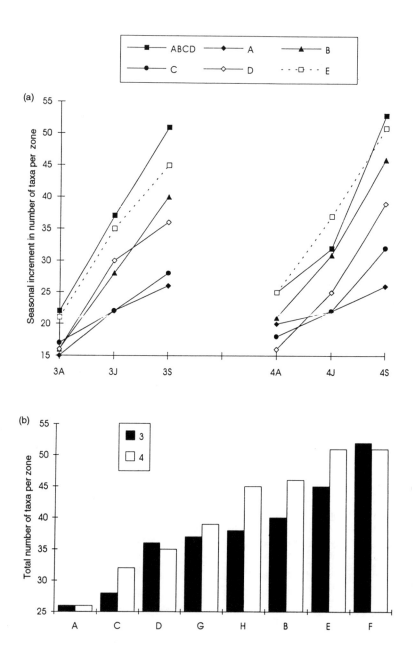

FIGURE 5.8.   (a) Seasonal increment in number of taxa per habitat zone. (b) Total number of taxa per habitat zone in channels 3 and 4

TABLE 5.3. The relationship between the variables, velocity (*V*), depth (*D*), occurrence of macrophytes (*M*) and the abundance of 11 common taxa, mean total abundance and numbers of taxa per sample. Multiple regression analysis on transformed (log *n* + 1) data; r^2 values with the probabilities associated with t-ratios for each variable in the multiple regression; (a) for all samples (*n* = 186) and (b) for mean values per habitat zone (*n* = 30), excluding the samples from zones F, G and H

Taxon	r^2	*V*	*D*	*M*
(a) *All samples*				
Valvatidae	9.9	0.256	0.000	0.133
Bydrobiidae	16.3	0.004	0.000	0.312
Sphaeriidae	15.8	0.081	0.006	0.278
Oligochaeta	35.7	0.558	0.210	0.000
Asellidae	14.1	0.000	0.450	0.001
Gammaridae	20.9	0.000	0.000	0.050
Baetidae	27.7	0.000	0.145	0.000
Ephemeridae	19.5	0.000	0.000	0.045
Hydropsychidae	20.8	0.000	0.000	0.144
Tanypodinae	26.0	0.003	0.029	0.114
Orthocladiinae	24.2	0.000	0.000	0.001
Total numbers	44.4	0.000	0.000	0.000
Numbers of taxa				
(b) *Zones*				
Valvatidae	34.5	0.122	0.005	0.924
Hydrobiidae	29.6	0.051	0.007	0.863
Sphaeriidae	52.5	0.366	0.078	0.066
Oligochaeta	52.9	0.258	0.481	0.003
Asellidae	31.4	0.161	0.192	0.283
Gammaridae	38.9	0.003	0.665	0.853
Baetidae	41.0	0.001	0.369	0.337
Ephemeridae	36.0	0.001	0.021	0.966
Hydropsychidae	29.4	0.003	0.059	0.997
Tanypodinae	64.5	0.013	0.312	0.010
Orthocladiinae	36.8	0.002	0.084	0.239
Total numbers	56.1	0.356	0.081	0.002
Numbers of taxa	53.3	0.000	0.713	0.033

Table 5.4 lists the ordination characteristics. All eigenvalues are low, indicating a low variation in taxon composition within the data set. The percentage variance accounted for by the three axes is low but this is to be expected in macrofaunal data sets (Verdonschot and Higler, 1989). The environmental variables describe the biological variation well as indicated by the high species—environment correlations. Table 5.4 also shows the results of an analysis of variance of ordination scores. Differences between channels were significant in April and June but not so in September for axis 1. Treatment differences were significant in June and September but not in April for both axes. A marginally significant channel/ treatment interaction for axis 1 suggested that the channels influence score values but that this influence varied with treatment. Similar but less significant results were noted for axis 2.

In April, before the artificial flow modification, all polygons are situated close together with a considerable amount of overlap. By June there is a pronounced division into habitat

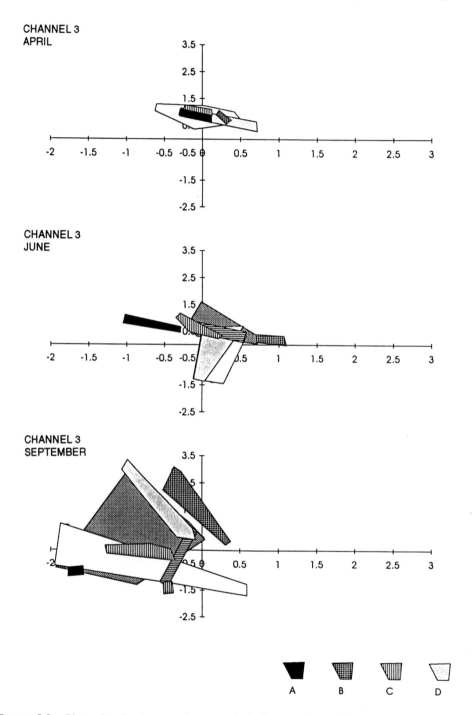

FIGURE 5.9.   Plots of ordination sample scores (axis 1 = *x* axis, axis 2 = *y* axis) derived from Canonical Correspondence Analysis of the entire sample site matrix. The channel and seasonal components have been separated from the total plot to illustrate spatial and temporal changes. The polygons surround the data points for each habitat zone

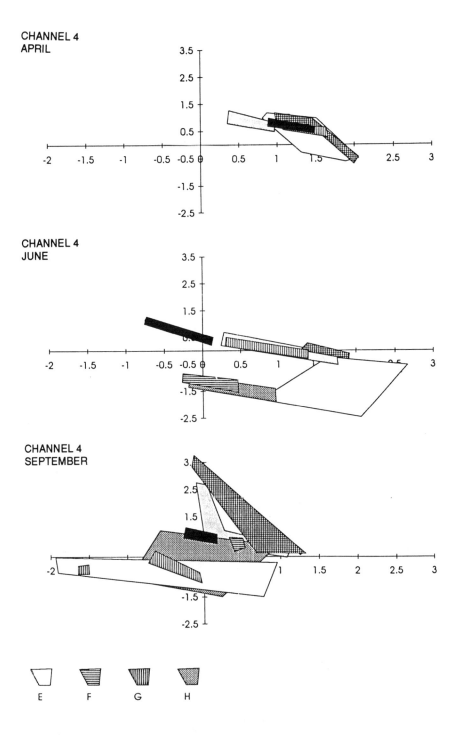

FIGURE 5.9. *continued*

TABLE 5.4. CCA ordination characteristics for (a) the first three axes and (b) results of ANOVA on ordination scores of axis 1 and axis 2. The degrees of freedom (df) for the F-statistic are given with the associated probability values (p). Significant values of $p < 0.05$ are in italic

(a) CCA ordination characteristics for the first three axes

Axes	1	2	3
Eigenvalues	0.078	0.034	0.024
Percentage of variance	4.1	5.9	7.2
Species—environment correlations	73.3	55.7	57.3

(b) ANOVA on ordination scores of axis 1 and axis 2

Source of variation	April			June			September		
	df	F	p	df	F	p	df	F	p
ANOVA on ordination scores of axis 1									
Channel	1	113.64	*0*	1	50.78	*0*	1	6.03	*0.019*
Treatment	4	2.15	0.095	4	22.19	*0*	4	4	*0.008*
Channel/treatment	4	2.09	0.102	4	2.69	*0.045*	4	2.45	0.06
Error	36			38			38		
Total	45			47			47		
ANOVA on ordination scores of axis 2									
Channel	1	3.59	0.066	1	3.76	0.06	1	0.16	0.694
Treatment	4	0.31	0.867	4	7.69	*0*	4	15.23	*0*
Channel/treatment	4	0.73	0.577	4	1.88	0.135	4	1.36	0.266
Error	36			38			38		
Total	45			47			47		

zones. The control section zone E in channel 3 constitutes an intermediate area between the slow (A) and fast (B) zones. The artificial samplers at F and the transect zone H also fall into this intermediate area. In channel 4 the zones are less similar to one another and this is reflected in their degree of separation. By September the control zone and the downstream zone H show a large amount of overlap but the other zones are fairly well separated. The main point to emerge from this analysis is the increasing range of community types. In April there was a very small range of community types and the faunal composition was relatively homogeneous. By the end of the study heterogeneity had increased noticeably. This was not a function solely of artificial flow modification as can be seen by comparing the areas occupied by the control zone E. In both channels there was a fourfold increase in the area covered by the control sites. This is despite a reduction in discharge from 2.1 to 0.34 m^3 s^{-1} in the adjacent river.

DISCUSSION

Many rather general statements have been made about the effects of low flow on lotic communities on the basis of studies which have had a broad focus (Armitage and Petts, 1992). It has therefore been difficult to identify specific effects except in extreme cases

(Giles, Phillips and Barnard, 1991). In this study the effort has been concentrated on a small part of the problem: short-term community change in a channel experiencing an approximately sixfold reduction in discharge between April and September.

It has been demonstrated that despite low discharge a relatively wide range of community diversity can be found, provided that habitat heterogeneity is maintained. It is also shown that the range of diversity can be achieved with relatively small changes in the physical environment.

These general effects were noted in both channels, which reinforces the validity of the observations. There were, however, some differences between the two channels and it is worth considering these in more detail. Despite their close physical and chemical similarity there were differences in the fauna and flora. The main environmental difference between the channels was the greater throughput of water in channel 4. Velocities were generally higher in this channel and this may have favoured the development of *Ranunculus* and *Elodea* at the expense of *Cladophora* which dominated channel 3. The deposition of organic fines is likely to have been higher in channel 3. Where these clog interstitial spaces in the gravel substratum leading to anaerobic conditions, rooted vegetation is destroyed (Bolas and Lund, 1974). It is possible that this may have contributed to the difference between the two channels but there is no direct evidence and it is also possible that the differences arose by chance. That is to say channel 3 may initially have had more *Cladophora* propagules than channel 4. Subsequent floral community structure would then have depended on species interaction. The faunal communities may reflect the differences in floral community (Wright, 1992) or in the associated hydraulic conditions (Dawson and Robinson, 1984; Marshall and Westlake, 1990). It is instructive that two adjacent "streams" receiving similar water and having similar physical characteristics should support differences in their faunal communities. It emphasizes the existence of much natural variation in the system. The detection of variation is also a function of scale and the differences are clearer because the study was finely focused. It is important to recognize these points when reporting the effects of stressors on the environment.

The role of macrophytes is important in creating a range of velocities (Marshall and Westlake, 1990) and this study has shown that, contrary to expectations, the control section showed nearly as much variability in faunal composition/diversity as the artificially modified section. This has important implications for the use of macrophytes to optimize low-flow conditions for faunal communities (Hearne and Armitage, 1993). Relatively simple management of weed beds could create a range of velocities when discharge is reduced.

Under natural flow conditions there is a macrophyte succession from *Ranunculus*, when velocities are high in spring and early summer, to *Nasturtium* in late summer and early autumn. The latter species is usually washed out by high winter flows. Where such flows are reduced the *Nasturtium* stands may remain and result in reduced velocities, increased siltation and the inhibition of the spring growth of *Ranunculus* (Westlake, 1967; Ham, Wright and Berrie, 1981; Westlake, 1981; Giles, Phillips and Barnard, 1991). Such events will radically alter the composition of the faunal community and prolonged low flow will result in exposure of the stream bed and invasion by marginal plants (Lewis, 1990; Giles, Phillips and Barnard, 1991).

The reduced depth and wetted area will lower the availability of suitable habitat, a relationship central to the development of instream habitat modelling techniques (Nestler, Milhous and Layzer, 1989). For fish the effects of habitat loss may be severe (Cowx, Young and Hellawell, 1984) and lead to loss of spawning habitat and "living room" (see Giles,

Phillips and Barnard, 1991, for a review). For invertebrates the reduction of habitat is not a major threat provided that the adverse conditions are not prolonged. Absolute densities may be lowered, the proportions of available microhabitat will be changed but diversity is maintained because the fauna is small, mobile and able to use refugia to survive the unfavourable period. In this study active colonization took place throughout the study period and no taxon was observed to die out. This indicates that niches were available for immigrants even at the lowest discharge.

The crucial factor in the maintenance of the stream integrity is the timing and nature of the disturbance (see Reice, Wissmar and Naiman, 1990, for a discussion of disturbance regimes, resilience and recovery of stream faunal communities). Streams and their biota have an inbuilt resilience to disturbance such as droughts and spates provided it follows long-established patterns to which the biota have adapted. Severe problems for the benthic community arise only when these patterns are disrupted. In the naturally disturbed stream a range of refugia are available when times are hard. These refugia later provide a source of recolonizing invertebrates which move out and re-establish themselves as suitable habitat becomes available. This study has shown that refugia can be created quite simply by artificial means but also that the macrophyte beds can furnish almost as wide a range of conditions providing that the macrophyte development is, itself, not inhibited by low flows. It is likely that prolonged low flows will require some management to maintain open waters. The old style of river-keeper would have manually cut weed in the smaller streams (Sawyer, 1985). Certain weed beds would have been left uncut and others trimmed to direct the flows from side to side and create areas of faster flow. This maintained habitat diversity and a return to such methods, in conjunction with the use of small hatches to control flow where appropriate, and a re-assessment of macrophyte cutting times (Hearne and Armitage, 1993) would help reduce the impacts of low flows.

This study did not have management as a primary objective; however, the information obtained does have relevance to conditions pertaining in streams during periods of low flow. Research proposals which involve detailed finely-focused work are not apparently attractive to many funding bodies and all too often a more general study is sponsored — usually more sites and less effort per site. It is clear, however, that detailed studies are needed. Often the effects of, for example, low flow may be specific to a particular type of river (Armitage and Petts, 1992) and an extensive survey while identifying this fact will not provide sufficient data to set rules for river-specific flow management. Rules for ecological management are based on information from a variety of sources and focused small-scale ecological studies have a valuable role in providing detailed data which can be used in formulating management policies and techniques.

ACKNOWLEDGEMENTS

I am most grateful to Rob Ewing who provided essential assistance in field and laboratory. Jon Bass helped with sampling, Isabel Pardo (University of Santiago) drew the vegetation maps, and Ralph Clarke gave statistical advice. I am grateful to these and to other colleagues at the River Laboratory for their advice and comments. The work was partly funded by the Natural Environment Research Council.

6

The Microflow Environments of Aquatic Plants — An Ecological Perspective

ALISON M. NEWALL

University of Reading, Reading, UK

INTRODUCTION

Aquatic plant habitats are varied and diverse due to the wide range of plant morphologies and colonization patterns associated with different plant species (Clapham, Tutin and Warburg, 1954). For example trailing plant species such as *Ranunculus* spp. provide a greater range of habitats and a different type of habitat to that provided by floating-leaved plants such as *Nuphar* spp.; while the former tend to form large dominant stands in suitable rivers, the latter tend to be found in smaller clumps, along with various other plant species. In comparison, emergent plants such as *Phragmites* spp. provide a different habitat again, and are found colonizing the slow-flowing shallow waters at the river's edge. This diversity could be used by river managers to maintain the river's natural diversity of flora and fauna for both conservation and aesthetic quality (Boon, Calow and Petts, 1992). The maintenance of a wide range of aquatic plants has the added advantages of generally reducing erosion (Stephens et al., 1963) and providing a food source — both in some of the plants themselves and in the fauna inhabiting them — for larger animals in the aquatic community (Blindow, 1987; Dionne and Folt, 1991). This physical diversity is difficult to quantify in terms of the environment experienced by invertebrates — the microflow environment. In this paper the microflow environments of aquatic plants have been characterized and quantified from field sites in six British rivers, coupled with microflow studies carried out in an experimental flume.

OBJECTIVES

The main objectives of the study were:

(i) to examine the microflow environments of aquatic plants of different morphological forms in their natural setting, the river;
(ii) to test the influence of plant parts and numbers of plants, using physical models in an experimental flume, on the microflow conditions; within, upstream and downstream of the plant stand.

The Ecological Basis for River Management. Edited by D.M. Harper and A.J.D. Ferguson. © 1995 John Wiley & Sons Ltd

The hypotheses tested were:

(i)   that plants of different morphologies affect microflows in different ways;
(ii) that fewer numbers of plants cause less flow variation (being defined as the range of velocities and flow parameters) than greater numbers of plants;
(iii) that greater numbers of plants cause a reduction of mean velocity when compared with fewer numbers of plants.

Various questions arise from these hypotheses and this research provides some answers. Questions that arise include:

(i)   Do trailing plants provide microflows of more diversity / variation, and if so would they therefore provide more microhabitats than emergent or floating-leaved species for use by invertebrates, fish and larvae?
(ii) How far upstream / downstream of a stand of plants does that stand affect the flows / microflows, if at all?
(iii) Is the relative plant depth significant in determining the microflow patterns within and around plant stands?

METHODS

The experimental work was carried out in a laboratory flume, 7.5 m long by 30 cm wide and maximum workable depth of 40 cm. The discharge was controlled by a valve with an ultrasonic doppler velocity meter indicating the percentage of the maximum inflow. Experiments were carried out at 15%, 60% and 85% which translate to roughly $0.0045 - 0.005$ m^3 s^{-1}, $0.019 - 0.0225$ m^3 s^{-1}, and $0.025 - 0.0285$ m^3 s^{-1}. The depth was controlled by use of an angled outflow gate, which for these experiments was kept at a constant height. Velocity transects were measured at three sections; upstream, middle (which was the experimental section) and downstream. A Nixon Streamflo 426 propellor current meter with a diameter of 8 mm was used. Readings were taken at intervals of 20 mm across the sections and at four heights in the water column; as close as possible to the bottom, and at 0.2, 0.4 and 0.8 of the depth. Measurements were taken at three different discharges.

Model aquatic plants were used along with some real plants taken from the River Loddon at Sindlesham, Reading. The models were built from plastic tubing to a similar morphology as *Potamogeton natans*, with leaves of a roughly oval shape. The real plants were *Potamogeton nodosus*, since *P. natans* was not readily available.

The first experiment used a single model stem, the second a stem with one leaf; the third a stem with two leaves; and the fourth a stem with four leaves and one submerged leaf form (considered one plant for these purposes, since any more caused the plant to be too dense and bulky). Two model plants were then used, followed by four real plants, eight real plants and finally 16 model plants.

In the field, data were collected using the Nixon current meter for the microflows within each stand of plants, and a Braystoke current meter for transects upstream and downstream of the stand. For the three river transects (across the total width of the river) velocity was measured at 0.4 of the depth, while in the stands of plants the microflow velocities were measured at 0.2, 0.4 and 0.8 of the depth, and readings were taken every 30 cm across the width of the stand. There were five such transects taken within the plants, with the exception of the Bristol Avon where three were taken.

RESULTS

Flume study

Figures 6.1−6.3 illustrate that the presence of obstructions, in this case model and real aquatic plants, influences the microflow patterns in the vicinity of the plants and downstream of them, but not upstream. (The sections were a distance of 1−1.5 m apart.)

Figures 6.2 and 6.3 show that the greater the number of plants, the greater is the deviation from the normal flow profiles and transects, and also that there is a definite mixing up of the flow profile: the normal increase in flow velocities upward from the bed is not seen. With 16 models (the greatest number of plants studied), the range of velocities is greatly increased.

Table 6.1 showing the ranges of flume velocities reveals three patterns associated with microflows. These are:

(i) for model plants the range of velocities increases with discharge, with a greater increase between 15% and 60% than between 60% and 85%;
(ii) that the range for each condition (i.e. set of three sections: upstream, middle and downstream) is greatest in the experimental (middle) section, and lowest in the upstream section;
(iii) that the ranges of velocities are considerably and consistently less for the real plants.

Flow descriptors are shown in Table 6.2. The Froude number, which describes the ratio of the inertial to gravity forces, increases with increasing discharge, although similar in value for different numbers of plants or models. In the empty flume it also increases in a downstream direction. Reynold's number, which describes the ratio of viscous to inertial forces, shows substantial increases with increased discharge, although there is a levelling out of the rate of increase with increased disturbance from plants under higher discharges. Reynold's values greater than 2000 indicate turbulent flow while values less than 500 describe smooth or laminar flow. Roughness Reynold's number shows a similar pattern. Shear velocity increases with increasing discharge. There is a decrease in the thickness of the laminar sub-layer with increasing discharge, as expected, with similar values for different numbers of plants.

Manning's *n* roughness coefficient shows a decrease with increasing discharge in the empty flume, which is the expected pattern. The pattern shown for the model plants is an increase of *n* with increasing discharge, while for eight real plants the value of *n* is roughly constant at around 0.003.

Table 6.3 supports the hypothesis that there is a significant difference in velocities between fewer plants and greater numbers of plants, with the fewer numbers resulting in higher velocities.

Field study

Table 6.4 shows ranges of velocities for the field study, and it is evident that trailing plants, e.g. *Ranunculus* spp. and *Potamogeton natans* are associated with a greater range of

82

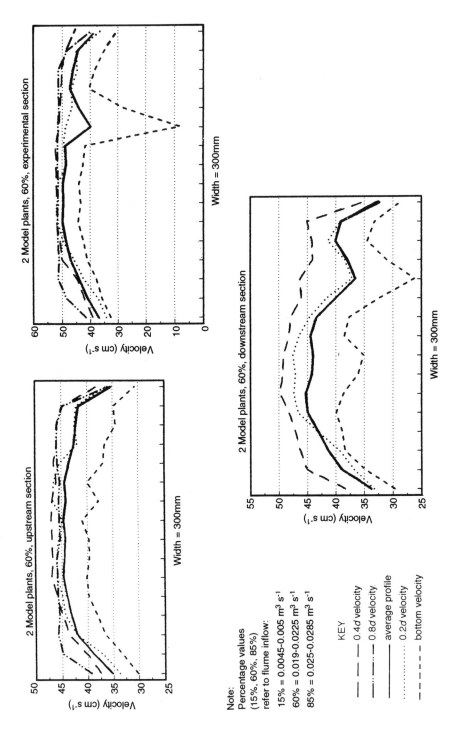

FIGURE 6.1.   Microflow velocity profiles and transects for two model plants

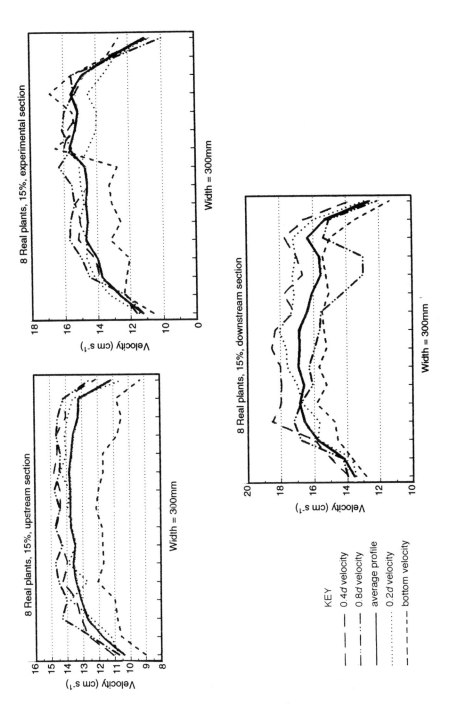

FIGURE 6.2. Microflow velocity profiles and transects for eight real plants

84

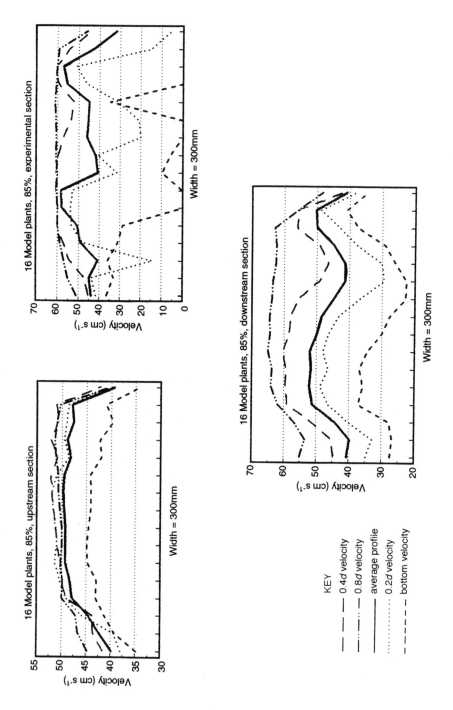

FIGURE 6.3. Microflow velocity profiles and transects for 16 model plants

TABLE 6.1. Ranges of flume velocities (all velocities in cm s^{-1})

	EMPTY FLUME		1 MODEL	
	Range	Overall range	Range	Overal range
15%				
Upstream	6.41		5.52	
Middle	7.03	7.32	11.14	11.14
Downstream	7.26		6.23	
60%				
Upstream	16.1		17.67	
Middle	18.44	20.07	39.99	42.52
Downstream	20.07		19.83	
85%				
Upstream	19.69		21.26	
Middle	22.04	25.61	54.76	58.29
Downstream	23.9		23.11	
	8 REAL PLANTS		**16 MODELS**	
15%				
Upstream	5.78		5.58	
Middle	6.84	9.52	15.48	16.52
Downstream	7.19		14.33	
60%				
Upstream	17.51		13.85	
Middle	21.31	24.63	45.77	48.48
Downstream	18.68		34.09	
85%				
Upstream	19.65		18.0	
Middle	19.7	21.09	55.4	59.25
Downstream	20.32		42.12	

microflow velocities than emergent or floating-leaved plants, e.g. *Phalaris* spp. and *Nuphar lutea*. It also highlights three patterns of the ranges of microflow velocities with regard to distance into the stand in a downstream direction:

(i) a decrease of the range of velocities with increased distance into the stand in a downstream direction (e.g. River Axe, 1991);
(ii) an increase of the range of velocities with increased distance into the stand (e.g. Millstream);
(iii) an increase into the middle of the stand, then a decrease to the downstream end of the stand, of the range of velocities (e.g. River Lymington).

Table 6.5 shows estimates of the forces exerted on stands of plants of different species, along with the discharges of the rivers in which they were studied. It is evident that the greatest forces are exerted on trailing, submerged plants, e.g. *Ranunculus penicillatus* in the River Axe (34.6 Newtons) and *Potamogeton natans* in the River Forth (31.7 N), and that the least force is exerted on submerged leaves of floating-leaved plants such as *Nuphar lutea* in the Bristol Avon (1.36 N) (the plants studied only had submerged leaves). However, these do

TABLE 6.2.   Experimental flume flow descriptors

	1	2	3	4	5	6	7	8	9
Empty flume									
Q	5077	5179	4926	20 699	20 759	20 917	29 233	28 150	30 403
Froude no.	0.148	0.151	0.154	0.361	0.362	0.365	0.407	0.427	0.442
$U*$	4.82	4.71	1.19	1.19	1.19	4.67	4.54	4.61	
Reynolds no.	17 815	18 464	18 043	79 307	85 960	77 904	111 316	107 239	117 842
$Re*$	—	n/a	—	—	n/a	—	—	n/a	—
δ	0.023	0.022	0.022	0.084	0.078	0.087	0.021	0.022	0.022
"n"	0.01	0.01	0.01	0.001	0.001	0.001	0.003	0.003	0.002
One Model (stem plus four leaves)									
Q	5184	5197	5094	22 787	21 129	22 019	31 139	27 327	27 791
Froude no.	0.162	0.163	0.159	0.379	0.351	0.366	0.441	0.397	0.404
$U*$	—	0.52	—	—	2.58	—	6.93	6.87	—
Reynolds no.	20 694	20 621	20 095	92 066	87 489	85 845	120 693	102 349	100 693
$Re*$		372			2559			7719	
δ	0.184	0.186	0.187	0.037	0.036	0.038	0.014	0.015	0.015
"n"	0.001	0.001	0.001	0.002	0.002	0.002	0.004	0.004	0.004
Two models									
Q	5283	5359	5325	21 076	23 001	20 130	27 362	28 931	29 542
Froude no.	0.158	0.161	0.16	0.35	0.382	0.335	0.415	0.439	0.449
$U*$	—	0.35	—	—	2.74	—	—	4.75	—
Reynolds no.	18 636	19 208	19 721	79 832	88 128	76 250	101 908	114 137	117 933
$Re*$		411			1891			5055	
δ	0.307	0.302	0.292	0.037	0.037	0.022	0.021	0.020	
"n"	0.0007	0.0007	0.0007	0.002	0.002	0.002	0.003	0.003	0.003

not correspond with maximum and minimum discharges; the river with the lowest discharge exerting the fourth greatest force (14.96 N), i.e. the River Lymington on *Callitriche hamulata*.

ECOLOGICAL SIGNIFICANCE OF THE RESULTS

Results shown in Figures 6.1–6.3 suggest that the presence of aquatic plants provides different environments from those found in open flow, which are likely to attract a different community of invertebrates and vertebrates. It also shows that the first hypothesis is supported, i.e. that more plants cause greater flow variation.

The figures show that downstream flows are affected by the plants. It cannot, however, be ascertained from this research how far downstream the effect of the plants is detectable. This is likely to depend on the dimensions and the species of the stand in question. Thus it must be considered when river management proposes to, for example, cut back plant growth, that the flow, and therefore the invertebrate and possibly vertebrate communities of the plants, are affected in a greater area than may be expected. Alternatively and more positively, if an aquatic plant habitat is to be created, it may be that a greater range of habitat is developed than just that within the stand of plants. The distortion of the velocity profile would also

TABLE 6.2. continued

	1	2	3	4	5	6	7	8	9
Eight real plants									
Q	4293	4559	5011	18 218	16 532	19 465	26 872	27 718	26 912
Froude no.	0.134	0.143	0.157	0.324	0.303	0.339	0.391	0.410	0.409
U^*	—	1.21	—	4.26	4.22	4.29	5.46	5.42	5.38
Reynolds no.	15 224	15 829	17 044	65 652	58 007	63 612	98 977	98 818	92 959
Re^*		1323			3998			8116	
δ	0.089	0.091	0.093	0.025	0.026	0.027	0.019	0.020	0.021
n	0.003	0.003	0.003	0.003	0.004	0.003	0.003	0.003	0.003
Sixteen models									
Q	4688	4414	4780	19 315	18 980	19 166	21 807	25 462	25 836
Froude no.	0.147	0.138	0.150	0.337	0.338	0.351	0.317	0.377	0.410
U^*	—	0.46	—	5.61	5.57	5.52	6.19	6.15	6.01
Reynolds no.	16 623	15 821	17 414	72 339	70 298	69 441	70 572	87 499	91 618
Re^*		297			3404			3804	
δ	0.235	0.233	0.229	0.018	0.019	0.019	0.019	0.018	0.018
n	0.001	0.001	0.001	0.004	0.004	0.004	0.005	0.004	0.004

1 = 15%, upstream   4 = 60%, upstream   7 = 85%, upstream
2 = 15%, middle   5 = 60%, middle   8 = 85%, middle
3 = 15%, downstream   6 = 60%, downstream   9 = 85%, downstream
Q = Discharge (cm³ s⁻¹); Re^* = Roughness Reynolds number; U^* = Shear velocity (cm s⁻¹); δ = Thickness of the laminar sublayer (cm); n = Manning's n

affect the types of invertebrates found in the habitat, possibly increasing the range of species found living in the water column, or otherwise causing a rearrangement of the invertebrates in the water column.

Results from Table 6.1 reinforce the previous findings of the influence the plants exert upstream or downstream, with ranges of velocities being greater downstream than upstream. The reduced range of velocities, for real plants, between 60% inflow and 85% inflow could be due to the "drowning out" effect. Thus, less of an approach area is presented to the flow thereby causing less resistance, with the stems and leaves becoming more compacted and as a result causing less flow velocity variation.

The greater range of velocities with trailing plants (Table 6.4) is ecologically significant since greater numbers of macroinvertebrates are found in macrophytes than are found anywhere else (Marshall and Westlake, 1978), and larger and more varied animal populations have been found to be associated with plants with more finely dissected leaves (Harrod, 1964) (cf. Rooke, 1984). More recently, however, Wright et al. (1992) found that emergent macrophytes provided the most faunistically diverse samples (62.7%) compared to submerged and floating macrophytes (33.3%) and non-macrophytic substrata (4%). If this is the case, it may be that the ranges of velocities are not the main, or only, determinants of the faunal communities associated with aquatic plants, as was suggested by Edington (1968) and tested in this research. The different patterns with distance downstream are thought to be determined by the density and species of the plant stand. A dense stand of *Ranunculus* spp. in the River Axe (1991) caused a decrease in the ranges, while a less dense stand of *Ranunculus* spp. and *Phalaris* spp. caused an increase, and a dense stand of *Callitriche* spp. in the River Lymington presented the third pattern.

TABLE 6.3.   Statistical results

Comparing average velocities for different numbers of plants in different parts of the flume

t-test: comparing velocities at the upstream, experimental and downstream sections separately
Hypothesis: average velocities are greater for fewer plants
Result: ACCEPTED for upstream and downstream sections

Upstream section only		t	p	d.f.
$p<0.05$	1 model > 8 real plants	1.73	0.0435	92
	1 model > 16 models	1.70	0.0465	90
$p<0.1$	stem > 8 real plants	1.4	0.08	93
	stem > 16 models	1.36	0.09	91
Downstream section only				
$p<0.05$	stem only > 16 models	1.99	0.0295	92
	stem +1 > 16 models	2.43	0.0085	87
$p<0.1$	stem +1 > stem +2	1.26	0.1	91
	stem +1 > 1 model	1.31	0.095	90
	stem +1 > 4 real plants	1.42	0.08	89
	stem +1 > 8 real plants	1.52	0.065	89
	2 models > 16 models	1.36	0.09	91

t-test: comparing velocities at all sections
Hypothesis: average velocities are greater for fewer plants or models
Result: supported in 17 cases, opposite supported in 1(*): ACCEPTED

$p<0.05$	1 model, pos1 > 16, pos3	2.05	0.021	90
	8 real pos1 < stem +1 pos3	−2.12	0.0185	90
	stem, pos1 > 16mod, pos3	1.71	0.045	91
	8 real, pos1 < stem, pos3	−1.67	0.049	93
	16 mods pos3 < stem +1 pos3	−2.09	0.0195	87
$p<0.1$	stem +1 pos1 > 16mod pos3	1.50	0.07	92
	stem +2 pos1 > 16mod pos3	1.55	0.065	92
*	1mod pos1 > stem +2 pos2	1.46	0.075	93
	1mod pos1 > 16mod pos2	1.54	0.065	91
	2mods pos1 < stem +1 pos3	−1.26	0.1	90
	2mods pos1 > 16mods pos3	1.26	0.1	93
	4 real pos1 > 16mod pos3	1.40	0.085	93
	8 real pos1 < stem +1 pos2	−1.39	0.085	93
	8 real pos1 < 2mods pos2	−1.40	0.08	93
	16mods pos1 < stem +1 pos2	−1.34	0.09	91
	16mods pos3 < 2mods pos2	−1.36	0.09	92
	16mods pos3 < stem pos3	−1.63	0.055	92
	stem pos2 > 16mod pos3	1.45	0.075	93

As the Reynold's number increments are reduced with greater discharge and more plants (Table 6.2), it appears that there is a threshold number of plants or size of stand above which the turbulence is little affected by increases of discharge. From this research it is thought that this threshold may be very low in terms of river communities, although it must be remembered that the results cannot be extended to plants of different morphologies or types. Under lower discharges the number of plants does not significantly affect the transitional nature of the flow. An increase in the number of plants presenting themselves to the flow under higher discharges causes a reduction of the turbulence. This may be due to the

TABLE 6.3. continued

Comparing average velocities for different numbers of plants in different parts of the flume			

Comparing models with real plants
t-test: comparing different positions and velocities
Hypothesis: there is a significant different between models and plants at each separate position and inflow velocity
Result: only supported for upstream section, velocity 1

$p < 0.1$	models, vel1 v plants, v1	1.73	0.086	152

t-test: Comparing positions across the channel
Hypothesis: that velocities in the middle of the sections will be greater than those at the edges
Result: supported in 4 cases, not in 1(*): ACCEPTED

$p < 0.05$	models, mid v plants, mid	2.26	0.025	201
	mod, mid > plants right	2.01	0.023	197
$p < 0.1$	mod, mid > mod, right	1.79	0.037	645
	mod, left < mod, mid	−1.34	0.09	580
*	plants, mid < plants right	−1.38	0.085	213

Pos1 = upstream, Pos2 = middle, Pos3 = downstream, Vel1 = 15%

channelling of the flow over and around stands of plants (Gambi et al., 1990; Hammerton, 1990). Watts and Watts (1990) suggest that different plant species have different effects on the turbulence of flow, such that *Oenanthe* spp. aggravate and *Potamogeton* spp. dampen the turbulence. It can thus be seen how stands of plants can present refuges from the main body of the flow in streams and rivers at different stages of the life cycle for invertebrates, fish and mammals.

The use of k, the relative depth, is not as useful with plants as for benthic microenvironments (as used by Davis and Barmuta 1989), since it is often found to be equivalent to the water depth. It is difficult to take an accurate value for k since the plants and models are by nature flexible and fluctuate in the water column. Since the calculation of the roughness Reynold's number and thickness of the laminar sub-layer employs k, the relative depth, the values vary depending on what subjective value is used. Roughness Reynold's number increases with increasing discharge regardless of how many plants there are, and the increase and range of values is greater for real plants (Table 6.2). The former is due to the bouyancy of the plants giving a higher value for k, while the greater range of values would be likely to provide a wide range of habitats within the plants, which may be utilized by aquatic insects and fish. It is also noticeable that the values are considerably greater than those described for benthic microenvironments, as described by Davis and Barmuta (1989), and therefore the use of the roughness Reynold's number is not really applicable to higher aquatic plants.

The presence of aquatic plants may encourage filter feeders more than other macroinvertebrates since the food can be supplied at a greater rate while shleter is provided in the stand, due to the increasing shear velocities with discharge (Table 6.2).

While the models cause Manning's n roughness to increase with increased discharge, the real plants allow the roughness to remain fairly constant (Table 6.2). This may be due to the advanced streamlining of the real plants. D. Watson and E.A. Keller (unpublished) found that the roughness due to aquatic plants increases nonlinearly with plant loading, and

TABLE 6.4. Ranges of microflow velocities

River	Minimum velocities (cm s⁻¹)	Maximum velocities (cm s⁻¹)	Range for transect (cm s⁻¹)	Species	Overall range for stand (cm s⁻¹)
River Axe 1991					
Transect 1 (upstream)	5.051	62.776	57.720	*Ranunculus penicillatus*	57.830
2	4.950	56.340	51.390		
3	5.051	55.180	50.130		
4	5.051	20.094	15.040		
5	5051	16.042	10.990		
Mill Stream					
Transect 1	5.203	10.876	5.670	*Ranunculus calcareus* and *Phalaris* spp.	22.480
2	4.950	11.838	6.890		
3	4.950	11.838	6.890		
4	5.001	15.789	10.790		
5	4.950	27.433	22.480		
Bristol Avon					
Transect 1	5.051	9.103	4.050	*Nuphar lutea*	7.040
2	5.001	10.319	5.320		
3	5.001	12.041	7.040		
River Lymington					
Transect 1	5.051	11.636	6.580	*Callitriche hamulata*	21.590
2	5.001	17.055	12.050		
3	4.950	26.536	21.590		
4	4.950	23.032	18.080		
5	5.001	10.471	5.470		

Brooker et al. (1978) found that Manning's coefficient was significantly correlated with plant biomass. Also, the streamlining of the plants must be inadequate at some point to allow the forces on plants to be enough to wash out plants in high discharges. Since *n* remains fairly constant with an increased discharge (in the range of discharges studied), macroinvertebrates may be able to use plants as a refuge without being washed away since the plants' resistance to the flow is not reduced.

Table 6.5 shows that trailing, submerged plants such as *Ranunculus calcareus* or *Potamogeton natans* are subjected to a greater force by the flow than floating-leaved plants, such as *Nuphar lutea*. This is significant in the morphology of the plant, in that a greater area of presentation to the flow would increase the force exerted on the plants, so streamlining becomes important; and also the location of the plants in the river would influence the flow velocities experienced and that would affect the forces exerted on the plants. In this context it can be seen that the flow characteristics and patterns influence the colonization of plants, with survival only being likely where the force is not so great as to wash the plants out of their substrate.

TABLE 6.5. Forces exerted on plants by the flow

	R. Axe 1991	Mill Stream R. Frome	R. Forth	R. Avon, Bristol	R. Dove[b]	R. Lymington	R. Axe 1992
$Q(\text{m}^3\text{ s}^{-1})$[a]	0.742	0.315	2.702	0.216	0.448	0.0768	0.427
$F(\text{N})$[a]	17.524	6.024	31.65	1.358	5.039	14.96	34.6[c] 228.36[d]
Plant(s)	R. pencillatus var. calcareus	R. cascareus + Phalaris spp.	Potamogeton natans	Muphar hutea	R. peltatus or fluitans	Callitriche hamulata	R. pencillatus var. calcareus

a Forces calculated after Dawson and Robinson's (1984) equations
b Calculated using Braystoke readings rather then microflow velocities
c Using $w = 2$ m, $v = 0.1511$ cm s⁻¹
d Using $w = 6$ m, $v = 0.1738$ cm s⁻¹

CONCLUSION

The main conclusions to be drawn from this research are that there is less flow variation with fewer plants; that trailing submerged plants provide a wider range of velocities and therefore of habitats than do emergent or floating-leaved plants; and that flow velocities are reduced with greater numbers of plants. Therefore stands of aquatic plants and the microflow environments and habitats they provide are important in the lotic community and must be maintained as an integral part of river systems.

7

The Ecological Basis for the Management of Flows Regulated by Reservoirs in the United Kingdom

D. TREVOR CRISP

NERC Institute of Freshwater Ecology, Barnard Castle, County Durham, UK

INTRODUCTION

There have been a number of general syntheses of published data on the downstream effects of river regulating schemes. These have included general reviews by Brooker (1981), Petts (1984) and accounts in the narrower UK context by Armitage (1980), Edwards and Crisp (1982) and Edwards (1984). The ecological impacts are generally divided into two categories: those impacts arising from physical alterations to the discharge regime and those impacts arising from changes during reservoir storage before release downstream. These impacts are also generally viewed on two scales: the immediate, local, ecological changes and the more extensive downstream consequences.

Four aspects of ecological impact within these two categories and scales are of importance to future management strategies for regulated rivers and are examined in this paper. These are:

(i) main downstream physical effects of impoundment and regulation in terms of water flow, water velocity, water depth, stream width and water temperature;
(ii) the transport and consumption of small organic particles;
(iii) the drift of invertebrates;
(iv) the effects of temperature and flow regime changes upon salmonid fishes.

The review draws heavily upon two UK regulation schemes for which relatively detailed ecological studies have been published: Cow Green reservoir on the River Tees, north-east England and Craig Coch and the Elan Valley reservoir, in the Upper Wye catchment, Wales (Edwards and Crisp, 1982).

DOWNSTREAM PHYSICAL EFFECTS

The main effects on flow regime immediately downstream of impoundments of different management types are summarized in Table 7.1. It is apparent that regulating reservoirs in

The Ecological Basis for River Management. Edited by D.M. Harper and A.J.D. Ferguson. © 1995 John Wiley & Sons Ltd

TABLE 7.1. Physical impacts of four general types of reservoir upon the downstream river

Reservoir type	Environmental impact				
	Modified amplitude of discharge regime	Modified seasonal pattern of discharge	Modified diel pattern of discharge	Decreased average flow	Increased average flow
Rivering regulating	+	+	+		
Water suppy	+	+		+	
Hydroelectric	+	+	+		
Inter-river transfer	+	+	+	+(donor)	+(recipient)

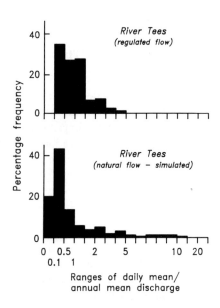

FIGURE 7.1. Percentage frequency distributions of values of daily mean discharge for the regulated and the natural River Tees. Note the irregularity of the scale on the *x* axis

particular will modify the amplitude of flow fluctuations and the seasonal patterns of flows, since their main task is to store water during wet periods (usually winter in UK) and release it during dry periods (usually summer). A good example of such modification of amplitude is shown in Figure 7.1. The virtual elimination of flows less than 0.1 or more than five times the annual mean value is apparent.

The manner in which changes in flow regime can influence bed movements (scour and deposition) and water velocity, depth and width immediately downstream of an impoundment are shown in Table 7.2. Each of these variables will fluctuate in a more regular fashion due to the elimination of highest and lowest flows by regulation. Analyses of data from extensively regulated UK catchments show that the main extensive downstream effect is the increase of minimum discharge (Bussel, 1979).

Annual temperature cycles, for rivers below three UK regulation schemes, plotted as monthly means, show similar patterns of reduction of annual amplitude and delay in the annual cycles as a result of the impoundment upstream (Figure 7.2). This effect is much more extreme below the two impoundments that show temperature stratification than in the third which shows temperature stratification only rarely and briefly. The damping effects of reservoir storage upon downstream temperature cycles alone appear to be appreciable (e.g. Cow Green), but small in comparison to both storage and stratification (Kielder Water and Elan Valley).

More detailed examination of the stratification pattern for Kielder Water (Figure 7.3) shows that in 1983, for example, stratification occurred from June to September inclusive. During this time the temperature of the water discharged to the River Tyne would have depended upon the choice of draw-off depth within the reservoir. Knowledge of the depth distribution of different temperatures could be used, in conjunction with the mixing of suitable quantities of water from appropriate draw-off depths, to give a mix of discharged water at any chosen temperature between the extremes represented in the reservoir water column. It is thus possible in reservoirs with multiple draw-off valves, to manage temperatures downstream of the dam so as to simulate, or even to improve upon, those of the natural river during periods of stratification in warm weather.

These temperature effects are likely to be fairly localized. There do not appear to be any simple and useful published models for prediction of temperature at different distances downstream of the release point (those which do exist seem unduly complex for useful application). It is apparent, however, that for reservoirs on the UK scale (up to 200×10^6 m^3 capacity) the downstream temperature effects probably become negligible within $10-30$ km of the point of release. The maximum observed effect of Cow Green reservoir, for example, upon the temperature of released water is a depression of 2.8°C in the July mean. Between the dam and Broken Scar, near Darlington (22.7 km downstream), tributaries raise the average discharge by a factor of 6.6. In terms of dilution alone, the temperature depression is therefore reduced to a mean 0.4°C at Broken Scar. In addition, the released water will rapidly adjust towards air temperatures as it travels down the river channel, and it is probable that this adjustment will be more rapid than that due to the entry of the tributaries. Management of the temperature of water released from reservoirs is therefore likely to be effective and necessary only on a local scale.

TABLE 7.2. Three types of modification to discharge regime and their consequent physical impacts

Discharge modification	Physical impact					
	Bed scour	Sediment deposition	Water velocity	Stream depth	Stream width	Wetted area
Reduced flow	−	+	−	−	−	−
Increased flow	+	−	+	+	+	+
Modified flow pattern	+ or −	+ or −	+ or −	+ or −	+ or −	+ or −

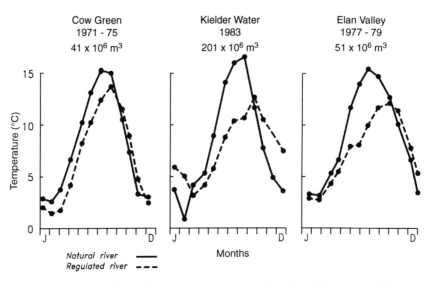

FIGURE 7.2.   Comparison of monthly mean temperatures of regulated river water downstream of
three UK reservoirs with the temperature of the natural river

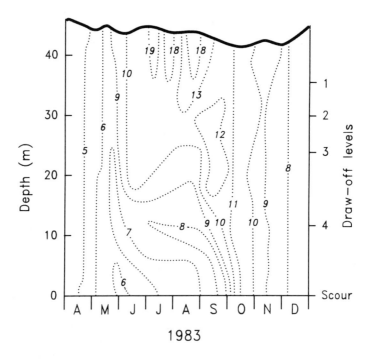

FIGURE 7.3.   Depth/time diagram of water temperature (°C) in Kielder Water, showing draw-off
levels

TRANSPORT OF SMALL ORGANIC PARTICLES

Streams draining upland areas may carry appreciable quantities of eroded peat in a fine particulate form (Crisp, 1966; Tallis, 1973). Peat is a relatively inert material before erosion, but its entry as fine particles into the aquatic environment with neutral or alkaline pH and calcium in solution is likely to lead to breakdown, which may cause release of some plant nutrients and conversion to particles such as bacteria of more value to filter-feeding animals. McLachlan, Pearce and Smith (1979) showed that eroded peat can be a significant food source for invertebrates in still waters.

One Tees tributary headstream transports such peat at a mean rate of 3 g dry wt s^{-1} (95 $\times$ 10^3 kg ann^{-1}). Of this, 80% is transported during spates which carry 33% of the annual runoff but occur for only 3% of the year (Crisp and Robson, 1979). Calculations on the basis of catchment ratios would give an estimated 4 $\times$ 106 kg dry wt ann^{-1} to Cow Green reservoir. Most of this (over 91% according to Armitage (1977) and over 95% according to more recent estimates) is retained and may be expected to support production of the reservoir biota. This is, thus, a potential, but unquantifiable, loss of food and nutrients to the downstream biota which in natural streams would be available at all times of the year, albeit in variable quantities.

Reservoirs discharge appreciable quantities of living fine particles, despite acting as sinks for incoming inanimate material, in the form of phyto- and zooplankton. Microcrustacea appear to be particularly important: at Cow Green reservoir, Armitage and Capper (1976) found that the material was 87% Cladocera (*Daphnia hyalina* and *Bosmina coregoni*) and 13% Copepoda (*Cyclops agilis* and *Cathocamptus staphylinus*). Most material (97% Cladocera and 90% Copepoda) was discharged during July to October inclusive. The total quantity was estimated as between 1 and 4 $\times$ 10^3 kg fresh wt (approximately 150 kg dry wt) ann^{-1}, 91% Cladocera and 9% Copepoda. Edwards and Crisp (1982) quoted a study showing 15 $\times$ 10^3 kg fresh wt ann^{-1} of microcrustacea in compensation discharges from a deep draw-off at Caban Coch reservoir to the River Elan (Hopper F.N., unpublished).

Such suspended microcrustacea are rapidly incorporated into the benthos of the downstream river. Armitage and Capper (1976) found that only 1−2% remained in the water column at a point 6.5 km downstream of Cow Green. The loss of suspended material appeared to be related to distance downstream in an inverse fashion, although the data were not adequate to demonstrate the relationship conclusively. In the River Elan, most of the microcrustacea also disappeared within 6.5 km of the discharge point. Inter-species differences in removal rate were observed, similar to the observations of Ward (1975) for a US impoundment.

Substantial numbers of microcrustacea occurred in trout stomachs downstream of Cow Green reservoir (Crisp, Mann and McCormack, 1978). Few were found in stomachs collected in May, when numbers were small in both reservoir and downstream drift. In August and October, microcrustacea were only found in stomachs (which were collected during daylight) when the uppermost valve being used for draw-off was 11 m or less below the reservoir surface. This implies that quantities discharged from the reservoir were dependent upon the depth distribution of the microcrustacea within the reservoir in relation to draw-off depths. Other studies, however, have provided conflicting results which indicate that further investigation of reservoir crustacea releases would be beneficial: substantial quantities of microcrustacea were detected during discharges from a "deep draw-off point" (Edwards and Crisp, 1982) and a "hypolimnetic release" (Ward, 1975).

There is thus the possibility that further investigations might yield a framework for the utilization of different draw-off levels in order to manage, within limits, the amounts of microcrustacea available to invertebrates and fish downstream of the dam, at least on a local scale.

DRIFT TRANSPORT OF MACROINVERTEBRATES

Most studies of invertebrate drift in natural streams have been concerned with what may be termed "behavioural drift". This phenomenon is reasonably predictable and occurs during low to moderate discharges. Relatively little attention has been paid to the "catastrophic drift" which occurs during both natural spates and regulatory releases and which can transport large quantities of both aquatic and terrestrial invertebrate material.

Anderson and Lehmkuhl (1968) found during autumn spates in an Oregon stream that there was an increase in insect transport,which was larger in terms of weight than numbers. Elliott (1971) and Zelinka (1976) subsequently suggested that there might be power law relationships between drift transport and either water velocity or stream discharge. Crisp and Robson (1979) showed for a north Pennine stream that empirical models could be fitted to relate animal transport and animal concentration to stream discharge (Figure 7.4). Models of this type, however, obscure the more complex patterns which occur during the course of individual spates. The following features become apparent:

(i) Mean animal concentration, in terms of both weight and numbers, peaks before the hydrograph (Figure 7.5) and this implies that numbers available to drift become depleted during the course of the spate either as a result of depletion of total numbers or because those present seek shelter in some way.

(ii) The mean weight per drift animal peaks with the spate, which reflects the presence of terrestrial casualties (Figure 7.6, cf. "discharge" and "terrestrial").

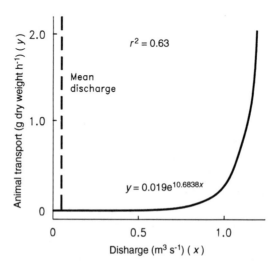

FIGURE 7.4.   Relationships between discharge (m³ s⁻¹) and animal transport (g dry weight h⁻¹) in a small north Pennine stream

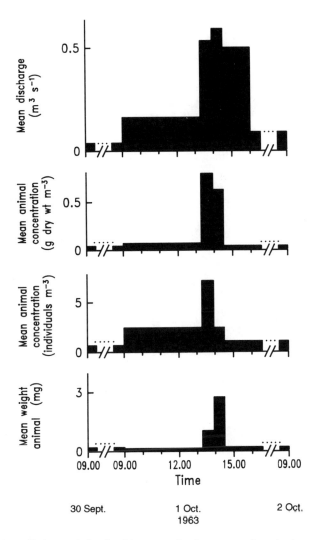

FIGURE 7.5. (a) Mean discharge (m³ s⁻¹); (b) mean animal concentrations (g dry weight m⁻¹); (c) number (m⁻³) and (d) mean weight per animal (mg), during the spate of 30 September−2 October 1963. Crisp and Robson (1979), reproduced by permission

(iii) The pattern of variation in drift concentration relative to discharge varies between different invertebrate taxa (Figure 7.6).

Armitage (1977) conducted a comparative study between the River Tees below Cow Green and the Maize Beck, an unregulated tributary, which may be used as an example of "behavioural drift" because discharges in the Tees on the sampling dates varied between 18% and 159% of the average daily flow but were mainly below 100%. The benthos and the drift of the Tees contained large numbers of Cladocera, Copepoda, *Nais* spp and *Hydra* sp., though the last two contributed little to the biomass. There was no clear evidence of difference between the total drift quantity between the two streams if these small animals are

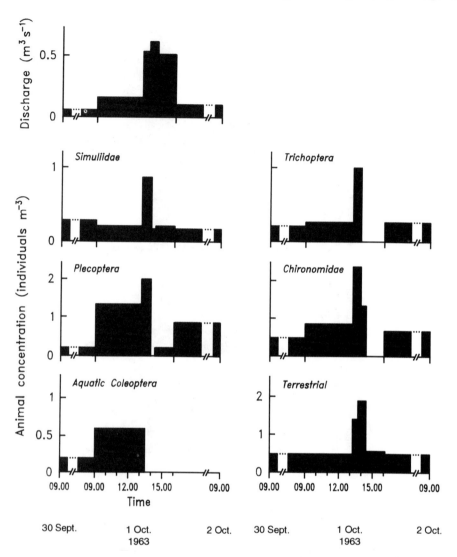

FIGURE 7.6.   Mean discharge (m³ s⁻¹) and mean numbers (m⁻³) for various invertebrate taxa during the spate of 30 September – 2 October 1963. Crisp and Robson (1979), reproduced by permission

excluded, but there were differences in drift composition. Plecoptera and Baetidae had similar benthic population densities at the two stations but were more abundant in the drift in the unregulated stream. In general, there appeared to be an inverse relationship in the Maize Beck between discharge and concentration of drift.

Brooker and Hemsworth (1978) conducted a study of freshet releases of two days duration in the Elan/Wye system which may be used as an example of "catastrophic drift". An increase in release discharge from 1.3 m³ s⁻¹ to 4.3 m³ s⁻¹ from Caban Coch reservoir to the Elan led to an increased flow in the river Wye below its confluence with the Elan from 2 to 5 m³s⁻¹. The arrival of the freshet gave rise to sharply elevated daytime values of total numbers and concentrations of drift animals, followed by a substantially increased night-time

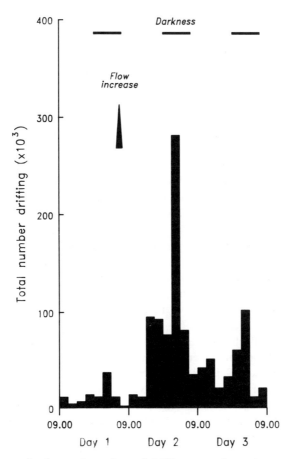

FIGURE 7.7. Changes in the total number of drifting macroinvertebrates. After Brooker and Hemsworth (1978)

peak (Figure 7.7). During the second day of increased flow the average and peak drift concentrations returned to pre-freshet values, but the total numbers transported were still elevated. On the two days of increased flow the numbers drifting were seven and three times higher respectively than the numbers drifting the day before release.

There were also several differences in detail of the drift patterns. For example, *Rheotanytarsus* sp. larvae responded immediately to the increased flow by a 30-fold increase in numbers drifting, but the quantity decreased fairly steadily as time passed, with no real evidence of diel fluctuations. Similar behaviour by Chironomidae was noted by Armitage (1977) and is apparent in Figure 7.6. In contrast, *Ephemerella ignita* showed a 10-fold increase in nocturnal drift peaks but little change in daytime rate.

It had previously been suggested by Mundie (1974) that the artificial enhancement of daytime drift could be a method of increasing the food supply to salmonid fish. Artificial freshets do appear to increase drift concentration and they may, therefore, have potential in the management of food supply to fish. Several aspects of such management possibilities require more detailed study however, such as the possibility that increased drift depletes the standing crop of benthic invertebrates.

EFFECTS OF FLOW REGIME AND TEMPERATURE ON SALMONID FISHES

In many rivers in the northern hemisphere, salmonid fish populations are the main economic resource, so it is hardly surprising that the impacts of flow changes upon salmonids are the subject of a considerable literature. Recent general reviews of environmental requirements of different life history stages and of human impacts upon them, chiefly within a UK context, can be found in Crisp (1989, 1993). The most important impacts of regulated flows for salmonids are the physical effects upon spawning sites and the temperature effects upon food, feeding and growth.

Impoundment modifies downstream flow regimes and, hence, river depth, water velocity, wetted area and patterns of transport of bedload and suspended solids (Tables 7.1 and 7.2). Depth, water velocity and wetted area are important in determining the choice and suitability of salmonid spawning sites (Jones and King, 1950; Peterson, 1978; Crisp and Carling, 1989) and also the balance of suitability of the habitat for juvenile salmon and trout (Kalleberg, 1958; Symons and Heland, 1978; Heggenes and Borgstrøm, 1991; Crisp, 1991; Crisp and Hurley, 1991a, b). Indirect consequences such as bed compaction, sediment deposition and bed movement may influence the viability of intra-gravel stages (Tautz and Groot, 1975; Hausle and Coble, 1976; Hall and Lantz, 1979). The deposition of fines within the gravel interstices may reduce intra-gravel flow and lead to asphyxiation of eggs or alevins within the gravel (Peterson and Metcalfe, 1981; Witzel and MacCrimmon, 1983; MacCrimmon and Gots, 1986; Marty, Beall and Parot, 1986; Olsson and Persson, 1986, 1988; Philips et al., 1986) and also to the entrapment of alevins within the gravel when they attempt to emerge (Philips and Koski, 1969). Bed movements during natural or man-made spates can lead to washout of eggs or alevins from the gravel, with subsequent high mortality. Washout has been considered a major cause of mortality in several populations of Pacific salmon (Wickett, 1952; Gangmark and Bakkala, 1960; Lister and Walker, 1966) and of sea trout in the River Dyfi (Harris G.S., unpublished). The ability of spates to disturb artificial eggs buried at depths of 5—10 cm, or in more exceptional spates up to at least 15 cm, in natural streams, has been noted (Crisp, 1989b). In some instances, large reservoir releases can wash away the gravel from downstream spawning areas. Most upland UK reservoirs effectively eliminate the very high flows capable of doing this but at least one example of gravel bed disturbance or loss by this means appears to have been observed in the UK (see discussion by Hey of Edwards and Crisp, (Hey et al., 1982)).

The more equable summer temperatures downstream of regulating impoundments will reduce the incidence of stressful, high temperatures (Edwards and Crisp, 1982) and modified seasonal temperature patterns may change growth rates (Elliott, 1981), incubation times (Crisp, 1981, 1988), and times of smolt migration (Solomon, 1978). However, the direction, magnitude and full consequence of these changes will vary from site to site. Temperature changes in the River Elan, for example, decreased the "scope for growth" of trout (Edwards and Crisp, 1982), whereas at Cow Green there was virtually no change in "scope for growth" and otherwise improved conditions were exploited by the fish through an increased population density rather than increased growth (Crisp, Mann and Cubby, 1983).

The oviposition time of salmonids is mainly determined by day length (Bye, 1984) which does not change when impoundment occurs, but embryonic development is influenced by temperature (Crisp, 1981; Jungwirth and Winkler, 1984; Humpesch, 1985; Crisp, 1988). It is, therefore, possible for temperature changes to upset the relationship between emergence and first feeding. A possible example of this was identified at Kielder Water (Crisp, 1989a) where temperature change had the potential to break the life cycle of the Atlantic salmon (*Salmo salar* L.) or, at least, to modify the balance of ecological advantage between salmon and trout. Such effects are only likely to occur on a local scale, however. A number of useful models for predicting the effects of temperature changes on UK salmonids exist and are summarized in Crisp (1992), but it has already been stressed that these effects are likely to be localized immediately below the dam.

Information on the effects of flow changes and consequent effects, in the UK, is more limited and often not quantitative (Milner et al., 1981). This leads to frequent application of quantitative relationships from North America, usually derived for different species in very different types of rivers, which may not be applicable in the UK and indeed, may not even be widely applicable in North America. These difficulties arise chiefly from the following causes:

(i)   There is not yet an adequate basis of physical science to allow accurate prediction of the locations, areas, or depths of gravel disruption that are likely to result from a spate of a given size in any given river section.

(ii)   There are difficulties in designing and executing appropriate experiments and/or observations, to relate river flow, via gravel composition, percentage fines and a complex of other factors, to the survival of intra-gravel stages. These difficulties are aggravated by variations in definitions and by marked small-scale spatial variations in intra-gravel flow in natural river gravels. Such problems can lead to large differences between studies, in the quantification of the relationships. For example, consideration of two studies on effects of intra-gravel flow and dissolved oxygen concentration on the survival of rainbow trout (*Oncorhynchus mykiss* (Walbaum)) embryos showed a difference of an order of magnitude in the apparent velocity (cm h^{-1}) of intra-gravel flow required to give 50% survival at dissolved oxygen concentrations of >7 mg litre^{-1}.

ACKNOWLEDGEMENTS

This paper was prepared as part of a contract with the Ministry of Agriculture, Fisheries and Food. The figures were prepared by Mr T. Furnass.

8

The Combined Impacts of River Regulation and Eutrophication on the Dynamics of the Salt Wedge and the Ecology of the Lower Ebro River (North-East Spain)

CARLES IBAÑEZ

Universitat de Barcelona, Barcelona, Spain

ALBERTO RODRIGUES-CAPITULO

Instituto de Limnología, La Plata, Argentina

and NARCÍS PRAT

Universitat de Barcelona, Barcelona, Spain

INTRODUCTION

The ecology of the Lower Ebro River has been deeply modified by human influence over the past century. River water has been increasingly used for industrial processes, cooling of nuclear power stations, urban supply and agriculture, with no treatment of the effluents. Additionally, more than 100 reservoirs have been built in the basin. River ecology changed further in the 1960s when a big dam, Mequinença, was built 100 km upstream from the Ebro Delta (Figure 8.1). Thus, the river flow has become more regulated and water quality has worsened, with a large increase in eutrophication.

Low flows (less than 100 m^3 s^{-1}) and the presence of a salt wedge in summer were known to be a natural phenomenon in the river, but nowadays the salt wedge is present for periods of the year other than summer, and can persist all year round in dry periods (e.g. 1989) due to river regulation. Low flows and eutrophication have probably enhanced processes leading to oxygen depletion in the salt wedge, affecting 25 km of the estuarine reach, with the disappearance of benthic organisms. During wet periods (usually the spring), when river discharge is higher than 300 m^3 s^{-1} (the mean annual discharge), the salt wedge is washed away in few days, and the estuary becomes a river. In the zone near the mouth, the influence

The Ecological Basis for River Management. Edited by D.M. Harper and A.J.D. Ferguson. © 1995 John Wiley & Sons Ltd

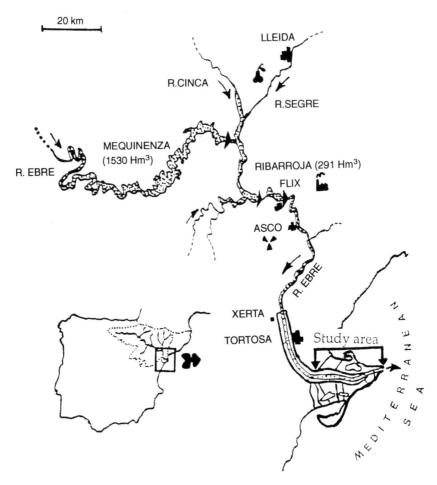

FIGURE 8.1.   General map of the lower Ebro River showing the reservoirs and the study area in the
Ebro Delta

of marine water can persist in flows up to 400 m³ s⁻¹, depending on the dynamics of the
sand bar at the river mouth and the sea level.

Future plans for river management have recently been prepared by the river authorities.
More abstraction of the river waters for irrigation purposes and new flow regulations (49
new reservoirs) are among the measures proposed, while eutrophication will not be
diminished immediately. This will affect water quality and fish productivity. The persistence
of the salt wedge will increase and the benthic communities of the estuarine reach will
continue to suffer severe environmental stress.

The benthos of the estuarine reaches of Mediterranean rivers has not been studied. Data
have usually been collected upstream from the zone affected by the salt wedge (e.g.
Battegazzore et al., 1992). Such information is also scarce in other rivers like the Rhine (Van
Urk, 1978; Admiraal, van der Velde and Cazemiek, 1992).

MATERIALS AND METHODS

During 1988−92 several surveys were undertaken of the last 30 km of the River Ebro (Figure 8.1) in order to study the dynamics of the salt wedge, relating hydrological to physicochemical processes (Ibañez and Prat, 1993). This is a continuation of an earlier study of the same river (Muñoz and Prat, 1989; Muñoz and Sabater, 1990). Here we present results concerning the effects of the river changes on the estuarine system, especially on benthic organisms.

In the field surveys we performed several profiles of conductivity, temperature and dissolved oxygen along the estuary (30 km), and water samples were also taken at different depths in order to determine the suspended material, chlorophyll and dissolved nutrients. The suspended material was determined by filtering 1 litre of water though a Watman GF/F (0.4 μm) filter. For chlorophyll, the same amount of water and the same type of filter was used; the pigments were extracted with acetone (80%). The readings were made at 664, 647, 630 and 430 nm. Dissolved nutrients (nitrates, ammonia and phosphates) were determined in a Technicon autoanalyser.

Samples of the benthos were taken from a boat at five different sampling stations along the river, from the river mouth to the city of Amposta, 30 km upstream. The sampling depth was around 5 m, to make sure that sediment was affected by salt wedge. The benthos was sampled in October 1992, after a long anoxic period that ended at the beginning of the month. At this moment, increased river discharge caused the retreat of the salt wedge. Two samples were taken with an Ekman grab at each point and organisms were separated using the sugar flotation technique (Anderson, 1959).

CHANGES IN THE RIVER

Patterns of the river flow

The original flow pattern of the Ebro River before the construction of reservoirs on tributaries was characterized by a strong increase in flow in spring followed by a minimum flow in summer and a further continuous increase in flow from September until March (Figure 8.2). After the construction of reservoirs in the tributaries (between 1951 and 1965) the river regime was similar to the original flow pattern, but with reduced flow in spring due to retention in the reservoirs. After the construction of large reservoirs in the main channel (1972−80), this trend increased. In dry years (e.g. 1980−90), water released by reservoirs was less than the discharge in earlier periods. In the last period for which data are available (1989−90), the mean monthly river flow for winter was similar to that previously recorded for summer, and was the lowest historical record of mean yearly discharge for the Ebro River.

Thus the effect of river damming has been a reduction in the mean annual discharge and, more importantly for river eutrophication and salt wedge formation, a decrease in the number and extent of flood events and the mean monthly discharge in spring. This decreasing of the mean flow in the Ebro River is thought to be a consequence of water losses by evaporation and filtration (in reservoirs, irrigation, industrial and urban uses), as well as some reduction of the rainfall in the Ebro basin.

Eutrophication

Eutrophication of the River Ebro has sharply increased in recent years. Monitoring data available from Ebro River water authorities (Confederación Hidrográfica del Ebro) show

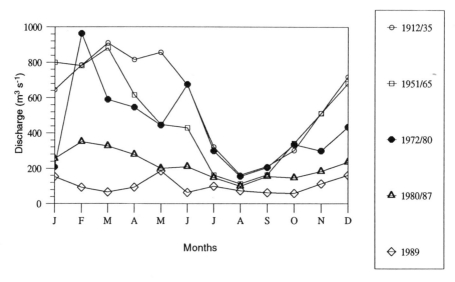

FIGURE 8.2.   Mean monthly discharge of the river at the city of Tortosa (40 km from the mouth) for
different periods along the present century. Data from the Confederación Hidrográfica del Ebro

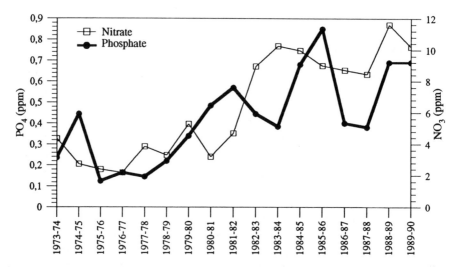

FIGURE 8.3.   Mean values of dissolved nitrate and orthophosphate in April at the city of Tortosa from
1973 to 1990. Data from the Confederación Hidrográfica del Ebro

that the mean annual orthophosphate and nitrate concentration for the Lower Ebro increased
from 0.2 mg litre^{-1} and 3.0 mg litre^{-1} in the 1970s to 0.9 mg litre^{-1} and 9.0 mg litre^{-1} at
the beginning of the 1990s respectively (Figure 8.3). Among the main causes are the
development of intensive farming, increased population in the basin and industrial
development as well as dam construction and reduced discharge of the river.

EFFECTS ON THE ESTUARINE REACH

Salt wedge hydrodynamics

The establishment of the salt wedge in the Ebro River and its advance and retreat largely depend on river discharge, but this is not a linear relationship because of irregular topography of the estuarine bottom (Figure 8.4). However, as can be seen in Figure 8.5, there is a good linear regression between river discharge and the depth of the interface (Ibañez and Prat, 1993). When the river discharge is higher than the mean annual discharge (about 300 m^3 s^{-1} for the last 10 years), the salt wedge cannot be established and the estuary works like a river. This usually happens in the spring, when the discharge is high due to the melting of the snow in the Pyrenees mountains, but sometimes can also happen during autumn or winter after a rainy period. When the river discharge is in the range of 250−300 m^3 s^{-1}, the salt wedge occupies the last 5 km, whereas with lower values it advances quickly until reaching a shallow point (3 m depth) at Gracia Island, 18 km from the mouth. The salt wedge remains arrested at this point until the river discharge is lower than 100 m^3 s^{-1}, and then it advances quickly until it reaches another shallow point (2 m depth) upstream of the city of Amposta, 30 km from the mouth. This point is nowadays the maximun extent of the salt wedge, though before the regulation of the river, the salt wedge was even further extended in some dry summers (Aragon, 1943).

These shallow points in the Ebro Estuary (including the sand bar at the mouth) also play a crucial role in the mechanism of retreat of the salt wedge. When the river discharge increases above some of the critical values which determine the stable positions of the salt wedge (Figure 8.6), the saline/freshwater interface reaches the bottom at the shallow point and the upper part of the salt wedge becomes isolated. This isolated part of the salt wedge is then progressively eliminated by mixing with the river water. A model based upon the regression equation between river discharge and the depth of the interface at several points along the estuary, together with the topography of the bottom, allows the position of the salt wedge and its general hydrological dynamics to be easily predicted.

This model can be used to assess the historical trends of the presence of the salt wedge in the Ebro Estuary, using the record of river discharge. The average duration of the salt wedge

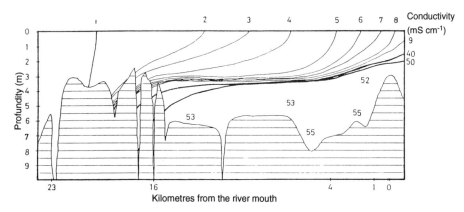

FIGURE 8.4. Typical structure of the salt wedge in summer in the Ebro estuary. Notice the irregular topography of the bottom

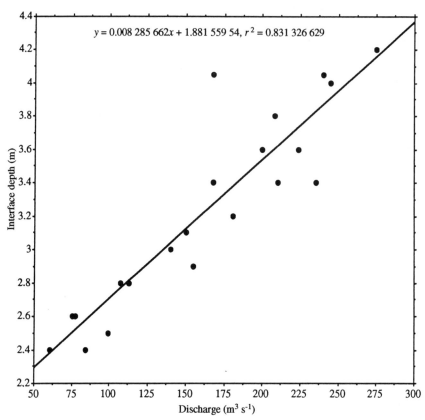

FIGURE 8.5.   Linear regression between mean daily discharge of the river at the city of Tortosa and
the depth of the interface 13 km from the mouth

for different periods in the past can be calculated, for example. Table 8.1 shows the average
number of months per year with the salt wedge for different periods in the last century (there
is a lack of data in the period 1935 – 50 due to the civil war in Spain). This shows that, as a
consequence of river regulation and reduced discharge, the number of months with salt
wedge has increased in recent years, especially during dry periods such as in the 1980s.

Physicochemical processes

When the salt wedge is present in the Ebro Estuary, the physicochemical structure of the water
column always shows sharp gradients at the interface (Figure 8.7), reflecting the big differences

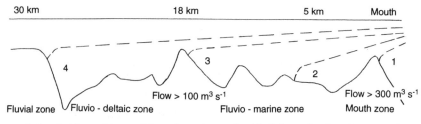

FIGURE 8.6.   Stable positions of the salt wedge and critical flows for its advance and retreat

TABLE 8.1. Predicted time of presence of the salt wedge in the Ebro estuary for different historical periods

Period	Mean annual flow (m³ s⁻¹)	Months with salt wedge (annual average)
1913–19	747	3,1
1920–26	519	4.4
1927–34	518	4.6
1951–57	389	6.9
1958–64	582	4.3
1965–71	516	5.1
1972–79	424	6.4
1980–88	318	7.7

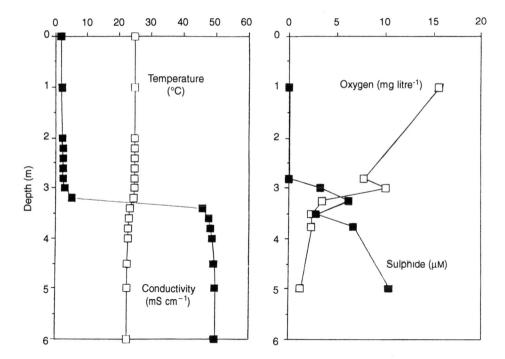

FIGURE 8.7. Profiles of conductivity, temperature, dissolved oxygen and sulphide in summer. Data from July 1991

in the water features between the upper layer (riverine origin) and the lower layer (marine origin). The high nutrient content of the surface waters and the presence of abundant light produce a high primary production (high chlorophyll levels), basically due to riverine phytoplanktonic species (Muñoz and Prat, 1989). This organic matter is sedimented and decomposed in the deep zone, where the lack of light and vertical mixing leads to oxygen

TABLE 8.2. Physicochemical features of the water column at stations 1, 2 and 3 in July 1991

Distance from mouth (km)	Station 1 1	Station 2 5	Station 3 13
Conductivity (mS cm^{-1})			
Surface	4.05	3.04	1.70
Upper interface	26.0	19.0	12.0
Lower interface	42.9	42.0	39.0
Salt wedge	47.9	48.4	47.1
Oxygen (mg litre^{-1})			
Surface	7.7	8.0	7.8
Upper interface	3.8	4.9	5.7
Lower interface	4.7	2.9	0.8
Salt wedge	7.2	3.6	0.7
TSS (mg litre^{-1})			
Surface	32.5	13.5	28.2
Upper interface	15.7	26.7	27.7
Lower interface	18.0	69.0	47.0
Salt wedge	67.0	52.2	49.7
Chlorophyll (mg m^{-3})			
Surface	65.9	76.8	21.5
Upper interface	55.8	33.2	39.6
Lower interface	50.8	3.2	14.2
Salt wedge	6.6	9.9	4.9
PO$_4$(μM)			
Surface	0.65	0.40	0.33
Upper interface	0.50	0.43	0.50
Lower interface	0.40	0.60	4.25
Salt wedge	0.38	0.50	3.65
NO$_3$(μM)			
Surface	188.1	233.3	238.1
Upper interface	228.6	221.4	214.3
Lower interface	183.3	42.8	33.6
Salt wedge	27.8	13.3	5.0
NH$_4$(μM)			
Surface	28.9	7.8	28.9
Upper interface	4.4	9.4	12.2
Lower interface	1.1	37.8	102.2
Salt wedge	14.4	128.9	110.0

depletion and the establishment of an anaerobic metabolism. The high ammonia and phosphate contents due to anoxia in the salt wedge are shown in Table 8.2. In general, there is a depletion of oxidized forms, e.g. nitrate, and an increase of reduced ones. The behaviour of the estuary in these conditions, which mainly occur in summer, is like that of a eutrophic lake. The longer the salt water remains in the estuary, the greater are the chemical and biological changes produced.

Benthic communities

In general, densities and specific diversity of the benthos are low or zero (Figure 8.8), indicating the existence of some stress factors. Close to the mouth, due to the presence of

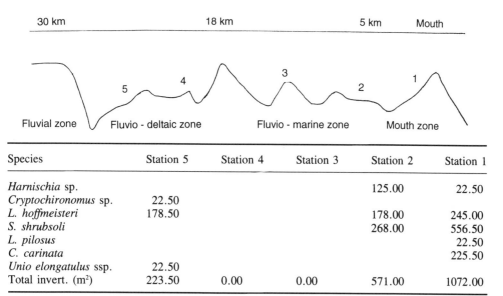

Species	Station 5	Station 4	Station 3	Station 2	Station 1
Harnischia sp.				125.00	22.50
Cryptochironomus sp.	22.50				
L. hoffmeisteri	178.50			178.00	245.00
S. shrubsoli				268.00	556.50
L. pilosus					22.50
C. carinata					225.50
Unio elongatulus ssp.	22.50				
Total invert. (m²)	223.50	0.00	0.00	571.00	1072.00

FIGURE 8.8. Location of the sampling points, and results of the survey carried out to determine the composition and densities of the benthic community in th Ebro estuary

oxygen and high salinity in the deep waters, several species typical of marine and estuarine waters are present (stations 1 and 2). The most abundant species were the Polychaeta *Streblospio shrubsoli* and the Oligochaeta (Tubificidae) *Limnodrilus hoffmeisteri*. The first of these is euryhaline, present also in the Alfacs Bay in the Ebro Delta (Martin, 1991). The second is more indicative of hypoxic environments like eutrophic lakes and reservoirs (Real, 1993). Other euryhaline species found were the Isopoda *Cyathura carinata* and the Amphipoda *Leptocheirus pilosus*, only present in station 1, where there were also low densities of the Chironomidae *Harnischia* sp. and some Nematoda.

In the zones which are anoxic during the summer (station 3 and sometimes station 4), no fauna were found. The effects of anoxia due to sewage pollution were studied in the Nechez River estuary (Harrel and Hall, 1991), showing the same results as in the Ebro, the absence of macroinvertebrates. In the upper edge of the estuary that is less often affected by anoxia, only several Tubificidae (*Limnodrilus hoffmeisteri*), some Chironomidae (*Cryptochironomus* sp.) and some Unionidae (*Unio elongatulus* ssp.) were present (station 5). This last species is typical of the lower parts of large rivers, but is low salinity tolerant (Altaba, 1992). It was found in a shallow sandy point rarely affected by the salt wedge.

The composition of the benthos in the riverine zone, upstream of the estuary, is more rich, abundant and constant (Muñoz, 1990), with the abundance of the ephemeropteran *Ephoron virgo* being the most remarkable. This species produces huge mating swarms in August and the begining of September (Ibañez et al., 1991).

DISCUSSION

The highly variable spatio-temporal dynamics of the salt wedge and the strong stratification of the water column are the factors that determine the ecological features of the estuarine reach of the Ebro River. Sharp changes in salinity and dissolved oxygen are the stress factors which prevent the establishment of either a real estuarine ecosystem or a fluvial one. Although these stress factors are natural in this kind of estuary, they have been enhanced by human-induced changes in the river (water control and eutrophication).

Benthos in the estuarine reach is a consequence of salt wedge dynamics and water quality. Only near the mouth (the last 5 km) is some benthos typical of estuarine or marine environments found. In the rest of the estuary, no benthos is usually found, though the upper reach can be recolonized by some fluvial species after a long period with no salt wedge. As a function of the benthic community, salt wedge dynamics and sediment type (Verdaguer, Serra and Canals, 1985), the estuarine reach of the Ebro River can be divided in three different zones (Figure 8.6).

The lowest zone is the mouth (the last 5 km), where the sand is the major component of the sediment. The water of the salt wedge is similar to the marine water coming in, characterized by the presence of dissolved oxygen, low dissolved nutrient concentrations and low concentrations of suspended solids. The salt wedge is present, on average, for about seven months per year. Some marine benthos and fish are present.

The middle zone, the fluvio-marine zone, exists from 5 km to 18 km (Gracia Island) upstream, where sand is mixed with abundant clay and organic matter, and the sediment shows a black colour. The water features of the salt wedge are deeply modified by the occurrence of long anoxic periods. The salt wedge is present, on average, for about six months per year. No benthos is found in the deep parts affected by the salt wedge.

The uppermost zone is the fluvio-deltaic zone, from Gracia Island to the end of the estuarine reach (30 km upstream, at the city of Amposta). The sediment here is similar to the previous reach but less rich in organic matter. Changes produced in the water of the salt wedge are the same as in the previous reach, but the salt wedge occurs, on average, for less than two months per year. The common occurrence of long periods with no salt wedge in this reach (several months, and sometimes more than one year) strongly increases the possibility of recolonization and reconstruction of the riverine benthos.

These conditions show clearly that the management of the river should be based upon the maintenance of flows higher than 100 m^3s^{-1} during long periods (except the summer), to allow the recolonization of the fluvio-deltaic zone by the riverine benthos. It is also necessary to maintain high flows (more than 400 $m^3 s^{-1}$) during some periods (at least the spring season) to wash the salt wedge away, in order to avoid the progressive accumulation of reduced, organic-rich sediment in the bottom of the fluvio-marine zone. To diminish the possibility of the occurence of anoxia in the longer term, it is also necessary for there to be a severe reduction of sewage inputs to the river, in order to lower the phytoplankton levels which cause the accumulation of organic matter in the salt wedge.

Nowadays, the Ebro River, despite its strong regulation, still has moderate discharges (up to 400 $m^3 s^{-1}$) in winter and spring. This is important not only to wash away the salt wedge and retain the riverine ecosystem but also to maintain the high biological productivity of the marine coastal ecosystems influenced by the river. Such requirements are not taken into account by the new Hydrological Plan of the Spanish Government. This future plan will greatly increase the river regulation (49 extra reservoirs) and the abstraction of considerable amounts of water (up to 50 $m^3 s^{-1}$). If this plan is carried out, the impacts on the lower Ebro River, the Ebro Delta and the coastal zone will be catastrophic.

9

The Utility of Stream Salinity Models in the Integrated Management of Australian Rivers

DAVID R. BLÜHDORN and ANGELA H. ARTHINGTON

Griffith University, Queensland, Australia

INTRODUCTION

Salinization of soil and water is causing major environmental and economic problems in Australia. The value of lost production due to salinization was estimated at A$100 million in 1990, while the environmental cost could only be conjectured (Macumber, 1990). The phenomenon is principally attributed to land and water management practices inappropriate to this continent's highly variable climate and stream discharge regimes (CSIRO, 1992).

Land-use practices, such as deforestation and irrigation, have contributed to the increase of soil and water salinity levels due to rising water tables and increased groundwater runoff (Evans, Brown and Kellet, 1990; CSIRO, 1992; McMahon and Finlayson, 1992). Water management practices have produced regulated rivers, created artificial storages, and disrupted the natural cycles of stream discharge. Additionally, point-source increases in stream salinity either create or exacerbate such problems. The potential environmental consequences of such disturbances include deleterious effects on soil structure (Close, 1990), the loss of species, adverse changes in biotic community structure (Anderson and Morison, 1989; Anderson, 1991), and the proliferation of exotic species.

Stream salinity is affected by catchment-wide land and water use as well as climatic, topographic, and geological characteristics. The generally arid climate of Australia, in which evaporation greatly exceeds rainfall has led, over a geologic time-scale, to the storage of large volumes of salt in sub-surface sediments. With the anthropogenic impact of rising water tables, this salt is increasingly being re-introduced into surface stream and soil systems (Macumber, 1990). The complexity of the interactions between these factors means that the problems created are not readily amenable to direct management. Salinity management depends on an integrated, catchment-wide approach which incorporates a knowledge of the effects of flow regime alterations, the regulation of point and diffuse sources, and the adoption of appropriate land-use and water management practices. An early step in the development of an integrated approach to stream salinity management is to understand the processes involved and to try to predict salinity responses under various circumstances. A recent study of Barker and Barambah Creeks, which form a sub-catchment of the Burnett

The Ecological Basis for River Management. Edited by D.M. Harper and A.J.D. Ferguson. © 1995 John Wiley & Sons Ltd

River system in south-east Queensland, has facilitated the analysis of a number of broadly applicable relationships between stream discharge and salinity.

STUDY AREA

The Barker–Barambah catchment, covering an area of 5 905 km², drains the northern slopes of the Great Dividing Range at the Bunya Mountains and the western slopes of the Coast and Brisbane Ranges (Arthington, Conrick and Bycroft, 1992). It comprises two principal streams: the largely unregulated Barambah Creek and the completely regulated Barker Creek. Figure 9.1 shows the location of the catchment and the streams.

Natural flows in Barker Creek are augmented by releases from Tarong Power Station (Figure 9.1), the water from which originates outside the catchment at Boondooma Dam on the Boyne River, another sub-catchment of the Burnett system. The augmented discharge in Barker Creek is utilized for local irrigation.

The principal regulatory structure on Barker Creek is the Bjelke-Petersen Dam, 1.3 km upstream from its confluence with Barambah Creek (Figure 9.1). Two irrigation areas and three urban communities are serviced by water from this dam.

The Barker–Barambah catchment has a negative water balance, where potential evaporation exceeds mean annual rainfall (Arthington et al., 1992). Under natural, unregulated conditions stream salinities ranging from $<300\ \mu$S cm^{-1} during periods of flood discharge to $>2\ 300\ \mu$S cm^{-1} during low discharge periods have been recorded for these waterways (Arthington et al., 1992). In general, the native biota are well adapted to this range of salinities.

In 1991 and 1992, the water quality of the Barker–Barambah catchment was measured at a number of locations over 15 sampling cycles. Salinity values, expressed in terms of electrical conductivity (EC), were measured and normalized to 25°C. Stream discharge data were available from three gauging stations adjacent to sampling sites and discharge at intermediate points was calculated by extrapolation (Blühdorn and Arthington, 1994).

Modelling stream salinity

The arrangement of regulated and unregulated streams, the dam, and the power station, facilitated the development of three EC–stream discharge relationships. The first, termed an "intrinsic flow effect", relates to EC values under conditions of natural discharge. The second, termed a "regulated flow effect", relates to EC values under regulated discharge conditions. The third, termed the "Tarong effect", models the effect of the point-source discharge from the Tarong Power Station combined with local irrigation effects.

The sampling sites were combined into three sub-catchment groups, and Figure 9.1 illustrates the location of the sampling sites and their associated groupings.

The first sub-catchment group (UBM) comprised those sites on Barambah Creek upstream from the influence of the Bjelke-Petersen Dam. This sub-catchment was unregulated and was used as the basis for the model representing the effect of unregulated flows on electrical conductivity throughout the catchment.

The second sub-catchment group (IRR) comprised the sites located downstream from the Bjelke-Petersen Dam and within the Byee Irrigation Area. These sites were subject to flow regulation from the dam.

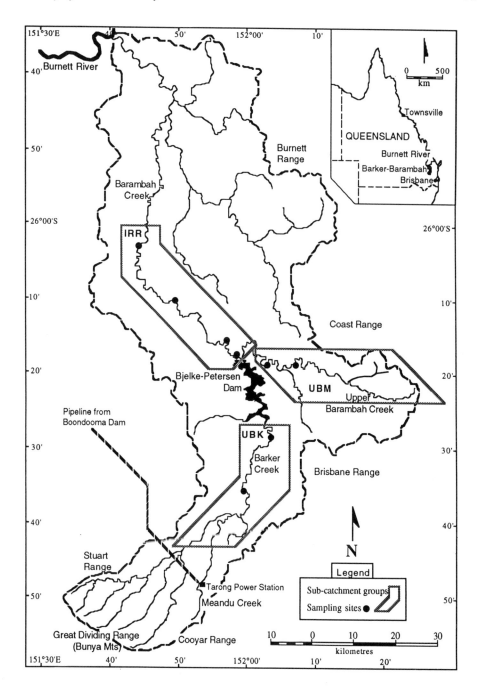

FIGURE 9.1. Map of the Barker–Barambah catchment showing the sampling sites grouped by sub-catchment

The third group (UBK) comprised the sites on Barker Creek upstream from the impoundment of the Bjelke-Petersen Dam, but downstream from the Tarong Power Station. Unregulated flows in this sub-catchment were augmented, and often dominated, by releases from the power station.

Intrinsic flow effect

Under unregulated conditions, electrical conductivity (EC) was found to be primarily a function of both surface and groundwater discharge. This intrinsic flow effect provided the foundation for the EC of all sites, although in some cases it was modified by other effects. The intrinsic flow effect (IFE) was estimated by analysing the relationship between EC values and discharge data for the unregulated part of the catchment (UBM) and also from other sites within the catchment when floods eliminated any modifying effects.

The derived IFE relationship was:

$$\text{EC } (\mu S \text{ cm}^{-1}) = 1837.4 \times \text{discharge } (10^6 \text{ litres day}^{-1}) -0.231$$
$$r^2 = 0.782, \quad \text{d.f.} = 50 \tag{9.1}$$

The limits to this function are an upper EC value of about 1850 μS cm^{-1} for discharges less than 10^6 litres day^{-1}, depending largely on local soil and baseflow salinity, and a lower EC value of about 300 μS cm^{-1} for flood discharges, depending on the conductivity of the suspended sediments. Figure 9.2 illustrates this relationship and the data points from which it was derived.

Thus, under natural conditions in this catchment, EC is inversely related to discharge. Floods (high discharge, low EC) flush the system at irregular intervals, reducing EC values to about 300 μS cm^{-1} throughout. Dry periods (low to zero surface discharge) result in localized high stream salinity, largely due to the effects of saline groundwater discharge.

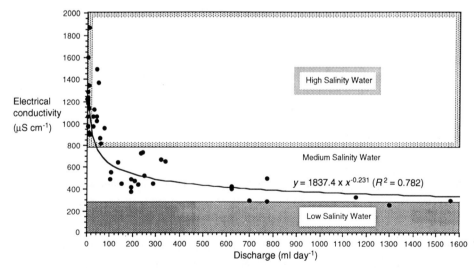

FIGURE 9.2. The intrinsic flow effect: electrical conductivity versus discharge under unregulated and flood flow conditions. The classification of EC values into low, medium and high salinities corresponds with the general guidelines for salinity of irrigation waters (ANZECC, 1992)

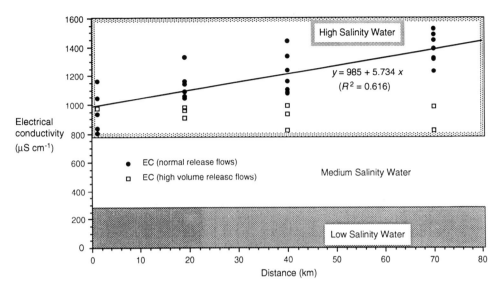

FIGURE 9.3. The regulated flow effect: electrical conductivity versus distance from the Bjelke-Petersen dam under conditions of normal irrigation releases. Data points for high volume releases are also given to illustrate the evenness of conductivity throughout the irrigation area under such conditions

Regulated flow effects

Stream discharge in the IRR sub-catchment (Figure 9.1) fell into three distinct categories. When natural discharge was high, or irrigation demand low, no flow releases were made and the discharge comprised water from the unregulated Upper Barambah Creek only. At other times, high volume releases ($>300 \times 10^6$ litres day^{-1}) were made to transport large volumes of water to the Bundaberg Irrigation Area, at the mouth of the Burnett River. Such releases occurred, on average, once per year, and lasted from two to four weeks. Finally, the IRR sub-catchment was subject to normal volumes of release flows ($<200 \times 10^6$ litres day^{-1}), to fulfil the requirements of local consumers.

EC data for the IRR sub-catchment group were then analysed to determine whether the IFE model fitted the regulated (IRR) sites. For those cycles where there were no release flows (periods of high natural discharge or low irrigation demand), the IFE model was applicable. However, whenever there was a release from the dam, the IFE model no longer fitted the data adequately.

Data recorded during high volume releases from the dam ($>300 \times 10^6$ litres day^{-1}) were analysed. These showed an even value throughout the irrigation area equivalent to the EC of the release water, i.e. from 800 to 1 000 μS cm^{-1}. Thus, high volume release flows reset downstream EC values in much the same fashion as natural floods, but the resultant EC value was considerably higher. Figure 9.3 shows the EC values recorded during high volume release flows.

The remaining IRR data, representing times of normal irrigation releases from the dam ($<200 \times 10^6$ litres day^{-1}), were also analysed. These data were largely independent of discharge and combined the EC level of release water with a function of distance from the dam, in the following manner:

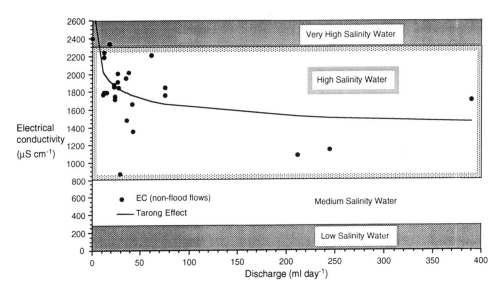

FIGURE 9.4.   The Tarong Effect: the solid line represents expected EC values (calculated from IFE equation +1 000) versus discharge. Also plotted are the actual, non-flood, electrical conductivity values recorded in Upper Barker Creek

$$\text{EC } (\mu\text{S cm}^{-1}) = 985 + 5.734 \times \text{distance from Bjelke-Petersen Dam (km)}$$
$$R^2 = 0.616 \quad \text{d.f.} = 25 \tag{9.2}$$

This model suggests that, under conditions of normal irrigation releases from the dam, EC values in the irrigation area are a combination of the EC of the release water plus the effects of natural and irrigation-modified baseflow. This baseflow effect acts in a longitudinally compounding manner to increase EC in proportion to stream distance from the source of the release water. Figure 9.3 illustrates this relationship between EC and distance under regulated flow conditions.

The Tarong effect

The EC values of Barker Creek (UBK) were significantly higher than those of the unregulated Upper Barambah Creek (UBM) from which the IFE model was derived. This difference was largely attributable to the flows released from the Tarong Power Station, which had an average EC value of 1770 μS cm^{-1} over the period February 1988 to May 1991. When combined with the natural discharge and local, irrigation-modified baseflow, the effect on the waters of Barker Creek was to increase EC values by an average of 1 000 μS cm^{-1} over those expected under the intrinsic flow effect model. Figure 9.4 shows the actual recorded values compared with those expected from the Tarong effect model.

Models

For the three sub-catchments, the following models illustrate the postulated contributory elements to salinity levels.

- Upper Barambah Creek (UBM):
  Intrinsic flow effect (equation 9.1)
- Byee Irrigation Area (IRR):
  Intrinsic flow effect (equation 9.1), with no release flow *or*
  EC value of release water only, under high volume release flows *or*
  EC value of release water + distance from Bjelke-Petersen Dam (equation 9.2) under
    normal release flows
- Upper Barker Creek (UBK):
  Tarong effect (intrinsic flow effect + 1 000 μS cm^{-1})

While these models greatly simplify the complex interactions between factors such as infiltration, surface discharge, and baseflow, they present the ultimate effects of these interactions in a format that is both easy to comprehend for the purpose of salinity management, and amenable to manipulation via the medium of release flows.

Effect on Byee Irrigation Area

As indicated earlier, the intrinsic flow effect is dependent on discharge, and high EC values could be reduced by a large flushing flow. However, from the IRR model under normal release conditions, it becomes apparent that EC is no longer dependent on flow. It is, instead, dependent on the EC of the release water and distance downstream from the dam.

Release water in any volume cannot be used to achieve the low EC values of high volume natural flows. Water from the Bjelke-Petersen Dam is usually within the range $800-1000$ μS cm^{-1}. This is equivalent to the EC of natural flows of $15-35 \times 10^6$ litres day^{-1}. Release water will, therefore, reduce EC values only if natural flows are less than 15 $\times 10^6$ litres day^{-1} (less, if the site is further away). Generally, the effect of release water will be to increase stream salinity in comparison to equivalent natural discharges. The median natural discharge for the IRR sub-catchment was 41×10^6 litres day^{-1} (Blühdorn and Arthington, 1994).

At the other extreme, release flows act to limit the natural rise in EC values as the dry season progresses. The highest EC value recorded under regulated flow conditions was less than 1600 μS cm^{-1}, in contrast to values greater than 2 300μS cm^{-1} recorded before the construction of the Bjelke-Petersen Dam (Arthington et al., 1992).

The effect of regulated flows on this section of the stream has been to reduce the variability and range of salinities, as well as to generally increase the salinity level of the water available for irrigation and urban use.

Factors affecting electrical conductivity in the Bjelke-Petersen Dam

There are two principal sources of water which supply the Bjelke-Petersen Dam via Barker Creek. These are flood discharges, and non-flood natural discharges combined with release flows from Tarong Power Station. Flood discharges provide large quantities of low salinity water over short periods of time. Combined natural and Tarong discharges provide lesser quantities of high salinity water over extended periods of time. The combination of these two sources has produced an impoundment with a relatively stable EC ranging from 800 to 1 000 μS cm^{-1}. This salinity level is fundamental to the effects produced by flow regulation on downstream sites (i.e. the regulated flow effect). It remains to be seen what effect a series of dry seasons will have on the EC of the impoundment. It may remain balanced by occasional flood inflows or it may increase as the proportion of high salinity water steadily increases.

ECOLOGICAL IMPLICATIONS AND MANAGEMENT POTENTIAL

The effects of flow regulation on stream salinity in the Barker—Barambah catchment have been to reduce the variability and the range of EC values recorded in the irrigation area. The effect of the point-source input from Tarong Power Station has been to increase EC values in Barker Creek by approximately 1 000 μS cm^{-1} over that expected under natural conditions. These effects have a number of potential ecological impacts.

The reduction in variability and range of salinity values represents a significant environmental perturbation with potential impacts on biotic community structure. Many native freshwater fish species are euryhaline (Allen, 1989) and adapted to these variable conditions. Without this variability, populations of such species may be environmentally disadvantaged. A reduction in these parameters may also provide suitable conditions for invasive species, which may be less euryhaline but more suited to the changed conditions. Arthington and Mitchell (1986) indicate that disturbed ecosystems are particularly susceptible to plant invasions.

Increases in mean salinity levels are likely to impact on stenohaline species, especially invertebrate and plant communities (Anderson, 1991). Such increases are also likely to adversely affect the structure of susceptible soil types under irrigation within the catchment (QDPI and QWRC, 1979). Significant rises in salinity will affect the use of this water for both irrigation and urban supplies. The models have indicated the changes to stream salinity brought about by the combination of anthropogenic impacts. With further refinements, they will allow future salinity conditions to be examined. However, their principal utility is in indicating those elements where management attention needs to be directed.

From these models it is evident that the pivotal element for managing stream salinity in this catchment is the impoundment of the Bjelke-Petersen Dam. One of the aims of an integrated management strategy should be scheme transparency (Blühdorn and Arthington, 1994). This is the ability to replicate inflow conditions at the outflow of the development scheme under management, so that the scheme is effectively transparent in terms of flow pattern and water quality. If the impoundment contained water of equivalent salinity to flood water, then almost the entire spectrum of stream salinities could be replicated at downstream sites.

Release flows from the dam would replicate natural flows by being combined with local baseflow. Under conditions of low volume release flow, the baseflow would be allowed to predominate, increasing EC values. Under normal release flow conditions the mixture of baseflow and release flow could be used to replicate the EC of equivalent natural flows. Under high volume flow conditions, the release flow would predominate, overriding the effects of baseflow and replicating the low salinities of natural high volume (flood) discharges.

Such a management strategy could be used to meet the ecological needs of the instream biota. Low salinity dam water would also be suitable for other consumers of water such as irrigation and urban use.

MODEL UTILITY

The utility of these simple models is that they are relatively easy to understand but provide all the information needed for management, or point to those areas which need further research. They have indicated the changes caused by anthropogenic impacts, those principal elements contributing to the changes, and they provide an element of predictability.

In the case of the Barker—Barambah catchment, the models indicate that the salinity of the impoundment of the Bjelke-Petersen Dam is pivotal to managing downstream salinity regimes for both environmental and consumer purposes. As well, they highlight the effect of the point source from the Tarong Power Station. With further refinement, and in conjunction with an integrated approach to management, less obvious impacts, such as diffuse salinity sources emanating from land-use practices, may be included in these models.

As developed, the models are broadly applicable to catchments of similar type to the Barker—Barambah catchment. Some modifications would be needed in more tropical catchments subject to monsoonal influences, or in locations where the climate and precipitation are not so variable.

ACKNOWLEDGEMENTS

This study was conducted as part of a Queensland Department of Primary Industries, Water Resources funded study into the effects of stream regulation by the Bjelke-Petersen Dam. The financial and logistical support of this department is gratefully acknowledged.

10

The Expert Panel Assessment Method (EPAM): A New Tool for Determining Environmental Flows in Regulated Rivers

STEPHEN SWALES and JOHN H. HARRIS

Fisheries Research Institute, Cronulla, Australia

INTRODUCTION

River flow regulation

Waters carried by streams and rivers are subject to a wide variety of uses and abuses. The manipulation and diversion of river flows for the purposes of hydro-power generation and agricultural land development constitute some of the most damaging environmental changes perpetrated by human societies in developed countries. There is now abundant literature documenting the adverse effects of river flow regulation on the aquatic biota (Ward and Stanford, 1979; Brooker, 1981; Lillehammer and Saltveit, 1984; Petts, 1984; Craig and Kemper, 1987; Petts, Armitage and Gustard, 1989), and rivers around the world continue to be dammed, diverted and regulated.

In most developed countries, there is a reasonably wide-ranging body of ecological knowledge on the status of biota in regulated rivers upon which to base management decisions. Even in those countries which lack a long history of research into the ecology of their regulated rivers (e.g. Australia), there is still extensive anecdotal and circumstantial evidence linking declines in aquatic communities, such as native fish in the Murray-Darling River, with the effects of river flow regulation (Walker, 1985; Cadwallader, 1986). One of the primary reasons today for the continued widespread degradation of environmental conditions in streams and rivers throughout the developed world is the inadequate input, transfer and application of ecological knowledge into decisions concerning the management of waters and their resources.

Environmental flow assessment

Instream flows provided for environmental reasons, sometimes called "environmental flows", are designed to enhance or maintain the habitat for riparian and aquatic life. They may be provided for preserving native species of flora and fauna, maintaining aesthetic quality, maximizing the production of recreational or commercial species for harvest, or

The Ecological Basis for River Management. Edited by D.M. Harper and A.J.D. Ferguson. © 1995 John Wiley & Sons Ltd

protecting features of scientific or cultural interest (Gordon, McMahon and Finlayson, 1992). A wide variety of methodologies have been developed for assessing the instream flow requirements of fish and other aquatic biota to assist in the development of environmental flow recommendations. Most of these methods have been developed and applied in North America where altered stream flows have jeopardized the continued survival and abundance of commercially and recreationally important fish species, particularly salmonids (Tyus, 1990).

Techniques for assessing the instream flow requirements of aquatic biota in rivers fall into three broad categories:

- historical discharge or "rule-of-thumb" methods, which are based largely on historical flow records and use a fixed proportion of flow;
- habitat analysis methods, which use a combination of hydrology and hydraulics and determine useable habitat by transect analysis and hydraulic simulation;
- instream habitat modelling methods, which determine habitat preference curves for species and model how changes in discharge affect habitat availability.

Within these general categories, a wide variety of different methods have been developed and applied over the last few decades (Richardson, 1986; Kinhill Engineers, 1988; Orth and Leonard, 1990; Gordon, McMahon and Finlayson, 1992). However, as yet there is no one tried and tested standard technique for assessing the instream flow needs of fish and other instream biota that is suitable for all situations. All of the techniques so far developed have their own particular drawbacks and limitations and new techniques continue to be developed for specific conditions and geographic regions. All of the flow assessment methods have their own proponents and critics. Gordon, McMahon and Finlayson, (1992) have reviewed the history of the development of instream flow methods.

Reiser et al. (1989) surveyed the use of different instream flow assessment methods by state and federal agencies in North America. The most commonly applied method (in use in 38 states or provinces) for assessing instream flow requirements was the Fish and Wildlife Service Instream Flow Incremental Methodology (IFIM) (Bovee, 1986). However, this technique has been widely criticised by fisheries scientists as being ecologically simplistic and lacking validation (Mathur et al., 1985; Scott and Shirvell, 1987; Orth, 1989; Armour and Taylor, 1991).

River management and professional judgement

In recent years there has been an increasing recognition of the need to establish more effective communications between river managers and scientists to improve river conservation and management (Boon, 1992). One of the ways in which this can be achieved is by more direct involvement of scientists in river management decisions. There have been some recent examples where expert-systems techniques have been successfully applied, such as the River Conservation System developed in South Africa by O'Keefe, Danilewitz and Bradshaw, (1987). In North America, Angermeier, Neves and Nielson (1991) recently described how the decisions of aquatic resource professionals were used to identify features of streams and rivers that conferred ecological and fishery values. Chaveroche and Sabaton (1989) have described how qualitative expert advice was used in formulating habitat use curves for brown trout in a river in France.

The first instream flow methods developed in North America in the early 1970s were based on the judgement of biologists (Gordon, McMahon and Finlayson, 1992). Reiser et al.

(1989) found that today many states in the USA still rely on professional judgement and so-called "rule-of-thumb" techniques when determining instream flow requirements. In river management, there is an increasing general recognition of the need for an interdisciplinary approach to protecting instream flows (Jackson et al., 1989; Hill, Platts and Beschta, 1991).

River regulation in Australia

In the rivers of semi-arid south-eastern Australia, the allocation of water for environmental purposes has become an important issue in recent years as river systems show indications of major environmental degradation (Barmuta, Marchant and Lake, 1992; Hart, 1992). The arid and unpredictable climate of Australia means that freshwater is scarce and by world standards the flow regime of rivers is highly variable (McMahon and Finlayson, 1992). Consequently, instream flow assessment methods developed overseas may not be applicable under Australian conditions and there is a need to develop other techniques (Arthington et al., 1992).

In New South Wales, as in most other densely populated parts of Australia, there are very few rivers whose flow regimes are not regulated. Most of the catchment of the Murray-Darling River system, at over 1×10^6 km^2 the largest in Australia, is contained within NSW. Since European settlement in the area in the 1800s, streamflows have been intensively regulated by dams and low-level weirs for the purposes of land irrigation, flood alleviation and, to a lesser extent, hydro-power generation (Harris, 1984; Walker, 1985).

NSW Fisheries has investigated a variety of techniques for assessing the instream flow needs of native fish in its streams and rivers. Richardson (1986) evaluated and compared four different instream flow methodologies (the Montana method, flow-duration curve analysis, transect analysis and the incremental method) on the Tweed River in northern NSW. All of the methods used were found to have shortcomings and it was recommended that other instream flow assessment techniques be developed which are more suitable for Australian conditions. In particular, it was recommended that there was a need for the development of a suitable reconnaissance planning technique for initial assessment of proposed developments.

The Expert Panel Assessment Method

The development of a reliable and accurate method for assessing environmental flow requirements is a priority in freshwater fisheries management. Ideally, the method used should be widely applicable, inexpensive and not require extensive field measurements (Orth and Leonard, 1990). Many of the techniques in use today, e.g. IFIM, are time consuming, expensive and require extensive field measurements.

An expert panel assessment method was developed by NSW Fisheries in response to these needs in the variable and unpredictable nature of streamflow conditions in Australian rivers. The method depends on the utilization of the professional experience of specialists in fluvial sciences to assess the suitability of instream flows for river ecosystem processes. The suitability of streamflows for the survival and abundance of native fish is taken as the primary criterion of the suitability of the discharge as an environmental flow. Fish communities are generally acknowledged to be a good indicator of overall environmental quality or river "health", and respond to direct and indirect stresses of the entire aquatic ecosystem (Fausch et al., 1990).

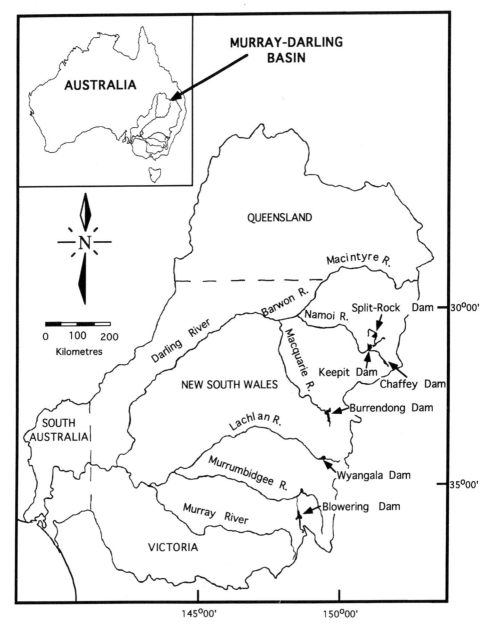

FIGURE 10.1. Map of the Murray-Darling River Basin and the location of water storages on major
tributaries

THE METHODOLOGY

The method was tested experimentally in 1992 at sites on regulated rivers below six
headwater storages on tributaries of the Murray-Darling River in eastern NSW (Figure
10.1). These storages operate largely to provide water for crop irrigation in downstream

river valleys during the summer growing period. Two expert panels were set up, each consisting of three specialists in the fields of freshwater fish ecology, river invertebrate ecology and fluvial geomorphology. In conjunction with NSW Department of Water Resources, arrangements were made in winter 1992 for the provision of experimental flow releases below each of the six storages. At each storage a range of four flows were released, representing the 80%, 50%, 30% and 10% flow percentiles, determined from flow-duration curves for each river.

River inspection sites for the expert panels were selected at distances of 3−5 km downstream from each dam. The two panels were asked to assess the suitability of the experimental flow releases below each dam as environmental flows, with the suitability of each flow for ecological processes affecting fish communities being used as the primary overall indicator of environmental quality. Flow assessments were made visually, with the specialists ranking flow suitability on a scale of 1 (poor) to 5 (excellent). The panels operated independently, with each panel providing scores based on a consensus decision produced through collaborative discussions within the panel.

The main criteria which the panels were asked to consider included the suitability of the flows for fish survival and abundance, including such aspects of life history as adult spawning requirements, fish passage, juvenile recruitment and survival, feeding and growth. The river invertebrate ecologist advised the panel on the suitability of each flow for the productivity of aquatic and riparian macroinvertebrates, while the fluvial geomorphologist advised the panel on the consequences of each flow for fish habitat, including river morphology, bank erosion and substrate stability. The panels were asked to assess the suitability of the flows on a seasonal and non-seasonal basis.

RESULTS AND DISCUSSION

Non-seasonal scores for flow suitability for fish, invertebrates and habitat quality varied considerably between storage sites and to some extent between panels (Figure 10.2). However, overall general trends of scores were similar within each site. Seasonal scores provided by the panel for spring, summer, autumn and winter showed a strong similarity between sites (Figure 10.3). In general, the overall trend was for the panel to prefer the lowest release (80% flow percentile) as a summer flow, intermediate flows (50% and 30% flow percentiles) as spring and autumn flows, and the highest releases (30% and 10% flow percentiles) as winter flows. The overall recommendation of the expert-panels was for a return towards a more natural pattern of seasonal flows.

The releases selected by the panels as suitable environmental flows were contrary to the current regulated flow regime in each river, with maximum releases being provided from each storage in summer to provide water for downstream irrigation requirements. Minimum releases are made in winter when there is little or no demand for irrigation releases and the storages are operated to mitigate downstream flooding and to store water for the summer release period. The regulated flow regime represents an inversion of the natural unregulated flow regime, which for most rivers in this area consists of low summer flows and high winter flows.

In general, for non-seasonal flows the preferred discharges were the mid-range flows (30% and 50% flow percentiles), while extreme low (80% flow percentile) and high (10% flow percentile) flows were least preferred. In general, habitat diversity in most rivers is at a maximum at intermediate flows and decreases as flows approach the extremes (Bain, Finn

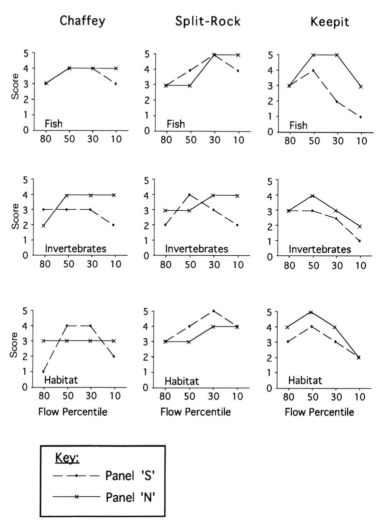

FIGURE 10.2.   Non-seasonal scores for fish, invertebrates and habitat quality made by the expert
panels at each flow percentile at sites below water storages

and Booke, 1988; Leonard and Orth, 1988). Consequently, the judgements of the expert
panels correspond with the streamflows at which maximal faunal diversity and abundance
might be expected.

The initial findings of the study suggest that EPAM can be a useful tool in assessing the
suitability of instream flows for native fish and as a means of assessing suitable
environmental flows. The technique does have drawbacks and limitations, however, like all
other instream flow assessment techniques currently in use. The main advantages of EPAM
relate to the direct communication of specialist knowledge from recognized experts in the
fields of fish biology, river ecology and fluvial geomorphology into river management
recommendations. Other major benefits include the incorporation of interdisciplinary
judgements and the fact that the approach does not require extensive field measurements.

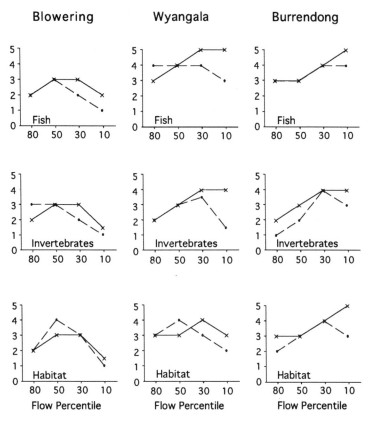

FIGURE 10.2. *continued*

However, the method is site-specific and does not lend itself to the production of a general model for use over a wide geographical area. EPAM is likely to be of most benefit when used as a tool in new approaches to environmental flow assessment, such as the holistic model described by Arthington et al. (1992).

EPAM may be criticized as being a largely subjective approach in that it does not involve field measurements of biota or habitat conditions. However, other more complex instream flow methods, such as IFIM, have been criticized as being simplistic in that they implicitly assume that the complex array of ecological processes occurring in streams and rivers can be modelled using a small range of isolated, non-interacting, physical habitat variables. Such models ignore ecological factors and other processes which cannot be easily incorporated into the models (Mathur et al., 1985; Scott and Shirvell, 1987; Orth, 1989). The use of expert opinion may be preferable in some situations, particularly in rivers such as those in south-eastern Australia, which show variable and unpredictable flow regimes.

Although a large number of instream flow assessment methods have been developed, there are few instances where the recommended flow regimes have been validated by assessing the response of fish and other biota to the recommended flow regime. One of the criticisms of techniques such as IFIM relates to the lack of evidence that fish populations respond to changes to available habitat (Scott and Shirvell, 1987). All instream flow assessment

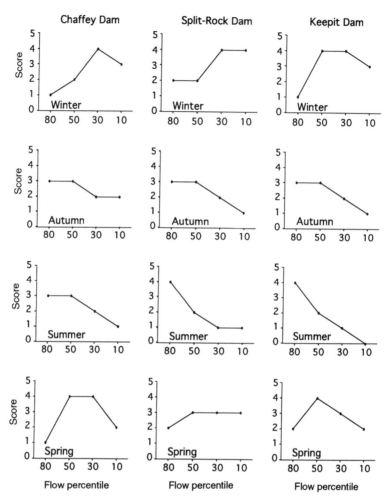

FIGURE 10.3.   Seasonal scores for experimental flow releases made by the expert panel at each study
site below water storages

methods must be considered as experimental until biotic community responses to flow
alterations are assessed and compared to the predicted changes.

One approach to validating instream flow methods and testing their effectiveness is to use
several different methods to assess the instream flow needs of biota and compare the results.
This procedure was carried out recently for native fish in the Peel River (Chaffey Dam),
New South Wales (Swales et al., 1994). Three instream flow methods were used: flow
duration curve analysis, habitat analysis and the expert panel method. The environmental
flow allocation derived using the expert panel method was similar to that obtained using the
flow duration curve approach. This result provides some validation of the expert panel
approach. Further tests of the expert panel method are planned for other rivers in south-
eastern Australia.

There is no single instream flow assessment method that is applicable to all circumstances.

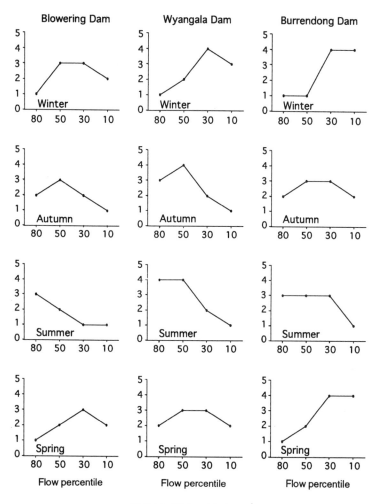

FIGURE 10.3. *continued*

Each method has its own advantages and disadvantages and the applicability and suitability will vary accordingly. In most situations it may be preferable to use a range of methods including both desk-top "rule of thumb" methods, such as flow duration curve analysis, for rapid preliminary assessments, together with detailed habitat-based field studies, such as transect habitat analysis, to provide a quantitative assessment of the effects of streamflow conditions on habitat availability.

The expert panel method might be most usefully employed as a field-based approach to assessing the suitability of a recommended flow regime, derived using a combination of "rule of thumb" and field-based techniques, as an environmental flow. This would allow the suitability of a particular flow regime to be evaluated prior to its actual implementation. Subsequently, to validate the approach taken, the response of biota and habitat conditions to the new flow regime should be monitored.

The use of expert panels in instream flow assessment provides an avenue whereby ecological knowledge can be incorporated into decisions concerning sustainable river management. The use of expert panels can form an effective bridge between freshwater scientists and river managers and so assist in the communication of ecological information which is urgently needed to improve river conservation and management.

ACKNOWLEDGEMENTS

Thanks go to the following expert panel members; W. Erskine, T. Hillman, P. S. Lake, B. L. Lawrence and J. Tilleard. The staff of NSW Department of Water Resources, in particular Hugh Cross, assisted considerably in project planning and organization. Funding support was provided by the Murray-Darling Basin Commission.

11

The Ecological Basis for the Management of Water Quality

RON EDWARDS

National Rivers Authority, Cardiff, UK

INTRODUCTION

It is perhaps mischievous to challenge the title of this book and ask to what extent river management has ever been based on ecological principles, either with the objective of protecting the quality or quantity of abstracted water or the maintenance of the physical, chemical and biological qualities of the rivers for instream uses. Until recently, ecology was a very minor player in influencing river management — except for fisheries. This was, in part, because of our ignorance of ecological processes and their impact on river management and, in part, the low priority we gave to aquatic wildlife conservation (Holdgate, 1979). But knowledge can replace ignorance and priorities can change.

There is, however, one function — flood defence — which has an inherently different design objective for rivers from those of fisheries, conservation and landscape, the flood defence engineer seeking to evacuate water from rivers through straight smooth channels of uniform slope, and nature trying to retain water within rivers through riffles and pools where slope is sufficient, to braided channels and then meanders in the flood plain (Yang, 1971). Taking account of ecological requirements in the design and maintenance of flood defence schemes eases the pain of nature's compromise.

There are at least three reasons why ecological understanding should permeate river management, particularly with respect to water quality, which this chapter seeks to explore:

(1) Ecological processes, both within the river and within the catchment, influence water quality.
(2) Ecological resources, e.g. fisheries and aquatic wildlife conservation, need protection.
(3) Ecological descriptions provide indicators of river quality, their diagnostic value being commensurate with our ecological understanding.

Rivers are open systems, frequently having short retention times, and their management must support not only intrinsic needs but also those of receiving water bodies of very different character, such as lakes and reservoirs, estuaries and the sea. These extrinsic needs are perhaps most clearly perceived with respect to nutrients which can induce blooms of

---

The Ecological Basis for River Management. Edited by D.M. Harper and A.J.D. Ferguson. © 1995 John Wiley & Sons Ltd

planktonic algae, and persistent toxins which accumulate in sediments and biota. Such issues of water quality, also serve to emphasize the interdependence of water quality and quantity, expressing itself in several ways through both catchment and instream processes.

Hydrological pathways through soils and shallow aquifers are influenced by changes in precipitation and evapotranspiration, these in turn being determined, in part, by seasonal factors. Such seasonal and hydrological changes can produce highly variable concentrations of determinands, such as nitrates (Likens, 1984). In some cases, variations in the water quality of the precipitation itself (e.g. pH) can induce major chemical changes in drainage waters (Bird, Walsh and Littlewood, 1990).

Several authors (Edwards, 1973a; Oborne, Brooker and Edwards, 1980) have sought to describe flow — concentration relationships in streams using simple equations, particularly

$$C = aQ^b$$

where C is the concentration, Q is the river flow and a and b are constants. Most determinands (e.g. PO_4-P, Ca, Mg, K, Na, HCO_3) generally have negative values of b indicating an increasing concentration with decreasing flow and, in the case of soluble phosphate, which in many rivers is derived principally from sewage effluent, the value of b may approach unity, reflecting a constant load. In contrast, the concentration of a few determinands generally increases with flow. Such behaviour can reflect the erosional properties of high flows within the catchment and river channel (suspended solids), the seasonal variation of flows coupled with variation in rates of biogeochemical cycling (NO_3) as well as in-river uptake processes by diatoms (SiO_2).

More complex models of flow — concentration relationships have been proposed which separate the flow into components (e.g. run-off, baseflow, effluents). These generally explain considerably more of the variance than the power equation described above (Oborne, 1981) and have some predictive value, particularly when coupled with the inclusion of seasonal influences on catchment and river processes (Webb and Walling, 1992).

Flow within rivers also exerts a wide range of effects on water quality and aquatic communities, principally through its impact on water velocity (or its reciprocal, retention time) as well as river depth and width. The primary impact route can either be:

flow --►aquatic community --►water quality

or   --►water quality --►aquatic community

an example of the former being the effect of retention time on diatom growth and its utilization of silica (Edwards, 1974) and of the latter being the effects of water velocity and depth on river aeration and its consequences to the oxygen economy and behaviour of aquatic organisms (Turner, 1992). Some of these relationships are explored elsewhere in this volume.

ECOLOGICAL PROCESSES AND WATER QUALITY

It is convenient to distinguish three groups of ecological processes which influence water quality, namely catchment, riparian and instream processes.

Catchment processes

Catchment budgetary studies have been carried out for a variety of purposes. At the one extreme they have involved large complex catchments, containing substantial populations, industrial discharges and areas of intensive agriculture (Garland and Hart, 1972), the purpose being to determine sources of materials and, sometimes, their subsequent fate within the river (Edwards, 1973b; Owens, 1970). At the other extreme they have involved small, uniformly vegetated and managed catchments with the receiving stream being merely a useful sampling point for catchment losses, the purpose being to compare precipitation and other inputs with catchment losses and to assess the impact of land use and management on these input — output relationships (see Likens et al., 1970).

Most of these studies demonstrate that virtually all the major chemical components of river water are derived principally from diffuse catchment processes and not direct effluent discharges — except in the most heavily populated and industrialized catchments (Garland and Hart, 1972), and that fluxes of many chemical components, such as combined nitrogen, are primarily dependent on ecological processes within the catchment or are influenced by catchment use and management. Possibly the most dramatic examples of the latter have been described for the Hubbard Brook catchments where, in one experiment, felling of the trees (but not their removal) caused, through the mineralization and oxidation of organic nitrogen and production of nitric acid, soil acidification and the leaching of major inorganic cations, including Al^{3+}. Sulphate losses decreased and, although the mechanisms are complex, the decrease seems in part due to the suppression of microbial mineralization of organic sulphur (Likens et al., 1970). The annual loss of particulate matter increased from 2.5 t km^{-2} ann^{-1} before felling to 38 t km^{-2} ann^{-1} three years afterwards, and was associated with the increased erodibility of the soil (Bormann et al., 1974).

In recent years the impact of acid precipitation on the water quality of streams and on aquatic communities has been widely studied (Stoner, Wade and Gee, 1984; Weatherley and Ormerod, 1987; Ormerod, Wade and Gee, 1987; Edwards, Gee and Stoner, 1990), and is the subject of the next chapter. The increased concentration of aluminium species, leached from the soil under acid conditions, is an important component of the toxic effects, particularly on fish communities (Brown and Sadler, 1989). Land use has a major modifying influence on acidification processes, coniferous forest exacerbating both acidification and aluminium leaching (Department of the Environment, 1990). Amelioration of aquatic damage using lime, either directly to water bodies (Weatherley, 1988) or to source areas of catchments (Hornung, Brown and Ranson, 1990), has been used successfully although costs can be high and a range of conservation impacts needs to be assessed.

We are now able to evaluate the impact of several catchment management practices and uses (e.g. coniferous plantations: Ormerod, Wade and Gee, 1987) on river quality and ecology, and reviews of catchment losses of specific chemicals (e.g. nitrate: Feth, 1966) have been published. It remains to translate understanding into action, although a useful start has been made through the production of forestry guidelines (Forestry Commission, 1991) and agricultural best-practices (National Rivers Authority, 1992g) to minimize aquatic damage.

Riparian effects

The structure of the vegetation canopy within the riparian zone influences river quality and aquatic communities in several ways (Ormerod, Wade and Gee, 1987), but most directly through the interception of radiant energy. This interception (and insulation) is reflected in differences in daily and annual temperature rhythms, which affect growth rates and life cycles, in the seasonal pattern (and quality) of organic matter available for microbial and animal utilization (Hynes, 1975), and in the daily and annual oxygen rhythms, greatly influenced by the site — aquatic or terrestrial — of organic production and utilization.

The temperature effects on salmonid growth can be significant, Weatherley and Ormerod (1990) predicting first-year growth rates of brown trout (*Salmo trutta*) up to 39% higher in moorland streams than in neighbouring forest streams. These authors also suggested that the voltinism of some aquatic insects, particularly mayflies, could be affected by riparian shading.

Of greater importance is the contrast in organic inputs to shaded (allochthonous) and unshaded (autochthonous) streams. The spatial patterns in the balance of these sources has been a major element in the river continuum theory (Vannote et al., 1980) which postulates predictable downstream patterns of the trophic structure of animal communities related to this changing nature of energy inputs within catchments from headwaters to the flood plain.

The development of aquatic macrophytes in unshaded rivers, although increasing habitat diversity and producing organic matter for heterotrophic consumption, can present problems — the most significant being to increase water levels and the incidence of summer floods. They can also profoundly affect pH and oxygen levels — the former, in poorly buffered waters, through the photosynthetic uptake of CO_2. Fish mortalities have occurred at high pH (although generally in lotic waters), and night-time oxygen levels, sometimes coupled with high summer temperatures, have caused major fish kills (Brooker, Morris and Hemsworth, 1977). It is now known that such low oxygen levels can also affect the drift behaviour of many invertebrates (Turner, 1992). Dawson and Kern-Hansen (1979) and Dawson (1989) have suggested that the growth of aquatic plants, where excessive and damaging to drainage or water quality, could be managed by the planting of riparian trees rather than the conventional cutting and removal procedures.

Where streams in afforested areas are acidic it has been suggested that marginal buffer strips, without trees, should be maintained. The basis of this management regime has been the presumed reduction of acidity and aluminium content of water flowing through these strips, a presumption which has not yet been adequately verified. The direct effect of the physical structure of stream margins, below the water-line, must also be considered in relation to habitat diversity and the ecology of fish and invertebrates: some invertebrate species occur predominantly in areas of tree roots (Ormerod, Wade and Gee, 1987).

Instream processes

Whilst rivers play a major role in global biogeochemical cycling by transporting elements, such as Ca, Mg, K, Na and Cl, which are often present well in excess of biological demands within the rivers, they also play a major transformation role for others, such as C, P and N, which are in relatively short supply in relation to ecological requirements (Newbold, 1992).

The concentrations of some elements (e.g. calcium) — despite their fluxes being relatively unaffected by biological processes (except where photosynthesis in poorly buffered waters may induce calcium deposition at high pH — influence the distribution of some species (Ormerod and Edwards, 1987).

Some of the earliest functional studies of elements transformed within rivers related to the kinetics of breakdown of organic material discharged in sewage effluents, studies which described such breakdown, and the associated consumption of oxygen, as first-order reactions (Streeter and Phelps, 1925). This description was used as the basis for predicting the distribution of oxygen within river systems. Later studies have shown that most rivers do not behave in this simple fashion (see Newbold, 1992). Except for the most organically polluted rivers, "natural" sources of particulate and soluble organic matter are of major significance and the break-down processes are complex, involving, in the case of coarse particulate material, a wide variety of invertebrate functional groups (Cummins, 1992) as well as heterotrophic microbes (Maltby, 1992). Except for the deepest rivers, most decomposition takes place on the river-bed or other surfaces (Edwards and Owens, 1965). In slow-moving rivers, fine organic sediments are formed where microbial oxidation of organic matter is greatly influenced by macroinvertebrate activity in aerating, mixing and stabilizing sediment (Edwards, 1964; Edwards and Owens, 1965; Krantzberg, 1985). In rivers where macrophyte growth is luxuriant, the ratio of plant surface to that of the river bed may exceed 50:1 during the summer and so extend the sites of microbial colonization and activity (Edwards and Owens, 1965).

Added to these complexities in heterotrophic pathways and rates, autochthonous production must also be considered. Although most of this production, particularly with macrophytes, is respired directly and imposes major night-time demands on the oxygen resources of some rivers (Edwards, 1968), about 10% ($\sim$ 100 g C m^{-2}) in productive shallow rivers decays during the autumn (Edwards and Owens, 1965).

In recent years the behaviour of carbon and other elements which cycle through biotic and abiotic phases, has been linked to considerations of downstream transport to produce the concept of spiralling, with turnover or cycling times being expressed as distances, i.e. turnover lengths. With respect to carbon it has been concluded that turnover lengths are often longer than the river, much of the carbon entering the river being refractory and reaching the sea (Newbold, 1992).

It is evident that there is scope to influence the decomposition of organic matter and its consequences, particularly to the oxygen economy of rivers, which are mediated through ecological processes. In addition to regulating allochthonous sources from the catchment, and particularly the riparian zone, aquatic plant management can influence both heterotrophic activity and organic production. Furthermore, sediment accumulation and behaviour, sensitive to flow patterns, can greatly affect decomposition processes and oxygen distribution within reaches (Edwards and Rolley, 1965).

The nitrogen cycle is more complex than that of carbon, with some states and processes being of direct concern to water quality managers. Ammonia is directly toxic at concentrations found in some industrial rivers (Alabaster et al., 1972) and consumes oxygen in its oxidation to nitrate, with benthic nitrification rates of ammonia of up to 2.5 g m^{-2} day^{-1} having been recorded (Newbold, 1992). It represents a significant cost in the chlorination of public water supplies and, with nitrate, is implicated as a limiting nutrient of algal blooms — principally marine. Nitrate is toxic to babies at very high concentrations and was, until recently, considered a possible factor in the causation of stomach cancer (Fraser, 1984). Nitrate can, nevertheless, be regarded as a resource, its reduction through denitrification at very low oxygen levels sometimes preventing or delaying the onset of completely anoxic conditions. Rates of denitrification up to 700 mg m^{-2} day^{-1} have been recorded from river muds, sufficient to explain losses equivalent to 50% of the nitrate sources in some lowland rivers in Britain during the summer months (Owens et al., 1972).

The accurate prediction of the loss rate of ammonia under a range of environmental conditions (through volatilization, nitrification and assimilation) or of nitrate (through nitrification and assimilation) is not yet possible. In consequence, management is focused on controlling inputs. With respect to point sources, ammonia is generally oxidized to nitrate, and nitrate reduced to nitrogen. Their use in agriculture and forestry is also being re-assessed and controlled in some areas (National Rivers Authority, 1990). The autochthonous production of combined nitrogen through fixation by some procaryotic organisms, particularly the blue-green algae, is widespread, but fluxes seem very low compared with other processes of nitrogen transformation. Whenever the control of these organisms is sought it is to avoid aesthetic and toxicity problems (National Rivers Authority, 1990) rather than those associated with excess nitrate.

ECOLOGICAL RESOURCES AND WATER QUALITY

In addition to the functional role of ecological processes in influencing aspects of water quality, ecological communities have a direct value in their own right. Wildlife appreciation — particularly riparian birds, mammals and some insects (such as dragonflies) — and fisheries for sport and food are dealt with in this volume but not specifically in relation to water quality.

Most early studies of the impact of water quality on fisheries related to the direct toxic effects of concentrations of specific poisons (or low concentrations of oxygen) which had acute lethal effects on a few standard fish species. The relevance of these studies to natural situations was questioned. In nature, pollution fields are complex (with several poisons at variable concentrations); several fish species are exposed throughout their life cycles and are dependent upon food-web tolerance and limitation of pollutant transfer. Bridging the gap between short-term laboratory dose—response tests and understanding the impact of water quality on fish populations, so as to manage and develop fisheries more effectively, is a process that is still continuing. Reviews by Sprague (1970, 1971) demonstrate early progress. The relevance of the short-term laboratory lethal test to describing the distribution of fish in polluted catchments was explored by Alabaster et al. (1972): these studies assumed an additive effect of poisons (Herbert and Shurben, 1964). Although physiological, biochemical and behavioural studies are now increasing our understanding of the mediation of toxic effects (Lloyd and Swift, 1976; Morgan and Kuhn, 1984; Whitehead and Brown, 1989; Norrgren, Wickland Glynn and Malmborg, 1991), genetical variability of species, and the impact of selective processes induced by water quality on the genetics of fish stocks, have not been adequately explored. Despite these weaknesses in our understanding of the biological processes, the recovery of fish populations in rivers, such as the Taff in Wales, of changing pollution status, has coincided with empirically derived water-quality standards described in the EC Directive 78/659/EEC as needed to support fish life (Mawle, Winstone and Brooker, 1985).

Considering a wider taxonomic base of relevance to conservation, Maltby and Calow (1989) have recently reviewed the use of bioassays in freshwater systems and, although invertebrates are more widely subjected to toxicity studies than fish (there are three times as many studies carried out on invertebrates, which is not surprisingly in view of their enormous taxonomic diversity and frequent use in monitoring), there is the same preponderance of short-term lethal tests. Inadequacies in the database have generally precluded conservation evaluations of specific toxic pollutants or predictions of the distribution of particular invertebrate or plant species in polluted river systems. Nevertheless, recognition of some of the weaknesses is now leading to studies on such aspects as the effects of pollution episodes (Ormerod, Wade and Gee, 1987; Turner, 1992) and the impact of water quality on reproductive and other physiological processes (Maltby, Naylor and Calow, 1990; McCahon and Pascoe, 1988a, b).

Behavioural responses, such as downstream drift, may be particularly critical with respect to the tolerance of some species and their absence from certain locations. Turner (1992) has demonstrated that short exposures to oxygen concentrations within the range $2-6$ mg litre^{-1} DO can result in the emigration of several benthic invertebrate species. Such concentrations are frequent at night in weedy rivers.

ECOLOGICAL ASSESSMENT

The impact of water quality on the biological communities in rivers, and the opportunity this affords to describe the severity of pollution through ecological assessment and classification, have been recognized for almost a century (Kolkwitz and Marsson, 1908). Early studies were linked specifically to sewage discharges and it is perhaps surprising that despite the substantially increased range and ubiquity of pollutants introduced during the 20th century — including synthetic biocides (not all conveniently discharged to the river at a constant rate via bank-side pipes) — these early classification systems have not been more widely and radically changed. Not only are they still generally based on community structure, whereas toxicity studies measure functional parameters which are generally determined for a single species (Maltby and Calow, 1989), but several essentially use the same sewage-sensitive communities to describe pollution status. In defence of conservatism in monitoring systems, temporal comparisons of quality can only be made where classification systems remain unaltered or where procedural changes are assessed and calibrated.

The importance placed on ecological classification systems has varied greatly between countries (see Newman, Piavaux and Sweeting, 1992) but in England and Wales, until recently they were frequently used as an adjunct to chemical classifications based on the intermittent sampling of a few sanitary determinands and, where the ecological and chemical descriptions differed, the chemical one took precedence. Proposals for statutory water-quality objectives redress this imbalance (National Rivers Authority, 1991c).

Despite the reluctance to change established monitoring systems, developments in ecological assessment methods have occurred and been widely reviewed (see Edwards, 1987). With respect to classifications based on community structure, two approaches have been adopted:

(i) to ignore the qualitative information within data sets and use only numerical information about numbers of species (or other taxa) and numbers of individual organisms within a sample to derive what are commonly referred to as diversity indices;
(ii) to produce numerical scores or named classes, based on diversity and perceived sensitivities to pollution of species or larger taxa within samples. These sensitivities are generally derived, somewhat arbitrarily, from field data of previous associations of organisms and pollutants (usually sewage discharges) rather than from experimental studies. Such biotic indices rarely use abundance data of individuals within a sample or within specific taxa, although a few indices (e.g. Chandler) combine the quantitative and qualitative approaches.

In recent years the relationship between community structure and water quality has been explored using multivariate statistical procedures and these are forming the basis of new classification systems, such as RIVPACS (Sweeting et al., 1992).

Diversity indices

Diversity indices are generally of the form:

$$fT.fN$$

where f is a function, and T and N are the number of taxa and individuals respectively within a sample. They clearly contain little community information and the numerical value of many such indices is dependent on sample size (Edwards et al., 1975). Furthermore, the density of individuals of even the most abundant and regularly distributed species varies widely, even within comparatively uniform habitats such as riffles, thus causing considerable variation in diversity indices of "replicate" samples and making inter-site comparisons difficult.

There was a fashionable use, some 20 years ago, of diversity indices based on Information Theory, particularly the Shannon-Wiener index, and although these indices utilized more community information than many others — particularly the numerical distribution of individuals between taxa — and had a seemingly theoretical base and legitimacy, they were widely abused and often found to be of limited value.

Biotic indices

These indices have been derived principally from associations either of attached micro-organisms or of benthic invertebrates, and for chronically polluted rivers either association is suitable. However, there is clear advantage with intermittent discharges in selecting communities having long recovery times. As pollution from continuous point-source discharges is reduced (either by regulations or economic instruments — or both), intermittent discharges become more significant and require effective detection systems: in this respect benthic invertebrates, many of which have annual life cycles, would seem to offer advantages over micro-organisms which rapidly recolonize areas after pollution incidents.

The important issue of recovery has been reviewed by Milner (1994). Distinction was drawn between the period required for invertebrates to reach maximum densities and to reach species equilibrium, a relevant consideration in relation to both diversity and biotic indices. Although the season when pollution incidents take place, as well as their character and severity, can affect both recovery rates and successional pathways, nevertheless, it was concluded that in most cases recovery is completed within one year. This was so even in the case of the Sandoz accident, described as "one of the worst chemical spills ever", which involved the release of toxic chemicals to the Rhine following a fire. Invertebrate recovery was complete within a year, and within two years, 40 of the 47 indigenous fish species had recolonized. Studies by Turner (1992) in which pollution episodes (high NH_3 and S_2^-, low O_2) were experimentally induced in small rivers, showed that benthic invertebrate communities recovered within three months, at least after summer episodes — the principal route being downstream drift. With such rates of recovery, conventional annual invertebrate surveys will clearly miss many pollution incidents.

Biotic indices, developed to maximize the "signal" between water quality and community structure, have frequently been derived from samples taken from only one "standard" habitat (or even an artificial habitat), the intention being to reduce "noise" created by sampling all habitats at sites of differing habitat diversity. In the UK, riffles have been sampled wherever possible. This habitat restriction, together with the limited taxonomic penetration required for many indices, limit the usefulness of such indices as descriptions of the conservation value of sites. Ormerod (1985) has shown the considerable differences in the perceived distribution of many invertebrate species within river systems, depending on whether marginal habitats or riffles are sampled.

Multivariate statistical approaches

In recent years multivariate techniques have been applied to faunal and water-quality data sets from several individual river systems. Classification methods have grouped those sites with similar faunas or those species which commonly occur together in space or time, and ordination methods have explored the relationships between faunal distributions and environmental variables. These methods have been critically compared for one polluted river system by Learner, Densem and Iles (1983). Wright et al. (1984) extended his study to 268 sites — all seemingly unpolluted — on a wide range of British rivers. From this study it has been possible to predict with reasonable accuracy (except for rare species in low abundance) the probabilities of macroinvertebrate taxa occurring at unpolluted sites in British rivers using only five physical and chemical variables (distance from source, mean substrate particle size, total oxidized nitrogen, alkalinity and chloride). On this basis the fauna of sites can be compared with that expected in the absence of pollution and, as a measure of pollution stress, the difference can be expressed using distance measurements (e.g. Jaccard's index). This approach shows promise as the basis for a river classification system (RIVPACS); nevertheless, there are still aspects which need resolution. Unlike most sampling protocols, collections are made from all site habitats, so confusing faunal/water-quality relationships by structural differences between sites. Furthermore, certain of the chemical variables, to which the fauna at unpolluted sites in the original data set were related, are frequently attributed to pollution (NO_3^-, Cl^- and the possibility of diffuse or intermittent pollution occurring at some of the unpolluted sites (e.g. farm wastes) cannot be excluded.

There is scope to extend the usefulness of systems, such as RIVPACS, beyond that provided by conventional distance measurements, by a detailed qualitative comparison of the "predicted" and observed faunal composition at polluted sites. Given an adequate eco-toxicological base, the qualitative differences could reveal the class of pollutants primarily responsible. Some progress has already been made with respect to organic pollution by farm wastes and a useful invertebrate key has been devised to assess the degree of damage from this form of pollution (Seager et al., 1992). The development of a strategy to extend such an eco-toxicological base is important as the methodological options are diverse (Edwards, 1987), the costs significant and the opportunity for the application of results, sensitive to the rationale (Maltby and Calow, 1989).

In contrast to structural indices, some functional indices have been used to assess pollution but application has not been sustained or extensive. At community level three aspects of functional organization have been distinguished: energy flow, material cycling and regulation (Matthews et al., 1982).

Material cycling has been discussed earlier in this contribution, particularly with respect to nitrogen. Most interest has been focused on aspects of energy flow, generally expressed through oxygen or carbon fluxes which are, to some extent, interdependent. Odum (1956) assessed photosynthetic (P) and respiratory (R) rates of stream communities from the analysis of diurnal oxygen curves: these methods have since been refined (Edwards and Owens, 1965). The degree of heterotrophy, sometimes substantially increased through organic pollution, was commonly expressed by P:R ratios. Sealed benthic microcosms have also been used to determine oxygen fluxes and furthermore, whilst these avoid the need to correct oxygen fluxes for reaeration through the river surface, they inevitably introduce a degree of artificiality, particularly in the measurement of benthic processes which dominate except in very deep rivers, and furthermore, some large components of stream metabolism, such as macrophytes, must be excluded. The introduction of ^{14}C labelled organics into microcosms to assess heterotrophic microbial activity has been widely used (Hall et al., 1990).

The impact of pollution on the energy balance of individuals and populations of macroinvertebrate species, such as *Gammarus pulex*, has also been studied and proposed as a field tool in pollution surveillance, the basis being a stress-induced decrease in the "scope for growth" (both somatic and reproductive) resulting from enhanced respiration and reduced feeding (Maltby, Naylor and Calow, 1990). Simplified field procedures for caged *Gammarus*, measuring only feeding rates, have proved more practical (McCahon and Pascoe, 1989; Crane and Maltby, 1991) although their use is likely to remain limited to complex catchments requiring sensitive diagnostic tools. Aspects of the copulatory behaviour of caged *Gammarus* have also proved useful in detecting low levels of some pollutants (Poulton and Pascoe, 1990). But sub-lethal tests on restrained animals may have only limited relevance if drifting behaviour is induced at even lower pollutant concentrations (Turner, 1992).

There is a widespread literature on the bioaccumulation of some pollutants, such as metals and biocides, by both plants and animals (Burrows and Whitton, 1983; Kelly and Whitton, 1989), and their transfer and biomagnification through food webs. This subject is covered in detail in Chapter 14. Despite the problems of relating contaminant levels in tissues with levels in both water and sediments to which many organisms are exposed (Bryan, 1976), and despite the dynamics of uptake and loss being important, particularly when pollutant concentrations are highly variable, nevertheless the techniques have been developed into standard monitoring procedures: aquatic bryophytes seem to be favoured for heavy metals (Whitton et al., 1991) but uptake levels by fish cropped for human consumption are of particular concern (Mason, 1991).

Sediments too are used to monitor the levels of accumulative pollutants in river systems and it is evident that in some instances the relation between tissue and sediment concentrations would support this approach (Bryan, 1976). Nevertheless, pollutant concentrations vary with particle size and composition. The behaviour of benthic animals, on which uptake may depend, is influenced by particle characteristics, particularly with respect to ingestion.

DISCUSSION AND CONCLUSIONS

In recent years there have been many promising advances in our understanding of ecological processes and their effects on water quality. We have also increased the profile of aquatic conservation and the use we make of ecological and eco-toxicological tools in protecting water quality and aquatic resources. There are four areas in which attention could profitably be directed in the future.

Widening horizons

Over the past two decades or so there has been an increasing recognition of the importance of riparian and catchment processes in stream quality. This lesson has been reinforced by the widespread effects of atmospheric pollution on lake and river acidification — mediated by land uses. The major scientific challenges are no longer the control of continuous point discharges but land-use and management practices which cause deterioration of river quality, and the improvement of atmospheric quality — particularly with respect to acid areas with poor buffering capacity. In recent years, there has been the emergence of policy instruments, such as guidelines for forestry and agricultural practices, indicative land-use strategies, zones where potentially polluting practices are regulated (protection zones) and atmospheric emission targets. A renewed focus on riparian processes and river corridor management would now seem worthwhile, to draw together the disparate scientific initiatives and studies which have been carried out.

Harnessing functional understanding of instream processes

Microbial ecology is now contributing greatly to our understanding of chemical transformations — particularly of nitrogen and carbon — and there has been a similar advance in our appreciation of the role which invertebrates can play in processing organic matter to accelerate microbial action (see Barnes and Minshall, 1983; Calow and Petts, 1992).

Measurements of component rates of transformation are being made and placed within the theoretical framework of nutrient spiralling (Newbold, 1992) but further progress needs to be made before we can reliably predict the rates of transformation of nitrogen and carbon under the wide range of conditions prevailing in rivers and before we can confidently manipulate processes to management advantage.

Spatial and temporal variability

Rivers are far more variable systems than we have generally acknowledged and that variability is now being recognized both in the spatial and temporal senses. Temporal changes in overall flow have been measured accurately for many decades and their consequences to water quality have been referred to earlier, but it is now becoming evident that the spatial pattern of water movement within rivers can be extremely variable and that this spatial variability has a major impact on ecological processes and challenges some of our traditional views on river plankton (Reynolds, 1992; see Chapter 13). The effects of velocity shear on organisms at the sediment−water interface were identified and described many years ago but flow through porous sediments, its variation and its effects on benthic and hyporheic communities and on water quality, have not yet been given the attention they deserve (Thibodeaux and Boyle, 1987; Turner, 1992).

Pollution episodes, not related to flow changes, have also been neglected, in part because biological and chemical descriptions before and during events were rarely possible. In consequence, stream recovery has been seen essentially as a spatial phenomenon, with named zones within an overall classification system. The development of continuous chemical monitors and automatic samplers has improved our ability to describe short-term events and confirmed the need for describing chemical conditions at night — particularly of oxygen. One might have hoped that the plethora of short-term toxicity tests undertaken over many years would prove useful in assessing the likely consequences of episodes, but unfortunately most of these tests did not extend to observations after the exposure period to establish whether mortalities or recovery from sub-lethal damage occurred.

The seasonal timing of pollution episodes is also critical in relation to both biological damage and the process of recovery. The impact of low oxygen concentrations during the winter is much less than during warmer seasons (Turner, 1992). The degree of coincidence of episodes with sensitive life stages can also be important: differences of sensitivity of life stages up to 1000-fold have been reported (Pascoe and Edwards, 1990). The course and speed of recovery are also dependent on season (Milner, 1994).

Utilizing bioassays

In the past, chemical measurements have been used to consent and monitor effluents and, together with descriptions of community structure, have been the basis of describing water quality. While this emphasis is unlikely to change, it is important that functional studies should complement and support such descriptions in two ways: firstly, by interpreting the significance of chemical conditions in relation to changes in community structure and, secondly, by their direct use in effluent and river monitoring. The process of interpretation has progressed considerably in recent years, particularly through the development of eco-toxicology, but the integration with genetical, behavioural and biochemical aspects of biology need to be further strengthened. The second process has progressed further with effluent than river monitoring, particularly through the use of fish-behaviour monitors and the introduction of toxicity consents for effluents of complex or variable quality. But functional monitors are being developed for rivers, and are being introduced selectively as an adjunct to traditional monitoring procedures.

12

Acidification: Causes, Consequences and Solutions

ALAN G. HILDREW

Queen Mary & Westfield College, University of London, London, UK

and STEVE J. ORMEROD

University of Wales College of Cardiff, Cardiff, UK

INTRODUCTION

The acidification of freshwaters has been one of the most intensively studied environmental problems of the last decade, with much research being relatively formally organized within applied and strategic programmes (USEPA, 1983; UKAWRG, 1989; Royal Society, 1990). Progress has been encouraging as a demonstration of the role of science in identifying a phenomenon, assessing its importance and distribution, revealing its causes, characterizing and modelling its consequences and suggesting various solutions, along with some indication of their likely efficacy. Acidification thus exemplifies an area in which scientists and managers have worked together with mutual benefits. Less of a cause for optimism is the realization that acidification of surface waters is widespread, its environmental costs severe and rectification expensive and uncertain. Further, there remains a great deal of more fundamental work to be done on the basic ecology of acidified waters, the results of which could eventually yield novel management options for the amelioration and management of acidification.

In this chapter we seek, first, to summarize the causes and ecological consequences of acidification of streams and rivers, but will then devote most space to a range of management options, organized around the concept of spatio-temporal scale in ecology. That is, there are things we can do over very large geographical areas and over substantial time periods, that are likely to be effective in the long term but extremely expensive, and there are small-scale, short-term options, that are cheaper but sometimes of more doubtful benefit.

CAUSES OF SURFACE WATER ACIDIFICATION

Not surprisingly for such an important phenomenon, the causes of surface water acidification have been extensively reviewed (e.g. Last and Watling, 1991; Radojevic and

The Ecological Basis for River Management. Edited by D.M. Harper and A.J.D. Ferguson. © 1995 John Wiley & Sons Ltd

Harrison, 1992), and only a brief recapitulation is necessary here. Essentially, anthropogenically mediated acidification occurs where the atmospheric deposition of strong acid anions (sulphate SO_4^{2-} and nitrate NO_3^-), accompanied by H^+ and NH_4^-, exceeds the buffering capacity of the soil; increased flux of sulphate through soil systems leaches calcium and magnesium, depleting these bases where their supply is not sufficiently maintained by weathering. Soil pH and alkalinity decline, and cations such as Al^{3+} begin to appear in runoff. Thus, while natural processes of acidification can be recognized, the two features necessary to explain the current global distribution of acidified waters are inherent sensitivity to acidification, and the presence and history of acidic deposition (Reuss, Cosby and Wright, 1987). Sulphate and nitrate deposition arise from the combustion of fossil fuels, whereas sensitivity in this context is greatest in regions dominated by granitic or siliceous bedrock. Affected waters in Britain have most recently been mapped by the Critical Loads Advisory Group (Battarbee, 1992) and include areas in mid- and north Wales, south-west England, parts of south-east England, the Pennines, Galloway and west-central Scotland. In some of these cases, afforestation with conifers has exacerbated acidification and, particularly, has enhanced the release of aluminium from soils to surface waters (e.g. Reynolds et al., 1988; Ormerod, Donald and Brown, 1989). This effect occurs mostly because of the "scavenging" of airborne material by the forest canopy, increasing the deposition of sulphate by around 50% of ambient, and nitrate by around 100% (Department of the Environment and the Forestry Commission, 1990).

In general, the deposition of sulphate over most of Britain has declined over the last 20 years, and we will return to this point below in the context of deposition control. By contrast, nitrogen emissions have continued to increase, making it inevitable that nitrogen will make a growing contribution to acid deposition in all its forms. Although it has been widely believed that nitrate is less mobile through upland soils than sulphate, being incorporated into plant or microbial production in the catchment, growing evidence shows that it can be abundant in runoff (Stevens et al., 1993). Research effort is currently directed at understanding the circumstances under which such nitrogen leakage occurs, and at understanding the consequences for surface waters. There are potential ramifications for primary producers, perhaps qualitatively if not quantitatively (Winterbourn, Collier and Graesser, 1988; Winterbourn, Hildrew and Orton, 1992). Also, there are some indications that nitrate will contribute to acidity particularly during episodes of winter rainfall or snowmelt; unfortunately our understanding of the true ecological effects of acid episodes is not yet clear enough to assess how such nitrate release will perturb stream biology (Ormerod and Jenkins, 1994).

CONSEQUENCES OF SURFACE WATER ACIDIFICATION

A summary of consequences

The biological responses to acidification can occur at a wide range of "levels of organization", from sub-cellular to global, and these are summarized in Table 12.1. The ecological consequences of acidification for streams and rivers form the focus of this paper: these include impacts on populations (size, density, range); communities (species richness, relative abundance, trophic structure); ecosystem processes (productivity, nutrient transformations); as well as very large-scale landscape and global patterns and processes.

Acidified streams are often fishless (Milner and Varallo, 1990) and other water-associated vertebrates may be absent or scarce along acid stretches. This has been best shown for the

TABLE 12.1. Potential responses to freshwater acidification at different scales

Scale of perturbation	Examples of effects
Sub-physiological	Tissue concentrations of mercury in fish
Physiological	Physiological dysfunctions and morphological reproductive impairment in fish and Amphibia; damage and malformation in fish; altered shell structure in some birds.
Behavioural	Time-activity use in birds; avoidance behaviour in fish and invertebrates
Individual	Changes in body condition, growth, and energetics in birds
Species	Presence/absence
Population	Density, biomass, recruitment, mortality, and survival patterns in fish, some birds and invertebrates
Community	Altered community structure in invertebrates and aquatic plants; reduced diversity; altered predator – prey relationships
Ecosystem	Altered quality of production, reduced decomposition
Landscape/biome	Cumulative responses across ecosystems (i.e. lakes and streams)
Global	Cumulative responses across biomes and continents

dipper (*Cinclus cinclus*; Ormerod et al., 1991), but there is also evidence for effects on the otter (*Lutra lutra*; Mason and Macdonald, 1987). There are very clear effects on populations and communities of macroinvertebrates (Sutcliffe and Hildrew, 1989; Ormerod and Wade, 1990), acid streams having fewer species, largely as a result of the reduction in the particularly "sensitive taxa" Ephemeroptera, Crustacea and Mollusca, but also in the diverse Chironomidae and Trichoptera. A few tolerant species, such as some Plecoptera, several chironomids and one or two Trichoptera, persist and may be rather abundant under acid conditions. Smaller invertebrates, including benthic copepods and Cladocera, are also much less species-rich in acid streams in southern England (Rundle and Hildrew, 1990) and in upland Wales (Rundle and Ormerod, 1991) although again, those species which do persist in acid streams may actually be very abundant (Rundle and Hildrew, 1992). Little is known of impacts on other taxa of smaller invertebrate animals, such as nematodes, ostracods, rotifers and protozoa.

Macrofloral assembleges in upland Welsh streams differ markedly with pH (Ormerod, Wade and Gee, 1987), the liverworts *Scapania undulata* and *Nardia compressa* being particularly characteristic of acid water, whereas the alga *Lemanea* and the moss *Fontinalis squamosa* were common in more circumneutral streams. Turning to the very important epilithic algal layers, there are very clear differences in species composition between acid and circumneutral conditions and, indeed, such differences are the basis of biomonitoring and pH reconstructions using diatoms (Battarbee, 1984; Juggins et al., 1989). *Eunotia* and *Tabellaria* are diatom genera particularly characteristic of acid water, whereas they are often joined by *Acnanthes*, *Gomphonema* and *Cymbella* in circumneutral streams (Stokes, 1981; Winterbourn, Hildrew and Box, 1985; Ormerod and Wade, 1990). It has also been reported that filamentous algae are particularly prevalent in acid streams (Stokes, 1981). Effects of acidity on biomass and productivity of epilithic algae, however, are far less clear. There have been reports of increased (e.g. Mullholland et al., 1986) and decreased (Collier and

Winterbourn, 1990) biomass and productivity in acid compared with circumneutral water, and Winterbourn, Hildrew and Orton (1992) found no consistent difference in the biomass of algae colonizing experimental substrata in a number of British streams of contrasting pH.

Decomposition of coarse particulate detritus has usually been found to be reduced in acidified streams (Hildrew et al., 1984; Mackay and Kersey, 1985; Chamier, 1987) and microbial activity is usually reduced under acid conditions, other circumstances being equal. Bacterial and fungal counts, however, are not always lower in acid waters, there being examples of decrease (Chamier, 1987; Groom and Hildrew, 1989), increase (Ormerod and Wade, 1990) and little difference (Simon and Jones, 1992) with declining pH in various situations. Simon and Jones (1992) did find that a gliding, filamentous, yellow-pigmented bacterium was absent from streams in the English Lake District with a pH < 5.5. This bacterium was incapable of growth in acid conditions in the laboratory. Clearly, inhibition of such bacteria, which characteristically degrade cellulose and other plant polymers, could contribute to the affect of acidification on decomposition.

Direct and indirect impacts

Such population and community level consequences may be the result of what we may term either direct or indirect impacts of acidification (Figure 12.1). Acidification is a complex chemical phenomenon involving a reduction in alkalinity and an increase in aluminium concentration (and often other metals) as well as increased hydrogen ion concentration (UKAWRG, 1989; Sutcliffe and Hildrew, 1989). Any of these chemical changes can have direct effects on aquatic biota by a range of mechanisms, from the sub-cellular to the whole organism level. These effects can be lethal or sub-lethal and could clearly account for most of the ecological changes observed in freshwater systems. Direct effects of acidification have been well described for a variety of organisms and reviews can be found for aquatic macrophytes (Farmer, 1990), macroinvertebrates and fish (Morris et al., 1989). Far less is known for algae, smaller invertebrates and microbes.

Indirect impacts of acidification can only arise as a consequence of a direct impact. When a species, ordinarily present under circumneutral conditions, is absent or reduced in abundance, or its behaviour or physiological condition is altered, other species with which it normally interacts may then be affected "indirectly" through the absence or modification of these interactions. These indirect impacts of acidification are of great ecological interest but are in many cases contentious and difficult to disentangle from other processes.

We can distinguish three main types of indirect effects. First, where an important predator is susceptible to acidification, any of its prey, physiologically resistant to acidification and normally regulated or controlled by the predator, may increase in abundance. This constitutes a "top-down" effect and occurs, for instance, where fish are absent and their vulnerable prey increase in abundance. Such vulnerable taxa may be large, predatory invertebrates, including the net-spinning caddis *Plectrocnemia* (Schofield, Townsend and Hildrew, 1988; Hildrew, 1992). Where increases of algal biomass have accompanied experimental acidification (Hendrey, 1976; Hall et al., 1980; Allard and Moreau, 1985), this has sometimes been attributed to reduced grazing pressure, a further example of a possible "top-down" effect.

A second type of indirect effect can occur where the acid-sensitive taxa are important food or prey in food webs for species not directly affected by the acidity. This is a "bottom-up" effect and a good example is the loss of dippers from acid streams *via* a reduction in their

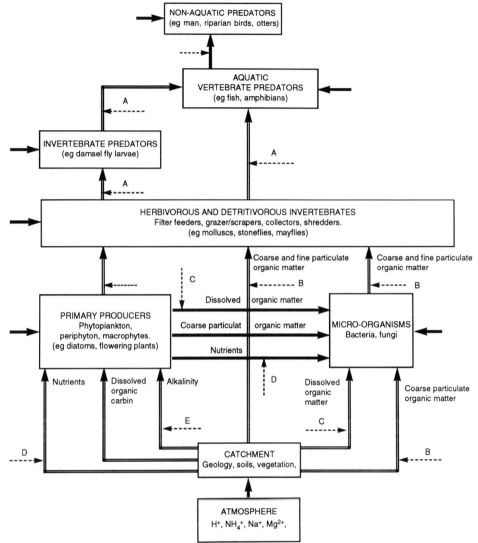

NOTES
(i)   Bold black arrows indicate trophic groupings where acidification processes will have direct effects (see text). Such effects may be lethal or sublethal.
(ii) Broken lines indicate possible indirect effects on groups and/or processes.
(iii) Indirect effects are labelled as follows:
    A Loss of fish resulting in increasing populations of larger invertebrate predators and decreases in abundance of other benthic organisms. Also there may be a reduction in food supply for vertebrate predators.
    B Reduced decomposition of coarse particulate organic matter affecting quality of food available for detritivorous invertebrates.
    C Complexation of dissolved organic water by metals may lead to a decrease in uptake of dissolved organics and hence a reduction in microbial productivity.
    D Aluminium may compete for phosphate thus reducing the availability of nutrients for primary production
    E Loss of alkalinity may lead to reduced availability of inorganic carbon, limiting photosynthesis.

FIGURE 12.1.   A diagrammatic representation of the potential effects, both direct and indirect, on freshwater ecosystems (after UKAWRG, 1989)

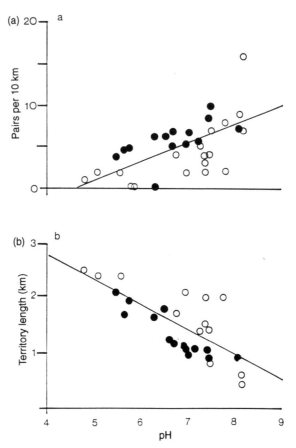

FIGURE 12.2.   The (a) breeding density and (b) territory length of dippers in relation to mean stream pH in Wales (open symbols) and Scotland (closed symbols) (after Vickery and Ormerod, 1991)

food, which is usually dominated by acid-sensitive stream invertebrates and fish (Figure 12.2). A reduction in microbial activity on decomposing particulate detritus also results in a reduction in its quality as food for detritivorous invertebrates, and perhaps contributes to reduced secondary production in acidified streams (Figure 12.3).

A third possible effect is that competitive relationships among species could be altered by differential susceptibiltity to acidification ("horizontal" effects). This is particularly contentious because of the controversy about the importance of interspecific competition in streams generally (Hildrew and Giller, 1994). There have been few attempts to test this hypothesis, though it might be noted that ducklings on acidified fishless lakes in Canada survived and grew better than those reared on lakes with fish, possibly because of reduced competition for large invertebrate prey (Bendell and McNicol, 1987).

Clearly, there are further categories of species interactions, including mutualistic, symbiotic, parasitic and phoretic associations, which could be implicated in the ecological changes wrought by acidification, but we know very little of them in running waters and it is unlikely that they will be of widespread importance (Hildrew, 1992). Whatever the mechanism by which the impacts of acidification occur, however, it is clear that the

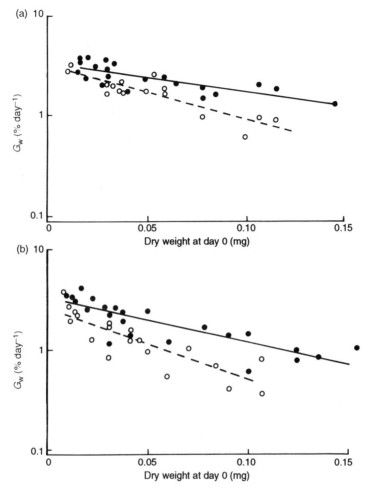

FIGURE 12.3. The growth (G_w) of a common, detritivorous stonefly (*Nemurella pictetii*) was a function of their initial size at the beginning of a growth experiment during which animals were fed on (a) decaying alder leaves or (b) beech leaves. Growth was slower, however, when animals were fed on leaves which had been conditioned in acid (open symbols, broken lines) rather than in a circumneutral stream (closed symbols, solid lines) (after Groom and Hildrew, 1989)

ecological changes are severe, both in terms of species composition, and thus of conservation, and in terms of ecosystem processes such as productivity.

We now turn to the ways in which ecological science can guide measures to ameliorate or prevent acidification, a real example of the "ecological basis for river management".

SOLUTIONS TO SURFACE WATER ACIDIFICATION

A question of scale

The ultimate causes of acidification are large-scale processes, involving the transport of pollutants over long distances and over regional and national boundaries. The progress of stream acidification is then modified by progressively smaller-scale processes, involving local geology, land use, transport of hydrogen ions and metals through soils and vegetation and across the riparian zone, in-channel phenomena such as downstream transport and exchange of materials with mineral and organic sediments and, finally, interactions at the scale of individual aquatic organisms and the biofilm. Thus, acidification can be seen in the same conceptual framework as other environmental factors in rivers, which are hierarchically arranged physical systems in which large external forcing factors (such as tectonics and climate) ultimately set conditions at progressively smaller scales from the whole river system to single substratum units (see, for example, Frissell et al., 1986; Hildrew and Giller, 1994).

The challenge to ecologists to understand how phenomena at different scales interact is of great contemporary interest (Levin, 1992). We have organized our discussion of management, therefore, into a series of measures requiring action at progressively smaller spatio-temporal scales which are summarized in Table 12.2. The approach envisages that strategies which might have a long lasting impact at large spatial scales would take longer to implement or to become effective than those applied more locally.

The supra-catchment scale

Among the potential solutions to the acidification problem (Table 12.2), only one can be effective at a supra-catchment scale, and involves reducing emissions at source. As the only management of the root cause of damage to a wide range of natural ecosystems, this option must remain at the heart of restorative measures. Difficulties arise, however, in understanding where, and by how much, emissions must be reduced to achieve the required goals and how long such improvements would take. Since we cannot determine the required emission reductions by trial and error, only a forecasting or modelling-oriented approach is possible.

Here, the "critical loads and levels" concept, although replete with practical problems, is proving useful as a policy tool (Battarbee, 1992). Stated simply, the concept is based on assigning the maximum tolerable load of pollutants that will not cause lasting damage to a given sensitive ecosystem (Figure 12.4). Some understanding of the sources and inter-regional movements of air pollutants will then provide guidance as to which emission sources should be restricted to achieve protection of particular systems or species. In

TABLE 12.2.   Scales of approach to the management of acidification

Spatial scale	Option	Temporal scale (y)	
		Implementation	Effectiveness
Supra-catchment	Reduce deposition	$10^1 - 10^2$	$10^1 - 10^n y$
Catchment	Manage land use	$10^0 - 10^1$	$10^0 - 10^1$
Sub-catchment	Liming	$10^0 - 10^1$	$10^0 - 10^1$
Riparian	Buffer strips	$10^0 - 10^1$	?
Habitat	Channel morphology	$10^0 - 10^1$	?
All scales	Do nothing	Environmental cost?	

Critical and Target Loads

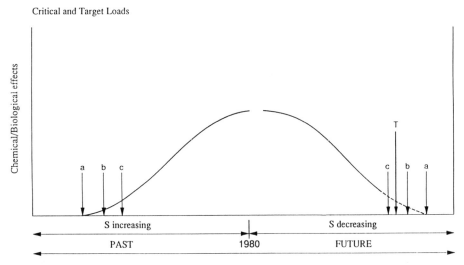

FIGURE 12.4. The critical loads concept. The critical load for a site is exceeded at point a; points b and c represent critical loads for species or systems with a different sensitivity. As sulphur deposition decreases in the future, a target load (T) might be chosen to protect a given species, or to enable full recovery (point a) (after Battarbee, 1992)

freshwaters, critical loads of sulphur and acidity to catchments are being set to avoid the risk of acid neutralizing capacity (ANC) in runoff falling below critical limits; an ANC of 0, for example, is a threshold at which streams become unlikely to support brown trout, *Salmo trutta* (Ormerod, in press). Parallel methods also relate sulphate concentrations in deposition to calcium concentrations in runoff and enable some prediction of change to sensitive ecosystem components such as diatoms (Battarbee, 1992).

Calculation of critical loads for freshwaters, albeit from a restricted survey, is now complete for each 10 km grid square in Britain and Northern Ireland (Battarbee, 1992). Values range from as low as 0.2 keq H^+ ha^{-1} ann^{-1} (i.e. very sensitive) to >2.0 keq H^+ ha^{-1} ann^{-1}. At present, sulphur deposition exceeds these loads by two to five times in the most sensitive areas. Thus, even after the UK's plan to reduce sulphur emissions by the year 2005 by 60% of 1980 values, 114 individual 10 km grid squares will still receive sulphur in excess of their critical loads (Reynolds and Ormerod, 1993). The more sensitive predictions provided for diatoms show that even larger areas will still be exceeded. All indications are, therefore, that emission reductions more substantial and expensive than the current 60% will be required to return Britain's acidified surface waters to something resembling their pre-acidification status. This is in agreement with predictions from earlier site-specific modelling exercises (e.g. Ormerod et al., 1988). Moreover, acidity due to nitrogen has yet to be incorporated into predictions, while rates of biological recovery are known only sketchily.

The logic of ameliorating acidification by reducing emissions is that we thereby remove the largest-scale external forcing factor to the system (the production and dissemination of strong acids in the atmosphere). However, this requires large-scale coordinated action to be sustained over a long period and at enormous economic cost, and even then we may have to wait for decades for an improvement. Subsequent sections ask whether we can intervene at smaller scale and cost with any chance of success.

The catchment and sub-catchment scales

Given that the emission option requires large-scale coordinated action, sustained over a long time-scale and with a long hysteresis between action and response (Skeffington, 1991), can we manage individual catchments, or parts of catchments, with any chance of preventing or reversing acidification in particular streams and rivers?

One prophylactic action at the catchment scale is to ensure that acidification is not exacerbated locally. The most prominent example comes from commercial forestry, where substantial local increases in acidifying depositions can result from conifer cover. The problem is particularly apparent in upland locations with large inputs from polluted fog and mist. Current National Rivers Authority guidelines prescribe acceptable levels of planting according to catchment sensitivity (Ormerod, Donald and Brown, 1989). Such prescriptions, of course, place the goal of protecting surface waters above the objectives of timber production.

At the sub-catchment scale, the most widely advocated symptomatic treatment for acidification has been the use of limestone (Howells and Dalziel, 1991). For example, spread onto hydrological source areas, lime applications have proved of value in increasing calcium concentrations and pH in runoff, thus reducing aluminium (Figure 12.5). Although some of the available data are equivocal, such chemical change is likely to benefit fish populations. At the same time, however, this technique has questionable aspects. First, it does not restore pre-acidification chemistry, creating instead new conditions with pH and calcium concentrations much higher than those prior to acidification (Ormerod et al., 1990). Some aquatic organisms are negatively affected, particularly some mosses and liverworts (Weatherley and Ormerod, 1992). Also, responses among invertebrate communities have been slow and partial, with limed streams gaining only a fraction of the communities found in streams which are naturally circumneutral (Rundle, S. D., Weatherley, N. S. and Ormerod, S. J. unpublished; Figure 12.6). The effective duration of catchment treatments is unknown, while repeated lime application to wetlands has ramifications for conservation which are almost certainly negative. Work is currently in progress to assess the wetland resources, such as mosses, insect communities and birds, that are at risk from liming, but we are still some way from knowing all the losses that are likely. Such unknowns prompt conservation organizations to advocate liming only in those isolated instances where important genetic resources are at risk from acidification, such as in the case of isolated populations of Arctic Char, *Salvelinus alpinus*.

The riparian scale

Vegetation and land use can have an important modifying effect on the course of stream acidification and, in particular, commercial coniferous afforestation in susceptible catchments seems to exacerbate the problem in British uplands receiving acid depositions. There are other deleterious effects of forestry on streams (Ormerod, Mawle and Edwards, 1987) and there have been recent attempts to devise and implement methods of riparian management in forests to ameliorate these impacts (Forestry Commission, 1988). These involve various kinds of buffer strips, either of moorland vegetation or of native broadleaf trees, either left at the time of planting or created subsequently by the removal of conifers.

Ormerod et al. (1993) recently assessed the effects of these measures by sampling macroinvertebrates in 66 first- to third-order streams in upland Wales and Scotland. The streams were categorised into six different types of management options (Figure 12.7). The

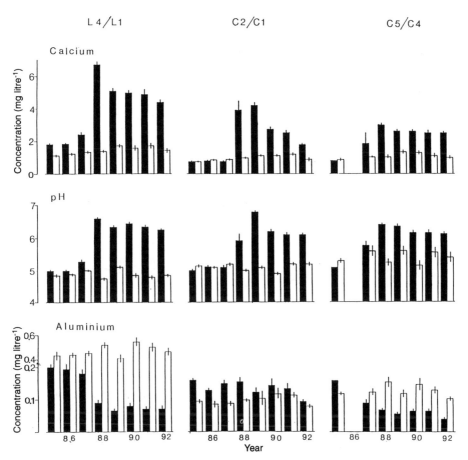

FIGURE 12.5. Mean annual values for calcium, pH and aluminium (with SE) in three Welsh streams whose catchments were limed in 1988 (L4, C2, C5: black bars), and in three adjacent references (L1, C1, C4: open bars)

effects on water chemistry were disappointing and aluminium concentrations were, on average, high in all treatments with conifers, whether there were buffer strips or not.

Chemistry was the most important determinant of the biota, and thus effects of riparian management on invertebrate communities were modest. There were benefits, however, and catchments of pure conifer, without any buffer strip, did have the lowest mean species richness. Our conclusion on these kinds of riparian management options, therefore, is that they do bring a modest environmental benefit but probably do little or nothing to ameliorate acidification on their own.

The stream channel scale

This last option is by far the most speculative, at least in terms of any substantial benefit to acidification. It is well known that many reactions involved in organic decomposition, particularly in anaerobic lake sediments, generate alkalinity (Schindler et al., 1986; Davison

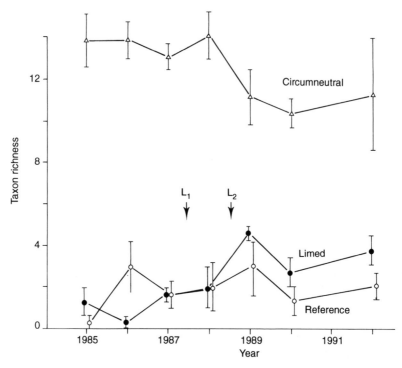

FIGURE 12.6. The mean (SE) species richness of 18 acid-sensitive invertebrates in limed, acidic reference and naturally circumneutral streams in upland Wales. L_1 and L_2 represent points of liming

and Woof, 1990) (Table 12.3). These processes have been put to good effect in the recovery and and management of acid lakes and lagoons (Davison, 1990; Davison et al., 1989). Could such reactions play a part in de-acidification in streams? Very recent research in some acid streams in the Ashdown Forest of southern England (Thomas, J. D. pers. comm. 1993) shows the de-acidifying effect of depositing leaf litter in a pool at low summer flows, where anoxic conditions are found just below the sediment surface. Nitrate and sulphate reduction seem to be the most important processes in involved. Crucial to the presence of anaerobic micro-environments, either in the hyporheic or in litter accumulations, in higher gradient streams would be a substantial external supply of leaf litter, such as would be obtained in a stream lined by broad-leaf trees, and the effective, long-term physical retention of that litter in large debris dams. The most important retention devices in natural stream channels are probably large pieces of dead wood, which decompose very slowly (Bilby and Likens, 1980). High gradient, native wooded streams in North America retain high litter standing stocks (e.g. Webster et al., 1992) and this was presumably also the original condition of upland streams in Great Britain. Management experiments in increasing litter standing crop and retention in upland, acid streams would seem worthwhile.

In this context, Dobson and Hildrew (1992) recently manipulated litter retention in some southern English streams, with the object of demonstrating food limitation for the feeding guild of shredding invertebrates (which chew up coarse particles of leaf litter). Not only did they increase litter standing crop in all but the most naturally retentive stream channel but they also reported substantial local increases in the populations of detritivorous

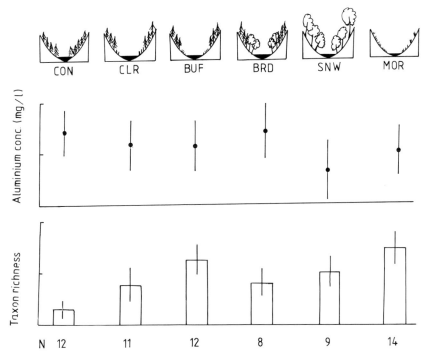

FIGURE 12.7. Ormerod et al. (1993) assessed aluminium concentration and macroinvertebrate taxon richness in upland streams in Wales and Scotland. Here, the streams are categorized in six land-use types: CON, conifers planted throughout the catchment including the riparian strip; CLR, conifer forest with a cleared riparian buffer strip 5–20 m wide; BUF, conifer forest with a buffer strip free of planting; BRD, conifer forest with a buffer strip of broadleaves left in place or allowed to regenerate; SNW, catchment entirely of semi-natural broadleaf woodland; MOR, moorland catchment

TABLE 12.3. Deacidifying reactions generating bases in freshwaters

Reaction	ΔAlk	
(1)	$2CH_2O + NO_3^- + 2H^+ \rightarrow 2CO_2 + NH_4^+ + H_2O$	2
(2)	$5CH_2O + 4NO_3^- + 4H^+ \rightarrow 5CO_2\ 2N_2 + 7H_2O$	1
(3)	$CH_2O + 2MnO_2 + 4H^+ \rightarrow CO_2 + 2Mn^{2+} + 3H_2O$	2
(4)	$CH_2 + 4FeO(OH) + 8H^+ \rightarrow CO_2 + 4Fe^{2+} + 7H_2O$	2
(5)	$2CH_2O + SO_4^{2-} + 2H^+ \rightarrow 2CO_2 + H_2S + 2H_2O$	2
(6)	$(CH_2O)_{106}(NH_3)_{16} + 106O_2 + 16H^+ \rightarrow 106CO_2 + 16NH_4^+ + 106H_2O$	1
(7)	$Ca^{2+}(sed.) + 2H^+(water) \rightarrow Ca^{2+}(water) + 2H^+(sed.)$	2
(8)	$CaCO_3 + 2H^+ \rightarrow CO_2 + Ca^{2+} + H_2O$	2

Δ Alk is the number of moles of base generated by each mole of inorganic component involved. Nitrogen in organic matter is represented by $(CH_2O)_{106}(NH_3)_{16}$ according to the Redfield ratio (reaction 6) (after Davison and Woof, 1990)

invertebrates. This is encouraging because it indicates environmental benefits, in terms of increased animal populations, diversity and production, whether or not there is any enhancement in the rate of natural de-acidification. In further field experiments in upland Welsh streams, Dobson and Hildrew (unpublished) were again able to enhance litter standing stock and shredder populations.

A final strand of evidence comes from recent observations by Lancaster and Hildrew (1993), who showed that there were prominent "dead zones" (see also Chapter 13, this volume) in a small, wooded stream in southern England with a complex, natural channel and a high standing stock of woody debris. Dead zones are areas of essentially non-flowing water in which solutes, particles and living organisms are retained, and thus provide conditions in which local anoxia could develop. We would speculate that rehabilitation of the fluvial morphology of small upland streams to something approaching their pristine condition, with copious fallen wood and litter, complex heterogenous flow and a high retention of water and litter, could thus have benefits in terms of both conservation and water quality. Whether enhancing natural de-acidification, by channel restoration, could make a quantitatively important contribution to the improvement of water quality in acidified streams, however, is presently open to doubt.

CONCLUSIONS — ACTION OR INACTION IN THE MANAGEMENT OF STREAM ACIDIFACTION?

Clearly, acidification has represented a substantial and widespread problem for aquatic ecosystems, with impacts on biological diversity, on economically important resources, and on valuable conservation assets. The restoration or protection of at least some of the affected (or potentially affected) waters is an important goal, and we have presented here a range of options through which such a goal might be achieved. Among them, the causative approach is important, but the magnitude, expense and time-scales involved in recovery should not be underestimated. Water undertakings, and others involved in the management of river catchments, may thus wish to appraise some of the alternatives at hand. Among them are options with a mix of environmental costs and benefits not only to aquatic systems, but also to other resources in affected catchments: for instance, the impacts of liming on conservation, or the loss to commercial forestry of not planting whole catchments. In other cases, such as manipulations of channel morphology, we currently have little information on which to base any judgment of success or gain. We believe, nevertheless, that our template for evaluating the options available, based on the concept of ecological scale, can provide a valuable tool in the formulation of future policy. Adding all the appropriate information through which each option might be compared, implemented or blended, is an important next step.

13

River Plankton:
The Paradigm Regained

COLIN S. REYNOLDS

NERC Institute of Freshwater Ecology, Ambleside, Cumbria, UK

INTRODUCTION

This chapter summarizes the outcome of recent investigations into the mechanics of maintenance and production of river phytoplankton. In an earlier presentation (Reynolds, 1988) introducing the programme of studies, I described the survival of plankton against constant removal by unidirectional washout as being paradoxical, arguing that without assuming continuous re-seeding in a non-stochastic manner, it was impossible to account for the reproducible cycles of phytoplankton abundance and species composition observable at given fixed points along our larger rivers. Thitherto, the criterion of suitability of a river as a medium to support the growth of planktonic algae, assuming no chemical factor constrained it, was the period of hydraulic residence: the river needed to be sufficiently long or to flow sufficiently slowly to permit the seven or eight divisions required to transform an inoculum of a few cells per millilitre to the order of 10^6 litre^{-1}. Even then, it was probably necessary also to embrace the view of Wawrik (1962) that reservoirs of standing water along the channel, in bays, pools, underbanks and blind side-arms, contributed to the between-bloom survival of fluvial populations.

This the earlier paper (Reynolds, 1988) referred to as "the paradigm of river plankton". It was, however, a paradigm that had not been explained mechanistically or even quantitatively supported, neither was it without paradoxical inconsistencies. The general models relating to dispersion, to dilution or to transport of entrained particles did not accommodate the observable phenomena (Day and Wood, 1976; Chatwin and Allen, 1985), whereas many proximal growth factors varied with flow. I concluded that "any relationship of the maintenance, growth and attribution of populations to fluvial discharge goes far beyond a single equation governing the rate of downstream transport" (Reynolds, 1988).

What, then, have the five years of investigations into the dynamics of phytoplankton in selected UK rivers revealed that is sufficient to establish the mechanisms by which populations are maintained? To what extent have we answered the challenge to the paradigm of the river plankton? The relevant findings have been reported fully to the sponsors of the research (UK Department of the Environment, UK National Rivers Authority: Reynolds,

The Ecological Basis for River Management. Edited by D.M. Harper and A.J.D. Ferguson. © 1995 John Wiley & Sons Ltd

Carling and Glaister, 1989; Reynolds and Glaister, 1992) and, in part, in several published papers (Reynolds and Glaister, 1989, 1993; Reynolds et al., 1990; Reynolds, Carling and Beven, 1991). This chapter assembles information drawn from these sources to address the specific issues raised above.

First, I seek to show which rivers, or parts of rivers, are likely to support a substantial planktonic community. Then, some evident trends in the spatial and temporal distribution of different kinds of algae are presented; next, the question is posed as to what hydraulic properties are necessary to account for the algal dynamics. Based on the case studies available, the adequacy of the mechanisms to account for those properties is then considered. Finally, the extent to which the population ecology of potamoplankton in specific rivers or riverine reaches is predictable or manageable is discussed.

OVERCOMING RELUCTANCE

"Standing with reluctant feet, where the brook and river meet"
(H.W. Longfellow, Maidenhood!)

During 1990 and 1991 a series of synoptic surveys was undertaken during which the phytoplankton was sampled at up to 71 stations on 18 UK rivers. The sites had been selected to achieve a good geographical spread and to cover a broad range of river types in England and Wales. These sites included upland, mainly steeply-descending headwater regions (e.g. of the Ure, Exe and of the Severn and Wye catchments), relatively steep downstream sections (of the Wear and Tees), as well as low-gradient rivers, both small (the Tame, upper Great Ouse, the Cherwell, Stour and Brue) and relatively large (lower parts of the Trent, Great Ouse, Thames, Severn and Wye). Their profiles are shown in Figure 13.1(a). The selection included stations linked by reaches of differing sinuosities ("sinuosity" being defined as the length of river divided by the length of valley between the nominated upper and lower limits of the reach concerned). It is also believed to have coincidentally embraced river reaches of differing alkalinity, of nutrient richness, certainly of suspended sediment content and even of anthropogenic pollution. Three surveys were carried out, in June, September and April, so far as possible to coincide with the early- and late-summer periods of algal abundance, as well as the anticipated spring bloom. Care was taken, however, to avoid any further generalization about the hydrological condition of the rivers as a whole, although it is fair to claim that, even in April, the stations were visited when far short of their bankful capacities.

The representations of algal mass, shown in Figures 13.1(b) − (d), are based on the measurements used to construct Figure 13.2: they correspond to the sum of the biovolumes of the planktonic diatoms, cyanobacteria, chlorophytes and cryptophytes but which exclude all non-planktonic forms. Five biovolume categories are inserted on the profile skeletons described in Figure 13.1(a), in order to give an immediate visual impression of the distribution of phytoplankton mass.

Two impressions may be readily gained. The more subtle of the two is the depressed mass evident among the September samples. It might have been anticipated that the generally low flows of the summer period and the conversely increased concentrations of plant nutrients would have supported the largest biomass of the year. In fact this does not always follow, especially if the community structure had developed to the extent that a filter-feeding zooplankton and/or benthos was able to exert some control over the phytoplankton crop. I have no evidence for or against this supposition but the observation serves as a timely

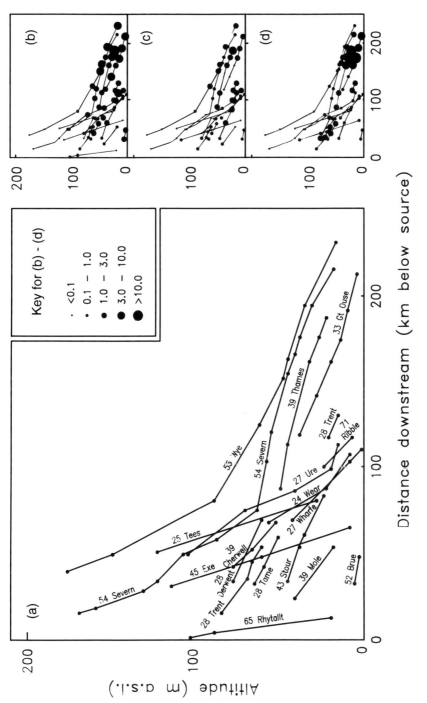

Distance downstream (km below source)

FIGURE 13.1. (a) The rivers and stations sampled by Reynolds and Glaister (1992) sketched as long profiles. The cumulative biovolumes of live suspended algae are marked on to skeleton profiles, representing surveys of (b) June 1990, (c) September 1990, and (d) April 1991. Note the tendency for large crops to be most prevalent in low-gradient reaches, especially of longer rivers

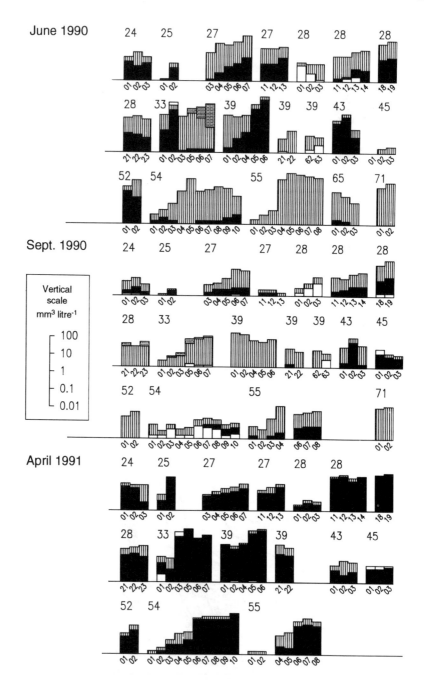

FIGURE 13.2. Summary of algal collections made by Reynolds and Glaister (1992), expresswed in biovolume and plotted on a logarithmic scale, in respect of the three survey periods represented in Figure 13.1. Differential shading, to represent non-planktonic mass (white), planktonic diatoms (black), cyanobacteria (stipple) and all other plankters (mostly chlorophytes and cryptomonads; vertical hatch), is, however, applied in arithmetic proportions. The identity of sites is appended

(Appendix 1)

reminder that it is dangerous to assume all populations of algae necessarily perform to the capacity of the supposed limiting factors.

The second, and more obvious impression survives the scrutiny of the timing of the surveys. It is, simply, that the greater masses (larger dots) occur predominantly towards the lower right-hand corners of the figure, although those of intermediate size are sometimes strung along the more gentle slopes towards the left-hand side (note especially those for the Thames, Tame and, in June at least, the Brue). Steeper gradients, without exception, carried low scores; high scores are always on gentle slopes (but gentle slopes do not always carry high scores!). The inference is very clear: a substantial biomass of phytoplankton is most likely to be carried in the low-gradient reaches of a river, the more so at increasing distances downstream. It has to be said at once that other aspects of the environment of long, lowland rivers (e.g. increasing depth and suspended sediment load) are likely to ensure that the tendency towards larger downstream biomass is not a continuous function. Nevertheless, the present data set does nothing to invalidate the suggested paradigm of river plankton or the view that it will be most evident in the rivers sufficiently long or slow-flowing to permit its development.

THE BOURNE OF TIME AND PLACE

"From out our bourne of Time and Place The flood may bear me far"
(Alfred Lord Tennyson, Crossing the Bar)

Exploring the same data set a little further, it can be readily shown that the downstream increase is not simply a matter of the same inoculum expanding uniformly: even the populations of each of the various species making up one or other of the preliminary subdivisions of the planktonic mass shown in Figure 13.2 change independently downstream, either more or less. Nevertheless, there is an immediate impression to be gained with regard to the relative abundance of diatoms, mainly of the (centric) genera *Cyclotella* and, especially, *Stephanodiscus*, in the vernal samples, and of chlorophytes (most particularly of the Chlorococcalean genera *Scenedesmus* or one of several *Chlorella*-like species) in June and September. Of still greater interest is the fact that in the April samples, to a limited extent, and particularly in the June collections, diatoms were often proportionately and absolutely a minor component of upstream stations in the Ure, Thames and Severn. Apart from suggesting a downstream replacement of green algae by diatoms, a property of river plankton noted by several other workers (Descy, 1987; Sabater and Muñoz, 1990; Stoyneva, 1991), the transition in their relative abundances seems to migrate up or down river with the season (Descy, 1987; Sabater, 1990; Reynolds and Glaister, 1993), presumably under the influence of discharge.

There is no experimental evidence to explain this "moving boundary" effect, though it is not difficult to suggest a plausible mechanism. Given that the phytoplankton species in rivers are selected for their investment in rapid growth rate, Reynolds (1988) has already identified the importance of high surface-area-to-volume ratios characterizing both the common diatom- and chlorophyte-genera of the phytoplankton. Elsewhere, I have argued (Reynolds, 1994) that the diatoms, by virtue of their accessory pigmentation and, sometimes, their superior ability to intercept light, are better adapted than chlorophytes to make the best use of short photoperiods in a well-mixed, turbid environment. This is so, whether or nor the turbidity is due to suspended silts or to algae. This feature of rivers suits some genera to the fluvial life or, at least, gives them a competitive advantage over other species. The green algae will not necessarily out-perform diatoms in the high-light environments of headwater streams, or in channels free of fine sediment, save through the loss-of-cells mechanism. Even these need not be critical, save in one important respect: the greater density of diatoms when compared to that of the green algae. Reynolds et al. (1990) have shown how non-motile, non-aggregating particles (including diatoms) are especially vulnerable to increased sinking losses in shallow water. If the selective advantage moves from small chlorophytes in clear, shallow waters to diatoms in deeper, perhaps muddier, middle reaches, then it is possible to recognize that either the channel deepening or greater turbidity associated with winter flow, or both, will bring the selective advantage nearer to the diatoms. Conversely, dwindling discharge, falling levels and reducing suspended sediment load (turbidity) will bring it back towards Chlorophytes. It is then not difficult to see that the point of equality of opportunity is itself discharge-sensitive and, hence, it will move up- or down-channel accordingly with seasonal fluctuations in flow.

Pending an alternative explanation for what is probably a general structural property of the planktonic community in rivers, it would be reasonable to suggest that the new knowledge and interpretation do not invalidate the general paradigm.

RISING UP IN WEARY RIVERS

"Even the weariest river winds somewhere"

(A.C. Swinburne, The Garden of Persephone)

Perhaps the most fascinating paradox that we have probed concerns the establishment of a fluvial population of planktonic algae and the apparent rate of their downstream development (Reynolds, Carling and Glaister, 1989; Reynolds and Glaister, 1992, 1993). What was first observed in the River Severn was that, for certain sections of river at least, the population of certain species found at the downstream station would not be simply greater than at the upstream one but be so by a factor that was difficult to explain in terms of direct growth of the population during the intervening period of fluvial passage. For instance, the populations of *Stephanodiscus* at Montford Bridge, were 1.3 to 1.8 times greater than the corresponding ones at Melverly, some 16.5 km upstream. Taking the velocity of the main channel to be $0.3-0.4$ m s^{-1}, the increase would need to be sustained by the equivalent of up to 1.75 cell divisions per day. On these occasions, the water was generally warm ($15-18°C$), but always turbid. There is no tributary of any significance on this reach and there is no additional storage capacity beyond that of the channel itself. Moreover, this was not an isolated case: similar observations were made at other reaches at other times and often involving more than one species (Figure 13.3). It has since been found to apply to many other of the rivers considered in the synoptic survey.

If we turn the relationship upside down, we could express the observed mean downstream increase as a function of the time of travel, if we knew the net in situ time-specific rate of population increase of the alga. Of course, the latter is much less easy to measure or to reconstruct but a series of relationships had emerged from many years of study of the ecophysiological dynamics of common species of phytoplankton which enabled maximum species-specific cell-replication rates to be calculated (Reynolds, 1989). When assembled in a small software program, a species-specific replication rate could be quickly derived for the water temperature and turbidity in the reach on each occasion. The time required to accommodate the observed increase over a fixed distance of river then defines the maximum mean velocity of flow (this has been calculated in Figure 13.3).

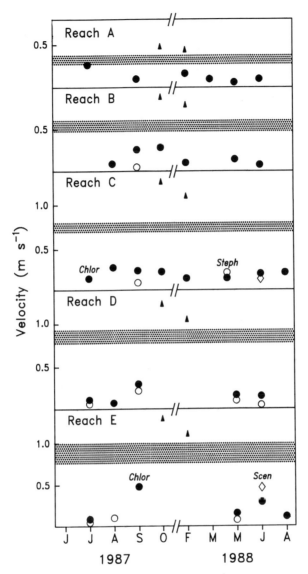

FIGURE 13.3. Mean velocities of downstream transport in the River Severn, calculated from the time of travel required to accommodate the observed growth between the upper and lower sampling points. The shaded areas represent the range of measured mean velocities in the main channel in each case. It is argued that the consistent differences may be explained only by intermediate channel storage. (The reaches are identified in Appendix 2)

The validity of the model equation has been established through its ability to simulate carefully measured or inferred growth rates of named algae in limnetic enclosures (the studies are summarized in Reynolds, 1986) and elsewhere (Hilton, Irish and Reynolds, 1992). The validity of their use may be doubted on more intuitive grounds, not the least being that algae do not always or even often perform to the physical capacity of a system, especially if nutrients or carbon availability are at rate-limiting concentrations. Moreover, we have not attempted to account for grazing or settling losses or cells along the way. Yet these failures only increase the anomaly — still more time is needed to accommodate the observed population increase net of limitations and losses. Our problem is to explain the fact that not even maximal cell-replication rates can account for the rate downstream increase, without supposing a large proportion of the population increase is supported in water flowing at substantially less than the mean rate.

A simple calculation shows how the mechanism might operate: suppose a river is discharging 100 m^3 s^{-1} through a channel with a cross-section of 200 m^2, its mean velocity is 0.5 m s^{-1} or 43 km day^{-1}. For an alga dividing once per day, we should reasonably anticipate a doubling of its mass every 43 km. Suppose the channel to be physically divided into a slow section taking 25 m^3 s^{-1} and a fast one carrying 75 m^3 s^{-1}, and suppose that the sections are exactly 100 m^2 in cross-section. The two sections are distinguished by their velocities, now 0.75 m s^{-1} and 0.25 m s^{-1}. Now, over 43 km, the fast-moving population could have increased by a factor of 1.4 but the slow-moving population will have made its second doubling, quadrupling its mass. When they are merged and mixed, the mean population would not have doubled over the distance but would have increased 2.7-fold.

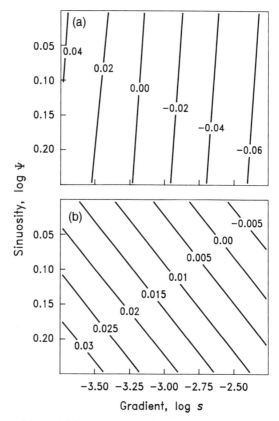

FIGURE 13.4. Contours of the multiple regression describing retentivity enhancement of downstream increase of phytoplankton in (a) the June 1990 survey of Reynolds and Glaister (1992), from data shown in Figure 13.2, and (b) from the March 1991 survey. Gradient is the key predictor at low discharges but sinuosity becomes relatively more important during the winter–spring period of greater discharges

It does not follow that this logic provides an explanation for these enhanced downstream growth increments, although the variability in the velocity field, with a variable fraction not flowing at all, is ultimately invoked as the key mechanism retarding the downstream progress of suspended populations, at least for long enough for them to increase, or decrease, relative to the population entrained in the adjacent mainflow. This, at first sight, anomalous behaviour has been observed by other stream scientists, in relation to solutes or fluorescent dyes, and has been explained in similar terms of intermediate channel storage (Bencala and Walters, 1983; Young and Wallis, 1987). Different authors appear to differ in exactly how the volume of non-flowing (dead-zone) water is aggregated but, so far as phytoplankton is concerned, it is essential to think of relatively large volumes of water in store exchanging in the order of a day or two or longer (Reynolds and Glaister, 1993).

It should be clear from this why I prefer the term "storage zone" to "dead zone" but the allusion to moribundity ties with the weariness of the flow in the section header: where the storage is greatest, the enhancement of the downstream increment in phytoplankton populations is potentially larger. It also follows that significant enhancement is likely to be indicative of protracted storage processing and, if identifiable, to make a statement about the reach of river concerned. From the data in their synoptic surveys, Reynolds and Glaister (1992) fitted significant multiple regressions to explain 32% of the variability in downstream phytoplankton performance in the June samples from British rivers and 41% of the April samples. In general terms, the weaker the slope, the greater is the retentivity and the greater is the potential downstream increase-enhancement: this was especially so during the low June flows (Figure 13.4(a)). In April, however, when discharges were two to four times greater, retentivity became more strongly predicted by reach sinuosity (Figure 13.4(b)). This is explicable, perhaps, in terms of diminishing areas of weak flow and the greater reliance upon the existence of channel forms protecting the storage zones surviving high winter flows.

LYING AT THE RIVER'S SIDE

"On either side the river lie ."
(Alfred Lord Tennyson, The Lady of Shalott)

It has not yet been possible to explain in detail the retentiveness of given fluvial reaches but is has been shown how individual storage zones might function (Reynolds, Carling and Beven, 1991) and how, in aggregate, they might account for the retentiveness upon which the downstream growth-enhancement hypothesis depends (Reynolds and Glaister, 1992, 1993). Within-channel storage has been demonstrated by means of *in situ* fluorimetry and by remote sensing of suspended chlorophyll (Reynolds, Carling and Beven, 1991): significant lenses of water are indeed held behind projecting bars or deltas adjacent to bank cavities or in other artefacts of non-uniform channel structure. They may retain water for quite sufficient time for it to become differentiated in terms of its suspended algal complement, despite a constant fluid exchange between the flow and the non-flowing store. The difference in algal concentration on either side of the intermediate boundary may then depend upon the *in situ* rates of cell increase for the species concerned, on either side of the boundary, and the rate at which water is exchanged across the boundary. The greater the former is and the longer the latter is, then the more marked may be the relative contrast in the standing crop of a given species either side of the boundary. Indeed, Reynolds, Carling and Beven (1991) identified

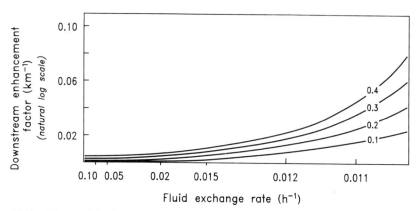

FIGURE 13.5. The model of Reynolds and Glaister (1993) to account for downstream growth enhancement as a function of relative storage zone volume (four cases: 0.1, 0.2, 0.3 and 0.4 times reach volume) and the rate of fluid exchange between flowing and non-flowing water

an asymptotic differential at which the growth increment within the store was identical to the quantity of cells removed per identical time step. Significantly differentiated storage-zone populations — values of between 1.5- and 43-fold greater concentrations have been measured — were associated with fluid exchange rates with main channel flow of between 0.02 and 0.006 h^{-1}. At each dilution by "inflow" to the store, there is a corresponding outflow from the storage zone, but this time charged with the enhanced algal population. To give any kind of contribution to the downstream populations, it is implicit that the aggregated proportion of the reach volume represents a quantitatively significant component of the channel flow, and that the rate of fluid exchange therewith is neither so rapid that new phytoplankton cells fail to be differentiated, nor so slow that new phytoplankton cells fail to be exported to the main flow in quantities sufficient to make a difference. A series of storage zones, together representing perhaps 10−40% of the reach volume and each exchanging with the main flow at rates of 1−2% per hour, would, by a saltating process, offer a collective opportunity to enhance, apparently, the performance of the autotrophic phytoplankton as it moves downstream. Reynolds and Glaister (1992, 1993) derived a family of predictive curves to describe the potential enhancement of population increase per kilometre of river, in terms of cumulative storage volume and the average fluid exchange rate. Their model is redrawn as Figure 13.5.

The question must surely arise, do rivers actually behave in this way? Reynolds and Glaister (1992) have presented evidence from one section of the Severn in Shropshire which supports the proposed mechanism. Using Daedalus thermal-line scanning, the surface temperature of a 10-km section of river was sensed remotely, with a maximum sensitivity of 0.1 °C and a pixel resolution of about 1 m. The full reconstruction (shown in part in Plate 2) assembled sectional images corresponding to the length of river between Dryton (NGR: SJ 581050) and Buildwas Park (SJ 632043), distinguishing a range of verified surface temperatures from ~ 16.5 °C to 20.5 °C. Having been taken at the end of a clear July morning, it was assumed that the most dynamic and most mixed water would have warmed little during the morning but that the surface veneer on non-flowing zones would, conversely, have heated significantly. Analysis of the percentage frequency of each colour of pixel in each sectional image revealed that some 82−94% of the river surface was

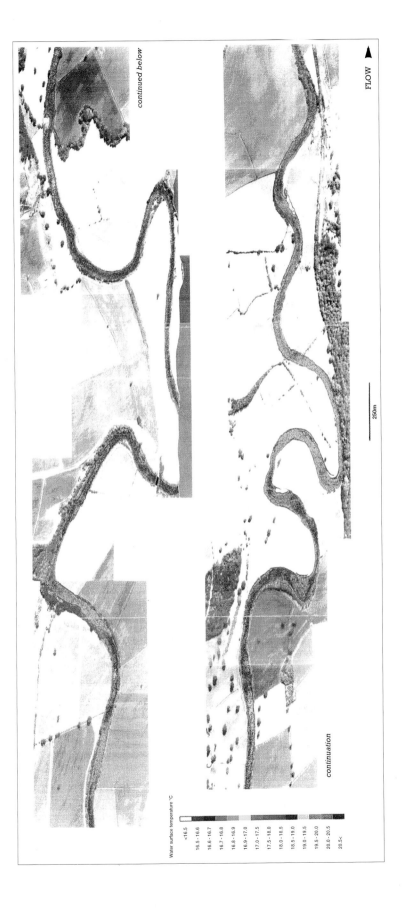

Water surface temperature °C

	<16.5
	16.5 - 16.6
	16.6 - 16.7
	16.7 - 16.8
	16.8 - 16.9
	16.9 - 17.0
	17.0 - 17.5
	17.5 - 18.0
	18.0 - 18.5
	18.5 - 19.0
	19.0 - 19.5
	19.5 - 20.0
	20.0 - 20.5
	20.5<

continued below

continuation

FLOW

250m

PLATE 2   Daedalus thermal line scan (TLS) over a section of the River Severn, from Cressage to Buildwas in Shropshire, showing remotely-sensed water-surface temperature differences of as little as 0.1 °C at a spatial resolution of about 1 m. The temperature scale (left-hand side) has been verified by ground truth measurements. (Surveyed on 10th July 1991; imagery is the property of the NERC, reproduced with permission)

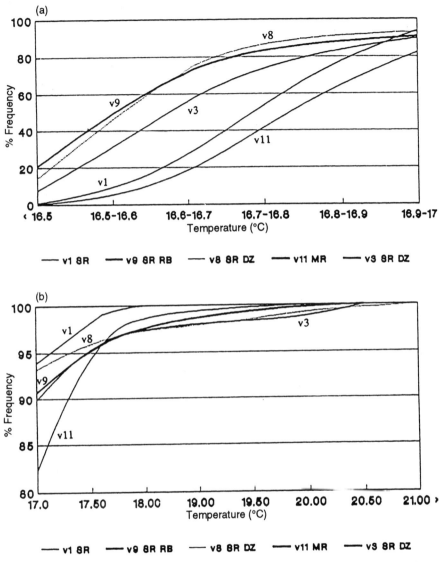

FIGURE 13.6. Plots of pixel frequency (density calibrated water temperatures) for selected squares making up the composite river run in Plate 2. In each case, some 82–94% of the river surface is between 16.5 and 17°C(a), and a further 6–18% is between 17 and 21°C(b). If the latter band is considered to be warmer because it is held in (temporary) storage, then 6–18% is the areal (not volumetric) function that is effectively the reach store

<17°C but that 6–18% was between 17°C and 21°C, much (1–4%) substantially above 17.5°C (see Figure 13.6). If these warmer bands are considered to identify water held in (temporary) storage, then 6–18% is the areal (but not volumetric) fraction of the effective reach store. This length of river happens to coincide almost exactly with the Cressage-Buildwas synoptic-study reach, for which positive enhancement factors of between 0.0048 and 0.0398 km^{-1} were recorded in each survey, the greatest being in the June sampling.

Reading back from Figure 13.5, the cumulative fluid exchange rate between river and a storage equivalent to between 0.06 and 0.18 of reach volume would fall in the range 0.010 and 0.017 h^{-1}. The hypothesis is not invalidated.

TO GO DOWN OR ROUND?

"Like the foam on the river . . . gone, and for ever!"

(Sir Walter Scott, The Lady of the Lake)

To conclude, the synopsis of the studies highlighted in this review is that it is possible to assert, with much greater confidence than previously, certain suppositions about the ecology of potamoplankton. First, it can certainly be native to rivers and, at times, become abundant. This is always more likely to be the case in longer and low-gradient rivers but the rate of downstream increase is most likely to accelerate where the headwaters level out to acquire the characters of a silted lowland river, probably with a distinct floodplain — literally, from the point where brook and river meet! Secondly, growth and loss dynamics will shape the size and structure of the potamoplanktonic assemblage but they will vary continuously down river and differ among species and among locations. In this way, the dominant species composition can alter downstream, generally from chlorophytes to diatoms but where there is a transition between them, it is a boundary which itself moves up- and downstream with changing discharges and changing seasons: there are fluctuations in dominance with both time and place!

Thirdly, it has been shown how fluvial behaviour within the channel contributes to the delay in the shedding of its water and which behaviour is "labelled" by the dependent downstream recruitment of phytoplankton: generally both are more prevalent in the quieter, low-gradient lowland sections. Fourthly, the proposed mechanism of hydraulic retention and growth enhancement in non-flowing storage zones, separated from the main flow by boundary zones across which fluid is exchanged, has been shown not only to operate in one length of lowland river but, apparently, to do so comfortably embracing the quantitative performance required to explain the otherwise paradoxical observations on downstream increase.

It is much too premature to regard these "discoveries" as being universally applicable to all other rivers or as providing a satisfactory explanation for the ecology of river phytoplankton. Nevertheless, they do provide a new perspective on the dynamics of river plankton that is adequate both to explain the paradigm and to suggest enough about how species might survive and enhance their downstream increase, at least in terms of distance, for the paradoxes to be dispelled. Phytoplankters do maintain themselves in flow-side storage zones, they do increase within them and they do augment the populations in the flow recruited through direct fluid exchange. Like Sir Walter Scott's foam, phytoplankters are inevitably carried away to the sea — ultimately; as is also frequently visible in the case of foam, however, not all slides off with the main flow but, particularly at the sides and in backwashes and lees, some of it rotates and swirls around, perhaps not indefinitely but seemingly quite indifferent to the nearby currents. Analogously, the same structures are probably also capable of maintaining stocks and supporting the growth of the river plankton.

ACKNOWLEDGEMENTS

I am grateful to the organisers of the Symposium for the opportunity to present this review. It is a pleasure to express my profound thanks to my colleagues Mark Glaister, for his cheerful support throughout our collaboration on rivers-related projects, and to Kirsty Ross, for her assistance in the preparation of the manuscript. The full financial support, formerly, of the UK Department of the Environment and, latterly, of the National Rivers Authority, for our work on phytoplankton is most gratefully acknowledged.

APPENDIX 13.1.   List of stations sampled by Reynolds and Glaister (1992) in the construction of Figures 13.1, 13.2 and 13.4

Station	River	Location	Station	River	Location
2401	WEAR	Sunderland Bridge	3921	CHERWELL	Clifton
2402	WEAR	Chester-le-Street	3922	CHERWELL	Gosford
2403	WEAR	Cox Green			
			3962	MOLE	Brockham
2501	TEES	Abbey Bridge	3963	MOLE	Downside
2502	TEES	Darlington			
			4301	STOUR	Sturminster's Newton
2703	URE	Wensley	4302	STOUR	Blandford
2704	URE	Masham	4303	STOUR	Wimbourne Minister
2705	URE	Tanfield			
2706	URE	North Bridge, Ripon	4501	EXE	Exebridge
2707	URE	Boroughbridge	4502	EXE	Tiverton
			4503	EXE	Thorveston
2711	WHARFE	Pool			
2712	WHARFE	Linton	5201	BRUE	Glastonbury
2713	WHARFE	Tadcaster	5202	BRUE	Mark
2801	DERWENT	Cromford	5401	SEVERN	Llanidloes
2802	DERWENT	Ambergate	5402	SEVERN	Caersws
2803	DERWENT	Allestree	5403	SEVERN	Newtown
			5405	SEVERN	Crew Green
2811	TRENT	Stone	5405	SEVERN	Montford Bridge
2812	TRENT	Gt Haywood	5406	SEVERN	Atcham
2813	TRENT	Wolseley Bridge	5407	SEVERN	Cressage
2814	TRENT	Handsacre	5408	SEVERN	Buildwas
2818	TRENT	Nottingham	5409	SEVERN	Bridgnorth
2819	TRENT	Gunthorpe	5410	SEVERN	Bewdley
2821	TAME	Hemlingford	5501	WYE	Llanwrthwl
2822	TAME	Tamworth	5502	WYE	Newbridge
2823	TAME	Alfrewas	5503	WYE	Boughrood
			5504	WYE	Brobury
3301	GT OUSE	Radwell	5505	WYE	Hereford
3302	GT OUSE	Bromham	5506	WYE	Holme Lacy
3303	GT OUSE	Barford	5507	WYE	Ross-on-Wye
3305	GT OUSE	St Neots	5508	WYE	Monmouth
3306	GT OUSE	Huntingdon			
3307	GT OUSE	Earith	6501	RHYTALLT	Pen-y-llyn
			6502	RHYTALLT	Bryn Afon
3901	THAMES	Abingdon	6503	RHYTALLT	Pont Peblig
3902	THAMES	Shillingford			
3904	THAMES	Henley-on-Thames	7101	RIBBLE	Ribchester
3905	THAMES	Marlow	7102	RIBBLE	Brockholes
3906	THAMES	Maidenhead			

APPENDIX 13.2. Reaches of the River Severn studied by Reynolds and Glaister (1993) for which the data are presented in Figure 13.3

Reach A: Mount Severn—Llanidloes
Reach B: Caersws—Newtown
Reach C: Abermiwl—Welshpool
Reach D: Green—Montford Bridge
Reach E: Cressage—Buildwas

14
Bioaccumulation of Pollutants and its Consequences

A. JAN HENDRIKS

Institute for Inland Water Management and Waste Water Treatment, Lelystad, The Netherlands

INTRODUCTION

The Dutch Rhine delta consists of several interconnected rivers and lakes (Figure 14.1) which serve as sources for industrial, drinking, fishing and irrigation water. In particular, surface water in lower areas of the Netherlands may contain up to 50% or more Rhine water.

FIGURE 14.1. Major rivers and lakes of the Rhine delta

The Ecological Basis for River Management. Edited by D.M. Harper and A.J.D. Ferguson. © 1995 John Wiley & Sons Ltd

The rivers and lakes are also used as media for shipping and recreation and as sinks for municipal and industrial effluents and agricultural runoff. The concern for the consequences of this exploitation has brought about the "Rhine Action Program". One of its main objectives is to reduce pollution and to restore destructed habitats of indigenous species (IRC, 1987).

As a first step, the countries concerned have agreed to reduce regular emissions of priority micropollutants (Vrijhof, 1984) by 50% during the period 1985 to 1995 (IRC, 1987). Unfortunately, even realization of higher percentages is no guarantee of achievement of the goal set.

Though priority pollutants are probably most hazardous, other compounds among the thousands estimated to be present may be important as well. Also, contaminants that are no longer tolerated might be substituted by other toxic compounds, such as possible replacement of chlorobiphenyls (Aroclor) by chlorobenzyltoluenes (Ugilec). Reduction does not necessarily lead to an immediate proportional decline in concentrations. Concentrations of compounds that occur naturally will decrease less than proportionally and it may take many years to reach new steady states for persistent chemicals in sediments. The average degree of contamination achieved after reduction may still be too high for indigenous species. Current estimations of "safe levels" are mainly derived from experiments in which one observes a limited set of standard species and effects.

In general, one may distinguish between non-persistent compounds that may harm the aquatic community after a large (accidental) release, and persistent microcontaminants that are emitted more regularly. In this chapter, we will focus on a special class of the last group: persistent compounds that accumulate in organisms. Their contamination is focused at the end of the river branches because they are largely adsorbed onto suspended solids that sediment there. Heavy metals and less persistent organics, such as polycyclic aromatic hydrocarbons and organophosphorus biocides accumulate in lower trophic levels after uptake from water (bioconcentration), whereas persistent organics, such as PCBs and DDT especially, accumulate in higher trophic levels after uptake from food (biomagnification).

We will discuss some of the monitoring and modelling efforts that were carried out in the Dutch part of the Rhine basin but the general perspectives that come out of this area apply to other sedimentation areas as well. The emphasis is on aspects of concern for management and policy-making rather than of research details.

MONITORING CONCENTRATIONS

Levels of relatively well-known compounds with accumulation potential can be monitored in organisms throughout the food web. If knowledge of accumulation in a particular river is scant, a few taxa that represent most of the trophic levels and habitats present may provide a first supplement to the general patterns discussed below. For time trends, measurements of a few species usually suffice. Heavy metals are preferably monitored in invertebrates while organochlorines can be conveniently measured in fish.

Hexachlorobenzene residues measured in aquatic Rhine delta taxa are shown in Figure 14.2 as an illustration. Though variability is present, residues in organisms are clearly proportional to those in sediments and suspended solids.

In time-trend assessment, two major patterns may be distinguished: residues of some compounds have decreased sharply in the past but seem to stagnate now, such as cadmium in

177

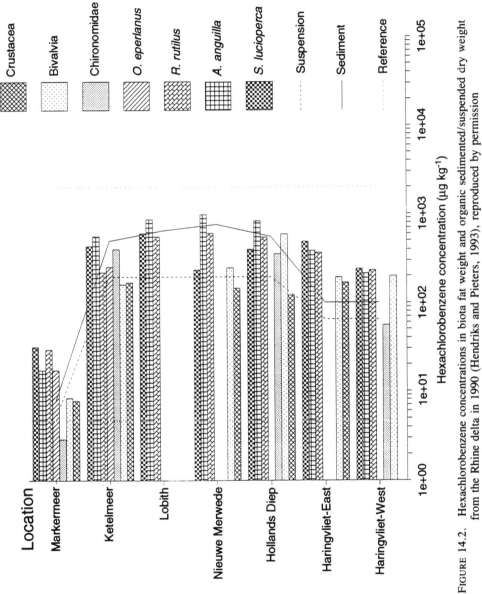

FIGURE 14.2. Hexachlorobenzene concentrations in biota fat weight and organic sedimented/suspended dry weight from the Rhine delta in 1990 (Hendriks and Pieters, 1993), reproduced by permission

TABLE 14.1.   (Proposed) consumption standards for animals (Stortelder et al., 1991). man (Van der
Valk, 1989) and birds and mammals (Romijn et al., 1991)

Microcontaminant	(Proposed) standards (μg kg^{-1} wet wt)		
	Animals	Man	Birds and mammals
Cd	5.0×10^1	5.0×10^1	1.6×10^2
Hg	1.0×10^3	1.0×10^3	4.0×10^2
HCB	2.0×10^2	5.0×10^1	5.0×10^2
PCBs	1.0×10^2	2.0×10^3	
PCB153	1.0×10^1	1.0×10^2	1.0×10^3
DDD+DDE	1.5×10^2	5.0×10^2	1.3×10^2
Dieldrin	3.0×10^2	5.0×10^1	1.0×10^2
HCH	2.0×10^2	1.5×10^2	1.6×10^2

zebra mussel (*Dreissena polymorpha*); concentrations of other compounds have remained at
the same level, such as PCB153 in eel (*Anguilla anguilla*) (Hendriks, 1993b). Similar
patterns have been observed for other compounds and compartments (Hendriks, 1993a;
Pieters and Hendriks, 1993).

A first impression of the impact of residues in invertebrates and fish on their consumers is
given by a comparison of levels with quality standards thought to be safe for the majority of
taxa. Such standards can be derived by different methods but usually reflect the lowest or the
five-percentile value of a set of concentrations that induce no response in laboratory studies.
Some standards that are used in the Netherlands are collected in Table 14.1 (Van der Valk,
1989; Romijn et al., 1991; Stortelder, Van der Gaag and Van der Kooij, 1991; Van der
Gaag, 1991; Van der Kooij et al., 1991). It is clear that, except for the best known
compounds, the scarcity of data severely restricts the outcome of any procedure used and
additional monitoring of response in (semi-)field studies is required.

MONITORING RESPONSE

The impact of calamities with microcontaminants on invertebrates (e.g. Van Urk, Kerkum
and van Leeuwen, 1993) and fish in the Rhine delta has been repeatedly demonstrated. The
long-term effects of toxic compounds on field populations are usually less easy to distinguish
from other factors that have an impact on populations. Effects on midge larvae
(Chironomidae) and worms (Oligochaeta) could be correlated with contamination of
sediments (Van der Guchte, 1990) but so far the compounds and mechanisms that caused the
response have not been identified. In cases of fish-eating birds and mammals, however,
correlations are supported by more in-depth (semi-)field studies in which biochemical,
pathological and ecological evidence is combined. Such studies have been carried out in the
Rhine delta for molluscivorous and piscivorous birds and mammals (e.g. Koeman et al.,
1973; Claassen and de Jongh, 1988; Scholten et al., 1989; Van den Berg et al., 1992;
Dirksen et al., 1993).

MODELLING CONCENTRATIONS AND RESPONSE

Since monitoring all compounds and species is virtually impossible, simple and intermediate models are being applied for hazard assessment on less well-known compound—taxon combinations. Generally, these estimations can be regarded to be a more compound- and taxon-specific refinement of the application of quality standards. An overview of simple and intermediate models for concentrations in and response by organisms is discussed in another paper (Hendriks, 1994) while specific elaborations for Rhine delta locations are presented elsewhere (de Vries, 1987; Dogger et al., 1992; Noppert et al., 1992).

Residues of accumulating compounds in organisms can be estimated from concentrations in water, suspended solids and sediments. Laboratory experiments and, to a lesser degree, field observations, have demonstrated that residues of persistent organic compounds in plants, invertebrates and fish are, at most, equal to the concentration adsorbed to organic solids. Accumulation of persistent organochlorines in molluscivorous and piscivorous birds (and probably mammals as well) is about one order of magnitude higher than residues in their food (Hendriks and Pieters, 1993, Hendriks, 1994). With these rules of thumb it is possible to estimate the microcontaminant burdens in other plants and animals with different habitats and diets.

The estimated exposure can be compared to levels that induce no response in organisms in laboratory experiments or in field studies carried out in other regions. Yet, as only a few species are studied frequently, no-response concentrations are seldom available for the species of interest. So, while exposure assessment discriminates between phylogenetic groups at a low level of taxonomy, no-response concentrations are often only available at the highest taxonomic level, e.g. algae, freshwater plants, worms, molluscs, waterfleas, insects, fish, amphibians, birds and mammals.

As an illustration, we have summarized some indicative exposure and no-response concentrations ratios for major taxa (see Table 14.2). The level of contamination is represented by the concentrations in water and suspended solids at Lobith in 1990. It should be noted that validation of these estimations with field data on concentrations in and abundance of organisms is poor so far. Yet, the in-depth studies mentioned above tend to confirm the estimations (for an overview on Rhine delta (semi-)field studies see De Wit et al., 1991).

CONCLUSIONS AND RECOMMENDATIONS

To obtain an impression of concentrations of accumulating compounds in water, suspended solids, sediments and organisms, regular monitoring programmes are carried out. Accumulation of heavy metals can be monitored well in invertebrates, preferably mussels, whereas concentrations of persistent organochlorines, like PCBs and DDT, are measured in fish. These monitoring studies have shown that contamination by some compounds, such as cadmium, has decreased substantially, while levels of other chemicals, like some PCBs, remain high.

Laboratory experiments and field studies have demonstrated that residues of persistent organics in aquatic plants and ectothermic animals are at most on the same level as the concentrations adsorbed onto organic matter in sediments or suspended if expressed in equivalent units. Residues in birds and animals may be up to one or two orders of magnitude higher as levels in their food. For heavy metals and less persistent organics a similar but somewhat more complicated approach may be followed.

TABLE 14.2. Preliminary estimates ratios of exposure and no response concentrations. Most taxa are threatened by one or several compounds

Organism group	Estimated exposure concentration/estimated no response concentration				
	Cd	Hg	HCB	PCBs	ΣDDT
Phyto-benthos	+	+			+
Zoo-benthos	+	+	+	+	+
Fish	−	+	+	+	+
Birds	−	−	−	−	+
Mammals	−	−	−	−	+

Estimated exposure concentration/no response concentration: $+ = <0.01$, $+ = 0.01 \ldots 1$, $- = >1$.

An initial impression of the chemical stress that an aquatic community has to suffer can be gained by comparison of measured or estimated concentrations with (proposed) quality standards. These concentrations generally reflect levels that induce no response in organisms in laboratory toxicity assays.

Obviously, only a limited set of species and responses can be tested. Therefore, current and future efforts should be focused on refinement to compound- and taxon-specific assessment. Preliminary estimates were qualitatively validated by the few in-depth field studies available. This will become increasingly necessary if the contamination decline in the Rhine delta indeed levels off near critical values, as is expected from the current trends.

Apart from refinement for relatively well-known accumulating compounds, tools for recognition of unknown accumulating compounds in field situations should be extended.

15

The Use of Macroinvertebrate Assemblages in the Assessment of Metal-Contaminated Streams

ANTHONY M. GOWER, GRAHAM MYERS, MARTIN KENT and MICHAEL E. FOULKES

University of Plymouth, Plymouth, UK

INTRODUCTION

Historically, the biological assessment of the effects of heavy metals in freshwater ecosystems has proved difficult. Metals can exist in dissolved, colloidal and particulate forms and can reach every part of an aquatic ecosystem; furthermore, the interactions and partitioning of the metal can vary with environmental conditions.

Methods used for characterizing the effects of metals on macroinvertebrates include identification of tolerant species, determination of species richness and other community parameters — with varying degrees of success. In recent years there has been an increase in the use of experimental streams (Winner, Boesel and Farrel, 1980; Clements, Cherry and Cairns, 1988; Leland et al., 1989) for analysing the biological effects of metals and some workers (Winner, Boesel and Farrel, 1980; Clements, Cherry and Cairns, 1988) have proposed a predictable graded response based on the proportions of major groups of insects within the community.

However, difficulties are still encountered in associating a particular concentration of a particular metal with biological characteristics and this has resulted from a tendency to over-generalize biological effects. Different taxonomic levels have been used in different studies, few interacting environmental variables have been considered and there has been extrapolation of effects from limited ranges of contamination and water quality.

We report here on our preliminary findings based on a survey in south-west England designed to explore relationships between macroinvertebrate assemblages in a number of streams with toxic metals and other environmental variables, using TWINSPAN (two-way indicator species analysis) for classification, and CANOCO (canonical correspondence analysis) for examining correlation with environmental factors.

STUDY AREA

South-west England is heavily mineralized, especially in the areas of the moorland granites and their metamorphic aureoles. Sampling sites (Figure 15.1) were located in catchments

The Ecological Basis for River Management. Edited by D.M. Harper and A.J.D. Ferguson. © 1995 John Wiley & Sons Ltd

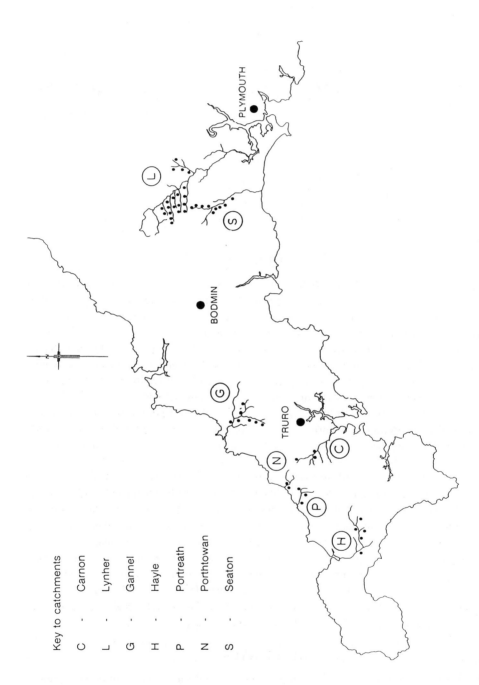

FIGURE 15.1.   Map of Cornwall showing sampling locations in the seven main catchments

Key to catchments

C - Carnon

L - Lynher

G - Gannel

H - Hayle

P - Portreath

N - Porthtowan

S - Seaton

around the edge of Carnmenellis, Goss and Bodmin moors in Cornwall, where a number of streams are affected by elevated metal concentrations resulting from past mining activity.

Mine drainage streams consist partly or entirely of adit (groundwater) flow and are characterized by a continuous and relatively constant level of contamination. Since many of these streams have been in existence for 100 years or more, conditions will have stabilized, allowing the development of tolerant communities. This, together with the fact that the streams are little affected by man's other influences, makes them particularly suitable for studying the long-term effects of metals on benthic macroinvertebrates.

METHODS

Conditions along metal-contaminated streams in the study area frequently change rapidly over relatively short distances due to the downstream processing of the metals, and sampling sites were carefully selected to enable direct comparison between different contaminated sites (i.e. both within and between streams) and between contaminated and control sites. Riffle macroinvertebrates were collected in October 1991 and April 1992 at each of 49 sites using a hand-net for three 1-minute kick samples. At each site the major habitats across the channel including the riffle and both margins/banks were proportionately sampled to ensure a representative collection of the community. Samples were preserved immediately in 4% formaldehyde and subsequently sieved and stained (using rose bengal). All taxa, including Chironomidae and Oligochaeta, were identified where possible to species.

Information was collected on a wide range of variables believed to have an important effect on the biota in Cornish mine drainage streams. The 35 physicochemical variables recorded at each site at two monthly intervals throughout the study period included hardness, alkalinity, pH, dissolved oxygen, sulphate, organic acids, substratum type, extent of chemical precipitation and the amount of algal growth. Concentrations of Fe, Al, Cu, Pb, Zn, Cd, Ca and Mg were determined by filtering (through acid-soaked 0.45 μm membranes) on site, acidifying with Aristar nitric acid and subsequent analysis by ICP-MS (inductively coupled plasma mass spectrometry).

Invertebrate assemblages have been classified using TWINSPAN (four "pseudospecies" were used after standardization to sampling site total). CANOCO (Ter Braak, 1988) is a method of direct ordination integrating species with environmental data. The resulting ordination diagram in which species, sites and environmental variables are represented can be used to demonstrate the relationships between the invertebrates and one or more of the environmental variables.

RESULTS AND DISCUSSION

Of the 35 variables initially used with CANOCO, 14 were selected for further analysis on the basis of their correlations with the ordination axes (Table 15.1); several of these variables are significantly ($p < 0.001$) correlated with axis 1, including Cu, Al, pH, alkalinity and UV absorbance at 254 nm. Figure 15.2 shows that the mean concentrations of copper recorded at the 49 sites varied from zero to 1.3 mg litre^{-1} with a full range of intermediate concentrations, enabling comparisons to be made across and within affected catchments. A total of 229 invertebrate taxa were collected during the course of this study, with 189 recorded in October.

TABLE 15.1.   Pearson product moment correlation coefficients of the 14 environmental variables used in the analysis with ordination axes determined by CANOCO, for the 49 stations

Variable		Axis 1	Axis 2
Al	log Al concentration	0.680***	0.097
Cu	log Cu concentration	0.834***	−0.305
Zn	log Zn concentration	0.415**	0.075
Cd	log Cd concentration	0.397**	−0.012
Pb	log Pb concentration	0.057	0.119
hrdns	log hardness	0.400**	−0.198
alk	log alkalinity	−0.619***	−0.284
Si	log silicate concentration	0.215	−0.403**
pH	pH	−0.677***	0.362**
uv254	absorbance at 254 nm	−0.643***	0.048
disch	log discharge	−0.160	−0.755***
ppt	precipitate, 10-point scale	0.365**	0.192
algae	log algal cover	0.473***	−0.170
slope	log slope	0.275	0.502***

** $p < 0.01$ *** $p < 0.001$

The sites/environment biplot and species/environment biplot, based on CANOCO analysis of the data from October 1991, are shown in Figures 15.3 and 15.4 respectively. The longer the environmental arrow, the greater is its influence on the community composition; and for interpretation purposes, each arrow can be extended backwards through the central origin. Thus sites or species with their perpendicular projections near to or beyond the tip of an arrow will be strongly positively correlated with and influenced by the environmental variable represented by that arrow. Those at the opposite end will be less strongly affected.

Communities in sites to the far right of Figure 15.3 are metal-tolerant and are associated with streams containing elevated concentrations of one or more of the metals, Cu, Al, Cd and Zn. The relationship with Cu is particularly strong, suggesting that this is a major determinant of community composition in these streams. It can also be seen that these sites are negatively correlated with pH, alkalinity and hardness. Other interesting groupings are: first, there are the three sites at the bottom of the diagram, still affected by metal contamination but occurring some way downstream — where discharge (and increasing pH) becomes increasingly significant. In general, separation on the second axis is in relation to slope and discharge. Secondly, there are the sites at the top left of the diagram, which are clean, with relatively undisturbed catchments (showing a negative correlation with silicate) in contrast to most other sites which have been disturbed by mining activity.

The Figure 15.4 biplot is restricted to species occurring at 10 or more sites for clarity of presentation, although all species recorded in October were included in the analysis. Metal-tolerant invertebrates appearing at the extreme right of the plot include the chironomids *Chaetocladius melaleucus* and *Eukiefferiella claripennis*, the empid *Wiedemannia* spp., the flatworm *Phagocata vitta* and the net-spinning caddis *Plectrocnemia conspersa*. The position of these species may be compared with that of the five invertebrates to the extreme left of the plot, i.e. *Eiseniella tetraedra*, *Tvetenia calvescens*, *Elmis aenea*, *Sericostoma personatum* and *Potamophylax* spp. which are usually absent from streams with elevated concentrations of heavy metals. The relative positions of the remaining species along the

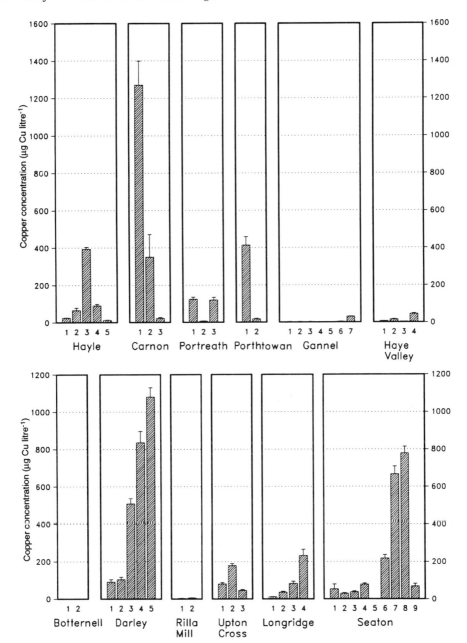

FIGURE 15.2.   Mean concentrations of copper in filtered water samples at the 49 sites from October
1991 to August 1992 (*n* = 6; vertical bars show standard errors)

primary axis reflect their sensitivity to heavy metals, thus demonstrating the potential
usefulness of the biplot in the analysis of invertebrate communities in relation to water
quality and other environmental conditions. Certainly, the occurrence of a group of species
at the far right of the plot within a reduced community is characteristic of metal-impacted

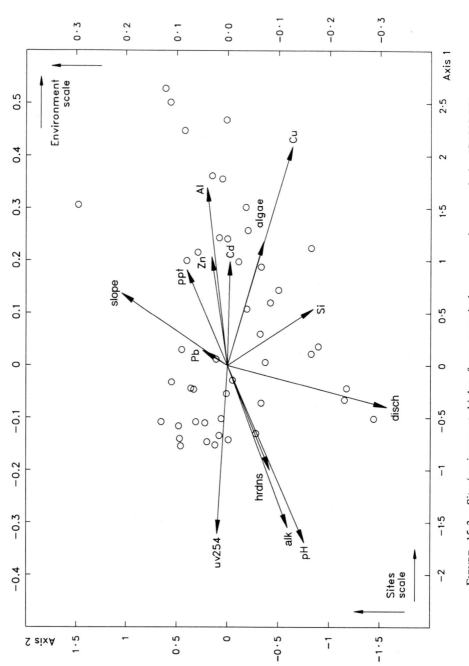

FIGURE 15.3. Sites/enviornment biplot from canonical correspondence analysis (CANOCO). Environmental variables are explained in Table 1

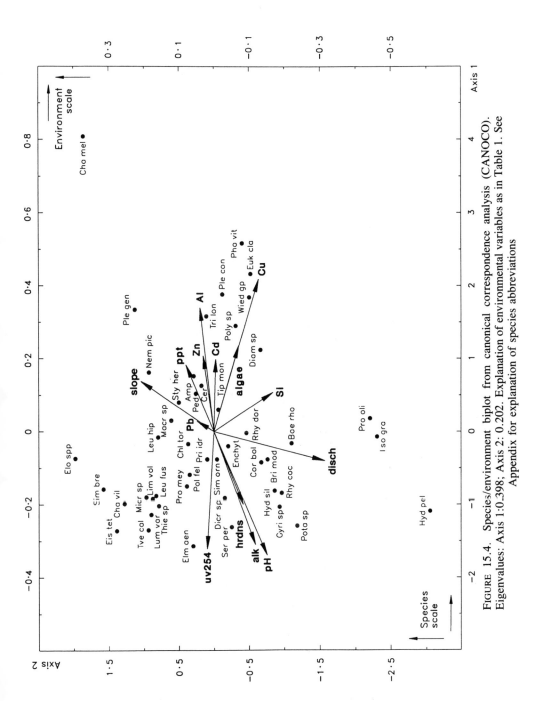

FIGURE 15.4. Species/environment biplot from canonical correspondence analysis (CANOCO). Eigenvalues: Axis 1:0.398; Axis 2: 0.202. Explanation of environmental variables as in Table 1. See Appendix for explanation of species abbreviations

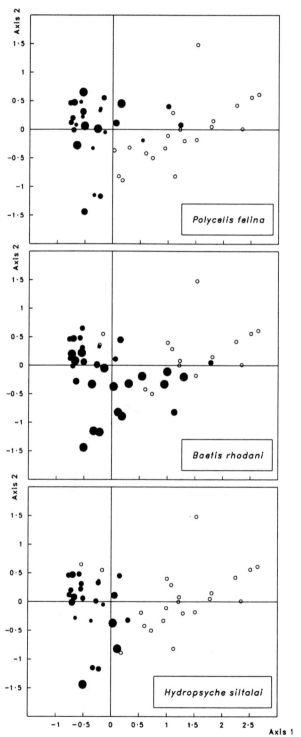

FIGURE 15.5.   Abundances of six invertebrate species superimposed on the sites/environment biplot:
• <1% community; • 1−4.9%; ●5−9.9%; ● ≥10%

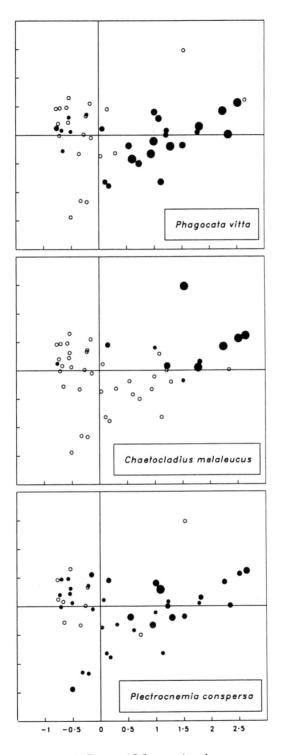

FIGURE 15.5. *continued*

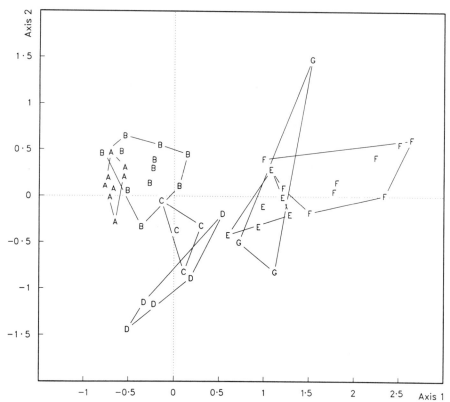

FIGURE 15.6.   TWINSPAN groups A to G superimposed on the sites/environment biplot

waters in south-west England.

Plots of individual species abundances superimposed on the sites ordination (Figure 15.5) provide further evidence that the distribution patterns of a number of species are significantly correlated with the metal gradient. These diagrams help to generate hypotheses about preferences and tolerances of the species. For example, the flatworms *Polycelis felina* and (the metal-tolerant) *Phagocata vitta* display clear separation along the primary axis. The net-spinning caddis *Plectrocnemia conspersa* is more widely distributed but is relatively more abundant in contaminated streams, where another net-spinner, *Hydropsyche siltalai*, is absent. The restricted distribution of the chironomid *Chaetocladius melaleucus* is also suggestive of both tolerance and competitive exclusion, with the latter being of particular importance. The mayfly nymph *Baetis rhodani* is a common and widely distributed species in south-west England and it is interesting to note that its distribution extends into streams of low level metal contamination where it may become prominent within the community.

TWINSPAN produced seven site groups and these are superimposed on the sites ordination in Figure 15.6. Site G separated at level 2 and sites A to F at level 3. Groups A and B which include all the relatively clean uncontaminated sites had the largest number of species with means of 52 and 37 respectively. Group E had the lowest mean species richness of nine. The overlap between the groups reflects a continuum of community variation along the copper-dominated gradient. However, the separation between the groups is sufficient to support the hypothesis that the effects of the variables along axis 1 are community based and the clear separation between groups A−D and E−G along the primary axis may suggest a metal concentration threshold at which there is a significant change in community structure and this is currently being investigated.

CONCLUSIONS

Preliminary analysis of our data by CANOCO confirms the considerable potential of this technique for studying community variation in metal-impacted streams. A particularly strong relationship with copper has been demonstrated and it is hoped to identify environmental thresholds above which there are significant changes in community structure. If effective use is to be made of macroinvertebrates in assessing the impact of metal contamination, species level identification is essential. The above examples show it is meaningless to generalize over the effects on Trichoptera and Platyhelminthes. Species richness is not on its own a reliable indicator of the severity of metal contamination, as sites with the highest metal concentrations did not have the lowest number of species.

ACKNOWLEDGEMENTS

This work was supported by a Natural Environment Research Council grant. Additional funding has been provided by the National Rivers Authority (South West Region). Confirmation of identification of selected specimens was kindly provided by Dr Patrick Armitage and John Blackburn (Institute of Freshwater Ecology) for oligochaetes and chironomid larvae and by Marc Ingelrelst (South West Water) for plecopteran larvae.

APPENDIX   Species abbreviations used in Figure 15.4

Amp	*Amphinemura sulcicollis*
Bae rho	*Baetis rhodani*
Bri mod	*Brillia modesta*
Cer	Ceratopogonidae
Cha mel	*Chaetocladius melaleucus*
Cha vil	*Chaetopteryx villosa*
Chl tor	*Chloroperla torrentium*
Cor bol	*Cordulegaster boltonii*
Diam sp	*Diamesa* sp.
Dicr sp	*Dicranota* sp.
Eis tet	*Eiseniella tetraedra*
Elm aen	*Elmis aenea*
Elo spp	*Eloeophila* spp.
Enchyt	Enchytraeidae
Euk cla	*Eukiefferiella claripennis*
Gyri sp	*Gryinus* sp.
Hyd pel	*Hydropsyche pellucidula*
Hyd sil	*Hydropsyche siltalai*
Iso gra	*Isoperla grammatica*
Leu fus	*Leuctra fusca*
Leu hip	*Leuctra hippopus*
Lim vol	*Limnius volckmari*
Lim var	*Limnodrilus variabilis*
Macr sp	*Macropelopia* sp.
Micr sp	*Micropsectra* sp.
Nem pic	*Nemurella picteti*
Ped	*Pedicia rivosa*
Pha vit	*Phagocata vitta*
Ple con	*Plectrocnemia conspersa*
Ple gen	*Plectrocnemia geniculata*
Pol fel	*Polycelis felina*
Poly sp	*Polypedilum* sp.
Pot spp	*Potamophylax* spp.
Pri idr	*Pristina idrensis*
Pro mey	*Protonemura meyeri*
Pro oli	*Prodiamesa olivacea*
Rhy coc	*Rhyacodrilus coccineus*
Rhy dor	*Rhyacophila dorsalis*
Ser per	*Sericostoma personatum*
Sim bre	*Simulium brevicaule*
Sim orn	*Simulium ornatum*
Sty her	*Stylodrilus heringianus*
Thie sp	*Thienemanniella* sp.
Tip mon	*Tipula montium*
Tri lon	*Trissopelopia longimana*
Tve cal	*Tvetenia calvescens*
Wied gp	*Wiedemannia* group

16

The Importance of Investigatory and Analytical Techniques in Biological Water-Quality Investigations

MAURIZIO BATTEGAZZORE

CNR-IRSA (Water Research Institute), Milan, Italy

ALESSANDRA GUZZINI

CNR-IRSA (Water Research Institute), Rome, Italy

ROMANO PAGNOTTA

CNR-IRSA (Water Research Institute), Milan, Italy

and ROBERTO MARCHETTI

CNR-IRSA (Water Research Institute), Rome, Italy

INTRODUCTION

Aquatic invertebrates for the assessment of running water quality have been widely used in various countries. Among the advantages of this approach compared to chemical analysis are the coverage of a wide range of impacts which affect the aquatic ecosystem and its relatively low cost. Single indices of water quality cannot be adopted everywhere, due to the different geographical distributions of the taxa. However, effort has been put into the comparison of various methods at the EC level which will hopefully lead to a standardization of the procedures in a forthcoming EC Directive. A step in this direction is represented by the RIVPACS programme (Wright et al., 1989), which evaluates water quality on the basis of a comparison between the real values of a biotic index (the Biological Monitoring Working Party — BMWP) score; and those predicted by a multivariate analysis involving physicochemical and geographical parameters which do not seriously affect quality. Less effort appears to have been put into a decisive aspect of biological water-quality evaluation, i.e. the standardization of macroinvertebrate sampling techniques.

The Ecological Basis for River Management. Edited by D.M. Harper and A.J.D. Ferguson. © 1995 John Wiley & Sons Ltd

For this reason, the first part of the study was undertaken on the rivers Tiber, Aniene and Po in order to evaluate the adequacy of three macroinvertebrate samplers in aquatic environments commonly adopted in water-quality investigations: the hand-net, the grab sampler and artificial substrates.

Another aspect that is not always considered as much as it deserves in the biological monitoring of rivers is the effect of general water-quality parameters on the taxa. This seems particularly true in highly polluted rivers, such as the Lambro, a highly polluted tributary of the River Po which is the object of the second part of the study. The main course of the river (also called the Northern Lambro) and its main tributary (the Southern Lambro) both collect the waste-waters of the intensively urbanized and industrialized area of Milan (about two million inhabitants). It has been calculated (Marchetti, 1988) that the River Lambro is responsible for about one-third of the entire load of pollutants from the Po basin, although accounting for between 1/10 and 1/30 of the volume of water flowing into the Adriatic Sea. The degree of pollution of the Lambro is heavy, and evidence of its progressive deterioration goes back to the 1930s (Baldi and Moretti, 1938).

Literature data show that the concentrations of the chemical parameters exceed even the limits set by the current legislation on industrial and domestic discharges. A law promulgated in July 1986 stated that the River Lambro was an area under high environmental risk. The discharges of the human activities in the area of Milan are not to date (1994) collected and treated but pass directly into watercourses which flow into the River Lambro. Only along the upper stretch of the North Lambro upstream from Milan have the domestic and industrial discharges been collected and treated.

In order to maximize the comparability between samples, and in order to evaluate the role of fundamental water-quality parameters on the taxa sampled along the River Lambro, the second part of the study was undertaken using only one sampling technique (artificial substrates).

STUDY AREAS AND METHODS

The rivers Tiber, Aniene and Po

Samples were taken at five stations along three watercourses (Figure 16.1 and Table 16.1): the Tiber River (three stations), its tributary, the Aniene River (one station), and the Po River (one station in the final stretch) in the Spring of 1989, between March 20 and June 20. The samplers adopted were a benthic 1 mm mesh hand-net (Maitland and Morris, 1978), a Petersen grab, and multi-plate samplers (Hester and Dendy, 1962; Figure 16.2). The multi-plate artificial substrates were used in all stations and represent the standard for comparing the relative effectiveness of the different techniques. The two indices EBI (Woodiwiss, 1978, modified by Ghetti, 1986) and BMWP score (Armitage et al., 1983) were calculated on the basis of the taxonomical lists.

Study of the River Lambro

In this part of the study, four sampling campaigns were undertaken in 1991 (February, May, July and October) at five sampling stations. Of these, four were situated along the main course of the river (North Lambro) and one was situated on the main tributary (South Lambro). A sketch of the study area is given in Figure 16.3.

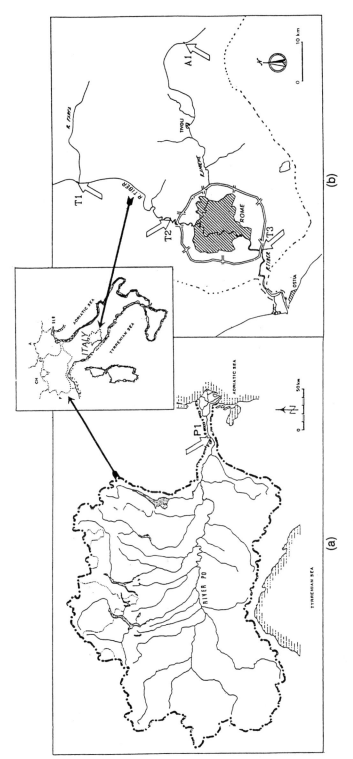

FIGURE 16.1. Study areas in the first part of the study: (a) the basin of the River Po and (b) part of the River Tiber basin are shown, together with sampling stations

TABLE 16.1. Some information on the watercourses Tiber, Aniene and Po

River	Station	Location	Type of bottom	River width (m)
Po	P1	Pontelagoscuro	Compact sand	300
Aniene	A1	Marano-Equo	Stones, sand, debris	20
Tiber	T1	Nazzano	Silt	60
Tiber	T2	Settebagni	Silt	50
Tiber	T3	Motorway bridge	Silt	70

Due to the difficulty of using traditional techniques (hand-net and grab), the sampling of macroinvertebrates was undertaken with multi-plate artificial substrates of the Hester-Dendy type according to the specifications of Petersen (1981). Samplers were submerged for about 30 days at a depth of about 1 m. Each sampling unit consisted of four multi-plate units (Figure 16.2). The number of samplers to be used for each sampling unit was decided on the basis of previous experiences (Battegazzore et al., 1992) in order to allow comparability of the samples.

Water samples for chemicophysical analyses were taken at the retrieval of the biological samples. Temperature, conductivity, dissolved oxygen (DO) and hardness were analysed for all samples. Following retrieval, samplers were stored at 4°C for the short time necessary

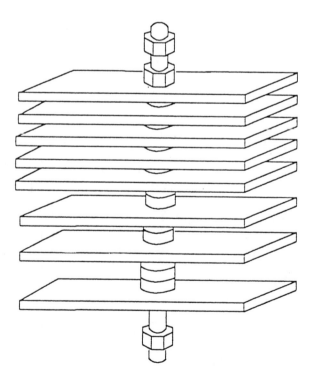

FIGURE 16.2. Hester-Dendy multiplate artificial substrate used in the present study (from Petersen, 1981, reproduced by permission)

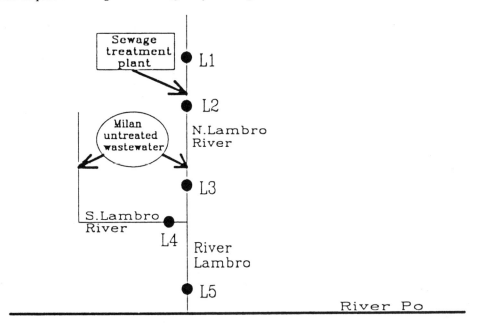

FIGURE 16.3. Simplified representation of the Lambro river system showing sampling stations and main polluting inputs

(about one week) for the sorting and taxonomical determination of the samples.

For all samples the following biotic indices based on macroinvertebrates, often used to assess water quality, were calculated: the Extended Biotic Index (EBI) by Woodiwiss (1978) as modified by Ghetti (1986) and the Biological Monitoring Working Party (BMWP) score (Armitage et al., 1983). These indices are typical non-quantitative biotic indices which stress the indicator value of the taxonomic groups found in the sample. Furthermore, Shannon-Wiener and Margalef diversity indices (Shannon and Weaver, 1949; Margalef, 1951) were calculated, together with the number of taxa *S*.

Finally, the ordination technique Canonical Correspondence Analysis within the CANOCO software package (Ter Braak, 1988) was adopted. The aim of this technique is to explain the spatial and temporal distribution of the taxa on the basis of the chosen environmental variables. The method requires a species matrix of the individual abundance and one of the values of the environmental variables. The linear combination of these variables allows the axis to be defined (axis 1) which determines the maximum dispersion of the species along the gradient of the same axis. Further axes, not correlated with the first one, can also be defined. Any two axes can be combined graphically in order to produce a biplot of the taxa, environmental parameters and samples. In this study, 29 taxa and six environmental variables were considered.

RESULTS AND DISCUSSION

The rivers Tiber, Aniene and Po

The lists of taxa are given in Table 16.2. The relative effectiveness of each technique in terms of the percentage of the taxa captured compared to the pooled number of taxa sampled

TABLE 16.2.   Taxa sampled in the rivers Po (Station P1), Aniene (station A1) and Tiber (stations T1, T2 and T3) in Spring 1989

	Sample									
	P1		A1		T1		T2		T3	
Taxa	Handnet	AS	Handnet	AS	Grab	AS	Grab	AS	Grab	AS
Plecoptera										
Isoperla sp.			X							
Leuctra sp.			X							
Dinocras sp.			X							
Ephemeroptera										
Baetis sp.			X			X		X		
Caenis luctuosa	X	X								
Caenis sp.				X						
Cloeon sp.								X		
Ephemerella ignita	X	X	X	X						
Ecodyonurus sp.			X			X				
Speorus sp.				X						
Heptagenia sulphurea		X								
Heptagenia sp.						X				
Rhythrogena sp.			X							
Habroleptoides sp.				X						
Diptera										
Brillia longifurca	X	X								
Cricotopus bicinctus	X	X								
Glyptotendipes sp.		X								
Polypedilum cultellatum	X	X								
Rheopelopia ornata		X								
Chironomidae			X		X	X	X		X	X
Ceratopogonidae			X			X				
Empididae			X							
Limoniidae			X							
Simuliidae			X	X				X		
Psychodidae			X			X				X
Trichoptera										
Ecnomidae						X				
Hydropsychidae	X	X	X	X	X					
Limnephilidae			X	X						
Policentropodidae						X		X		
Rhyacophilidae				X						
Odonata										
Calopteryx haemorroidalis	X	X								
Calopteryx sp.						X				
Ischnura sp.						X		X		

TABLE 16.2. *continued*

	P1		A1		T1		T2		T3	
Taxa	Handnet	AS	Handnet	AS	Grab	AS	Grab	AS	Grab	AS
Coleoptera										
Elmintidae			X	X						
Helichus substriatus	X									
Limnius sp.		X								
Haliplidae						X				
Dytiscidae			X			X				
Heteroptera										
Aphelocheirus aestivalis				X						
Hemiptera fam. gen. sp.						X				
Crustacea										
Echinogammarus veneris	X	X								
Echinoqammarus sp.					X	X	X		X	X
Gammarus sp.			X					X		
Asellus aquaticus		X				X		X	X	X
Proasellus sp.										X
Hirudinea										
Dina lineata			X							X
Erpobdella sp.						X				X
Helobdella stagnalis	X	X				X		X		
Hemiclepsis sp.						X				
Piscicola geometra	X	X				X		X		X
Oligochaeta										
Stylaria lacustris	X									
Naididae			X	X		X		X		X
Tubificidae gen. sp.	X		X		X	X	X	X	X	X
Enchytreidae gen. sp.										X
Lumbriculidae gen. sp.			X							
Lumbricidae gen. sp.			X							
Haplotaxidae gen. sp.			X							X
Mollusca										
Ancylus fluviatilis			X			X				X
Bythinia sp.				X		X		X		X
Physa sp.				X		X		X		X
Lymnea peregra		X								
Lymnea sp.						X		X		
Valvata piscinalis								X		
Pisidium sp.			X					X		
Theodoxus fluviatilis		X								
Tricladida										
Dugesia sp.						X		X		X

AS = Artificial substrates

TABLE 16.3. Numbers of taxa samples with three different techniques in the rivers Po, Aniene and Tiber

				Sample					
	T. Po	(P1)	R. Aniene	(A1)	R. Tiber	(T1)	R. Tiber	(T2)	R. Tiber	(T3)
Pooled no. of taxa	21	(100%)	32	(100%)	27	(100%)	17	(100%)	17	(100%)
No. of taxa sampled										
Handnet	13	(61.9%)	26	(81.1%)	—	—	—	—	—	—
Artificial substrate	18	(85.7%)	14	(43.8%)	27	(100%)	17	(100%)	17	(100%)
Grab	—	—	—	—	3	(11.1%)	3	(17.6%)	5	(29.4%)
Exclusive no. of taxa										
Handnet	3	(14.3%)	18	(56.2%)	—	—	—	—	—	—
Artificial substrate	8	(38.1%)	6	(18.8%)	24	(88.9%)	14	(82.4%)	12	(70.6%)
Grab	—	—	—	—	0	(0%)	0	(0%)	0	(0%)
No. of taxa in common	10	(47.6%)	8	(25.0%)	3	(11.1%)	3	(17.6%)	5	(29.4%)

TABLE 16.4. EBI and BMWP score values for all sampling techniques in the rivers Tiber, Aniene and Po

	Station									
	P1		A1		T1		T2		T3	
Inex	Handnet	AS	Handnet	AS	Grab	AS	Grab	AS	Grab	AS
EBI	7	8	12	8	4	11	4	8	4	7
BMWP score	44	74	104	76	9	101	9	53	17	36

AS = Artificial substrates

in each station is shown in Table 16.3. The values of EBI and BMWP score calculated from the taxonomic lists of each sample are given in Table 16.4.

While the hand-net seemed most adequate in the relatively small stony-bottomed Aniene tributary (56.2% of the taxa in this station were collected by this technique only), artificial substrates were more efficient in the larger Tiber and Po rivers (between 38.1% and 88.9% of the taxa were sampled exclusively by this technique). The grab sampler used in the silty Tiber River never gave exclusive or even complementary information compared to artificial substrates.

The EBI and BMWP scores at the three stations along the Tiber River for the artificial substrates both showed a trend of progressively decreasing quality, which was not paralleled by the grab samples. The artificial substrates, therefore, proved to give a much better representation of the situation. Elsewhere, the hand-net gave a higher quality evaluation than the artificial substrates in the small Aniene tributary, but a lower one in the Po River.

Study on the River Lambro

The complete taxonomic lists are given in Tables 16.5−16.8. The most abundant groups were the Chironomidae and the Oligochaeta, with Tubificidae prevailing numerically at stations L3, L4 and L5. The presence of the oligochete *Monopilephorus limosus*, in previous years also found in a small tributary of the Lambro (Erséus and Paoletti, 1986), and previously believed to be limited to the densely populated areas of China and Japan, was found on various occasions in both branches of the river. Other groups such as the mayflies (Ephemeroptera) or the caddisflies (Trichoptera) were present, but in general the richness of taxa was rather poor.

The trends of the average values of all the quality indices, together with the number of taxa, are given for each station and sample in Figures 16.4−16.8, together with their intervals of variation for each station. Stations L3 and L4, both below the Milan area and situated on the northern and southern branches of the river, respectively, were generally those showing lowest quality and diversity. Moreover, with the only exception of Shannon's H' (Figure 16.7), they also showed the smallest temporal variability. On the other hand, the first two stations were the ones showing generally better quality and higher diversity.

For all parameters the two stations downstream of Milan (L3 and L4) were the poorest in quality, and the two upstream stations (L1 and L2) were the best (Figures 16.4−16.8), but some uncertainty emerged among the parameters as to which of the two upstream locations

TABLE 16.5. Abundances of the taxa sampled in the five stations of the River Lambro in February 1991

Taxa	L1	L2	L3	L4	L5
Ephemeroptera					
Baetidae					
Baetis sp.	3	2			
Trichoptera					
Hydropsychidae					
Hydropsychidae gen. sp.	21	28			
Diptera					
Chironomidae					
Chironomus riparius		106	3	11	269
Cricotopus bicinctus		18			
Eukefferiella hospita		1			
Micropsectra atrofasciata	4	17			
Phaenopsectra flavipes		12			
Polypedilum laetum		88			
Rheopelopia sp.		4			
Parachironomus longiforceps		1			
Paratrichocladius rufiventris		1			
Psychodidae					
Anthomydae					
Limnophora sp.		2			
Odonata					
Coenagrionidae					
Pyrrhosoma nymphula					4
Platycnemididae					
Platycnemis pennipes					2
Isopoda					
Asellidae					
Asellus aquaticus	5	38			8
Hirudinae					
Erpobdellidae					
Erpobdella testacea	14	5			2
Erpobdella octolucata					2
Erpobdella sp.					3
Glossiphonidae					
Helobdella stagnallis			1		
Oligochaeta					
Enchytreidae					
Lumbricillus rivalis					1
Naididae					
Nais elinguis	1	859		4	18
Tubificidae					
Limnodrilus sp. (immat.)	2	4	271	20	54
Limnodrilus hoffmeisteri			112	1	186
Limnodrilus profundicola			5	3	31
Limnodrilus undekemianus	1		223	58	53
Monopilephorus limosus				21	
Tubifex blanchardi			68	4	26
Tubifex tubifex			129	10	74

TABLE 16.6. Abundances of the taxa sampled in the five stations of the River Lambro in May 1991

Taxa	L1	L2	L3	L4	L5
Ephemeroptera					
Baetidae					
Baetis sp.	4	1			
Trichoptera					
Hydropsychidae					
Hydropsychidae gen. sp.		2			
Diptera					
Chironomidae					
Chironomus riparius	2	118	557	14	4
Cricotopus bicinctus	2	192			1
Cricotopus triannulatus		234			
Eukefferiella hospita		12			
Micropsectra atrofasciata	1	37			
Nanocladius bicolor	1				
Orthocladius sp.		53			
Polypedilum laetum		34			
Polypedilum nubeculosum	1				
Rheopelopia sp.		1			
Parachironomus longiforceps					1
Potthastia gaedii		7			
Tipulidae					
Tipulidae gen. sp.		2			
Anthomyidae					
Limnophora sp.		20		3	
Isopoda					
Asellidae					
Asellus aquaticus	45	15			1
Hirudinea					
Erpobdellidae					
Erpobdella testacea	17	10			
Erpobdella sp.		1			
Glossiphonidae					
Glossiphonia sp.		1			
Helobdella stagnalis	1				
Oligochaeta					
Enchytreidae					
Enchytraeus albidus	1	14			
Lumbricidae					
Eiseniella tetraedra		6			
Eisenia foetida?		1			
Naididae					
Nais elinguis	3	155			
Tubificidae					
Limnodrilus sp. (immat.)	3	5	80	26	43
Limnodrilus hoffmeisteri					103
Limnodrilus profundicola				7	
Limnodrilus udekemianus	2	3	207		79
Monopilephorus limosus				4	
Tubifex blanchardi		2	43		
Tubifex tubifex	8		119	2	401
Tricladida					
Dugesia sp.	1				

TABLE 16.7. Abundances of the taxa sampled in the five stations of the River Lambro in July 1991

	Stations				
Taxa	L1	L2	L3	L4	L5
Diptera					
Chironomidae					
Chironomus riparius		2312	36	29	9
Dicrotendipes sp.	4				
Polypedilum laetum	2	130		4	
Polypedilum scalenum	1				
Anellida					
Hirudinea					
Erpobdellidae					
Erpobdella testacea	2				
Oligochaeta					
Lumbricidae					
Eiseniella tetraedra	2	1			
Naididae					
Nais elinguis					
Tubificidae					
Tubificidae gen. sp. (immat.)		144	50	47	10
Limnodrilus hoffmeisteri	2		9	13	1
Limnodrilus profundicola				18	
Limnodrilus udekemianus		5	8	21	6
Monopilephorus limosus			2	5	
Psammoryctides bavaricus					7
Tubifex blanchardi		54			1
Tubifex tubifex			10	2	16
Nematoda fam. gen. sp.					

presented the best quality. This is a matter of some importance, due to the outflow of the upper Lambro basin treatment plant occurring between these two stations. Some parameters (in particular the two diversity indices) seemed to indicate a better quality at station L1, above the treatment plant. Other parameters (in particular, the number of taxa and the EBI) seemed rather to indicate a slight increase in quality between stations L1 and L2. It is presumable that, while both groups of parameters responded to the toxic and oxygen-consuming discharges of the Milan area, the second group also responded (in an opposite way) to the non-toxic nutrient enrichment due to the treatment plant discharge.

The picture of the biological water-quality trend in the sampled stations was confirmed by the results of physicochemical analyses given in Table 16.9. For instance, in the case of dissolved oxygen (Figure 16.9), average values dropped between stations L1 and L2, but not to such low levels as occurred between station L2 and the following ones. Station L4 on the South Lambro tributary was also extremely low in oxygen content. Evidently, the oxygen drop at station L2 did not influence the trend of water quality measured by the EBI and BMWP score negatively. The very low values downstream of Milan are indicative of near-anoxic conditions for long periods of the year. The abrupt increase in conductivity occurring at station L2 due to the treatment plant discharge seemed to have an influence on the trophic conditions of the river rather than causing drastic quality variations.

TABLE 16.8.   Abundances of the taxa sampled in the five stations of the River Lambro in October 1991

Taxa	Stations				
	L1	L2	L3	L4	L5
Diptera					
Chironomidae					
Chironomus plumosus					2
Chironomus riparius		400		7	147
Dicrotendipes sp.					1
Glyptotyendipes sp.				2	
Polypedilum cultellatum	3				
Polypedilum laetum		194			
Crustacea					
Asellidae					
Asellus aquaticus			4		
Hirudinea					
Erpobdellidae					
Erpobdella testacea			2		
Oligochaeta		49			
Lumbricidae					
Eiseniella tetraedra		6			
Eisenia foetida		4			
Naididae					
Nais elinguis	1				
Tubificidae					
Tubificidae gen. sp. (immat.)	5	10	99	278	940
Limnodrilus hoffmeisteri	2	1	500	141	244
Limnodrilus profundicola					
Limnodrilus udekemianus			439	38	114
Monopilephorus limosus		2			
Tubifex blanchardi		22	387	107	289
Tubifex tubifex	1		566		77

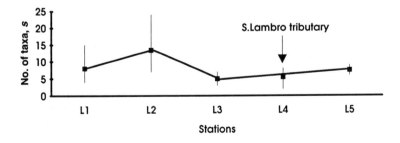

FIGURE 16.4.   River Lambro: trend of average values of the number of taxa (*S*) showing variation intervals

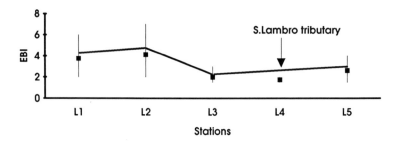

FIGURE 16.5.   River Lambro: trend of average values of the EBI showing variation intervals

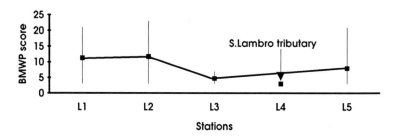

FIGURE 16.6.   River Lambro: trand of average values of the BMWP score showing variation intervals

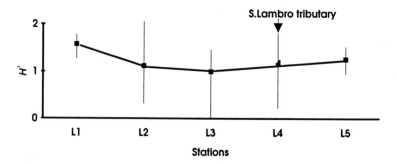

FIGURE 16.7.   River Lambro: trend of average values of Shannon's diversity index (*H'*) showing
variation intervals

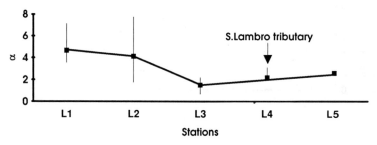

FIGURE 16.8.   River Lambro: trend of average values of Margalef's diversity index (α) showing
variation intervals

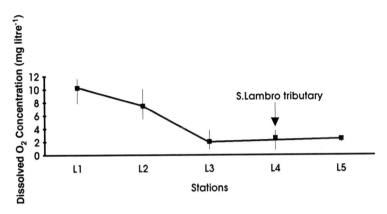

FIGURE 16.9.   River Lambro: trend of average values of dissolved oxygen concentration (DO)
showing variation intervals

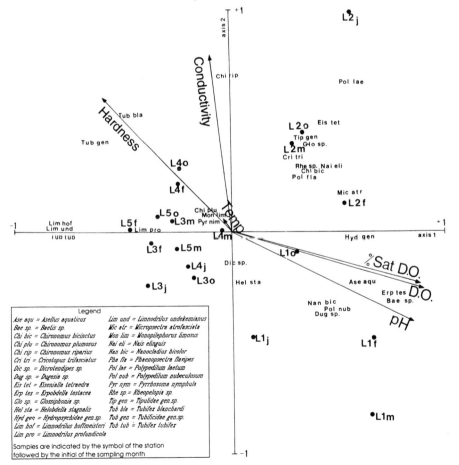

FIGURE 16.10.   Canonical Correspondence Analysis of 29 taxa and six enviornmental variables for all
samples taken in the River Lambro

TABLE 16.9.   Some chemicophysical measurements undertaken in the River Lambro together with the biological sampling

Date	Stations	Temperature (°C)	pH	Conductivity (μS cm^{-1})	DO (mg litre^{-1})	% saturation	Hardness (°F)
February	L1	5.5	8.33	473	11.64	94.0	14.3
1991	L2	6.4	7.87	716	10.10	83.0	15.8
	L3	8.7	7.56	534	3.82	33.1	19.5
	L4	8.9	7.53	728	3.16	27.6	17.5
	L5	8.4	7.52	578	2.70	23.2	19.0
May	L1	16.5	8.56	394	11.35	118.6	n.m.
1991	L2	14.0	8.01	731	5.48	53.9	21.2
	L3	14.6	7.70	626	1.93	19.2	24.0
	L4	14.8	7.76	677	3.73	37.2	19.0
	L5	15.5	7.77	597	1.93	19.5	23.3
July	L1	25.4	8.15	282	7.90	98.0	12.7
1991	L2	24.3	7.88	950	5.68	68.7	25.1
	L3	19.5	7.56	595	0.91	10.0	26.1
	L4	20.5	7.60	617	2.72	30.5	22.9
	L5	22.1	7.63	604	2.86	33.0	23.1
October	L1	16.7	8.08	364	10.12	106.2	17.1
1991	L2	17.5	7.89	571	8.82	93.5	22.3
	L3	16.5	7.53	597	1.61	16.7	26.0
	L4	16.6	7.39	866	0.79	8.3	24.2
	L5	17.0	7.48	605	2.53	26.4	26.1

nm = Not measured

The multivariate analysis (Figure 16.10), showed good agreement with the above considerations and gave an overview of the situation. The first axis is well correlated with both dissolved oxygen and its percentage saturation, and separates the two upstream stations from all the others. The second axis is well correlated with conductivity and hardness and separates the samples taken at station L1 from those at station L2. It appears evident that the group of species including *T. tubifex*, *T. blanchardi*, *L. hoffmeisteri*, *L. profundicola* and *M. limosus*, together with samples taken at stations L3, L4 and L5, were typical of low oxygen and intermediate conductivity levels. Another group, including the taxa *Baetis* sp., the Hydropsychidae, *A. aquaticus* and *P. nubeculosa*, together with the stations of the most upstream station L1, were typical of high oxygen levels accompanied by intermediate conductivity levels. Finally, the remaining group, including the taxa *C. trifasciatus*, *C. bicinctus* and others, together with the samples taken at station L2 downstream of the sewage plant discharge, was typical of relatively high values of both conductivity and DO.

CONCLUSIONS

The importance of the choice of the sampling technique and of multivariate analysis for the evaluation of running water quality on the basis of macroinvertebrates and for the interpretation of the relative data emerged clearly in the two phases of this study.

   In the first part of the study, the grab did not prove to be adequate for macroinvertebrate sampling aimed at water-quality evaluation. The artificial substrates adopted in this study were more appropriate than the active sampling techniques in the large rivers Tiber and Po, and helped to retrieve taxa which were not sampled by the handnet in the small stony river Aniene. Therefore, the combined use of artificial substrates and the hand-net sampler (where appropriate) in different environmental conditions would allow a more adequate comparison of the variations in water quality.

A limitation regarding the use of any artificial substrate is the relatively long time needed for colonization. However, this would not apply in the case of long-term water-quality monitoring and management involving methods such as the EBI and the BMWP score and non-quantitative sampling. From a methodological point of view, both indices seemed to give adequate and comparable water-quality evaluations; the BMWP score has the advantage of allowing the evaluation even of small differences among sample scores.

From the point of view of the sampling technique, artificial substrates used for water-quality evaluation offered an advantage compared to other techniques, due to the constant selection. From a methodological point of view, the combined use of biological and chemical methods, integrated by appropriate multivariate analysis, is extremely useful and is highly recommended for future studies on the environmental quality of running waters. This conclusion supports the one drawn in a study conducted by the EPA (1990) with the aim of comparing chemical and biological methods for water-quality evaluations.

The second part of the study indicated that the two stations along the River Lambro upstream of Milan, although showing a modest quality level, were the less polluted among the study sites. The anthropic impact of the urban−industrial area of Milan was clearly evident in the stations immediately below, while some sign of recovery, however small, was indicated at the final station.

The multivariate analysis of the 29 taxa and six physicochemical parameters sampled on all sampling dates and at all locations gave a global picture of the relative importance of the parameters as possible influences on community structure. A clear distinction among three groups of taxa emerged, related mainly to dissolved oxygen, conductivity, pH and hardness: a first group typical of the less polluted upper reach (station L1), a second group typical of high conductivity and hardness levels downstream of the treatment plant discharge (station L2), and a third group typical of the final, polluted stretches (stations L3, L4 and L5).

Nevertheless, it is possible that other parameters (such as specific pollutants) could have an important influence on the community structures. Therefore, further investigations ought to be carried out in this direction. In the framework of the restoration of the Po River basin (IRSA, 1990), the present study could represent a basis for the evaluation of the effects of the restoration measures, including the construction of sewage treatment works serving the Milan area, which are to be undertaken in the coming years.

ACKNOWLEDGEMENTS

The authors wish to thank Prof. Bruno Rossaro of the University of L'Aquila (Italy) and Dr Andreina Paoletti of the University of Milan (Italy).

17

Alleviating the Problems of Excessive Algal Growth

Irene Ridge, Judith Pillinger and John Walters

The Open University, Milton Keynes, UK

INTRODUCTION

Algae are an important component of river ecosystems, providing food and (for filamentous types) shelter to a wide range of organisms. Over the last 20 years, however, excessive growth of certain species has given rise to increasingly severe problems because of high nutrient inputs coupled with low summer flows. Offending algae are mostly filamentous or unicellular greens and blue-greens (cyanobacteria), the former developing mainly in lower order streams and shallow reaches and the latter in slow-flowing, deeper reaches. In the relatively short rivers of the British Isles, problems from filamentous algae are commonest, with phytoplankton problems arising chiefly in lakes and reservoirs. In larger continental rivers, however, nuisance phytoplankton are increasingly common especially where temperatures are high: in 1992 a state of emergency was declared in Australia after massive cyanobacterial blooms developed in the Murray River system.

The problems caused to humans by excessive algal growth include blocking of pumps and filters during water abstraction, release of toxins from cyanobacteria (NRA, 1990) and serious interference with leisure use and aesthetic enjoyment of rivers. Ecological problems are less clearly defined because it is virtually impossible to disentangle the direct effects of high nutrient levels from those of excessive algal growth but, in general, there is reduced growth of macrophytes, lower fish production and lower invertebrate diversity. Eutrophication is, however, a perfectly natural phenomenon and current problems are related to extreme eutrophication (hypertrophy) or a switch from oligotrophy to eutrophy caused by recent human activities.

It is in these clearly "unnatural" situations that efforts to reduce excessive algal growth should be concentrated but the mechanisms available to river managers are limited. Reducing nutrient inputs and hence eliminating the conditions which give rise to algal problems is a primary objective but, in the short term, may not be attainable. Even if terrestrial inputs decrease, dredging sediments in which phosphates have accumulated may

The Ecological Basis for River Management. Edited by D.M. Harper and A.J.D. Ferguson. © 1995 John Wiley & Sons Ltd

be necessary to effect any real improvement. Another approach is biomanipulation where the aim is to reduce algal crops by increasing or protecting the populations of herbivores (Moss, 1988). At sites in the Norfolk Broads, this technique combined with nutrient reduction has reduced phytoplankton very effectively (Moss, 1990) and the main problem hindering more widespread application of biomanipulation is often a lack of relevant ecological information about individual systems. Special problems are posed by filamentous algae because they are grazed by relatively few species but there is scope for biomanipulation even here: in Californian rivers, chironomid larvae that wove filamentous algae into tufts caused a marked reduction of algal biomass provided that invertebrate predators were controlled by higher order carnivores (Power, Marks and Parker, 1992). Mechanical removal of filamentous algae is the human equivalent of biomanipulation and is used successfully in some situations. However, it cannot be used in very shallow water, causes considerable disturbance to the system and leaves behind unsightly mounds of algae which are not easily disposed of; biomanipulation would clearly be preferable on ecological grounds.

A third approach to algal control is direct inhibition by chemical means, but commercially available algicides are beset by problems of low selectivity and potentially damaging side-effects. Their use may be justified in occasional emergencies but cannot be regarded as an ecologically sound tool for routine river management. The finding that regular applications of herbicide in drainage ditches led to the emergence of resistant strains of *Vaucheria* (Cave, G., pers. comm.) further reinforces this conclusion. However, one chemical system circumvents the problems described above and this depends on the release of natural inhibitors from rotting barley straw. The inhibitors are short-lived, highly selective against algae and are usually algistatic rather than algicidal so that small and probably beneficial amounts of algae persist during treatment. The applicability of this system to rivers and preliminary work which suggests that similar inhibitors are released from a range of other types of plant litter are discussed below. Managing rivers so as to maximize inputs of these natual inhibitors can then be regarded as another, environmentally acceptable approach to ameliorating algal problems.

BARLEY STRAW

The first clear demonstration that barley straw rotting in water inhibits algal growth in the field and does so by the release of antialgal substances was made in the late 1980s (Gibson et al., 1990; Welch et al., 1990). The inhibitors were released one to three months after placing straw in water and were shown to be active against a range of filamentous and unicellular green algae. Later work confirmed that they are also extremely inhibitory to cyanobacteria (Newman and Barrett, 1993) but Pillinger (unpublished) observed that diatoms apparently are not inhibited. Since 1990, barley straw has been used at sites in the UK and Eire to control both filamentous algae and phytoplankton in both lotic and lentic waters (Ridge and Barrett, 1992). Its use in river systems has not been extensive but is clearly a practical, relatively cheap option that appears to be environmentally safe and merits wider consideration.

A primary requirement for the successful use of barley straw is to maintain aerobic conditions and this is relatively easy to achieve in rivers by judicious positioning on inflow streams or in parts of the main channel where there is moderate current velocity. The main points to resolve concern anchorage, straw containment and dose rate. In drainage ditches where there was little or no flow, whole, 20-kg bales anchored to the bank at 20-m intervals successfully controlled *Vaucheria* (Cave, G., unpublished) but, where flows are stronger, the straw must usually be enclosed in some kind of netting. This should be well anchored but with sufficient slack to allow straw to remain submerged if water levels fall and it must be re-filled with straw or removed once the straw has decomposed to avoid any danger of entangling birds and mammals. Whole bales can be enclosed but, even with constantly high flows, the interior of the bales is likely to be poorly aerated and ineffective. Smaller quantities of straw loosely packed in netting are, therefore, preferable and are less likely to cause oxygen depletion in conditions of low flow and high temperatures.

The dose rate for rivers obviously depends on volume flow and must be to some extent a matter of trial and error. The original field trials were in a canal with an average summer flow of 5×10^3 m^3 day^{-1}, where approximately 140 kg of dry straw placed in a lock controlled *Cladophora glomerata* over a distance of at least 100 m (Welch et al., 1990; Ridge and Barrett, 1992). In static waters, an unpublished survey by P. R. F. Barrett (Aquatic Weeds Research Unit) suggests that the minimum dose rate may be as low as $3-5$ g m^{-3}, although rates of $20-50$ g m^{-3} provide a wider safety margin. In all situations in the UK, barley straw is effective only if in position at least two months before algal problems are anticipated; it remains effective only if constantly submerged in aerated water and must be replaced when fully decomposed.

OTHER FORMS OF PLANT LITTER

Based on the presumption that barley straw is unlikely to be unique in releasing anti-algal substances during decomposition, we have tested other types of plant litter. Two types of litter from trees were particularly effective: brown-rotted wood (BRW) and leaf litter.

Brown-rotted wood

Brown-rotted wood is the residue left after attack by brown-rot fungi, which degrade cellulose and other wall polysaccharides but not lignin. BRW thus comprises a lignin-enriched fraction with the lignin in a relatively undegraded form. Samples of BRW from various species of deciduous trees were tested against *Chlorella pyrenoidosa* using a three-day bioassay in buffered medium at pH 8.0 (Gibson et al., 1990). All samples inhibited algal growth by 90% or more at a dose rate of 10 kg m^{-3} (Figure 17.1) but hawthorn (*Crataegus monogyna*), birch (*Betula pubescens*) and a mixture of birch and beech (*Fagus sylvatica*) were the most effective at 2 kg m^{-3}. At this lower dose rate, however, brown-rotted elm strongly inhibited the cyanobacterium *Microcystis aeroginosa*. The algal inhibitors released from brown-rotted hawthorn were stable for at least nine days (Figure 17.2). Wood was removed from algal medium by filtering through glass fibre (GFC) immediately after autoclaving and this liquor was still inhibitory to *Chlorella* after standing for nine days at room temperature in the light. For reasons which cannot be explained, inhibitory activity was lost after only six days if the GFC-filtered liquor was filtered through a microbial filter.

In 1992, brown-rotted wood was tested in a field trial at Campbell Park Fountain, Milton Keynes, in collaboration with Dr M. Street (Milton Keynes Parks Trust) (Figure 17.3). The fountain basin (50 m in diameter) typically contains dense populations of the unicellular green alga *Scenedesmus obliquus* in summer. BRW from several species of deciduous trees

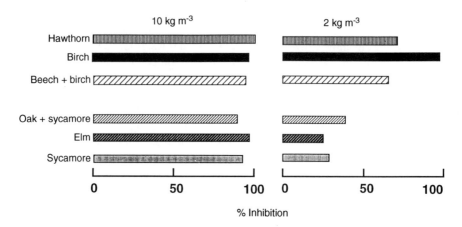

FIGURE 17.1.   Inhibition of algal growth by brown-rotted wood from deciduous trees. Wood samples were autoclaved with algal medium and tested against *Chlorella* using a 3-day bioassay based on cell counts. Results are expressed as % inhibition of *Chlorella* growth relative to controls

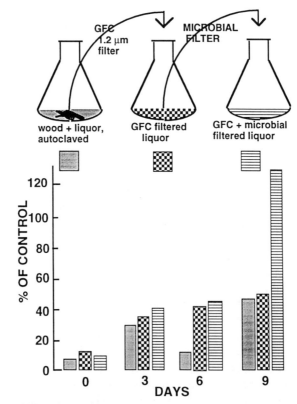

FIGURE 17.2.   The stability of anti-algal activity released from brown-rotted hawthorn. Wood was autoclaved with algal medium which was then filtered successively through GFC and a microbial filter. Medium plus wood and the filtered liquors were tested in three-day *Chlorella* bioassays (cell counts) immediately after preparation and after standing at room temperature in the light for the times shown

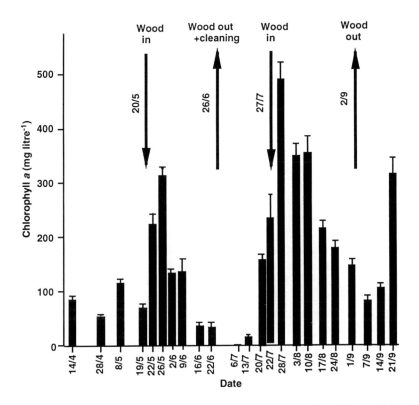

FIGURE 17.3. Algal density, measured as chlorophyll *a* concentration, during a field trial in 1992 using brown-rotted wood. Wood enclosed in sacks was placed in Campbell Park Fountain, Milton Keynes, at a rate of 50 g m^{-3} on the dates shown and removed when algal density had declined. Ten samples were removed weekly for chlorophyll *a* determination; vertical bars = SD

was enclosed in nylon mesh tubes which were placed in woven plastic sacks in the central pump chamber, providing a dose rate of approximately 50 g m^{-3}. Algae were monitored by weekly determinations of chlorophyll *a*. In the absence of a "control" fountain, wood was added and removed twice during the summer and after each addition (in May and late July) algal density declined after a lag of two to three weeks and increased after wood removal (in early July and September). The increase in algae after wood removal in early September provides the most convincing evidence that *Scenedesmus* was inhibited by brown-rotted wood. We cannot rule out the possibility that nutrient depletion contributed to algal decline in June and August, although it is unlikely to be the full explanation since summer blooms are usually sustained from June to September and densities as low as those observed from 16 June to 2 July would not be be expected. Similarly the increase in algae in July may be related to nutrient addition, since the fountain was drained and re-filled after wood removal (a matter beyond our control). Assuming, however, that some if not all of the algal decline in June and August is attributable to BRW, it is notable that a lag of two to three weeks occurred after wood addition; the relatively weak flow of water over wood in the central fountain chamber may account for this lag. Wood that had been removed from the fountain was still strongly inhibitory to *Chlorella* in laboratory bioassays. Clearly, further work is required to determine the efficacy of BRW as a source of algal inhibitors in field conditions but these preliminary results suggest that this is the case.

Leaf litter

Two kinds of leaf litter, freshly fallen leaves of oak (*Quercus robur*) and semi-decomposed conifer litter (mixed *Pinus* and *Picea*), showed strong anti-algal activity in *Chlorella* bioassays (Figure 17.4). Inhibition was greatest and most consistent when litter was autoclaved with algal medium but clear, although more variable, inhibition occurred when litter was placed in sterile media and allowed to stand overnight before inoculation. Conifer litter is highly acidic but this did not explain its anti-algal effects here since the pH of buffered media did not fall below 7.0 during bioassays. No field trials have been carried out but a simulated pool system was set up in an unheated greenhouse (temperature $4-20°C$ during the experiment) using a glass tank containing 218 litres of continuously aerated tap water. Dry, freshly fallen oak leaves (5.5 mg cm^{-3}) were placed in the tank, stirred daily and samples of both leaves and liquor (pH $6.9-7.0$) tested regularly in *Chlorella* bioassays. The leaves were dried in a desiccator under reduced pressure, chopped finely and autoclaved with algal medium (1 kg m^{-3}) before testing, whereas bioassays with the liquor were carried out after direct addition of algal nutrients and without autoclaving. Results to date (Figure 17.5) show that leaves have remained inhibitory for 90 days, with a brief period of lower activity around eight days after the start of the experiment. By contrast, tank liquor stimulated algal growth for the first four days but then became strongly inhibitory, remained so for 60 days, and showed a decrease in activity between 60 and 90 days.

The tank experiment demonstrated that algal inhibitors were released readily from oak leaf litter under conditions which approximate to those in the field. Furthermore, the inhibitors were stable in solution for several days: when liquor was removed from the tank after 58 days, filtered to remove leaf fragments and kept in the laboratory at room temperature for up to six days before addition of algal nutrients for bioassays, it showed no reduction in anti-algal activity. Oak litter inhibited *Microcystis* similarly to *Chlorella* in bioassays but has not been tested against other species. Our tentative conclusion, therefore, is that leaf litter from oak, and probably other species, may reduce algal growth in aerated pools for at least three months and this effect could extend downstream for an appreciable distance, given the stability of the inhibitors released.

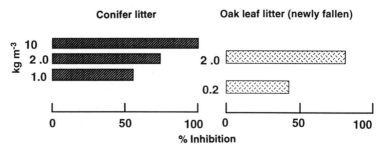

FIGURE 17.4.   Inhibition of algal growth by freshly fallen oak leaves and semi-decomposed conifer litter. Dry, coarsely chopped samples were autoclaved with algal medium and tested in three-day *Chlorella* bioassays (cell counts). Results are expressed as percent inhibition of *Chlorella* growth relative to controls. All treatments gave significant inhibition ($p < 0.001$, Student's t)

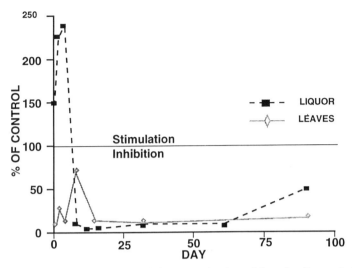

FIGURE 17.5. The anti-algal activity in oak leaves and released into the liquor during a 90-day incubation in aerated tap water. Freshly fallen, dry oak leaves were placed in a large tank of water (5.5 mg cm^{-3}) and both leaf and liquor samples were tested at intervals in *Chlorella* bioassays (cell counts). Results are expressed as percentage of control growth

THE NATURE OF ALGAL INHIBITORS

For a proper understanding of river ecosystems and to optimize the use of plant litter for algal control, it is essential to know how the inhibitors released from litter work, how they are produced and what they are. We have concentrated initially on the question of the chemical nature of algal inhibitors. The possibility that algal inhibition by rotting barley straw might be attributable to antibiotics produced by specific fungal decomposers was considered. However, the mycoflora of different samples of straw varied widely and although two species of cleistothecial ascomycetes, which were the dominant isolates from some samples, did indeed inhibit the growth of *Chlorella* on agar plates, these fungi were not detectable in all samples (Pillinger et al., 1992). It is unlikely, therefore, that the general anti-algal effects of decomposing straw can be explained by anti-algal properties of specific fungi. More likely is the release of algal inhibitors from straw itself during decomposition. Consistent with this view, algal inhibitors are also released from fresh, undecomposed straw after grinding to a fine powder, although quantitation in bioassays is hampered by the concurrent release of growth stimulants.

There are several reports in the literature of the anti-algal effects of phenolics (e.g. Dedonder and van Sumere, 1971; Hussein Ayoub and Yankov, 1985) and since all three kinds of litter which we studied release polyphenolic materials, these seemed a likely source of algal inhibitors. Barley straw, for example, is rich in cell wall phenolics and comprises up to 15% lignin. However, simple phenolics, which are active in the concentration range 10^{-3} to 10^{-5} M, cannot account for the observed anti-algal effects of straw because they are not present in sufficient quantities. Lignin is a more promising source and has been shown to solubilize from barley straw (Kivaisi et al., 1990). Preliminary studies of the liquor from brown-rotted wood also indicate that a major source of anti-algal activity is a lignin fraction which can be precipitated from aqueous solution by fivefold excess of ethanol (Pillinger,

Gilmour and Ridge, 1993). Consistent with this view, samples of white-rotted wood in which lignin had been extensively degraded showed little or no activity in *Chlorella* bioassays. Oak leaves are a well-known source of tannins and their presence was confirmed in this study using protein precipitation assays. It is possible, however, that these tannins are active against algae by virtue of their general polyphenolic structure rather than specific tannin properties (i.e. protein precipitation). Under bioassay conditions of high pH (8.0) and good aeration, phenolic hydroxyl groups may be oxidized (Pillinger, Cooper and Ridge, 1994) and authentic tannic acid does not cause protein precipitation. Such conditions were obtained not only in bioassays but also in field situations where barley straw was active, where brown-rotted wood appeared to be active (Campbell Park Fountain) and in the tank containing oak leaves, which simulated an aerated pool. Proper chemical characterization of the algal inhibitors from plant litter is still urgently needed but the isolation and characterization of single compounds is not imminent. For the forseeeable future, plant litter itself rather than any purified substance(s) must be used for algal control and the many other uses of litter in aquatic systems indicate that this is a better strategy on ecological grounds.

IMPLICATIONS AND RECOMMENDATIONS

There is no universally applicable solution to the problem of excessive algal growth in rivers. Rather, short-term amelioration measures, which must be tailored to particular situations, need to be combined with a long-term strategy involving management of whole catchments. Of the short-term measures available, use of easily obtained plant litter, particularly barley straw, is probably the cheapest and, in environmental terms, the safest option. Application of deciduous leaf litter, netted as described for barley straw, is also worth considering (not conifer litter because of its acidifying effects). The artificial application of brown-rotted wood is not recommended since this is both difficult to obtain and a valuable ecological resource in terrestrial systems. Barley straw and deciduous leaf litter can have additional beneficial effects in river systems in addition to controlling algae. Both provide a source of food and shelter for a wide range of invertebrates, and dense populations of shredders such as *Gammarus* have been observed in straw mats (Strect, 1979). There is no evidence so far for any adverse effects of barley straw on macrophytes, amphibians or invertebrates but it would certainly be wise to investigate what effects there are on sediment chemistry and community structure at sites where straw has been used for five years or more.

The most ecologically sound approach to algal control in the long term needs to combine top-down, bottom-up and "sideways" approaches: biomanipulation, nutrient reduction and enhanced litter inputs, particularly of leaves and brown-rotted wood. Past management practices have tended to minimize litter inputs, mainly by removing bankside trees and "tidying up" riparian strips. But leaf litter and coarse woody debris are valuable habitat components, with a clear association between the presence of woody debris and increased species diversity (Harmon et al., 1986; O'Connor, 1992); the possibility that they act also to restrict the growth of nuisance algae may, therefore, be regarded as a bonus. Undoubtedly, efforts to minimize nutrient inputs throughout river catchments need to be maintained but there are always likely to be accidental, point inputs. Judicious management which encourages both litter inputs and algal grazers where feasible will not only buffer river systems against excessive algal growth but will also promote richer, more diverse communities.

18

The Ecological Basis for the Management of the Natural River Environment

DAVID HARPER, COLIN SMITH,

University of Leicester, UK

PETER BARHAM

National Rivers Authority, Peterborough, UK

and RICHARD HOWELL

National Rivers Authority, Cardiff, UK

WHAT IS THE NATURAL ENVIRONMENT?

A common theme running through the legislative development of almost all countries of the world in the past two decades has been that of concern for the natural environment. The UK is probably typical, however, in that the legislation requires those who manage our rivers to "conserve the flora and fauna" and "natural beauty" without specifying exactly what this means. In practice, the objectives of modern river management are to carry out the statutory operational functions whilst maintaining and, if possible, enhancing the natural river environment. The objective of ecological advice is to define the natural environment precisely for the river manager, and indicate options for its maintenance, enhancement or restoration, i.e. ecological effectiveness combined with cost-effectiveness.

A definition of the natural environment is sometimes difficult to produce, because in rivers, as in most ecosystems close to human activities, the natural environment has been extensively modified. Rivers have been used for centuries as transport routes, water supply sources, effluent disposal units and power sources. Wholly natural rivers are rare, usually located long distances from human habitation, and often in mountainous terrain. The majority of rivers, particularly those in the lowlands, are subjected to pressures of human agriculture or urbanization and are no longer natural, although they may still have considerable ecological interest. Many are physically and chemically degraded although in most countries most degraded rivers are capable of restoration, with only a few beyond recovery (Boon, 1992).

The three most important levels of human influence are illustrated in Figure 18.1;

The Ecological Basis for River Management. Edited by D.M. Harper and A.J.D. Ferguson. © 1995 John Wiley & Sons Ltd

FIGURE 18.1.   The three states of human impact upon rivers and the consequences for management to enhance the natural environment

maintenance and enhancement of rivers that are in the middle category of human impact are particularly cost-effective. We need the "wild/pristine examples", however, to tell us what many rivers once were. Preservation of such examples, both in their own right and as models for restoration of more impacted rivers is considered more fully by Boon in Chapter 19.

It is obvious that the hydraulic regime of a river determines its geomorphology, whilst the underlying geology of the catchment and its soils determine the river's chemistry. Petts et al., Carling and Brookes have shown this in Chapter 3. The two abiotic influences combine to determine the unique combinations of animals and plants in communities of the channel, riparian zone and floodplain. Assessment and selection of pristine rivers for conservation often emphasizes these physical features, but they are no less important in semi-natural rivers with detectable human influence. We do not, however, have the time or the knowledge to base ecologically sound river management on physical features alone; yet neither have we the knowledge to use biological information exclusively, with the exception of certain key species such as the otter (see Chapter 23).

This chapter is an introduction to the practical problems of managing the natural environment, some aspects of which will be explored in greater detail in subsequent chapters. Our goal is to maintain, to enhance and to restore the ecology of river systems, as appropriate; to do this we need ecological — both physical and biological — information for sound management decisions.

THE NATURE OF ECOLOGICAL INFORMATION

There is a basic dilemma between the ecological information needed for effective management, and the ecological information that is often available. Management is usually directed at the structure of the ecosystem, but has greatest impact upon its function. Ecological information may exist about the structure, but is less likely to be available and thus more likely to be extrapolated for the function (Figure 18.2).

Understanding ecological function is more complex and expensive than description of ecological structure, so the former is usually confined to research organizations and the latter is more common in management organizations. The two are not mutually exclusive however, and more often than not an understanding of structure needs to precede an understanding of function (Figure 18.3). Coupled with this is the interpretation which can be made of structural information; not just in a comparative fashion such as in diversity indices (e.g. Hughes, 1978) and water quality (Hellawell, 1986) or conservation assessment (Wright

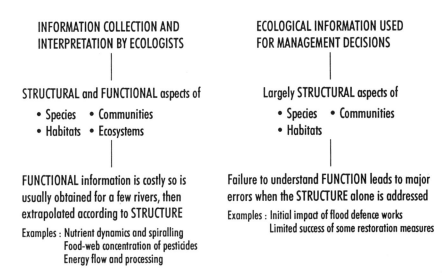

FIGURE 18.2.   The broad differences between collection and use of ecological information

et al., 1992), but also in making inferences about function, such as the state of succession (Hills et al., 1992) or of food-web structure from the presence of different feeding-guild members (Cummins and Klug, 1979).

The most important links between structure and function come from classifications of

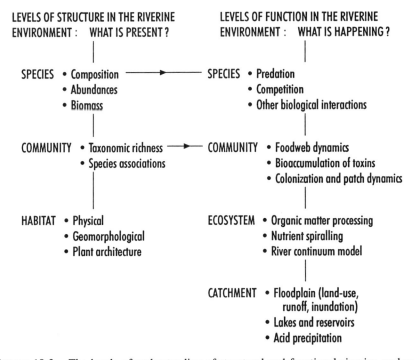

FIGURE 18.3.   The levels of understanding of structural and functional riverine ecology

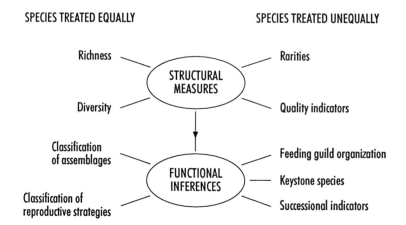

FIGURE 18.4. Examples of the use of structural information from species analyses to produce inferences about river ecosystem function

species assemblages. Their use has advanced rapidly over the past 10—15 years through the development of robust methods of multivariate analyses such as TWINSPAN and CANOCO (Figure 18.4), applied for example to plants (Ellenberg, 1984; Holmes, 1989; Rieley and Page, 1990) and invertebrates (Wright et al., 1984). These provide bases for an interpretation of community function (Cummins, 1974) and the identification of distinct habitats (Harper, Smith and Barham, 1992).

Accurate identification of distinct habitats in the aquatic and riparian environment has an intuitive value, since habitats link the impacts on the natural environment and its inhabitants (Figure 18.5). Moreover, habitats have considerable management potential for the simple reason that they can be recognized visually on the river bank whereas many species and most ecological functions cannot.

In the UK the main system of conservation recording, assessment and classification — the River Corridor Survey — was developed in the 1980s to a habitat-based recording model (Nature Conservancy Council, 1984, 1990; National Rivers Authority, 1993). Further developments since then within the NRA also use habitat-based systems for conservation

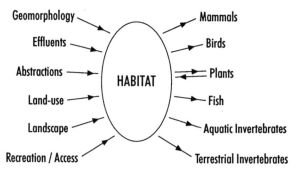

FIGURE 18.5. The concept of habitats as the centre of the natural river environment (modified after NRA, 1992)

assessment ("Post Project Appraisal" methodology and "River Habitat Survey" methodology). The basic principle underlying all these survey techniques is similar: a surveyor walking slowly along the river bank records visible features on a check sheet. RCS check sheets are supplemented by an annotated map, and most survey techniques can be extended to include detailed species-based information using standardized methodology (Figure 18.6). The information thus gained is used in a variety of ways (see Chapter 19).

The habitat-based approach is probably of greatest value in the field of physical restoration. Very few lotic restoration techniques are directed solely to species (certain programmes for mammals, such as beaver reintroduction in the Czech Republic, or for birds such as kingfishers, are exceptions); almost all of them are based upon a combination of geomorphological and habitat units (e.g. Kern, 1992; Petersen, Petersen and Lacoursière, 1992; Chapter 28 this volume), specified in a "building-block" approach to stream reconstruction.

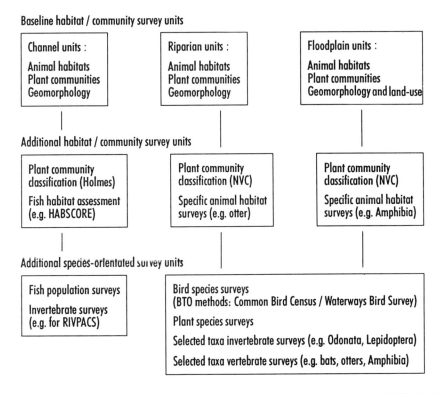

FIGURE 18.6. The layering of detail available in UK river survey techniques. NVC: National vegetation classification; BTO: British Trust for Ornithology

HOW EFFECTIVE IS THE HABITAT UNIT AS A TOOL FOR UNDERSTANDING AND MANAGING THE NATURAL ENVIRONMENT?

The effectiveness of habitat-based recording of the natural environment depends upon two factors. The first one is the ability of surveyors to consistently see and recognize habitat units on the river bank or in the channel. The second is the ecological value of the units being recorded; a list of habitats that are visually distinct is not necessarily the same as a list of habitats that are ecologically distinct. Harper, Smith and Barham (1992) called the former "potential habitats" and the latter "functional habitats". The value of such functional habitats as tools for ecological management depends upon objective demonstration of their ecological distinctiveness.

It has long been recognized that distinct habitats exist in streams, but few studies have shown the existence of a definitive list. Some studies have looked at the species abundance and richness of a single habitat (Williams, 1984, Strommer and Smock, 1989), or compared a number of habitats (Percival and Whitehead, 1929, Cummins and Lauff, 1969; Rooke, 1984; Suren, 1991) but have not started with a division of the whole. Some attempts have been made to partition the total habitat of the stream channel, to ensure that species inventories are complete in biological surveys (e.g. Brooker, 1982; see Table 18.1). Samples have then often been pooled for species identification, losing information on habitat selectivity (Ormerod and Edwards, 1987: riffle, margin and slow run). An alternative strategy has been to take samples from habitat groups (Rutt, Weatherley and Ormerod, 1989: riffles, margins) and then to interpret the results in terms of detailed habitat structure.

Barmuta (1989) studied the macroinvertebrate distribution between visually distinguishable classes of physical substrate in an Australian upland stream. Distinct community differences were found between erosional and depositional substrate types, with a large proportion of the variation accounted for by velocity, mean particle size and depth. Bournaud and Cogerino (1986) studied particulate and vegetative habitats on submerged banks of a canalized reach of the River Rhône. They concluded that the *a priori* definition of 12 potential habitats (*microhabitats prospectés*; Table 18.2) was validated by macroinverte-

TABLE 18.1.   Habitat types on the River Teifi (Brooker, 1982)

Habitat	Physical description
Riffle	High current velocity, disturbed surface
Fast run	Similar current velocity to, but deeper than riffle
Slow run	Similar to fast run, but with reduced current velocity
Pool	Discrete area between faster reaches; velocity reduced, depth variable
Slack	Shallow bankside area of much reduced current velocity, generally silty
Backwater	Area of minimal current velocity, partially isolated from channel during low flow
Tree roots	Submerged fibrous system of alder, ash, sycamore and willow in deep water
Grass roots (*Phalaris*)	Submerged fibrous systems of bankside stands
Ranunculus penicillatus	Extensive stands in regions of low current velocity, usually at margins of channel
Callitriche spp.	Extensive stands in regions of low current velocity, usually at margins of channel
Potamogeton natans	Extensive stands in regions of low current velocity, usually at margins of channel

TABLE 18.2. Habitat types on the River Rhône (Bournaud and Cogerino, 1986)

Boulders (25 – 100 cm ϕ)	Silted gravel	Branches <5 cm ϕ
Stones (3.2 – 25 cm ϕ)	Silted sand	Branches >5 cm ϕ
Gravel (0.2 – 3.2 cm ϕ)	Excavation (bare cavity under boulders, roots, etc.)	Fibrous roots
Sand (600 μm – 0.2 cm ϕ)	Roots <5 cm ϕ	Algae

brate distribution, subject to varying overlap within three wider habitat classes of erosion, sedimentation and vegetation.

Smith, Harper and Barham (1991) studied the macroinvertebrates of 42 "potential habitats" on the River Welland in eastern England, from which analysis using TWINSPAN (Hill, 1979) showed 20 "functional habitats". Some interpretation of the functional habitat list was required; for example "rocks in pool" was not included because it was an artificial feature; and sand was characterized only by absence of species, so scored only as the sole physical substrate at a site. Functional habitats were similar on a neighbouring river with different water chemistry, macrophyte and macroinvertebrate species, which indicated potential for a broadly-applicable list of functional habitats (Harper, Smith and Barham, 1992).

Broad studies such as these suggest macroinvertebrate assemblages indicate a set of functional habitats composed of substrate particle size classes and macrophytes. Macroinvertebrates should be a good taxonomic group to use for functional habitat definition because of this combined geomorphological/botanical dependence, but also because they have sufficient taxonomic diversity to test their robustness in such a definition. Habitat definition for macroinvertebrates involves combinations of substrate and plant types.

Habitats based upon substrate particle size

Particle size is probably the physical habitat variable for which most data are available: Leland, Carter and Fend (1986) found information for each of 21 common taxa in a Californian stream. The benthic fauna differs between substrates of dissimilar particle size (e.g. Doeg et al., 1989; Smith, Harper and Barham, 1991). Differentiation of linear or nonlinear community responses to particle size has been difficult in many studies, due to *a priori* definition of substrate size categories. Some cases have suggested that a series of discrete benthic community types exist, in terms of associated substratum particle size (Thorup, 1966; Reice, 1974). For the most part, however, a gradual change in species composition has been shown with the transition from fine to coarse sediment (Rabeni and Gibbs, 1980; Sheldon and Haick, 1981; Barmuta, 1989). Discrete communities may not exist in relation to particle size *per se*, but the latter is discontinuously variant in the stream channel. Transitions between riffle and pool regimes of substrate are often spatially abrupt, even though depth and flow rate can be normally distributed (Singh and Broeren, 1989). This habitat patchiness might produce community patchiness even for monotonous variation of species with substratum.

Some taxa are strongly associated with cobble substrata in streams (e.g. crayfish: Capelli and Magnuson, 1983; Miller, 1985; Elminthidae: Brown, 1987) and the high abundance of invertebrates on riffle substrata (typically cobble-based) has been established for a long time

(Wene and Wickliff, 1940; Pennak and Van Gerpen, 1947). In most cases the highest species richness is also associated with coarse sediment (Pennak, 1971; Cummins, 1975; Hart, 1978; Gore and Judy, 1981). Williams and Mundie (1978) looked at macroinvertebrate utilization of artificial gravel beds, with 11.5 mm, 24.2 mm and 40.8 mm diameter. They found maximum abundance in 24.2 mm gravel, while diversity was greatest in the largest substratum. Beds dominated by large particles generally include a range of finer sediment and organic matter which encourages both abundance and diversity of species (Hynes, 1970). Williams (1980) observed such a result with experimentally manipulated substrata, including a heterogeneous substratum with an upper layer of coarse material. These are the conditions which occur in established riffles, through the process of armouring (Jain, 1990).

Smith, Harper and Barham (1991) found that gravel (of *c*. $0.5-2.0$ cm ϕ) was a distinct macroinvertebrate habitat on the Rivers Welland and Wissey in eastern England. They also found differences between gravel at the head and tail of riffles, which were not explicable solely in terms of drift. Gravels are important as fish spawning sites, notably of salmonids but also of some key coarse fish such as chub *Leuciscus cephalus* (Wheeler, 1978).

The "hyporheic zone" of interstitial spaces in cobble- or gravel-based stream beds is an important habitat for invertebrates (Stanford and Gaufin, 1974). Waringer (1987) found Trichoptera larvae down to 1 m in a gravel bed, with maximum numbers of early-instar *Sericostoma* at $20-60$ cm. The habitat value of the hyporheic is generally reduced by large amounts of fine sediment (Nuttall, 1972; Boles, 1981) although organic matter has been found to be beneficial (Williams and Mundie, 1978; Milner et al., 1981). Distinct communities can be found in the hyporheic zone (e.g. of Limnohalacaridae: Husmann and Teschner, 1970), especially when the stream geomorphology produces zones of upwelling and downwelling (Dole-Olivier and Marmonier, 1992).

Sand is usually poor as a habitat in terms of both abundance and diversity for most invertebrates larger than 1 mm (Pennak, 1971; Bournaud and Cogerino, 1986; Smith et al., 1990). Wagner (1984) modified a portion of stream bed to make it homogeneous sand: the numbers of most taxa declined although some (e.g. Ptychopteridae, *Centroptilum luteolum*) became more abundant. The specialized meiobenthos of mostly smaller animals may be very abundant, though still species-poor (Whitman and Clark, 1984; Soluk, 1985) and not extending to such depths as in the gravel hyporheic environment ($15-30$ cm: Strommer and Smock, 1989). Sand is usually the least stable of riverine sediments on the time-scale of macroinvertebrate life cycles (Peeters and Tachet, 1989) but deposits associated with flow obstructions such as woody debris can accumulate organic matter (Newbold et al., 1981). They then support richer invertebrate communities (Anderson and Day, 1986) and may become vegetated.

Brown and Brussock (1991) found that riffle macroinvertebrates of the Illinois River were more species-rich and abundant than those of bedrock-dominated pools. Gore (1985) also stated that pools do not provide large amounts of suitable substratum for macroinvertebrates. This contrasts with the silt of pools studied by McCulloch (1986) and Smith et al. (1990), which held comparable or greater species richness and biomass to riffles. The main difference is probably between lowland and upland streams; and the key habitat feature is detritus-rich silt, stable for much of the summer, rather than pools *per se*.

Pools with a bedrock substrate support few individuals of few macroinvertebrate species (Logan and Brooker, 1983; Brown and Brussock, 1991). Boulders or bare rock in flowing water, however, provide an important habitat for filter-feeding species (Freeman and

Wallace, 1984; Huryn and Wallace, 1988). Smith-Cuffney and Wallace (1987) found that production of *Parapsyche cardis* was higher on bare rock than in pebble riffles, with drift items in the range of caddis catchnets four to ten times as abundant on the bare rock. Boulders increase the surface area available for epibenthic species, especially if the surface of the rock is pitted. Chironomids such as *Corynoneura* and *Thienemanniella* are often found in rock fissures (Cranston, 1982; Cranston, Oliver and Sæther, 1983).

Thin films of water on bare rock (e.g. seepages and beside waterfalls) are a specialized habitat of smaller macroinvertebrates, the "hygropetric zone" (Vaillant, 1953, 1954). Harpacticoid and cyclopoid copepods (Gurney, 1932, 1933; Harding and Smith, 1960), psychomyiid caddis larvae (Alderson, 1969; Jenkins, 1977) and Diptera larvae such as Thaumaleidae (Smith, 1989) and Chironomidae (Cranston, 1984) are typical inhabitants.

Habitats based upon aquatic plants

Plants act as habitat features for stream macroinvertebrates in several main ways:

- The living tissue is a food resource for species which shred, mine or pierce the plant. Some invertebrates also use leaf or stem segments as case material.
- Macrophytes provide an extension of the physical substrate; and a large surface for periphyton, grazed by many invertebrates.
- Instream plant litter is a food resource for detritivores, with a similar process of decomposition to allochthonous material.
- Both aerial and submerged portions are used as sites for oviposition; and some plants provide a route for emergence of insects.
- Macrophytes can provide a refuge from predation and adverse flow conditions.

Some invertebrates also obtain oxygen from roots, in otherwise anoxic sediment.

There are often statistical correlations between the species richness of macrophytes and macroinvertebrates (Jackson et al., 1979; Palmer, 1981; Ormerod, Wade and Gee, 1987) but they need not indicate causal relationships (e.g. Friday, 1987, showed the importance of pH to both). Stronger evidence has been obtained in studies of selected taxa. For example, Cuppen (1983) found that two of three *Hygrotus* spp. (Dytiscidae) were more abundant in macrophyte-rich waters; and Jeppesen et al. (1984: cited by Sand-Jensen et al., 1989) found that areal densities of simuliids and chironomids increased several-fold in the presence of macrophytes. A seasonal correlation between abundances of *Potamogeton pectinatus* and four invertebrate species was observed by Bergey et al. (1992). They showed definite ecological relationships between the invertebrates and the phenology (growth, canopy and senescence phases) of the macrophyte.

Direct consumers of living plant material are usually a small proportion of the macroinvertebrate community, with most use of senescent plants (Soszka, 1975). Dvorak and Best (1982) found direct consumers formed 0.6% of invertebrate abundance in Lake Vechten; and that together with miners and filterers, consumption was 0.03% of daily primary production. In streams, the use of living plants as food may be even less, and restricted to lentic habitats (e.g. Trichoptera: Elliott, 1969; Mackay and Wiggins, 1979; and Ephydridae: Berg, 1950). Mining chironomids (particularly Chironominae) are most abundant on emergent plants (Dvorak and Best, 1982), although they are usually filter feeders rather than direct consumers of the host plant (Walshe, 1948). A specific community of nematodes is found in the roots/rhizomes of aquatic plants (Prejs, 1977) and there are individual species associations (Prejs, 1986). The larvae of many Hydroptilidae pierce plants (filamentous algae) for food (Wallace, Wallace and Philipson, 1990) but this mode of feeding is uncommon (Rooke, 1986). Plant segments are used for case material by many Trichoptera (Wallace, Wallace and Philipson, 1990) and some Lepidoptera (Hasenfuss, 1960: cited by Verdonschot, 1992).

Vegetation is important to periphyton grazers such as Naididae (Learner, Lockhead and Hughes, 1978, Bowker, Wareham and Learner, 1985), chironomids (Tokeshi, 1986), chydorid Cladocera (Fairchild, 1981) and gastropods (Lodge, 1985). Diatoms of the periphyton are also needed as case material for some chironomids (Fairchild, 1981). The proportions of periphyton types (e.g. filamentous green algae and diatoms) can differ between macrophyte species, producing a diverse environment for selective grazers (Lodge, 1986). Macroinvertebrates make much more use of the periphyton than of the macrophytes. Kairesalo and Koskimies (1987) found consumption by oligochaetes and gastropods was $22-45\%$ of daily periphyton production (cf. 0.03% of macrophyte production: Dvorak and Best, 1982). Cattaneo and Kalff (1980) estimated that epiphyte production was almost as much as that of the macrophytes, which makes grazing of epiphytes an important link between primary producers and the animal community (Cattaneo, 1983).

The seasonal abundance of smooth substrate which some macrophytes provide is of benefit to macroinvertebrates such as some leeches and gastropods. Lodge (1985) studied the distribution of 13 gastropods and 10 macrophytes: he proposed that restriction of *Acroloxus lacustris* to *Nymphaea alba* and emergent species was due to its need for a broad substrate for attachment and locomotion. Rooke (1984) found no community difference between *Potamogeton amplifolius* (broad leaves, low habit) and the stone substrate, suggesting that it was used as an "extension" of the stream bed.

Dytiscidae use a range of macrophyte species for oviposition, either on the surface of shoots and roots (*Agabus*: Jackson, 1958) or within the shoots (*Ilybius*: Jackson, 1960), and many Odonata lay their eggs on or within macrophytes, with varying degrees of specificity (Corbet, 1980). Oviposition was suggested as a particular value of mosses in faster water by Glime and Clemons (1972). Plants provide a passage to the water surface for emerging insects (McGaha, 1952; Gaevskaya, 1966: cited by Rooke, 1984) and Rooke (1984) found that plants supported a higher proportion of species with aerial life stages than did stones.

The intricate structure of some submerged macrophytes (particularly mosses) can provide a refuge from predation and flow (Malmqvist and Sjöström, 1984; Wellborn and Robinson, 1987). Emergent vegetation is chosen as shelter by some *Gerris* species in response to wind or wave action (Spence and Scudder, 1980; Spence, 1981). Macrophytes can also be an important predation refuge for young fish (Hart, P.J.B., pers. comm.). The larvae and pupae of some Diptera insert their spiracles into the roots of aquatic plants for respiration in anoxic sediments (Keilin, 1944; Houlihan, 1969).

Most of the macroinvertebrates associated with aquatic macrophytes are found across a variety of species (Dvorak and Best, 1982), though with some degree of preference (Harrod, 1964). Broad but incomplete habitat tolerance has also been shown by Rooke (1984, 1986a), Iversen et al. (1985) and Engel (1988). Ecological affinities do not always reflect systematics: *Ranunculus penicillatus*, *Potamogeton pectinatus* and *Zannichellia palustris* seem more similar as habitats for invertebrates than *Potamogeton pectinatus*, *P. perfoliatus* and *P. natans*. Wright et al. (1992) discussed the value of macrophyte growth forms (habits) for effective study and for their relevance to issues of stream management. The use of

TABLE 18.3.   Taxon richness on plants and substrata (River Welland; Smith, Harper and Barham, 1991)

Cladophora sp.	49	*Potamogeton lucens*	35	*Schoenoplectus lacustris*	45
Enteromorpha sp.	39	*P. perfoliatus*	38	*Glyceria maxima*	52
		P. pectinatus	39	*Sparganium erectum*	31
Potamogeton natans	48	*Ranunculus penticillatus*	45	*Carex acutiformis*	43
Nuphar lutea	44	*Myriophyllum spicatum*	36		
Nymphaea alba	24	*Elodea canadensis*	64	Riffle substrate	68
				Gravel	46
Fontinalis antipyretica	71	*Agrostis stolonifera*	54	Sand	31
		Rorippa amphibia	49	Silt	49
		Phalaris arundinacea	63		

Submerged macrophyte samples did not include roots/rhizomes and underlying substrate

ecological, rather than taxonomic, plant categories is needed to study pattern between streams with dissimilar macrophyte communities.

Most aquatic macrophytes are readily categorized according to their habit; as emergent, submerged or floating-leaved. Many species (e.g. *Sparganium emersum, Butomus umbellatus, Oenanthe fluviatilis*) have leaves which are either submerged or emergent — *Sagittaria sagittifolia* can have linear submerged leaves, long-petiolate floating leaves and sagittate emergent leaves — but one habit usually dominates. Submerged species have been further categorized according to their topology; into those with broad leaves and those with fine or dissected leaves. Marginal herbs have various architectures, mostly different from those of emergent monocotyledons and they are also associated with a distinct set of depth and flow conditions. Mosses and macroalgae typically have growth forms distinct from angiosperms. There are thus seven categories, at least, which may be expected to serve as functional habitats:

- emergent species (e.g. *Sparganium erectum, Glyceria maxima*)
- floating-leaved species (e.g. *Potamogeton natans*, floating leaves of *Nuphar lutea*)
- submerged species with broad leaves (e.g. *Potamogeton perfoliatus*, submerged leaves of *Nuphar lutea*)
- submerged species with fine or dissected leaves (e.g. *Potamogeton pectinatus, Myriophyllum spicatum*)
- mosses (e.g. *Fontinalis antipyretica, Rhynchostegium riparoides*)
- macroalgae (e.g. *Cladophora glomerata, Enteromorpha intestinalis*)
- marginal species (e.g. *Rorippa* spp., *Phalaris arundinacea*)

Wright et al. (1992) found greater invertebrate family richness on emergent plants than on submerged and floating-leaved plants, which in turn were richer than the substrate, over a large number of British rivers. Their results were based on macrophyte samples which included the underlying substrate, on the premise that its habitat characteristics are modified by the plant. Data from Smith, Harper and Barham (1991; see Table 18.3) show that for one river at least, the invertebrate richness associated with macrophytes (not including rootstock/ substrate) was usually about equal to that of the substrate, except for sand. Communities of silt with or without macrophyte rootstock were quite similar; the greatest qualitative contribution of macrophytes occurred in the water column (Figure 18.7). Categorization of plants according to growth form was supported by their results: although invertebrate taxa were mostly found on several macrophytes, a number were restricted to each of submerged, emergent and floating-leaved categories (Figure 18.7). Habit-based macrophyte groups have also been used in more detailed studies of invertebrates. For example, Cuppen (1983) found that *Hygrotus decoratus* and *H. versicolor* were most strongly associated with emergent and submerged macrophytes respectively.

Krecker (1939) suggested that plants with dissected leaves consistently support more invertebrates than those with broad leaves. "Fine-leaved" plants might provide more surface area for growth of periphyton (Dvorak and Best, 1982; Lodge, 1986) and attachment of invertebrates (Lodge, 1985), capture more fine particulate matter from the flow (Gerking, 1957; Rooke, 1984), and offer more protection from predation or turbulence (Malmqvist and Sjöström, 1984). Experimental evidence (reviewed by Cyr and Downing, 1988) is equivocal; many investigations have supported Krecker's hypothesis but some, including Cyr and Downing themselves, found no systematic benefit of fine-leaved species. Data from Smith, Harper and Barham (1991) show some macroinvertebrate species restricted to each of fine- and broad-leaved submerged macrophytes (Figure 18.7). A large proportion of the 26 restricted species were uncommon but among the taxa found only on broad-leaved plants were gastropods (*Planorbis carinatus, P. planorbis, P. contortus, P. albus*), flatworms (*Polycelis* sp., *Dugesia lugubris, Dendrocoelum lacteum*) and leeches (*Helobdella stagnalis, Erpobdella octoculata*), all of which might be expected to prefer such a surface.

Mosses can support high invertebrate densities (Maurer and Brusven, 1983; McElhorne and Davies, 1983; Brusven, Meeham and Biggam, 1990) and species richness (Egglishaw, 1969; Thorup and Lindegaard, 1977). Suren (1991) demonstrated that in upland New Zealand streams the species associated with bryophytes and gravel were different, and Schwank (1984) also found highly specialized communities of smaller invertebrates such as nematodes and rotifers. The complex structure of a moss cushion is a refuge from predation and flow for small species and immature stages (Malmqvist and Sjöström, 1984). Fine sediment and organic matter accumulate in mosses (even in strong flows) providing physical substrate and a food resource.

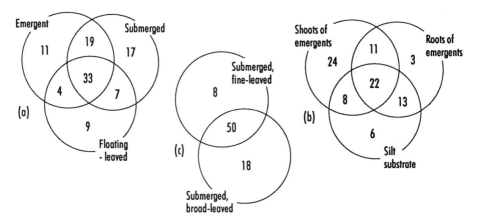

FIGURE 18.7.   Taxon richness and distinctiveness of habitats on the River Welland (Smith, Harper and Barham, 1991)

Some herbivores, such as *Nemoura*, are reported to feed on both mosses (Hynes, 1970) and associated detritus/periphyton (Frost, 1942) but most can be expected to use the periphyton. Glime and Clemons (1972) found fewer species and individuals on a plastic imitation of *Fontinalis*. Mosses (with liverworts) also retain water and provide mechanical structure to the hygropetric habitat.

The macroinvertebrate assemblage of *Fontinalis* in riffles on the River Welland was distinct from that of other macrophytes or the substrate (Smith, Harper and Barham, 1991); and more species-rich (49, 53) than *Ranunculus penicillatus* (33, 38) or *Potamogeton pectinatus* (22, 35) in similar flow conditions (run, riffle). Other bryophyte species are particularly important in upland or colder streams such as the River Tees (Holmes and Whitton, 1981) where they grow throughout the year in contrast to most other aquatic plants (Kelly and Whitton, 1987).

Macroalgae such as *Cladophora* and *Enteromorpha* occur naturally in streams, but their overgrowth is the most visually obvious consequence of the eutrophication of lowland rivers. It may also be the most ecologically important: by shading of substrates and other macrophytes; by alteration of the physical environment; and by quantitative modification of trophic relationships in the stream. *C. glomerata* was associated with a high abundance of limited species (pool, *Lymnaea pereger*; run, *Ephemerella ignita* and Orthocladiinae spp.) on the River Welland (Smith, Harper and Barham, 1991). Macroalgae may provide refuge from predators (Dudley, Cooper and Hemphill, 1986; Holomuzki and Short, 1988) and are an oviposition site for *Ilybius* (Balfour-Browne, 1950). There are some reports of macroalgae as a food resource for invertebrates (Gray and Ward, 1979; Behmer and Hawkins, 1986) and fish (Greger and Deacon, 1988). Feminella and Resh (1991) found that selective grazing on *Cladophora* by the caddis *Gumaga* had a significant effect on algal succession in a Californian stream. Generally, however, *Cladophora* is not an important food item (Patrick et al., 1983). *Cladophora* is a substrate for epiphytes, which take advantage of its low mucilage production (Chapman, 1964: cited by Learner, Lochhead and Hughes, 1978), and are subsequently grazed by macroinvertebrates (Dodds, 1991).

River margins are less thoroughly studied than the wholly aquatic environment, but may be the first areas to recover habitat complexity in managed channels. Shallow areas may be selected by small fish for food (Bardonnet, Gaudin and Persat, 1991) or as a refuge from predation (Harvey and Stewart, 1991). Schiemer and Spindler (1989) found that shallow margins on the Danube supported more fish fry than neighbouring revetted sections. Semi-aquatic macrophytes contribute to the physical richness of marginal areas for invertebrates, through a variety of habit and position in relation to the water level. Dvorak (1970) found that a marginal stand of vegetation supported a community ranging from semi-terrestrial gastropods to aquatic Heteroptera, varying with distance from the shoreline in a pond. Smith, Harper and Barham, (1991) found a large number of macroinvertebrate species in samples from lotic marginal macrophytes (Table 18.3: *Agrostis*, *Rorippa*, *Phalaris*). Their analysis of the data suggested that marginal plant species may form more than a single functional habitat.

Terrestrial invertebrates utilize the aerial parts of marginal and emergent vegetation in much the same way as they do herbaceous riparian vegetation. In many cases, the mature stages of otherwise aquatic invertebrates also use this vegetation zone for feeding, mating and oviposition. Reed beds have a distinct conservation interest for both birds and invertebrates (Ditlhogo et al., 1992). Some emergent species provide habitats in their own right: *Phragmites australis* hosts a unique faunal and floral community (Skuhrávy, 1978);

and the specific fauna of *Cyperus papyrus* changes with the stage of development of the umbel (Thornton, 1957).

Studies of the marginal emergent vegetation of Czech fish ponds have shown distinct faunal changes corresponding with vegetation changes (*Glyceria, Sparganium, Carex*) in the transition to dry land (Dvorák, 1978). This kind of evidence suggests that the identification of distinct vegetation communities can implicitly give coverage of invertebrate communities. Murdoch (1963) has shown that the carabid beetles inhabiting marshy areas divide into two species groups: those inhabiting vegetated areas which leave a litter layer overwinter (the plant species seems unimportant); and those inhabiting open ground.

Much of the information about the role of marginal and emergent vegetation as invertebrate habitat is vague. The feeding, territoriality, mating and oviposition of many dragonflies depends upon this vegetation (Hammond, 1983) but usually is not species-specific. For example, detailed studies of the habitat use of ovipositing *Leucorrhina intacta* recorded "shallow water over submerged vegetation" and "emergent vegetation" as preferred sites for oviposition (Wolf and Waltz, 1988). Erman (1983) came to similar conclusions about other aquatic insects at different stages of their life cycle — the habitat requirements are for shallow edges and variety in plant structure. The larvae of aquatic insect species sometimes, as an exception, had more precise nocturnal feeding requirements.

Many species utilize separate habitats for life stages, some including quite specific terrestrial requirements: the water-lily beetle *Galerucella nymphaeae* develops entirely on the upper surface of water-lily leaves, but the adult overwinters under the bark or litter of pine trees (Kouki, 1991). Combinations of habitats may also be important; some damselflies select only emergent macrophytes adjacent to fast-flowing water as sites for oviposition (Gibbons and Pain, 1992).

Aquatic habitats based upon riparian terrestrial vegetation

Riparian trees (e.g. *Salix, Alnus, Acer*) or dense growth of other vegetation (e.g. *Phalaris, Carex*) can produce a matrix of exposed roots, especially where the toe of the bank is scoured. Tree roots can be an important habitat for specialized species, such as some Trichoptera (Jenkins and Cooke, 1978; Wallace, Wallace and Philipson, 1990) and Ephemeroptera (Jenkins, 1975). Jenkins, Wade and Pugh (1984) suggested that some apparently rare species may be more common, but unsampled, among tree roots. Although Rhodes and Hubert (1991) did not find qualitative differences in the fauna between undercut banks and mid-stream habitats, the former supported a fivefold greater abundance.

Aquatic macroinvertebrates are often significantly associated with leaf litter with evidence for individual species (*Eisenia spelaea*: Omodeo, 1984) as well as species groups (Ephemeroptera: Hearnden and Pearson, 1991) and communities (Egglishaw, 1964, 1969; Arunachalam et al., 1991). Ingestion of litter by benthic animals was established by early work (Slack, 1936; Jones, 1950) yet until recently, its distribution and functions had received little attention (Macan, 1961, 1962). Interest in litter and its role in the stream "economy" began in the late 1960s (e.g. Kaushik and Hynes, 1968; 1971). There are three main potential values of litter for aquatic macroinvertebrates:

- direct food resource for the "shredder" feeding guild (*sensu* Cummins, 1973; Cummins and Klug, 1979);
- indirect food resource, as a site for production (via micro-heterotrophs) and capture of fine particulate organic matter (FPOM);

● physical substrate, increasing the available surface area, especially when leaf packs accumulate; and introducing large-scale structure to fine sediment.

Egglishaw (1964) found that the distribution of many riffle macroinvertebrates was influenced by litter abundance, and that similar results could not be obtained using artificial (rubber) leaves. Richardson (1992) also found that shredders were abundant on *Alnus* leaf packs but absent on artificial (polyester cloth) packs. Differences in litter breakdown rates in fine and coarse mesh bags were attributed to shredders by Rounick and Winterbourn (1983), although shredders are not always important to litter processing (Matthews and Kowalc zewski, 1969; Reice, 1978), especially in its later stages (Kaushik and Hynes, 1971). Gut analyses have confirmed coarse detritus as a frequently important dietary item of benthic species (Minshall, 1967, Coffman, Cummins and Wuycheck, 1971). There is also indirect evidence for the importance of litter as a food resource: *Gammarus pulex* became food-limited in the summer months in a Cotswold stream, ending with the leaf fall (Gee, 1988).

The abundance of collectors in leaf packs was related to FPOM by Short, Canton and Ward (1980), and differences in non-shredder abundance between natural and artificial leaves were accounted for by variation of trapped FPOM (Richardson and Neill, 1991; Richardson, 1992). The fine matter created *in situ* by processing of leaf litter may be of higher food value than general stream FPOM (Ward and Cummins, 1979), promoting the value of litter as a habitat for collectors. "Conditioning" of leaves by decomposers also increases the value of litter to shredders (Bärlocher and Kendrick, 1973; Cummins, 1974; Webster and Benfield, 1986). Leaf litter species vary in their complement of fungal and microbial decomposers. Readily-decomposed species may support the most macroinverte-brates (deciduous species more than coniferous; Short, Canton and Ward, 1980), while both abundance and diversity increase with the progress of conditioning (Dudgeon, 1982). Mackay and Kalff (1973) found caddis (*Pycnopsyche*) fed preferentially on leaf species that decayed quickly, especially those attacked by fungi.

It is intuitively clear that litter could act as a physical habitat feature and this has been shown in still water (Street and Titmus, 1982), but experimental evidence in streams is hard to obtain. Litter is the case material of many caddis larvae (e.g. Limnephilidae: Mackay and Kalff, 1973), especially in later instars (Wallace, Wallace and Philipson, 1990), but the minimum tolerable availability of case material has not been studied. Absence of shredders from artificial leaf litter (Egglishaw, 1964; Richardson, 1992) is strong evidence for the role of litter as a food resource, but does not disprove the value of litter as a physical substrate *per se*. Without food, the animals are unlikely to be found in an otherwise favourable environment.

Rounick and Winterbourn (1983) suggested that the retention of leaf litter was important in New Zealand streams, where poor shredder communities could be found despite input of litter. Riffles and backwaters were more efficient than pools and chutes in litter retention in South African streams, and supported highest shredder densities (Prochazka, Stewart and Davies, 1991). Speaker, Moore and Gregory, (1984) noted that accumulation in riffles (by cobbles and debris) is more permanent than accumulation in pools, due to scouring of pools during floods. Coarse woody debris is another important focus for litter retention (Bilby and Likens, 1980; Speaker, Moore and Gregory, 1984), especially in smaller streams. Bilby (1981) found removal of debris dams from a second-order stream produced a fivefold increase in the export of organic matter. The flexibility of leaves affects their retention by debris and coarse bed-material (Young, Kovalak and Del Signore, 1978), compounding the inter-specific differences in litter food value (Herbst, 1980; Dudgeon, 1982). Leaves entering the stream before senescence may require prolonged retention: Stout, Taft and Merritt, (1985) found a 26-day lag between immersion and breakdown of fresh *Alnus* leaves, which are otherwise most quickly processed (Anderson and Grafius, 1975; Sedell, Triska and Triska, 1975). The leaf-fall of deciduous trees is often followed by winter floods; and in coastal (or otherwise short) streams, the brief retention time may not permit leaf processing (Malicky, 1990). Buried leaves may be a temporary store of organic matter because of their slower decomposition (Herbst, 1980); they may also store nutrients during winter, as shown in marshes (Brinson, 1977; Morris and Lajtha, 1986). Macrophytes provide an additional, instream source of organic matter (Westlake, 1975; Fisher and Carpenter, 1976) which can function similarly to allochthonous litter. Macrophyte "litter" is rapidly decomposed and in unshaded streams (thus complementary to riparian sources) can contribute a large proportion of productivity (Anderson and Sedell, 1979). There has been comparatively little research on the role of senescent macrophytes, even though they may be the major source of litter in streams with managed corridors.

Coarse woody debris plays a major role in the geomorphology of pristine streams (Sedell and Froggatt, 1984; Triska, 1984). Debris is generally less abundant and more localized in large than in small streams (Keller and Swanson, 1979; Wallace and Benke, 1984; cf. Keller and Tally, 1979; Robison and Beschta, 1990) but rivers larger than any in Britain were structured on the scale of 100 km / 100 years by accumulation and break-up or debris dams (Triska, 1984). The first and most consistent steps in historical river management have been removal of debris and riparian deforestation. The full realization of debris-driven processes is now limited to smaller streams in old-growth forests (Grier and Logan, 1977; Robison and Beschta, 1990), but stream hydrology and geomorphology can be influenced by debris of lesser abundance (Gregory, 1992). The hydrograph is smoothed for light and moderate flood events in the presence of debris dams (Gregory, Gurnell and Hill, 1985). Accumulations of debris may be a cause of local scour but Gregory (1992) found that removal of debris increased overall sediment transport and erosion.

Debris has received considerable attention as an ecological channel feature, especially in North America, where most of the information was obtained for an extensive review by Harmon et al. (1986). Benke et al. (1984) showed that macroinvertebrate biomass and production were higher on debris (snags) than in benthic habitats, for a south-eastern USA river. Many of the invertebrate species studied by O'Connor (1991) in an Australian stream were restricted to debris samples. Some species, usually Diptera larvae, exploit debris directly as borers (Dudley and Anderson, 1982, 1987; Anderson, Steedman and Dudley, 1984). Accumulations of debris influence retention of leaf litter (Speaker, Moore and Gregory, 1984) which is an important food and habitat resource for benthic invertebrates (Egglishaw, 1964; Cummins et al., 1973; Prochazka, Stewart and Davies, 1991). Bass (1986) found that species richness of Chironomidae, though not abundance, was higher on debris or leaf litter than on the underlying sand.

Habitats based upon exposed substrata

Several studies have shown that bare substrata, disturbed by winter floods and colonized by a range of plant communities from pioneer to early scrub successions, have distinct invertebrate assemblages. Plachter (1986) found 48 species of Carabidae (Coleoptera) in four distinct riverine habitats in Germany:

- wet, plant-free areas of gravel bars near to the water's edge;
- dryer, but also plant-free centre of the bars, farther away from the edge;
- isolated stands of vegetation surrounded by gravel;
- fine sand or sand-clay slopes at the outer edge of bars covered with grasses and less frequently inundated.

Fowles (1990) found habitat partitioning of carabid species with particle size on shingle banks in Wales. The importance of the habitat mosaic of shingle banks is not confined to beetles; evidence also exists regarding wandering spiders (Uetz, 1977) and ladybirds (Majerus and Fowles, 1989), many species of which seem to be national rarities in the UK. Some may be on the edge of their range and confined to this one habitat because it is ephemeral and hazardous, so reducing competition. Anderson (1969, 1983), in Norway, classified carabid riparian habitats in more detail, in terms of particle size (including vertical heterogeneity), degree of moisture and plant influence.

A wider riparian habitat classification was developed by Lott (1992) from a preliminary study of terrestrial beetles (mostly Carabidae and Staphylinidae) on the floodplain of the River Soar. There was some overlap in species composition between all sites, but the four main habitat types were: "pioneer" sand and shingle banks disturbed by floods; more stable sand and silt bars; undisturbed meander cut-offs with a well-developed litter layer; and permanently waterlogged sites. At least the first three habitats represent stages in a temporal succession, as well as being spatially distinct.

Some indication exists in the literature that more steeply-inclined bare sites have their faunal importance. A survey of bees in Cheshire (Whitfield and Cameron, 1988) suggested that species richness along the River Dean (26 solitary species, nine social and parasitic species) was related to the suitability of bank soil for nesting, and to several nectar sources such as riparian shrubs. The animal communities of steep solid substrates (rocks and boulders) are hardly studied, but these are frequently rich in bryophyte and fern communities (Rieley and Page, 1990) and in the riparian zone may often be in damp conditions; a specialized moss fauna of microinvertebrates such as tardigrades occur in such circumstances.

DISCUSSION

In the UK, ecological information based upon functional habitats forms the firm framework for river survey techniques concerned with the physical integrity of the natural environment (river conservation). Most river conservation effort has been directed at maintaining the state of channel and riparian habitat in the face of land drainage works, but over the past few years serious concern has been directed at problems of low flows due to groundwater depletion and flood-water regulation. In Chapter 1, Petts et al., demonstrate the integration of species-based investigations (PHABSIM) with biotic indices to quantify ecological effects of low flows and to set ecological targets for flow restoration. Mantle and Mantle (1992) described the broad ecological effects of low flows on chalk streams in southern England; it is clear that most of the changes that have occurred in the wetted channel could also be described in terms of functional habitat shifts.

A comparison between the wholly submerged and the winter flooded/summer exposed substrata indicates the state of our present ecological knowledge of habitats and the use to which this can be put. Instream habitats have been well studied and can be both clearly

TABLE 18.4.   Summary list of instream functional habitats

Cobbles (more than 64 mm ϕ)	Dominant substratum in some high-energy streams, or elsewhere in riffles
Gravel	Dominance with above, and where cobbles have been removed (lowland)
Sand (less than 2 mm ϕ)	Point bars, patches in riffle−pool transition, or dominant in some streams
Silt	Deposited in pools, slacks, margins or off main channel
Macrophytes, emergent	Significant aerial portion, e.g. *Sparganium* (usually grasses, rushes, reeds)
Floating-leaved	Leaves lying on water surface, e.g. *Nuphar* and some *Potamogeton* species
Submerged, broad-leaved	Include strap-like leaves of for example, *Butomus* and *Sparganium emersum*
Submerged, fine-leaved	Include fine leaves (e.g. *Zannichellia*) or dissected leaves (e.g. *Ranunculus*)
Mosses	Aquatic types, e.g. *Fontinalis*, *Rhynchostegium*
Macroalgae	"Cott", usually *Cladophora* and *Enteromorpha* on lowland rivers
Marginal plants	Rooted around (e.g. *Phalaris*) or below (e.g. *Rorippa*) normal water level
Leaf litter	Deposited in pools, slacks, margins or as "leaf packs" in riffles
Woody debris	Fallen trees, logs, substantial branches and driftwood
Tree roots	Fine exposed roots or the fibrous clumps of, for example, *Alnus*, *Salix*, *Acer*
Exposed rock	Used instream by some filterers; and in wet places (hygropetric zone)

defined and relatively easily recognized (Table 18.4). By contrast, riparian flooded substrates are known to have important habitat mosaics but have to date been inadequately studied to produce a definitive functional list; rather we can produce lists of important habitat types, such as shingle banks, vegetation layers, etc. (Table 18.5), together with habitat modifiers such as plant litter or vegetation cover (Table 18.6) (Parkyn, Harper and Smith, 1992). Within the floodplain our understanding is at a similar level. We can produce a habitat classification based upon the three-dimensional structure of vegetation layers for animals, with modification for particular plant features such as nectar sources or Lepidoptera food plants (Parkyn, Harper and Smith, 1992). The vegetation itself can be used to indicate the habitat structure provided by soil character and water-table depth; and this is the basis of the recognition of plant alliances in phytosociology. About 31 alliances can be recognized in the riparian and floodplain zones, ranging from remnants of alluvial forest to assemblages of ruderal species characteristic of disturbed ground (Spence, 1964; Ellenberg, 1978; Rieley and Page, 1990). At present, these provide the most effective way of recognizing plant habitats.

Functional habitat description has been less-well used in water-quality studies, where the physical environment may remain intact despite gross changes in the chemical environment. In such circumstances the UK approach has been to progress from the use of indicator taxa from the 1960s−1980s to the use of taxonomic assemblages as a classificatory and predictive tool in the RIVPACS system (see Chapter 19). However, water-quality changes

TABLE 18.5.   Summary list of important riparian habitats

Mature trees (each species)	Value for invertebrate diversity and biomass, eipihytes, nesting sites
Mature shrubs (each species)	Nectar and fruit feeding stations for insects and birds
Tree features: holes, pollards	Additional value nesting for birds, roosting for bats
Patches of continuous low scrub	Value for passerine bird feeding and nesting, otter cover
Leaf litter and dead wood	Important for decomposer invertebrate community and small mammals
Field layer vegetation zones	Invertebrate architectural diversity: ground, turf, short and tall
Field layer: flowering plant species	Additional nectar value for lepidoptera and hymenoptera
Emergent plants	Invertebrate community, mating zygoptera, nesting birds
Marginal (rosette) plants	Terrestrial invertebrates and oviposition for aquatic
In-channel exposed sediment	Ground beetles and spiders, nesting wading birds, waterfowl
River banks (soft, near vertical)	Bees: classify by soil type and elevation
Rock cliffs and river banks	Epiphytic bryophyte and fern community with associated fauna

TABLE 18.6.   An example of the habitat mosaic: exposed substrata (after Anderson, 1969)

Upper substrate	Lower substrate	Elevation	Vegetation	Debris	Shading
Cobble	Sand	Never flooded	None	None	Open
Stones	Silt	Intermediate	Grasses/herbs	Leaf litter	Part shade
Gravel		Low-lying	Shrubs	Woody	Dense shade
Sand					
Silt					

do cause habitat changes through plant species shifts (e.g. broad-leaved *Potamogeton* spp. → fine-leaved spp. → macroalgae) and substrate shifts (e.g. cobbles become silt-covered); hence there is scope for consideration of functional habitats in water-quality monitoring, alongside biotic indices.

Restoration is the area of management activity where functional habitats have been most used. It is easy to see how a prescription for restoration — which may be drawn up by a biologist but executed by the driver of a piece of heavy machinery — is much smore robust if based upon visual "building blocks" rather than the names of species. In the developing management methodology of the National Rivers Authority, future planning and resource allocation will increasingly be channelled through catchment management plans (CMPs: see Chapter 36), which will incorporate the whole spectrum of river uses. Targets for ecological quality and measures of success will need to be established. Even simple comparative measures such as the numbers of functional habitats per unit length of river channel would enable targets of ecological quality to achieve a status equivalent to those of water quality, or water resources.

There will always be a contrast between the long-term needs for research into the true functioning of ecosystems and the immediate need for comprehensive structural information. Hildrew and Ormerod (Chapter 12) provide an excellent example for research in the case of acidification, where our knowledge and our response to one aspect of water quality management has been largely driven by research into the changes in ecosystem function. This contrasts strongly with the river management requirements described by Brookes (Chapter 2), where physical improvements to watercourses are evaluated on a rolling programme which needs a simple yet robust ecological input on a rapid time-scale. Functional habitats currently provide the most immediate solution to practical problems, yet utilize information about ecological functioning within their structural cloak.

19

The Relevance of Ecology to the Statutory Protection of British Rivers

PHILIP J. BOON

Scottish Natural Heritage, Edinburgh, UK

INTRODUCTION

The process of building principles of ecology into a framework of legislation is never straightforward. Not least of the problems is the difficulty of reaching consensus on areas such as definitions, objectives, classifications, and environmental standards, thus extending the period required for bringing legal measures into force and diluting their effectiveness. Nevertheless, the quantity of legislation that incorporates at least an element of ecology increases from year to year, including such recent examples as the EC Directive on the Conservation of Natural Habitats and of Wild Fauna and Flora (European Communities, 1992) and statutory requirements for Environmental Impact Assessment (European Communities, 1985).

In 1981, the passage of the Wildlife and Countryside Act in Britain empowered the Government's statutory adviser on nature conservation — the Nature Conservancy Council (NCC) — to select areas of land or water containing plants, animals, geological or landform features of special interest, and notify them to owners and occupiers as Sites of Special Scientific Interest (SSSIs). This procedure has now been taken over by the three successors to the NCC: English Nature (EN), the Countryside Council for Wales (CCW), and Scottish Natural Heritage (SNH). Following site survey and evaluation, SSSIs are notified (in addition to the owners and occupiers) to the appropriate Secretary of State, the local planning authority, and (in England and Wales) to the water and drainage authorities. Owners and occupiers also receive a description of the site including its importance for conservation (the "citation"), a map of the site boundaries, and a list of activities deemed likely to cause damage (Potentially Damaging Operations, PDOs). Once a site has been notified, the effectiveness of the SSSI system in protecting it relies as much on persuasion and negotiation as it does on restrictions and penalties. (Further information on SSSI procedures is given in booklets published by the statutory conservation agencies (e.g. Scottish Natural Heritage, 1994).

Compared with most other types of wildlife habitat in Britain, relatively few rivers have been notified as SSSIs in their own right. This is partly due to the high level of resources needed for survey and communication with riparian owners and occupiers, as well as a

---

The Ecological Basis for River Management. Edited by D.M. Harper and A.J.D. Ferguson. © 1995 John Wiley & Sons Ltd

continuing debate over the usefulness (or otherwise) of SSSI designation in protecting riverine habitats (Boon, 1991, 1994).

Since the NCC was reorganized, the approach to river notification has proceeded rather differently in each of the three new conservation bodies. In Scotland, the whole area of site designations is under review and the particular case of river protection has yet to be addressed fully. In Wales, too, the subject is being reconsidered, and although CCW is committed to the "renotification" of the River Wye under the 1981 Act (at present only notified under the National Parks and Access to the Countryside Act, 1949), it is investigating other less costly means of river conservation (Duigan, C. A. pers. comm.). In England, however, river notification is proceeding more rapidly, and in 1992 the Council of English Nature approved a new programme involving 16 rivers. Because of the resource implications of notifying more than 4000 owners and occupiers, this programme will not be completed until 1997 (Withrington, D. K. J. pers. comm.).

To what extent does river SSSI notification rely on a study of ecology? This chapter sets out to examine the relevance of ecological information, and the constraints in applying ecological principles, to each of the four principal phases of the statutory protection process: survey and classification, evaluation and selection, site boundary definition, and management and monitoring.

SURVEY AND CLASSIFICATION

"Ecology", as "the scientific study of the interactions that determine the distribution and abundance of organisms", has both structural and functional elements (see Chapter 18); thus, ecologists are chiefly interested in *where* organisms are found, *how many* occur there, and *why* (Krebs, 1972). Ideally, all three elements are needed to build a strong scientific basis for nature conservation, but in practice rather more emphasis has been given in Britain to the first, and rather less to the second and third.

Survey and classification are the two prerequisites of conservation assessment. If management options are to be formulated for particular rivers it is essential that those rivers can be put into context and their importance for conservation evaluated (Boon, 1992a). The assembling of basic information on habitats, species distributions and community structure is essential to this process, and inasmuch as this is part of ecology at its most fundamental level, so the first step in statutory river protection is an ecological one.

Over the past 15 years the NCC and its successors have devoted considerable resources to three principal types of river survey. Much of the early work centred on surveys of aquatic plants and led to the production of a classification system for British rivers (Holmes, 1983, 1989). This was based on data from more than 1000 sites on over 200 rivers, with the 56 end groups derived from TWINSPAN analysis amalgamated into 10 main community types for the purpose of selecting representative sites for SSSI selection (Nature Conservancy Council, 1989). A further period of survey (1988–91) increased both the geographical coverage throughout Britain, and the size of the total data set (to more than 1500 sites on over 300 rivers; Figure 19.1); re-working these data has produced some subtle, but significant, improvements to the classification (Table 19.1, Holmes and Rowell, 1993). Site classification and the prediction of community types is now extremely consistent, and clearly relates both to geographical location and to catchment geology. Thus, a framework is in place for selecting SSSIs, ensuring (in theory) that the best examples of all major types of rivers can be considered for statutory protection.

FIGURE 19.1. Distribution of sampling sites in British rivers for macrophyte surveys (1978–82, 1988–91) (open circles) and invertebrate surveys (1986–91) (closed circles) carried out by the Nature Conservancy Council

TABLE 19.1. Classification of river community types found in Britain (revised by Holmes and Rowell (1993) from the version previously published in the SSSI selection guidelines (Nature Conservancy Council, 1989))

Type	Group(s)	General description
I	A1	Lowland rivers with minimal gradients; predominantly in south and east England, but may occur wherever substrates are soft and chemistry enriched
II	A2	Rivers flowing in catchments dominated by clay
III	A3	Rivers flowing in catchments dominated by soft limestone such as chalk and oolite
IV	A4	Rivers with impoverished ditch floras, usually confined to lowlands and mainly in England
V	B1 and B2	Rivers of sandstone, mudstone and hard limestone catchments in England and Wales, with similar features to those of Type VI
VI	B3 and B4	Rivers predominantly in Scotland and northern England in catchments dominated by sandstone, mudstone and hard limestone; substrates usually mixed coarse gravels, sands and silts mixed with cobbles and boulders
VII	C1 and C2	Mesotrophic rivers where bedrock, boulders and cobbles are the most common components of the substrate; usually downstream of Type VIII
VIII	C3 and C4	Oligo-mesotrophic, predominantly upland rivers where boulders are an important component of the substrate; intermediate, and often between Types IX and VII
IX	D1 and D2	Oligotrophic rivers of mountains and moorlands where nutrient and base levels low; bedrock, boulders and coarse substrates dominate
X	D3 and D4	Ultra-oligotrophic rivers in mountains, or streams flowing off acid sands; substrates similar to Type IX but often more bedrock

This bias towards botanical work began to be addressed in the mid-1980s with a complementary programme of invertebrate survey carried out for the NCC by the Freshwater Biological Association (later the Institute of Freshwater Ecology (IFE)). Limited resources restricted the scope of this work to 123 sites on 45 rivers throughout Britain (Figure 19.1) (Wright et al., 1992), most of which had already been identified as important for conservation on the basis of their aquatic plant communities. Samples were collected using the standard methods employed by IFE thereby ensuring compatibility with the RIVPACS system for site classification (Wright et al., 1989).

Before rivers are notified as SSSIs, a wider habitat (or "river corridor") survey is frequently carried out, mapping the main habitat features in the channel, riparian zone and floodplain (e.g. wet woodland, fen, marsh) which may deserve inclusion within the SSSI. Several surveys of this type have been undertaken by the NCC during the past 10 years, some of which have been used in subsequent river notification (e.g. River Blythe, West Midlands), while others remain on file to be used in site "casework" but not yet in site notification.

While these processes of survey and classification are fundamental to conservation, their contribution to ecological knowledge is generally limited to the broad inferences that correlative statistical techniques permit. Regrettably, many studies are terminated there, and add little to the understanding of complex relationships between species populations, and between populations and their physical or chemical environment; yet these are the areas so necessary for ecologically focused river management.

EVALUATION AND SELECTION

In 1989, the NCC published a set of guidelines for selecting biological SSSIs (Nature Conservancy Council, 1989). The guidance given for running waters encourages evaluations based on a wide range of features, but accepts that in many cases selection will mainly concentrate on plant communities. Even then, a certain amount of basic ecological information is essential, for without that it is difficult to set the particular features of interest for which the site was notified within a broader context. However, until recently this tends to have been the exception rather than the rule. An examination of citations prepared by the NCC (Figure 19.2(a)) shows that very few provide a well-rounded description of the river and many fail even to mention such fundamental characteristics as substrate, flow regime, natural water chemistry or general descriptions of biological communities. In comparison, recent draft citations prepared by English Nature (Figure 19.2(b)) in readiness for the new programme of river notification, have done much to improve the situation.

The SSSI guidelines are based on widely accepted conservation criteria and emphasize features such as "naturalness", "representativeness", "diversity" and "rarity" which were discussed in detail in *A Nature Conservation Review* (Ratcliffe, 1977). Their relevance to river evaluation in Britain has been considered by Newbold, Purseglove and Holmes (1983) and by Boon (1991, 1992a), and similar criteria have also been applied to rivers elsewhere (e.g. North America: Rabe and Savage, 1979; Australia: Blyth, 1983; New Zealand: Collier and McColl, 1992).

By definition, attributing "value" involves a degree of subjectivity, in which individual preferences and philosophical viewpoints overlay the scientific information on which judgements are based. The real challenge in river assessment is not to eliminate the subjective, but to increase standardization and repeatability in the collection and handling of data used in evaluation. At the same time, the aim should not merely be to produce a broad-based description of rivers while still continuing to select sites using a narrow range of criteria, but to find ways of incorporating more ecologically relevant information throughout the assessment process. This is one of the main objectives of a project now under way in Britain, directed by Scottish Natural Heritage in conjunction with the other statutory conservation agencies and river protection bodies in the UK (Boon et al., 1994). Known as "SERCON" (System for Evaluating Rivers for Conservation), this project is examining the way that rivers are assessed at present, and is taking a fresh look at how such assessments might be conducted with greater consistency and objectivity. Initially, SERCON will concentrate on the evaluation of habitats and species, but because the role of both SNH and CCW (as well as the National Rivers Authority) extends to other features of the countryside it is hoped that parallel systems can be developed later to assess rivers, for example, on their landscape or recreational use.

The key characteristics of SERCON (PC-based and using a scoring method) have been derived from the River Conservation System (RCS), designed for rivers in South Africa (O'Keeffe, Danilewitz and Bradshaw, 1987). However, the system now under development is different from the RCS in many ways, having the capacity to link with databases, create its own databases, and provide information on screen in textual, graphical or photographic form. The primary data input for SERCON is derived from a wide range of attributes considered important in conservation assessment. Some attributes (e.g. altitude at source, river length, stream flow stability) are there to set the river in context, and do not contribute directly to the evaluation process. Others (e.g. relating to data on biological communities,

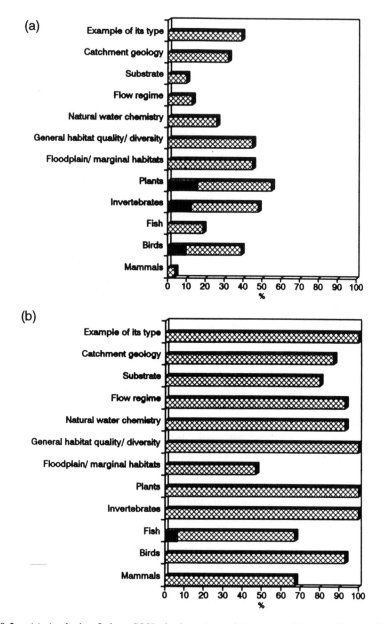

FIGURE 19.2. (a) Analysis of river SSSI citations ($n = 31$) produced by the Nature Conservancy Council, showing the percentage containing information (however brief) on various riverine features. (b) Analysis of citations (as for (a)) ($n = 15$) prepared by English Nature for proposed river SSSIs. (Solid black = percentage of citations not providing species details)

rare species, diversity of fluvial features) provide the raw data for assessing conservation value, while the assessment of threats and impacts on the river (e.g. acidification, inter-river transfer) takes place in a separate part of the system. However, unlike the RCS, the information on each attribute is not fed directly into the assessment, but is combined in various ways to produce scores for a range of conservation criteria such as representativeness, physical diversity, and species rarity.

SITE BOUNDARY DEFINITION

An SSSI must, by law, have a clearly defined boundary marked on the map accompanying the letter of notification. This is perhaps the main area in which the legislation, and the practical difficulties of applying it, do not encourage the inclusion of well-established ecological principles.

For river conservation to be effective it has to take account of five dimensions (Boon, 1992a). Three of these are spatial and refer to the intimate connections (structurally and functionally) between communities and their surroundings — longitudinally (upstream/downstream), laterally (channel/floodplain) and vertically (substratum/hyporheic) (Ward, 1989). The other dimensions are temporal, reflecting changes in time in geomorphology and community structure, and conceptual, in which reference to spatial and temporal dimensions must be fitted into a framework of philosophy, policy and practice. Of these five, the longitudinal and the lateral are of most immediate importance in defining SSSI boundaries.

Changes from source to mouth in physical, chemical and biological characteristics have been observed from the early days of stream ecology, leading to classifications based on distinct riverine zones (e.g. Illies and Botosaneanu, 1963) or more recently on the river continuum (Vannote et al., 1980). Although some doubts have been expressed over the usefulness of the river continuum concept as a universal paradigm (Winterbourn, Rounick and Cowie, 1981), it has done much to advance both the theoretical study of stream ecology and the practical understanding of how running waters respond to man-made disruptions to the continuum. For example, dam construction has been shown to break the normal downstream sequence of physical, chemical and biological conditions, a phenomenon termed the "serial discontinuity concept" (Ward and Stanford, 1983). The logical conclusion from ecological studies such as this is that the upstream and downstream boundaries for an SSSI should be the river's source and its mouth.

In contrast with the many early studies on downstream sequential patterns in rivers, it is only comparatively recently that concerted efforts have been made by ecologists to study the lateral connections between river channels and their wider corridors. There is now a substantial body of information worldwide showing the importance of terrestrial–aquatic ecotones within river corridors. For example, riparian zones alongside rivers play an integral part in ecological functioning by exerting hydrological control, acting as filters of diffuse pollutants and providing an external source of organic carbon (Pinay et al., 1990). Further away from the river channel, periodic inundation of floodplains may modify instream patterns of nutrient cycling (Pinay et al., 1990), support highly productive fisheries (Welcomme, 1979), assist in maintaining biodiversity and add to the visual quality of landscapes (Petts, 1990c).

There can be little doubt that an "ecological boundary" for a river SSSI should include an entire corridor of water and land from source to mouth, extending laterally to encompass the riparian zone and contiguous wet areas. An "ecologically optimal boundary" should be wider still as the realization has grown that streams and their valleys are functionally inseparable (Hynes, 1975). Information collected over the past 10 years on the hydrological and chemical impacts of afforestation, or runoff from agricultural land and urban developments, merely confirms that the appropriate unit for river management and for river protection is the entire surface water catchment. This point has been made frequently by workers in conservation and ecology (e.g. Boon, 1992a; Naiman, 1992; Newson, 1992), and has been recognized by national and international organizations around the world (e.g. Economic Commission for Europe, 1988; IUCN/UNEP/WWF, 1991), not least at the recent International Conference on Environment and Development, held in Rio de Janeiro (UNCED, 1992).

Yet, the results of ecological research, their implications, and the aspirations of conservationists can rarely be fully accommodated by statute. Attempting to notify whole river catchments as SSSIs would present immense difficulties: the time and expense of contacting many hundreds of owners and occupiers; the potentially huge numbers of consultations over PDOs and the requirement for many costly management agreements; and not least the problem of including within a site boundary large tracts of land which may not, in their own right, be valuable for conservation.

The SSSI guidelines (Nature Conservancy Council, 1989) can only offer broad advice on boundary definition. Although they do state that "where a catchment is still natural or semi-natural and especially where it is small and discrete, the whole of it should be included within the SSSI", they add that normally "only a much narrower 'buffer zone' can be incorporated". In practice, river SSSIs tend to be variable lengths of one or a few streams within a catchment, with their lateral boundaries extending either to the top of the banks or further out to incorporate adjacent wet habitat. Thus, "the main core of interest may vary from less than 1 m high or wide in narrow, shallow rivers to as much as 100 m high or wide in flat flood-plains". Also, in recognition of the important role of riparian vegetation (especially as buffer zones for pollution prevention), it is recommended that this should be included within site boundaries where possible.

Research on riparian buffer zones has grown steadily over the past decade, not least on their capacity to act as nutrient filters (e.g. Peterjohn and Correll, 1984; Haycock and Pinay, in press; Haycock, Pinay and Walker, 1993). These studies are now beginning to work their way through into practical river conservation planning in many different countries (e.g. Poland: Kajak, 1992; Sweden: Petersen, Petersen and Lacoursière, 1992; Australia: Bunn, 1993; USA: Osborne and Kovacic, 1993). The need now is for a concerted effort to draw together new developments in functional ecology and apply them to the practical problems of conserving river systems. For example, determining the optimum width of a buffer strip must be strongly influenced by the results of sound scientific research if the inclusion of these areas within the lateral boundaries of statutorily protected sites in Britain is to be defended. This need has been recognized in other countries as well. Extensive tree planting programmes are currently under way in some Australian catchments, yet the view has been expressed that much more information is required on the functioning of buffer zones of different widths and composed of different species before further replanting schemes are undertaken (Bunn, 1993).

Unfortunately, in Britain financial realities may well overshadow the conclusions of ecological research. For example, it seems likely that the cost to English Nature of implementing the new programme of river notification will mean that most lateral site boundaries will not extend further than the channel and its banks (Withrington, D. K. J., pers. comm.).

MANAGEMENT AND MONITORING

Biological survey enables rivers to be classified, evaluated against an agreed set of criteria and notified as SSSIs. Ensuring that the site retains its conservation value, however, requires a commitment to monitoring, an understanding of a wide range of ecological impacts, and an appreciation of the consequences of various river management strategies. River notification may provide a mechanism whereby potential problems can be discussed and damage minimized, but it is certainly not guaranteed to prevent the threat from arising in the first place. Table 19.2 lists an assortment of specific cases on river SSSIs (or proposed SSSIs) dealt with by the statutory conservation bodies. This illustrates three fundamental points relevant to the theme of management and monitoring.

First, there is the sheer range of impacts encountered (including many not shown in the table), over the full spectrum of scales from supra-catchment to instream impacts (Boon, 1992a). Second, and related to the first, there is the array of ecological concerns that such impacts generate and the scope of ecological knowledge required to address those concerns. This includes the physical and chemical characteristics of running waters, and their influence on plant and animal distribution, survival, growth and life cycles, feeding and breeding behaviour, together with biotic interactions such as competition and predation (Table 19.2).

Third, there is an expectation by some staff within the conservation bodies and others outside that definitive answers are always available, preferably instantly. Two questions illustrate the point, both asked by regional staff of the Nature Conservancy Council in letters sent to advisers at the NCC headquarters. One concerned a proposal to remove gravel from the bed of a river SSSI: "Are there any species found in rivers which are intolerant of such disturbance?" The other, strangely precise in its expectations ("How many sump washings will cause a problem at low stream flows?"), referred to an increase in vehicle traffic across a small woodland stream, and the practice of some drivers of using the crossing point as a convenient place for car maintenance.

While there may never be detailed research results available for every conceivable type of aquatic impact, these questions emphasize the need for river management strategies (including management of SSSIs) to be informed by sound ecological principles (Boon, 1992b). This in turn obliges those with responsibility for SSSI notification (and a continuing overview of their management) to keep abreast of developments in river ecology, and to ensure that unanswered questions are addressed by new research work. Both requirements are impractical except to a very limited extent: staff, time and money are in short supply compared with the size of the task. Moreover, many of the ecological dilemmas facing river managers are too complex to be solved by short-term studies.

For example, in 1988 the Nature Conservancy Council commissioned a three-year project on the effects of low flows caused by abstractions from rivers. The main objective was to find ways of recommending ecologically acceptable minimum flows, based on the responses of macroinvertebrates. No overall patterns or coherent trends in faunal distribution were observed when sites above and below abstraction points were compared, and no straightforward relationships were found between the type of stream, the intensity of abstraction, and the amplitude and direction of the faunal changes (Petts et al., 1991; Armitage and Petts, 1992). Only those sites having severely depleted flows at the time of survey had degraded fauna, and then only when discharge, depth and velocity were drastically reduced. A new approach at assessing impacts based on "ecological profiles" of individual species seemed promising, but no time was available at the end of the project to

TABLE 19.2. A selection of casework on river SSSIs (or proposed SSSIs) handled by NCC/SNH HQs, to illustrate the broad range of anthropogenic impacts experienced, and the ecological concerns they generate

Category of impact	Specific casework item	Examples of ecological concerns
Fishery management	1. Stocking with rainbow trout and brown trout	Impact on native fish stocks and other biota (predation; competition for food and space; genetic interactions)
	2. Introducing grayling (non-native to catchment)	Impact on native fish stocks and other biota (predation, competition for food and space)
	3. Introducing barbel (non-native to catchment)	Impact to native salmonids (damage to spawning redds; competition for food and space)
Abstraction	4. Abstracting water for fish farming	Flow reduction over stretch of river between intake and outflow — impacts on plants and invertebrates through altering flows, substrate composition, water quality
	5. Temporary (1 year) removal of water, diverted to pipeline	Uncertain — flow reduction in river minimal (1–2%)
Modification to aquatic habitats	6. Gravel cleaning and "pot-hole" creation for salmon	Effects on macroinvertebrates (increased drift, alteration to flow regimes)
	7. Replacing frequent summer weed-cutting by one severe autumn cut	Impact on invertebrate habitat and life cycles
	8. Dumping road-rubble into river channel	Damage to plant and animal habitats. Lethal or sub-lethal effects of polyaromatic hydrocarbons, heavy metals
Construction works	9. Construction of new road scheme with drainage to river	Increases in road runoff, e.g. suspended solids, oil, salt
	10. Routing an oil pipeline across a river	Varying concerns depending on which of three options chosen, but mainly increases in suspended solids and sedimentation
	11. Construction of business parks in catchment	Increased access to river bank; increased amounts of runoff from hard surfaces
	12. Replacing a ford by an Irish bridge	Increased stream disturbance (e.g. input of particulates) during construction
	13. Building a railway bridge across a river	As for No. 12.
Pollution	14. Discharge from sewage treatment works	General effects of increased BOD, ammonia, suspended solids, nutrient levels on biota
	15. Discharge from three storm sewage outfalls	As for No. 14, with additional concerns over impact of stormwater runoff
	16. Discharge of effluent from rainbow trout farm	General effects on biota of organic and inorganic pollutants, including fish farm chemicals
	17. Leachate from landfill site flowing into river	Effects of complex mixture of organic solvents and heavy metals on invertebrates
	18. Discharge of effluent from yoghurt factory	Increases in BOD, turbidity; particular concern over impact on migration of the smelt, a threatened fish species
Recreational developments	19. Opening a canoe slalom course	Increased visitor pressure on river (bankside and aquatic habitats); alteration to flow regime by course construction
	20. Water-skiing	Physical damage to aquatic vegetation, disturbance to birds and otters, bank erosion, fuel spills

TABLE 19.3. Water quality standards proposed by English Nature for inclusion in the Special Ecosystem Use Class

Target level	A: BOD (mg litre^{-1})	B: NH$_3$-N (mg litre^{-1})	C: PO$_4$-P (mg litre^{-1})
A/B/C/1	1.5	0.06	0.02
A/B/C/2	2.0	0.13	0.06
A/B/C/3	2.5	0.20	0.11
A/B/C/4	3.2	0.34	0.20
A/B/C/5	4.0	0.60	1.00

Figures for BOD and NH$_3$-N are 90 percentiles; those for PO$_4$-P are annual means

develop this further.

One way of utilizing scarce research funds more effectively is for organizations with shared interests (e.g. the statutory conservation bodies, National Rivers Authority (NRA), or River Purification Boards (RPBs)) to co-operate together. In the wider realm of SSSI management such collaboration is absolutely essential. A good example is the partnership approach between English Nature and the NRA in developing so-called "special ecosystem standards" as part of a new system of statutory water-quality objectives (Department of the Environment/Welsh Office, 1992). For many years the NCC, and its successors, have expressed concern at the deterioration of high quality rivers, including SSSIs. In particular, nutrient enrichment is perceived as an insidious threat, especially in lowland regions, a point emphasized in the recent report by the Royal Commission on Environmental Pollution (1992). English Nature has now put forward proposals aimed at maintaining or improving water quality in river SSSIs, comprising a five-point classification based on three chemical variables (BOD, NH$_4$-N, PO$_4$-P) (Table 19.3), derived from the levels of these substances recorded in SSSI rivers in 1990 and 1991. However, this approach is essentially pragmatic, and English Nature has now commissioned further work to investigate the ecological relationships between phosphorus levels and plant communities in rivers (Gibson, M. T., English Nature, pers. comm.).

The problems of obtaining and interpreting ecological information for SSSI management are no less daunting for SSSI monitoring. It is one thing to maintain a regular check on potential water pollution problems (the responsibility of statutory pollution control authorities), or to monitor the effects of a particular development project, but monitoring the condition of river SSSIs demands information of a different kind; in particular sufficient observations of habitats and species to establish whether the criteria for SSSI status are still being met. As yet, the statutory conservation bodies have not developed monitoring protocols or methodologies to achieve this, although they are acutely aware of its importance.

POSTSCRIPT — THE EUROPEAN CONTEXT

On 5 June 1992 the European Commission notified the United Kingdom of the Council Directive on the Conservation of Natural Habitats and of Wild Fauna and Flora (European Communities, 1992), commonly known as the Habitats and Species Directive. This requires

Member States to convey to the Commission by June 1995 a list of sites that contain habitats and species included on Annexes I and II.

Annex I contains a list of running water habitats (although only some are found in Britain). Annex II includes 11 British species found in rivers for at least part of their life cycle: otter (*Lutra lutra*), sea lamprey (*Petromyzon marinus*), river lamprey (*Lampetra fluviatilis*), brook lamprey (*Lampetra planeri*), allis shad (*Alosa alosa*), twaite shad (*Alosa fallax*), Atlantic salmon (*Salmo salar*), spined loach (*Cobitis taenia*), bullhead (*Cottus gobio*), freshwater pearl mussel (*Margaritifera margaritifera*), and the North Atlantic stream crayfish (*Austropotamobius pallipes*).

By June 1998 the Commission will agree with each Member State a draft list of sites of community importance, and as soon as possible thereafter (and within six years at most) the Member States will designate each site as a Special Area of Conservation (SAC).

The UK Government has announced its intention "that any sites notified as being of Community importance will already have been notified as SSSIs before being given the additional designation of Special Area of Conservation under the Directive" (Hansard, House of Lords, 1993). This has two implications: first, that the statutory conservation bodies (which are at present gathering information on potentially important sites) will need to review with some urgency their positions on river notification. Second, more autecological data may be needed for those species listed in Annex II, both for justifying proposals for potential SACs, and for assisting with subsequent management of any sites ultimately designated as such. While the ecological requirements of some species (e.g. otter, Atlantic salmon) have been well researched, for others (e.g. allis shad) far less is known (Maitland and Lyle, 1992).

Obtaining and collating information is one thing; synthesizing it and applying it is quite another. The gap between ecologist and river manager will only be bridged when ecological information is appropriately focused, easily understood, simple but not simplistic, unequivocal and authoritative, rapidly accessible, accepted and used (Boon, 1992b). This is essential if statutory river protection, whether in a British or a European context, is to function effectively.

ACKNOWLEDGEMENTS

I would like to express my thanks to David Withrington (English Nature) and Dr Catherine Duigan (Countryside Council for Wales) for supplying SSSI citations, and to Dr Terry Rowell (Conservation Data Services) and Jackie Graham (Scottish Natural Heritage) for producing the diagrams.

20

Integrating River Corridor Environment Information through Multivariate Ecological Analysis and Geographic Information Systems

ANGELA M. GURNELL

University of Birmingham, Birmingham, UK

DAVID SIMMONS

Geographical and Multimedia Solutions, Southampton, UK

and PETER J. EDWARDS

Geobotanisches Institute ETH, Zurich, Switzerland

VEGETATION, HYDROLOGY AND FLUVIAL PROCESSES AT THE LANDSCAPE SCALE

Vegetation composition reflects the characteristics of the surrounding environment. Indeed a major focus of plant ecological studies is to develop an understanding of the way that plant species and communities are adjusted to environmental controls and thus the degree to which changes in environmental conditions may affect the character of the vegetation. Vegetation is relevant to catchment and river management studies in three main ways: through the direct effect of the vegetation on hydrological and fluvial processes; through the modification of the effect of vegetation on these processes as a result of vegetation management; and through the adjustment of vegetation to environmental gradients.

Vegetation has a wide range of direct effects upon hydrological and fluvial processes, which have been illustrated by representative catchment, small plot and river channel studies (e.g. Clarke and Newson, 1978; Hino, Fujita and Shutto, 1987; Gregory and Gurnell, 1988; Dawson, 1989). These effects are sufficiently significant for them to have been incorporated to varying levels of complexity in hydrological models including, for example, the use of afforested percentage to estimate catchment water yields (Calder and Newson, 1979) and the use of forest age/growth to estimate canopy interception storage capacity, stress thresholds, proportion of roots in the topsoil horizon and thus to estimate the effects of forests on catchment water yields (Schulze and George, 1987).

The Ecological Basis for River Management. Edited by D.M. Harper and A.J.D. Ferguson. © 1995 John Wiley & Sons Ltd

Vegetation cover is one of the most easily manipulated components of a drainage basin or river channel, and because of the direct influence of vegetation on hydrological and fluvial processes, numerous studies have illustrated the enormous impact of vegetation change on these processes (e.g. Bosch and Hewlett, 1982; Trimble, Weirich and Hoag, 1987; Watson, 1987; Gregory and Davies, 1992; Gurnell and Midgley, 1993). Vegetation management through such processes as cutting, burning or grazing can lead to changes in biomass and, in some circumstances, to immediate or gradual changes in the vegetation type. All such changes have hydrological, fluvial geomorphological and thus catchment and river management implications.

In addition to the direct impact of vegetation on hydrological and fluvial processes, semi-natural vegetation communities are closely adjusted to environmental controls, including the hydrological (particularly soil moisture and river flow) regime. For example, Walker (1985) described the link between tundra vegetation and environmental gradients (including soil moisture, cryoterbation and temperature gradients) at three spatial scales in the Prudhoe Bay region of Alaska. Hupp and Osterkamp (1985), Hupp (1986) and Olson and Hupp (1986) have also related vegetation to fluvial landforms, soils and bedrock in catchments in Virginia, USA. In addition, Gregory and Gurnell (1988) have reviewed the role of vegetation in determining river channel changes. As a result of these close relationships with controlling processes, vegetation communities can indicate hydrological and fluvial processes and thus provide a basis for their extrapolation across areas for which process measurements are unavailable (e.g. Gurnell et al., 1984; Gurnell and Gregory, 1987).

It is this property of vegetation, to act as a surrogate index for environmental processes, that provides a potentially powerful tool in landscape-scale studies. In the context of river survey and management, the application of multivariate ordination and classification techniques to macrophyte species within rivers in England and Wales by Holmes (1989) has shown that at the national scale macrophyte communities reflect climate, rock type and water quality. Similar links between environmental controls and ecological characteristics underpin the development of systems such as RIVPACS (Moss et al., 1987) and PHABSIM (Milhous, Updike and Schneider, 1989), which relate to invertebrate and fish communities of rivers and their association with site characteristics. If such interactions exist between vegetation and environmental characteristics of rivers, then it should be possible to use them to characterize aspects of the environment from the plant composition of the river and its corridor.

This chapter presents the results of a pilot project which explores the potential for characterizing the river environment from the riparian and in-channel plant species that are present. A linked series of analyses of data from a tributary of the River Thames are used. The spatial pattern of the results of multivariate ordination and classification of the riparian and in-channel species are presented and the known environmental tolerances of the species are used to provide possible explanations for the observed spatial patterns. The spatial distributions of potential environmental controls are then analysed, to assess whether or not they match the inferences on controls derived from the species data.

The significance of this pilot study is that the character of a river system is explored from entirely secondary sources. In particular, it represents a novel use for the very substantial existing information available in River Corridor Surveys (RCS) for rivers in England and Wales. Indeed, complementary methodologies are currently being developed for abstracting quantitative information from RCS maps so that they too can be used to explore environmental characteristics at the whole catchment scale. Such approaches are essential

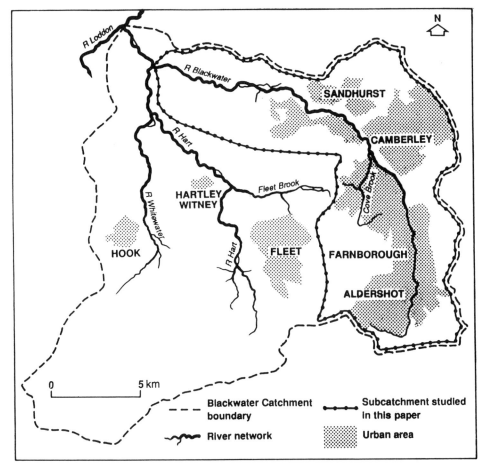

FIGURE 20.1. The River Blackwater Catchment

for landscape-scale studies where specialist, comprehensive field surveys are generally unavailable and would be prohibitively expensive and time consuming to commission.

The volume of data generated for this one pilot catchment was so large that it necessitated storing information in a purpose-designed database. GIS (Geographic Information Systems) coupled with a relational database manager proved to be a key analytical tool in this work, permitting efficient data storage, retrieval, integration, modelling and visualization.

DATA ASSEMBLY AND PREPARATION FOR A PILOT CATCHMENT STUDY

The River Blackwater (Figure 20.1) was selected to provide a test of a methodology based upon the identification of river environmental gradients from plant species information. It is a short river (approximately 30 km) within a small sub-catchment (146 km²) of limited relief (45−147 m OD) and which is strongly influenced by man's activities. The small size and relief of the catchment limit the potential variability of climate, the underlying Tertiary clays and sands provide a restricted geological variability, and the heavy disturbance of the river

TABLE 20.1.   Spatial entities and associated attributes used in the River Blackwater GIS

Entity type	Entity	Attributes
Points	Water-quality sample points	Levels of selected water-quality parameters
	Sewage	Level of discharge consent
Lines	River network	Riparian and in-chanel vegetation species lists
	(subdivided into 500 m reaches)	Surveyed channel width and depth
		(to vegetation limit)
		Surveyed bank height and width
		(to vegetation limit)
Polygons	Catchment boundary	
	Maximum limit of known flooding	
	Rock type boundaries	Enclosed rock type classes
	Floodplain land use	10 land use types, vegetation
		species lists for selected polygons
	Floodplain adjacent to	
	500 m reaches	
	Enumeration district voronoi	Property statistics
Raster data	Thematic mapper data	
	OS 50 m resolution DTM	

channel and predominance of urban and intensive agricultural land use are all likely to restrict the range of riverine plant species, thus providing a severe test of the sensitivity of this approach. A GIS for the River Blackwater catchment was assembled using the Intera Tydac Technologies SPANS product run under OS/2 version 2 extended edition on an IBM PS2 model 80 platform. A SPANS study area that included the watershed of the River Blackwater and its tributaries was constructed. Compatability was maintained with SPANS work carried out by the National Rivers Authority, Thames Region, in the preparation of their draft Blackwater Catchment Plan, so that data and output could be readily transported between the systems. Table 20.1 lists the entities and associated attributes used in the present study.

A two-volume RCS for the River Blackwater, commissioned by the National Rivers Authority, Thames Region, provided the basic data source for this study. Information in the RCS was relevant to two types of spatial entity: the River Blackwater itself, which was subdivided into 59 reaches of 500 m in length; and the river's floodplain, which was divided into polygons containing single classes of land use. For each river reach two plant species lists were provided, one for the riparian zone and one for in-channel vegetation. Field measurements of bank height and width (above the vegetation limit) and channel width and depth (below the vegetation limit) were also available for many of the reaches. Each floodplain polygon was allocated to one of 19 land use classes and, where the land use was of conservation interest, a plant species list was provided using an aquatic, grassland or woodland record card, as appropriate. Numerical codes were established for each plant species and their Latin and Common names according to the National Vegetation Classification (Rodwell, 1993). Two additional sets of vegetation-related information were assembled for the tolerances of the riparian and in-channel species, respectively, to particular environmental characteristics. These data sets were developed from the habitat requirements of river plants listed in the appendix of Newbold, Purseglove and Holmes

(1983), where a five-point scale is used to relate each plant species to each habitat characteristic. The habitat characteristics include altitude (lowland, transition, upland), soil or river substrate (peat, soft mud, clay, silt, sand, gravel/pebbles, cobbles/boulders, rock fissures), flow velocity (fast, moderate, slow), channel depth and width, water chemistry (base-poor, -neutral or -rich water; nutrient-poor, -moderate or -rich water; acid, neutral or alkaline soils), shade and disturbance tolerance.

The vegetation data used in the study were managed as relational tables under control of the OS/2 Query Manager. A data table was thus constructed to hold the reach-based survey information (e.g. channel width) and the DECORANA axis scores and TWINSPAN classes (see below) relating to in-channel and riparian vegetation. A code-to-name translation table was also available to support the extraction of plant species lists for particular sites. A tolerance table contained rows for each species and columns identifying scores against different habitat requirements. A similar structure was implemented to manage floodplain habitat polygons and their species lists.

Two raster data sets were used to provide contextual information on the catchment as a whole. Four tiles of the Ordnance Survey 50 m resolution DTM enabled detailed investigation and visualization of catchment topography and definition of watersheds for any location on the river network using the diffusion algorithm supplied within SPANS. Data for Landsat Thematic Mapper bands 3 (red) and 4 (infra-red) for 9 August 1984 were imported as a Normalized Difference Vegetation Image (NDVI) linearly stretched to the range $0-100$, with image rectification and pre-processing carried out by IDRISI. The NDVI was evaluated and density sliced on the basis of carefully selected training areas to provide estimates of the spatial distribution of arable, grassland, woodland and heathland, and urban land use. Bare (harvested) arable fields and dense urban areas were difficult to distinguish using single-date satellite imagery alone. To provide such differentiation, the 1981 census information of property density at Enumeration District level was used to generate a property density map based on voronoi boundaries. A SPANS matrix overlay procedure on the property density and NDVI maps was then used to apply the logic that differentiated between urban and arable land uses. Additional polygon and point data sets were supplied by the National Rivers Authority, Thames Region. Polygon-based information included the catchment and rock type boundaries, and the boundary of the maximum limit of known flooding. Point entities included the locations of sewage outfalls (for which the daily volume of the discharge consent was supplied) and water-quality sampling points (for which three years of observations of a variety of water-quality parameters were provided).

DATA ANALYSIS

The assembled data were used to explore three aspects of the river environment: the spatial pattern of plant communities as defined by multivariate ecological analysis; the spatial distribution of environmental controls inferred from the known tolerances of the plant species; and the spatial distribution of potential environmental controls estimated from observations of these controls or from surrogates.

Multivariate analysis of plant species information

The riparian and in-channel plant species lists associated with the fifty-nine, 500 m reaches of the River Blackwater were separately subjected to ordination and classification

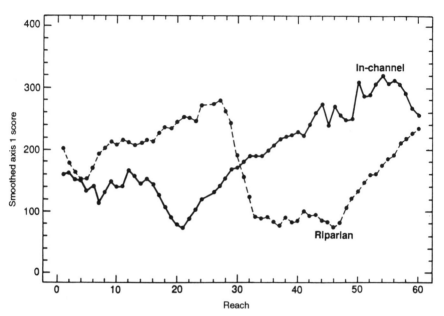

FIGURE 20.2. The downstream pattern in riparian and in-channel vegetation DECORANA axis 1
scores for river reaches. The trend is identified by a three reach running mean

techniques. Ordination techniques of vegetation analysis produce a spatial arrangement of
samples (river reaches) to reflect their similarity. The ordination approach to vegetation
analysis is usually used to relate species and environmental factors either by including
information on environmental variables in the ordination or by interpreting the axes that are
estimated from the vegetation data in terms of possible environmental controls. Detrended
Correspondence Analysis (DECORANA: Hill and Gauch, 1980) was used to achieve the
ordination of the riparian and in-channel plant species. In each case, the reach scores on the
first axis of the ordination generated a clear downstream pattern (Figure 20.2), which was
not apparent in the lower order axes, though in neither case was it the simple downstream
trend that might be expected. Both riparian and in-channel axis 1 scores show clear changes
in direction with a distinct change occurring around reach 21 for the in-channel species and
around reaches 36 and 46 for the riparian vegetation. TWINSPAN (Hill, 1979) is a
classification method which produces a classification of both plant species and sites or
samples in the form of a two-way table. It proceeds by progressively splitting groups of
samples into two, to give a hierarchical, dichotomizing classification. In much ecological
research, TWINSPAN is used for identifying plant species that occur in association with one
another, but in the present case the grouping of river reaches (i.e. the samples) was of more
interest.

The method for splitting the groups of samples begins with an ordination of the data of the
same kind used in DECORANA. This primary ordination is divided at its centroid and used
to identify differential or indicator species (i.e. those species that are preferential to one side
of the dichotomy). A refined division is then constructed using only these indicator species,
and this provides the basis for the further classification of the samples into groups. The
classification proceeds by progressively splitting the samples or river reaches into two

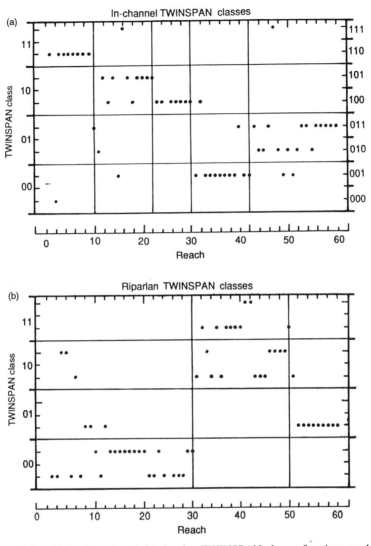

FIGURE 20.3. (a) In-channel and (b) riparian TWINSPAN classes for river reaches

groups. In the present analysis this process of subdivision was allowed to proceed to the third level, so producing eight classes of river reach. Figure 20.3 plots these eight classes of reach for both in-channel (upper diagram) and riparian species (lower diagram) against their reach number. As with the axis 1 DECORANA scores, the TWINSPAN classes form a clear, if complex, spatial sequence. In the case of the in-channel vegetation, classes 000 and 111 contain only one and two reaches respectively and so were omitted from further consideration. Of the remainder, the downstream sequence of classes was 110, 101, 100, 001, 01 (i.e. 010 and 011 combined), with breaks between classes coinciding approximately with reaches 10, 22, 30 and 42. In the case of the riparian vegetation, the downstream sequence of classes was 00, 10/11, 01, with breaks coinciding approximately with reaches 30 and 50.

TABLE 20.2. In-channel vegetation: average scores for each TWINSPAN class based upon plant species habitat requirements

	TWINSPAN class ordered from upstream (left) to downstream (right)				
	110	101	100	01	001
Substrate					
Peat	1.57	2.04	2.01	1.98	1.68
Clay	4.47	4.28	4.31	4.18	4.59
Silt	3.73	3.82	3.82	3.81	4.10
Sand	1.97	2.06	2.03	2.27	1.97
Gravel	2.26	2.36	2.00	2.68	1.76
Velocity					
Negligible/slow	4.65	4.58	4.53	4.07	4.49
Moderate	3.28	3.28	3.01	3.21	3.13
Fast	1.47	1.67	1.72	2.36	1.61
Depth					
>1 m	1.71	1.71	2.23	2.45	2.20
0.5−1 m	2.73	3.08	3.67	4.13	3.41
0.1−0.5 m	4.34	4.41	4.26	4.28	4.32
<0.1 m−dry	3.97	3.10	2.76	2.65	2.72
Width					
>10 m	3.39	3.63	3.70	4.04	3.85
5−10 m	4.15	4.13	4.17	4.07	4.21
<5 m	4.13	3.91	3.56	3.24	3.56
Water chemistry					
Base-poor	1.65	2.41	2.02	2.16	1.38
base-neutral	4.26	3.71	4.01	4.04	3.92
Base-rich	4.02	3.80	4.06	4.19	4.39
Nutrient-poor	1.65	2.15	1.81	2.00	1.32
Nutrient-moderate	4.07	3.71	3.97	3.89	3.82
Nutrient-rich	3.97	3.78	4.11	4.19	4.25
Tolerance					
Shade	2.50	2.56	2.40	2.37	2.32
Disturbance	3.68	3.56	3.65	3.53	3.58

Mean values are derived from allocating in-channel species within the reaches of a particular TWINSPAN class a score of 5 to 1 according to whether they thrive, are common, occasionally colonize, are rare or absent in these conditions (from Newbold, Purseglove and Holmes, 1983)

Habitat requirements of river plants and their apparent environmental gradients

Using the data on the habitat requirements of river plants given in Newbold, Purseglove and Holmes (1983), plant species were allocated scores of 1 to 5 for each environmental characteristic according to whether the plant is absent (1), rarely found (2), occasional (3), common (4) or thrives (5) in association with that environmental characteristic. The species lists for each reach gave a series of scores for each environmental characteristic so that an average tolerance score could be calculated for each reach, both for in-channel and riparian vegetation. The average score was then calculated for each environmental characteristic for each TWINSPAN class (i.e. that group of reaches belonging in the same class) that had a distinct spatial location. Tables 20.2 and 20.3 provide the results of this cross-tabulation and

TABLE 20.3. Riparian vegetation: average scores for each TWINSPAN class based upon plant species habitat requirements

| | TWINSPAN class ordered from upstream (left) to downstream (right) | | | |
	00	10 and 11		01
Soil				
Peat	2.28	2.49	2.16	2.19
Soft mud	2.57	2.51	2.61	2.57
Clay soils	4.21	4.10	4.19	4.35
Sandy soils	3.59	3.77	3.97	3.30
Loose shingle	2.20	2.46	2.39	1.89
Rock fissure	2.52	2.92	2.61	2.22
Slope				
$>60°$	3.71	4.10	3.90	4.03
$30-60°$	3.50	3.97	3.55	3.84
$<30°$	2.90	3.33	2.77	2.43
Velocity				
Negligible/slow	4.16	4.05	4.26	4.00
Moderate	3.78	3.90	3.94	3.89
Fast	2.79	2.95	2.77	3.03
Chemistry				
Acid	1.62	1.87	1.55	1.78
Neutral	4.07	3.82	3.97	4.03
Alkaline	3.63	3.51	3.74	3.41
Tolerance				
Shade	3.21	3.51	3.39	2.92

Mean values are derived from allocating riparian species within the reaches of a particular TWINSPAN class a score of 5 to 1 according to whether they thrive, are common, occasionally colonize, are rare or absent in these conditions (from Newbold, Purseglove and Holmes, 1983)

aggregation process for the in-channel and riparian vegetation, respectively.

By identifying the highest scores for each of the environmental characteristics, the in-channel plant species information indicates that the river has a clay substrate, but with a tendency towards the coarser end of the clay range (higher scores on sand and lower scores on clay) in the middle reaches (10 to 42) of the river. Flow velocity is slow but decreasing scores on "slow" and increasing scores on "fast" downstream suggest some increase in velocity in a downstream direction. Flow depth is 0.1 to 0.5 m but with increasing scores on $<0.1-0$ m upstream and on $0.503-1$ m and >1 m downstream suggesting a deepening of flow downstream. Flow width is $5-10$ m but increasing scores on >10 m downstream and <5 m upstream suggest channel widening downstream. Base status of the water is neutral upstream but changes to rich downstream. Nutrient status is moderate upstream, increasing to rich downstream, with the changeover apparently occurring around reach 10. There is relatively low shade tolerance which decreases slightly downstream, and moderate disturbance tolerance.

The riparian species information indicates predominantly clay soil with a tendency towards the coarser end of the clay fraction (decreased clay scores and increased sand scores) in the middle reaches (30 to 50). Bank slopes are steep with the $>60°$ class receiving the highest scores. Flow velocities are slow and soils are neutral. Shade tolerance is moderate with the highest scores occurring in the middle reaches (30 to 50) of the river.

The data on habitat requirements presented by Newbold, Purseglove and Holmes (1983) were derived from a national survey. The various manipulations of the data described above using the species lists from the RCS are clearly effective in identifying the general character of the River Blackwater as a whole: a low-velocity, narrow, shallow and disturbed river developed on alluvium derived from clay to sand-based rock types. However, the clear, if not particularly strong, spatial trends that have been inferred from the vegetation species data are very surprising given the small size of the river and catchment and the limited range of apparent environmental gradients operating within the catchment. The validity of these trends was, therefore, explored by estimating the actual environmental gradients or surrogates for them using the River Blackwater GIS.

The spatial distribution of potential environmental controls on plant species

In identifying the actual environmental gradients, it is the local (at-a-site and immediately adjacent) properties that are most likely to influence the riparian vegetation, whereas the in-channel vegetation is also likely to be influenced by characteristics of the catchment area draining to that river channel reach. Therefore, in using GIS to define actual or surrogate environmental gradients, three spatial units were used:

(i)   the catchment area upstream of selected reaches;
(ii) the area of floodplain adjacent to each reach as defined by the maximum limit of known flooding and perpendiculars to the river channel generated from the reach end points;
(iii) the area defined by the river reach itself.

The first two of these spatial units were generated within SPANS. The identification of sub-catchments used a new SPANS module for diffusion analysis to ascertain from the DTM the areas flowing to, or "uphill" of specified reaches. The area of the floodplain adjacent to each reach was established from a voronoi map based on the reach centre points cut to the floodplain boundary. This effectively gave perpendiculars from the reach end points to the floodplain margins.

The following variables or surrogates were used to establish the environmental gradients that had been indicated from the plant species habitat requirements:

River bed and bank materials

The likely character of the river substrate and bank soil type was established from the distribution of rock types within the catchment. The proportion of different rock types within the catchment area was calculated from area cross tabulation between upstream catchment and rock-type polygons. Figure 20.4(a) plots the underlying rock type for each reach and Figure 20.4(b) plots the percentage of the catchment area under five different rock types draining to selected reaches. Figure 20.4 shows that the upper and lower reaches of the river are on London Clay, whereas the middle reaches are on the coarser Lower and Middle Bagshot Beds. The indication from the in-channel TWINSPAN classes of some coarsening of the substrate between reaches 10 and 42 corresponds closely with the zone where the river

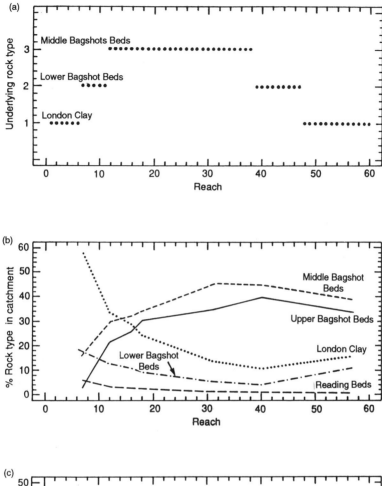

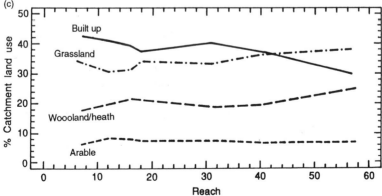

FIGURE 20.4. Downstream pattern in: (a) underlying rock type; (b) percentage of different rock types within the catchment area to individual reaches; (c) land use based on interpretation of Thematic Mapper data, plotted against reach number

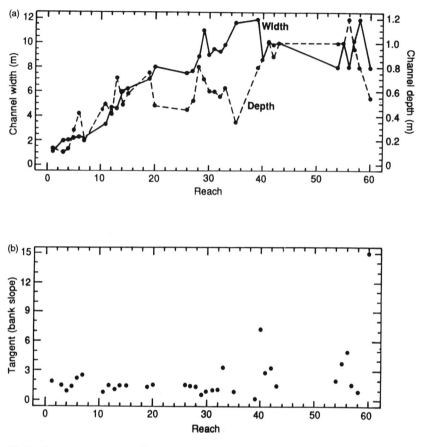

FIGURE 20.5.   Downstream pattern in: (a) channel width (m) and depth (10^{-1} m); (b) tangent of the bank slope, plotted against reach number

is underlain by Middle Bagshot Beds (Figure 20.4(a)). In contrast, the indication from the riparian TWINSPAN classes of some coarsening of the soil between reaches 30 and 50 appears to correspond with the pattern generated by rock-type percentages within the catchment area (Figure 20.4(b)), possibly reflecting the fact that the river banks are developed on materials from the upstream catchment rather than on the local underlying rock type. These explanations, although attractively simple and logical, must be treated with caution because of the heavy development of the River Blackwater floodplain for gravel extraction. The high percentage of open water on the floodplain (Figure 20.7) results from flooded gravel pits and is indicative of the level of development that has occurred.

Channel size

The inferred sizes and trends in river channel depth and width (based on the vegetation limit and, therefore, representative of the "bankfull" channel) and bank slope (based on the bank above the vegetation limit) from the in-channel and riparian TWINSPAN classes are confirmed by the available surveyed cross-section data. Figure 20.5(a) plots river width and depth as increasing downstream with size corresponding to the predicted $5-10$ m width range throughout the catchment and meeting the $0.1-0.5$ m depth range over much of the channel length. The tangent of the bank slope shows no clear spatial pattern apart from an increasing variance downstream, but the average tangent of the slope angle was 2.26 (66°), which falls within the predicted $>60°$ class.

Water chemistry

The likely pattern of water chemistry was estimated from three sets of information. The mix of rock types in the catchment (Figure 20.4(b)) provides a control on background water quality, but this is moderated by land use which was estimated through area analysis of the classified Landsat TM image for the sub-catchment areas draining to eight reaches (Figure 20.4(c)). A more direct indication of the urban impact on water quality can be gained by accumulating the discharge consent levels for sewage outfalls contributing to flow from varying sub-catchment areas. Thus SPANS was used to establish which of the discharge points fell within selected sub-catchments. However, since river discharge, which increases with the catchment size, dilutes effluent discharged to the river, a Discharge Consent Index (DCI) was derived by dividing the accumulated discharge consents in each sub-catchment by catchment area to produce an indication of the potential concentration of pollution within each reach. The resulting pattern in the DCI is presented in Figure 20.6(a). Average determinations for selected water-quality indices over three water years (1989–90, 1990–91, 1991–92) for the water-quality sample points located on the reaches used to construct Figures 20.4(a) and 20.6(b) show a similar spatial pattern to Figure 20.6(a), with relatively high water quality upstream of reach 10, which then deteriorates rapidly to reach 20. The water-quality surrogate of DCI and the observed water-quality indices confirm the inference from the in-channel TWINSPAN analysis that the water becomes increasingly nutrient-rich downstream. Figure 20.7 plots percentage of different land-use types on the floodplain adjacent to each river reach. These data were derived by an area analysis of the land-use polygons occurring within each floodplain reach polygon. Figure 20.7 presents running mean values plotted over five reaches in order to highlight the spatial trends occurring at the scale of the whole river. The increased scores for riparian shade tolerance between reaches 30 and 48 correspond with an increase in woodland cover of the floodplain. The slight decrease in inferred shade tolerance of the in-channel species downstream is presumably a result of the increase in channel width offsetting increasing tree cover on the floodplain.

This section has described the downstream patterns of single environmental characteristics or their surrogates and has established the degree to which these patterns correspond to those inferred from the vegetation species. Clearly this is a very simplistic approach which would be greatly improved by first analysing the strength of bivariate correlation between the pattern exhibited by the vegetation and by the environmental characteristics, and secondly by estimating multivariate relationships so that the relative importance of the interactions could be assessed. However, problems arise in developing either of these approaches with the current data set. From the above discussion, tables and figures, it appears that the vegetation is influenced by water quality and yet there are only eight water-quality sampling points on the River Blackwater. It would be a straightforward but scientifically questionable process to interpolate water-quality information for all 59 river reaches using a variety of modelling procedures within the GIS and then to estimate correlations between vegetation and

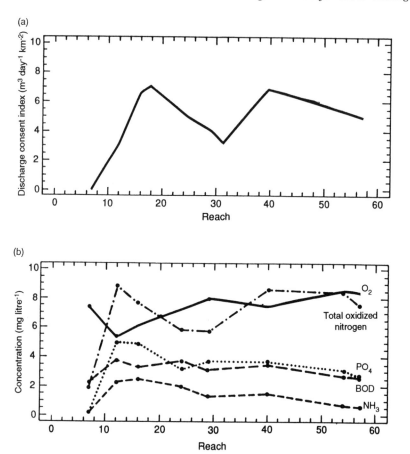

FIGURE 20.6. Downstream pattern in: (a) the discharge consent index; (b) mean of sampled water quality 1989–92

environmental properties over the entire sample of reaches. As Grayson et al. (1993) noted in the context of constraints to hydrological modelling in GIS, "There is a perception that GIS can generate information via interpolation but this is only possible under certain specific circumstances." In the present context, the only simple and valid approach to assessing and comparing the strength of the relationships between vegetation and individual environmental controls is to reduce the analyses to the eight reaches within which water quality is monitored. Given the small sample size, Spearman's rank correlation coefficients were estimated between the riparian and in-channel DECORANA scores and the various water-quality, channel-dimension, land-use and rock-type variables for these eight reaches. No significant correlations were found between the riparian scores and any of the environmental variables but highly significant ($p < 0.02$) correlations were estimated between the in-channel vegetation scores and dissolved oxygen concentration, catchment land use (built-up, wood/heath and grassland) and the underlying rock type. With a sample of eight reaches, multivariate analysis to establish the relative strength of the interactions is impossible, but such an approach would be the logical next stage if a larger geographical area were to be considered with more water-quality sampling points.

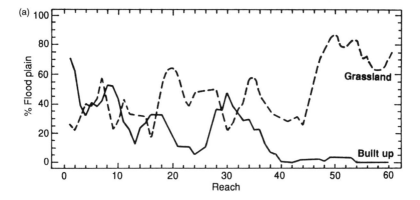

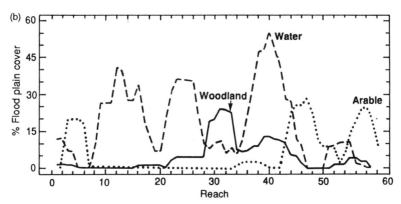

FIGURE 20.7. Land use on the floodplain adjacent to each channel reach. The diagram presents five reach running means of: (a) % built up and % grass cover; (b) % woodland cover, % arable and % open water

ERRORS AND GIS

It is tempting, having designed and assembled a GIS for the River Blackwater catchment, to undertake a range of analyses using the spatial modelling functions that are available. In the above analyses only very simple modelling procedures have been used because it was felt that the data did not support more sophisticated analyses. It is important to be aware of the vast range of sources of error that can be easily forgotten once a GIS has been developed. Burrough (1986) gives a thorough review of such errors, but in the present context the following error sources were thought to be most significant.

Data age

A variety of data sources was used in compiling the database, and, as is often the case with GIS applications, the dates of the surveys varied. For example, the river corridor survey was undertaken in 1986, land use was estimated from 1984 TM data, and water quality was based on surveys between 1989 and 1992. In addition to the varying age of different data sets, it is important to be aware of the fact that data standards change through time so that information from different dates on the same variables may not be strictly comparable, though the documentation of the data sets is often insufficient to indicate the nature of any difference and thus permit inter-calibration.

Areal coverage

Information is collected with respect to spatial units of different type (points, lines, areas) and spatial scale. Integration of information from these different spatial units is a straightforward process within a GIS but the results of such integration may be neither scientifically justifiable nor meaningfully interpretable. For example, integration of information for different-sized river reaches by overlaying will result in the smallest reaches having the strongest influence on the resulting pattern. Thus a simple GIS function (polygon overlay) can produce spurious results unless a purpose-designed overlaying strategy is developed.

Varying spatial scales of source information

Information on different aspects of the Blackwater catchment are available at different spatial scales, and are underpinned by differing densities of data points. For example, the 50 m resolution grid of the digital terrain model is derived by mathematical interpolation from the contours on the 1:50 000 scale Landranger maps; TM data are available as reflectance values for a ground resolution of 30 m^2 pixels; vegetation species lists relate to surveys of entire 500 m reaches whose boundaries are estimated by the surveyor in the field using a 1:10 000 scale map; river channel dimensions relate to the same 500 m reaches as the vegetation data but are only measured for one or two "representative" cross-sections within the reach. Clearly there is great variability in the positional accuracy and the representativeness of features associated with these different data sets.

Relevance of the information to the specified purpose

Ideal data sets are not always available because of the dependence on secondary sources, and therefore surrogates must be used. In the present study, TM data has been analysed with population census data to provide surrogate land use information, and information on rock type has been used to infer the character of bank and bed material in the river corridor. Whenever surrogate information is used it will not provide a perfect spatial representation of the variables of interest. Thus there will be an additional source of error to those listed above when surrogates are employed, though surrogate indices may offer invaluable data compression as well as simply filling gaps in the available data.

Potential errors in processing

In addition to the errors associated with the various secondary sources, errors will be accumulated as the data are entered into and processed through the GIS. In particular, digitizing errors can exaggerate errors in boundary positions that are already present because of the different scales and sampling designs associated with the source materials. Decisions are made when entering data into a GIS about the spatial resolution with which the data are to be held and these decisions can have significant effects on the results of subsequent spatial analysis.

These comments on errors illustrate that GIS is a very powerful tool that can be easily misused. Research is needed into error generation in general (e.g. Kemp, 1993) and, in the context of river corridor analysis, into the rationale for reach modelling in particular (Clark et al., 1993). It is accepted, for example, that the spacing of river geomorphological features is a function of the size of the river, and that river ecology is closely related to river geomorphology. Following from this, it is necessary to develop research on three fundamental issues:

(i) What is the appropriate base spatial unit for river corridor GIS applications?
(ii) What is the effect of combining data sets of different quality and spatial resolution?
(iii) How may we aggregate or split data associated with adjacent spatial units so that when we overlay different types of information we produce a spatial pattern that is not an artifact of the spatial units from which it is derived?

CONCLUSIONS

This paper has reported on the combined use of ordination and classification of plant species information; inferences on environmental gradients from plant species habitat requirements; and the application of visualization, diffusion modelling and area analysis techniques within a GIS to illustrate and quantify actual environmental gradients or their surrogates. These techniques were applied to data from a pilot catchment which was specially chosen to minimize the potential for obtaining striking results. It is a testimony to the information content of plant species information that environmental gradients have been inferred from the analysis of these data, which have largely been confirmed from the observed environmental gradients or their surrogates. Although it is impossible to identify exact causes for the gradients and classes identified from the plant species because of the limited size of the water-quality data set, the full range of inferred controls have been corroborated by observations of those controls. Throughout this analysis, the volume of data has necessitated careful data storage in a well-designed database structure. The power of some simple GIS functions in establishing catchment, floodplain and channel characteristics and their spatial distribution is also well illustrated by this pilot study. The Blackwater study shows the general power and value of combining a wide range of survey information in addition to integrating the more common types of map data within a GIS as a tool for understanding and, therefore, managing river channels in a more integrated manner.

The extent of the analyses presented has been limited by scientific caution rather than by the analytical power of GIS. Fundamental issues have been raised regarding errors associated with this particular GIS application and regarding research issues relevant to the development of the river corridor GIS in general. If river corridor GIS approaches are to develop as a tool for integrating secondary information at the landscape scale in a scientifically acceptable manner, then issues of data quality, scale, coverage and relevance must be addressed and reach modelling techniques must be developed that provide valid information on river corridor environments rather than artifacts of the data and data-handling methodology. On the other hand, as new sources of data become available (e.g. from the NERC Countryside Survey and from RCS), there will be considerable scope for developing integration models of this kind.

ACKNOWLEDGEMENTS

A brief description of the River Blackwater study has been published in Gurnell et al. (1993). Diagrams and text from this paper are reproduced here with permission of the International Association of Hydrological Sciences. The authors would like to thank the National Rivers Authority, Thames Region (particularly D. Mills, C. Woolhouse, A. Driver and J. Eastwood) for provision of data.

21

Practical Aspects of Restoration of Channel Diversity in Physically Degraded Streams

COLIN SMITH

University of Leicester, UK

TOM YOUDAN

National Rivers Authority, Kettering, UK

and CLAIRE REDMOND

National Rivers Authority, Peterborough, UK

INTRODUCTION

More or less distinct sets of stream invertebrates are associated with different bed materials and macrophyte types. Most fish and aquatic macrophytes also have definite depth, velocity and substratum preferences, while the birds and mammals which are found on watercourses have physical requirements of the channel. Details remain largely unquantified for macrophytes and macroinvertebrates, but we know that such preferences are widespread (chapter 18). Although habitat types differ in the numerical abundance and species richness of associated fauna and flora, biological diversity is clearly dependent on physical diversity in the river channel environment, subject to the effects of chemical water quality.

Many British lowland rivers were channelized during the early years of this century, following adoption of the 1930 Land Drainage Act in particular. During and after the Second World War, with a sustained drive for agricultural self-sufficiency, more work was carried out to extend the length of the grazing season on river floodplains or to bring them into arable production. Channelization work was planned only to carry the designed flood discharge and to provide the required floodplain drainage. A trapezoidal channel section was used, which minimized the land taken in enlargement. Channels were straightened, thereby increasing their gradient, putting land into production and simplifying maintenance. The bed was usually lowered, encouraging drainage of riparian land which was also routinely cleared of trees and river-bank hedges.

The Ecological Basis for River Management. Edited by D.M. Harper and A.J.D. Ferguson. © 1995 John Wiley & Sons Ltd

Legal protection of fisheries alone has existed for centuries (see Chapter 32). Inclusion of broad environmental duties in legislation is a more recent phenomenon, post-dating the major developments in land drainage and flood defence. Provisions of the 1968 Countryside Act and the 1973 Water Act, for example, were typical of a "reactive" approach, which sought primarily to limit the impact of otherwise-planned engineering works. Until the 1980s it was also hard for biologists to pursue active conservation with scant resources, given their first priority of extensive water-quality monitoring. The environmental role of the National Rivers Authority (NRA) for England and Wales has been recognized since its inception in 1989, and was consolidated in the 1991 Water Resources Act. The NRA now has a clear duty for development of conservation value which goes beyond the traditional role of responding to potential damage. The resources to pursue its duty have, to a significant extent, been provided.

There has historically been a conflict between river conservation and engineering, since professional perceptions of a "good" river and its function seemed inevitably opposed. A strengthened conservation input to channel management might have been expected to aggravate the conflict; but in practice it has meant that ecologists aware of flood defence needs — and calling on others' expertise for stream restoration projects — have found substantial common ground with river engineers. Initiatives for channel restoration are not motivated exclusively by issues of wildlife conservation or amenity value. Channelized streams have often been unstable, with the need for expensive mitigation measures or maintenance. Scouring of the bed or bankside and around structures is accompanied by subsequent deposition of sediment elsewhere, which both lead to continued management costs. Prolific weed growth in widened, unshaded channels — especially those of eutrophic lowland rivers — can also require annual weed-cutting to maintain channel capacity.

This paper discusses restoration measures using the case study of a lowland stream in the Anglian Region of the NRA; Harper's Brook, which is a tributary of the navigable River Nene in Northamptonshire. The work was carried out for conservation and flood defence reasons, with input from both disciplines in its planning and execution.

PROBLEMS AND OBJECTIVES

Harper's Brook is typical of many lowland streams in mid-eastern England. It has been channelized extensively, and some parts which lie within the River Nene floodplain have been diverted to allow extraction of gravels. Previous alteration of the channel has created both operational and environmental concerns for its continued management, which together provided a "shopping list" of desired improvements.

The lower reaches of Harper's Brook have supported prolific summer weed-growth. This itself reduced channel capacity; and the emergent plants acted as a focus for sediment deposition, with widespread formation of vegetated bars in mid-channel. In order to contain the designed flood, frequent maintenance has been necessary using hydraulic excavators equipped with weed-cutting baskets. There has been concern for the effect on an adjacent nature reserve of weed-cutting operations and bank-top re-location of the material so gained.

The bed of the brook is mainly gravels overlain with sand. Most larger stones have previously been removed and there is little transport of coarse sediment from upstream. Such coarse material as the stream contains is only poorly sorted by annual floods in its enlarged channel, which has a more or less uniform cross-section. Many of the aquatic macrophyte, invertebrate and fish species appropriate to the Harper's Brook favour conditions of flow / depth / substratum which are not presently offered by the channel.

In common with many streams in the region, the banks of the brook are uniformly steep to below the summer water level. Herbaceous marginal plants are restricted though in some places a small shelf has formed where *Phalaris arundinacea* lines the margin. Neighbouring streams with more physically varied margins support extensive stands of plants such as *Veronica beccabunga* and *Myosotis scorpioides*, which themselves provide habitat for invertebrates and young fish.

In the agricultural landscape which comprises much of the region, streams are important conservation features for more than the channel itself. Patches of trees and rough ground are potentially linked by the river corridor to provide a substantial resource. The brook has been largely cleared of bankside trees and hedges; and in the sections which were diverted there was certainly no planned vegetation of the riparian zone to replace losses.

Several basic objectives for future management of Harper's Brook were established, in response to the operational and environmental concerns for the stream:

- reduce the frequency of weed-cutting operations
- increase the heterogeneity of substratum / flow / depth
- provide a more varied channel margin
- provide a more diversely-vegetated riparian zone

MANAGEMENT METHODS

An enhancement programme for Harper's Brook was carried out in two phases. The first stage, implemented in 1992, was the introduction of a riffle sequence to the channel. The second stage, in 1993, was to increase physical diversity of the margins and partially re-vegetate the riparian zone.

Although Harper's Brook has a low gradient, riffle−pool sequences are found on similar streams in the locality, provided that coarse material is available. A riffle system has therefore been created to increase heterogeneity of the substratum and associated flow conditions. The basic capacity of the channel is reduced by restricting its cross-section at a number of places, but if seasonal deposition and vegetation of bars can be avoided or reduced then this will reduce the maintenance burden.

An ideal approach would be to distribute coarse bed-material widely and trust the stream to sort this during flood events. For Harper's Brook, however, the material was actually introduced in artificial riffles because the channel section is enlarged and sufficient floods to sort distributed material might be rare. On a natural river, riffles are typically 5−7 channel widths apart, but we did not know if this would be appropriate for such a modified channel. On the neighbouring River Welland, with a similar history of channelization, the position of introduced riffles has been aided by a correlation of riffle-spacing and discharge determined at sites with an established riffle sequence. Such correlation was not sought for the Harper's Brook because sites were apparent where the coarser fractions of existing gravel had accumulated and these were used for the introduction of material. There are two sections without riffles where suitable sites were unclear, but these will provide a future comparison with the populated sections and also indicate the extent to which the riffle material is transported.

The channel is 2−3 m wide over the reach in which 27 riffles were introduced. Existing bed material was removed and the riffles were constructed to about 0.5 m height and 7−8 m length with subsequent replacement of the finer borrowed material. The riffles were shaped to trail downstream, to produce some retention upstream and give a broken flow over the riffle.

The existing bank profile was modified in several areas, to provide a variety of bank slopes. Embayments were cut into the bank, together with a series of shallow berms at or above normal water level. Material won in shallowing the bank slope was used to create shallow scrapes in lakes on the adjacent nature reserve. The enhancement scheme used access to the south (nature reserve) bank but it has been decided to carry out future maintenance from the north bank only, thus reducing direct impact on the reserve. To permit this, the toe of the north-bank berm was reinforced with stone.

Alders and willows have been planted on the toe and bankside, with a mix of oak, ash, hawthorn, elder, blackthorn and dog rose in rough groupings on the top of the bank. Most of the tree planting is on the south bank, which might decrease the light available for excessive growth of submerged macrophytes, though there has been no quantitative investigation of this likelihood.

COSTS

Conservation work by the NRA and its predecessors has formerly been funded only "on the back" of management for other purposes such as land drainage and flood defence, albeit latterly as a prerequisite for such work. Resources are now available for pro-active conservation management but the process has a greater likelihood of support and success if, as on the Harper's Brook, projects are jointly promoted by ecologists and engineers in river management. The maintenance burden of this stretch was formerly about £250 per year for weed-cutting and £2000 for *c.* 7-yearly dredging.

Each riffle required 20 t of introduced material. No gravel was presently being extracted close to Harper's Brook but gravel rejects (*c.* 30–150 mm diameter) were obtained from workings 10 km distant. These gravels cost £3–4 per tonne at the pit which increased to £5 per tonne with delivery to the stream. The cost of material was therefore £100 per riffle; this would increase rapidly for larger streams though the Harper's Brook is typical of many streams in the region.

The livestock on the north bank of the stream had effectively been held in place by the channel, but with riffle construction this could no longer be guaranteed. An extra cost of the riffle system was therefore fencing, of a high standard, to prevent livestock access to these new potential fords. Fence timber and its transport cost about £100 per riffle — equivalent to the materials cost for the riffle itself.

The main outlay for the riffles was their installation, though this element of the cost would rise more slowly for larger streams than the cost of materials. Hire of the machine and its operator cost £25 per hour and there was additional labour to finish the riffle and place the fencing. The total cost per riffle was £500, which over the managed length amounted to about £6500 km^{-1}.

The cost of the fencing clearly increases with its required length, which depends on distance from the riffle edge; but if the fencing was too tight to the riffle then it might be compromised during floods and would need to be repositioned with shifts of the riffle material. A compromise was used which will probably involve some annual maintenance, and indeed some attention was required after the first winter.

The cost of trees was £500 km^{-1} and they were planted free of charge by the local Wildlife Trust, though planting would normally cost about £500 km^{-1}. Fencing to protect the young trees was the major cost — about £1500 km^{-1}. These costs would rise far less with stream size than would the instream enhancement measures.

Bank re-profiling cost about £5000 km^{-1}. As for the tree planting, however, costs would rise, at most, linearly with stream size. In both cases it is also worth noting that these are indicative costs for a "worst-case" stream with a more or less featureless river corridor.

DISCUSSION

There are initial signs that the introduced riffles may be successful. One of them was washed out during a flood but the material accumulated some distance downstream. This incurs an initial maintenance cost (the fences) but suggests that the stream is competent to sort riffle material once it has been provided, and also that the majority of the riffles are appropriately sited. Biological monitoring of several riffles is being carried out and they are being colonized by taxa such as Elminthidae and *Baetis* for which the channel was previously hostile. Physical monitoring of the morphology, stability and grain-size distribution of introduced riffles which will provide strong input to future methodology, is being undertaken as a university postgraduate dissertation.

Bank failure and bed erosion have also been a problem for channelized streams in the region. They have usually required local bank reinforcement and the construction of weirs, which themselves require maintenance. Unless streams can be re-meandered (as they have been with success in Denmark and the Netherlands), there will remain a need for structures to reduce the gradient of these straightened channels. Work elsewhere in the region is attempting to replace some of the weirs with riffle material; although they still need to be modified to retain the specified head, at least the substrate and flow conditions can provide physical habitat.

Costs arise in many ways, and may include measures to mitigate the effect of restoration measures on other functions of the river corridor. Riparian land use and the support of its owners or occupiers is very important to river management, never more so than in conservation-orientated restoration. Harper's Brook was a fortunate case since it was bounded to the south by a nature reserve, and to the north by a Site of Special Scientific Interest and by a sympathetic farmer. The land taken by re-profiling of the south bank would normally have been lost to agricultural production; the reserve also provided a ready use for the material so won; and the possibility of local bank scour during adjustment of the stream was accepted. In work on the River Welland in the adjacent catchment, a landowner transported material from farmyard to river on an *ad hoc* basis, reducing costs for the work.

As a result of their straightening and deepening over past years, we have inherited many monotonous rivers and streams. These often do not properly fulfil their function as wildlife habitats, and yet require a high level of maintenance to properly fulfil their function as carriers of flood water to the sea. If we can reintroduce physical diversity without adversely affecting their capacity to discharge flood water, then we can improve them as resources for conservation, amenity and efficient flood defence.

In some cases we can aim towards restoration based on a known history, but otherwise it must be valid to replace degraded channels with a close approximation. In either case, the pace of river restoration is increasing, with a growing background of experience and case histories to draw upon. There is a need for information to be collated and made available so that river restoration practitioners do not have to re-invent the successful methods, nor repeat the same mistakes.

22

Protection, Reclamation and Improvement of Small Urban Streams in Portugal

MARIA DA GRAÇA SARAIVA, PAULO PINTO,
JOÃO EDUARDO RABAÇA, ANDRÉ RAMOS and MARTA REVEZ

Instituto Superior de Agronomia, Technical University of Lisbon and University of Évora

INTRODUCTION

River corridors are major landscape elements, with very important ecological, amenity, scenic and recreation values. They establish a spatial and functional network in river basins and catchments, based on their pattern, geomorphological and biological characteristics, human uses and activities (Forman & Godron, 1986; Ahern, 1991; Boon, 1992a). Protection, conservation and enhancement of these features is an important environmental goal (Gardiner, 1991), particularly relevant in semi-arid and dry areas, where water can be a scarce resource for part of the year (Large & Petts, 1993).

Urban land uses in the areas surrounding river corridors largely affect their structure and functions, causing important environmental impacts (NCC, 1984; House, Ellis and Shutes, 1991; Fruget, 1992); for example, a reduction in the diversity of channel and bank due to maintenance and river drainage improvement schemes, and the disruption or destruction of natural and semi-natural adjacent habitats for urban use or development.

In Portugal, river corridors adjacent to urban areas show some degree of degradation. In the southern part of the country, those problems can be increased, due to the Mediterranean conditions (Saraiva, 1993a). Torrential flow with high flow variation, bank erosion and degradation, vegetation clearance due to urban, agricultural and overgrazing practices, pollution by urban and agricultural sources and a general situation of neglect with Authorities' "backs turned", are common problems.

The project described in this chapter is in progress in the southern part of Portugal, on the Sado River, to restore and improve two of its tributaries, flowing through the city of Évora as intermittent watercourses. There is a high flow variation between dry and wet seasons, and during summer they are usually dry.

The main purpose of the project was to improve the physical and visual conditions of river corridors, in order to create favourable conditions for the development of fauna and flora. The following components were incorporated:

- survey and elimination of pollution sources;

The Ecological Basis for River Management. Edited by D.M. Harper and A.J.D. Ferguson. © 1995 John Wiley & Sons Ltd

- survey and characterization of vegetation and aquatic life;
- cleaning and dredging of the river bed and bank stabilization;
- landscaping and enhancement of riversides and adjacent areas;
- bankside tree planting;
- creation of paths and riverside walks with information displays containing information about vegetation, fauna and general environmental values;
- public information and educational programmes for riverside users, to protect and enhance the environmental value of river corridors.

Location and preliminary survey

The city of Évora (Figure 22.1) lies just to the south of the divide of three main southern river basins of Portugal: the rivers Tejo, Sado and Guadiana. The project area, consisting of two tributaries of the River Sado in the suburbs of Évora, is in a granitic zone with rather flat slopes, but showing some important erosion effects. These tributaries are the River Xarrama, in the northern and eastern parts of the town, and the River Torregela, at the west side of the town (Figure 22.2). Both watercourses cross urban, suburban and rural areas, and are generally in poor condition in terms of water quality, vegetation clearance, bank erosion and stabilization.

A general survey of the catchment area was initially undertaken, covering geology, topography and slope, aspects, land use, hydrology and pollution sources, to identify the main problems that affect this area. More detailed characterization of physical and biological aspects of the watercourse in different zones was then developed. Three main zones, all along the River Xarrama, were identified according to the topography and land use: an urban zone, a peneplain zone and a rocky slope zone.

The main management problems identified were:

- torrential flow and high flow variation;
- bank erosion and degradation;
- vegetation clearance due to urban, agricultural and overgrazing practices;
- pollution by urban and agricultural sources;
- general situation of public abandon.

CHARACTERIZATION

Characterization of the following was undertaken:

- pollution sources and chemical water quality;
- biological diversity, including aquatic macroinvertebrates and birds;
- landscape survey, with the definition of reach types and the analysis of existing vegetation.

Pollution sources and chemical water quality

The water chemistry of the rivers Xarrama and Torregela (Saraiva et al., 1991; Simões et al., 1992) suggests a number of water-quality problems. The high levels of phosphorus (4.47 mg P litre^{-1}) and nitrites are usually a consequence of pollution from domestic effluents and animal farm activities (Saraiva, 1991). Very low values of dissolved oxygen (4.5 mg litre^{-1}) with high BOD and COD (448.9 mg O_2 litre^{-1}) also indicate high levels of organic inputs. In some places, the high values of silica are related to erosion of the

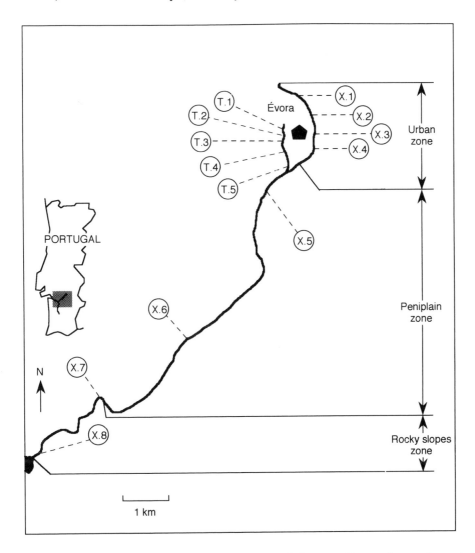

FIGURE 22.1. Study area, different topographic zones of Xarrama River and macroinvertebrate collecting sites on Xarrama River (X.1 to X.8) and Torregela River (T.1 to T.5)

underlying surface geology. Industrial effluents, although reduced in number, are responsible for high levels of pollution mainly in the River Torregela. Domestic effluents are very frequent which cause some pollution problems, despite their reduced charge.

Biological characterization

Vegetation

Table 22.1 shows that the aquatic and riverside vegetation is highly ruderal. In several places

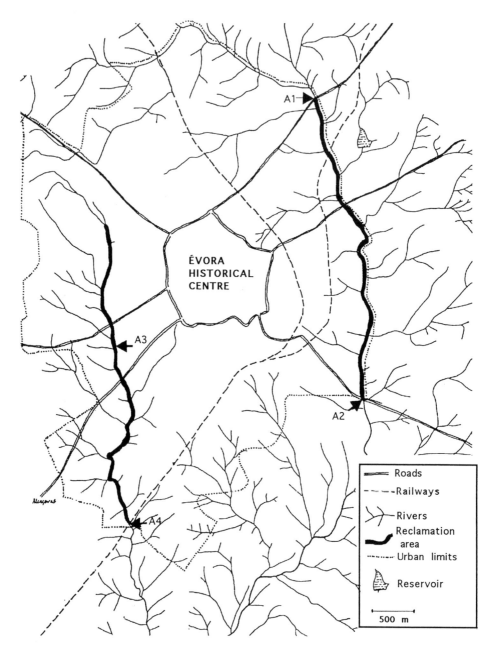

FIGURE 22.2. Reclaimed areas indicating the sites where the discharge was evaluated (A1 to A4)

banks and the river bed are invaded by species like *Rubus ulmifolius*, *Phragmites australis* and *Arundo donax*. High diversity of grass species occurs, often connected with agricultural activities. Tree and shrubs have been extensively removed by erosion and/or hand-cutting.

TABLE 22.1.   Most abundant plant taxa in the urban zone of Xarrama and Torregela rivers

Trees	*Populus alba*
	Populus nigra
	Fraxinus angustifolia
	Alnus lutinosa
Shrubs	*Salix* sp.
	Laurus nobilis
Grass	*Cyperus rotundos*
	Hordeum vulgare
	Briza maxima
	Amaranthus sp.
	Foenicum vulgare
	Torilis arvensis
	Fumaria oficinalis
	Anchusa azurea
	Mentha sp.
	Lavatera cretica
	Equisetum palustre
	Medicago arabica
	Rumex sp.
	Silybum sp.
	Senecio vulgaris
	Avena sterilis
	Papaver rhoesas
	Anacyclus radiatus
	Echium plantagineum
Aquatic vegetation	*Ranunculus tripartidus*
	Scirpus sp.
	Typha latifolia
	Poligonium persicaria
	Oenanthe crocata
	Juncus sp.

Benthic macroinvertebrates

Macroinvetebrates, sampled from points along the rivers Xarrama and Torregela using a 5 min hand-net sample (Hellawell, 1978), show a clear dominance of Chironomidae and Oligochaeta (Tables 22.2 and 22.3), which indicate high levels of organic pollution (Tachet, Bournard and Richoux, 1980; Pinder, 1986). The decrease in the number of species and individuals from spring to autumn, usual in these ecosystems (Pinto, 1988), in this particular case is increased by the organic waste deposited on the river bed during the summer.

The Torregela River has lower numbers of taxa and individuals than the Xarrama River. The highest densities· of Tubificidae also indicate this river to be more highly polluted (Brinkhurst, 1982). The high densities of Mollusca for the urban zone in spring are related to an increase of primary productivity induced by domestic organic pollution (Simões et al., 1992).

The longitudinal zonation of the trophic structure of the Xarrama River (Figure 22.3) shows two different situations in the urban zone. From spring to autumn a replacement of

TABLE 22.2. Macroinvertebrate taxa present in Xarrama River in the spring of 1991

Taxa	Collecting sites							
	X.1	X.2	X.3	X.4	X.5	X.6	X.7	X.8
Nematoda	3	1	1	0	0	2	0	0
Bithynia sp.	2	1	0	0	0	0	0	0
Physa sp.	0	1	4	6	0	0	0	0
Planorbis sp.	115	82	120	0	0	0	0	0
Ancylus sp.	0	0	22	5	0	0	3	0
Naididae	13	0	0	0	0	105	0	380
Tubificidae	0	27	10	49	26	134	5000	415
Lumbriculidae	0	0	0	1	0	0	0	0
Erpobdella sp.	0	0	0	6	0	0	1	2
Athyaephyra sp.	0	0	0	0	0	0	0	3
Procambarus sp.	0	0	0	0	0	1	4	0
Baetis sp.	6	0	3	16	0	11	1	48
Caenis sp.	0	2	0	0	0	0	1	7
Siphlonuros sp.	1	0	0	0	0	0	0	0
Cloen sp.	2	2	0	0	0	0	0	0
Thraulus sp.	0	0	1	0	0	0	0	0
Isoperla sp.	15	0	0	3	0	1	0	0
Perlodes sp.	0	0	0	0	0	1	0	0
Nemoura sp.	2	0	0	3	0	0	0	0
Plea sp.	0	1	2	0	0	0	0	0
Paracorixa sp.	0	1	2	0	0	0	0	0
Lacophilus sp.	0	0	0	2	0	1	0	0
Agabus sp.	12	17	2	0	0	0	0	0
Ditiscus sp.	1	0	0	0	0	0	0	0
Besorus sp.	1	0	0	0	0	0	0	0
Driops sp.	2	0	2	0	0	0	0	0
Stictonetes sp.	1	0	0	0	0	0	0	0
Ilybius sp.	1	0	1	1	0	0	0	0
Ceratopogonidae	276	150	400	250	8	100	700	46
Simulidae	551	86	10	739	0	112	8	10
Orthocladinae	2758	250	1300	477	1	200	86	129
Chironomus gr. *thumni*	0	100	0	23	35	49	595	0
Chironominae	0	0	0	0	0	36	0	0

scraper/grazers by collectors is observed. Cattle faeces deposited on the river bed during the dry period which, when inundated, increase the autotrophy of the river, may be the principal cause. Downstream of the urban zone the longitudinal pattern is similar for both seasons: the scraper/grazers tend to decrease while the collectors tend to increase.

The scores of the Belgium Biotic Index (BBI) (De Pauw & Vanhooren, 1983) emphasize a great decrease of the water quality from spring to autumn (Table 22.4). During the summer, the absence of flow allows the deposition of organic waste from human, agricultural and animal farm activities on the river bed. The first autumn rainfalls are not sufficient to clean the river bed, thus there is a decrease in the water quality. The very low values of the BBI in the Xarrama River at site X.5 are a result of the deposition of urban waste several hundred metres upstream (Simões et al., 1992).

TABLE 22.3. Macroinvertebrate taxa present in Torregela River in the spring and autumn of 1991

Taxa	Collecting sites					
	Spring					Autumn
	T.1	T.2	T.3	T.4	T.5	T.5
Physa sp.	0	0	0	6	0	0
Planorbis sp.	0	2	0	0	1	0
Ancylus sp.	6	1	0	0	θ	0
Tubificidae	89	1298	120	53	500	97
Lumbriculidae	0	2	0	0	0	0
Erpobdella sp.	1	0	0	0	0	1
Asellus sp.	0	0	0	0	3	0
Procambarus sp.	0	0	0	0	0	1
Baetis sp.	1	0	0	0	0	0
Cloen sp.	0	3	0	0	0	0
Lacophilus sp.	3	9	5	0	0	0
Ditiscus sp.	6	29	0	2	4	0
Potamophilus sp.	0	4	0	0	0	0
Disticidae n.i.	2	4	0	0	0	0
Ceratopogonidae	0	0	0	0	8	0
Simulidae	0	2	0	0	0	0
Tipulidae	1	2	0	0	0	0
Psycodidae	3	0	0	0	0	0
Orthocladinae	4	60	400	100	6	0
Chironomus gr. *thumni*	814	1890	400	100	6	16
Chironominae	0	0	150	0	0	0
Colembola	0	2	0	0	0	0

Birds

A point count technique (Blondel, Ferry and Frochot, 1981) was used to assess the breeding bird fauna of Xarrama River: 21 stations with a counting period of 10 min were set up along the watercourse, from the beginning of the urban zone (c. 8 km from headwaters) to the Vale do Gaio reservoir located near the mouth. Visits were carried out during the early morning under good weather conditions, and three observers with similar skills were involved. Additional information came from non-systematic visits to several stretches.

In all, 33 diurnal breeding bird species (aquatic, riparian and terrestrial species associated with the riversides) were recorded in the studied zones.

The low flow of the river and the lack of medium water bodies seem to be the main factors influencing the aquatic breeding bird fauna. Apart from a few pairs of mallards (*Anas platyrhynchos*) and moorhens (*Gallinula chloropus*) detected in places where suitable shore vegetation occur, no other aquatic bird species was recorded as a breeder in the stretches studied. Main species associated with fluvial habitats were: common sandpiper (*Actitis hypoleucos*) and kingfisher (*Alcedo atthis*), both found downstream of the urban zone; grey and white wagtails (*Motacilla cinerea* and *M. alba*), recorded only in the uppermost zone of the river, a 14 km stretch upstream of the Vale do Gaio reservoir where several rocky

SPRING

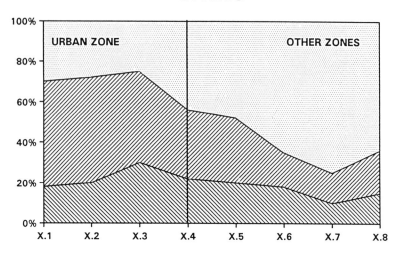

AUTUMN

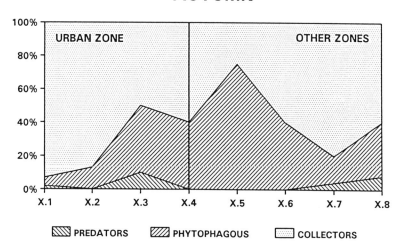

[legend] PREDATORS   PHYTOPHAGOUS   COLLECTORS

FIGURE 22.3.   Trophic structure of the macroinvertebrate communities in Xarrama River in the spring and autumn of 1991

outcrops are present in the channel bed and riversides; and Cetti's warbler (*Cettia cetti*), a species recorded in all sections of the river.

Only the moorhen and Cetti's warbler were found in the urban zone, although with different distribution patterns. Other breeding species were mostly terrestrial passerines associated with spots of good vegetation growth in the riversides: wren (*Troglodytes troglodytes*), melodious warbler (*Hippolais polyglotta*), blackcap (*Sylvia atricapilla*), Sardinian warbler (*S. melanocephala*), and nightingale (*Luscinia megarhynchos*).

TABLE 22.4. Scores of the Belgium Biotic Index (see text) in the spring and autumn of 1991 for the collecting sites

	Spring	Autumn
Xarrama River		
X.1	9 (I)	4 (IV)
X.2	5 (III)	2 (V)
X.3	6 (III)	4 (IV)
X.4	9 (I)	2 (V)
X.5	2 (V)	0 (V)
X.6	5 (III)	4 (IV)
X.7	5 (III)	4 (IV)
X.8	5 (III)	4 (IV)
Torregela River		
T.1	4 (IV)	
T.2	6 (III)	
T.3	2 (V)	
T.4	4 (IV)	
T.5	3 (IV)	2 (V)

I, No pollution; III, moderate pollution; IV, high pollution; V, very high pollution

Along the peniplain zone, the Xarrama River flows through an important bird area (Grimmet & Jones, 1989) with scattered holm oaks (*Quercus rotundifolia*), wheatfields and dry pastures. Within this area, a stretch of the river corridor (*c.* 1.8 km) with a line of poplar trees (*Populus nigra*) in the bankside, provides nesting places for white storks (*Ciconia ciconia*) and black kites (*Milvus migrans*). In the early 1980s, there was also a colony of grey herons (*Arderea cinerea*) nesting in a eucalyptus tree (e.g. Candeias, 1981), which fell down in the winter of 1984/85.

Landscape

Both watercourses show severe problems at the landscape level due to public and official neglect. The Torregela River flows mainly through an urban fringe, and almost all the catchment area is impervious due to urban development. There are general problems of litter, waste and sediment deposits, and pollution by illegal sewer linkages and sources. The Xarrama River has a larger river bed with a trapezoid cross-section as a result of former river regulation schemes. It flows mostly along rural arable or pasture. The low slope, however, induces sediment deposition, colonized by opportunistic vegetation. A low-flow course within the regulated one is formed during spring flows.

A detailed survey has been developed along the selected reaches to identify uses, vegetation and main problems along the river bed, banks and floodplain and suggests enhancement actions to be implemented in the next stages.

REHABILITATION

Following the survey and evaluation phases, a set of action plans was established (Figure 22.4), addressing the existing problems and the potential river corridor values (Saraiva,

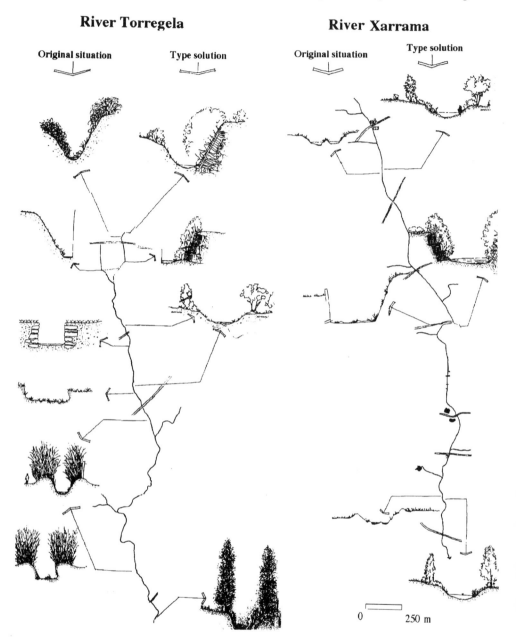

River Torregela

Original situation   Type solution

River Xarrama

Original situation   Type solution

0   250 m

FIGURE 22.4.   Reach types and implementation schemes

1993b). Some specific target landscape conditions (Aukes, 1992) were raised. In urban areas, along the Torregela, where land availability through municipal ownership did not represent a constraint, enhancement, landscaping and recreational improvements were the main objectives. Different landscape projects were implemented, with specific design schemes for bank stabilization and plantation (e.g. Figure 22.5), recreation and amenity

FIGURE 22.5.   Cross-section scheme of solutions for bank stabilization

areas, creation of paths and walks along riverside and integration of neighbourhood urban facilities. Provision of information panels along these paths is foreseen.

In rural and transition reaches, such as along the Xarrama, availability of space was the main problem, as land here is privately owned. Using legal powers and collaboration with landowners, a 10 m wide buffer strip along the bank for planting, protected by a temporary fence to avoid disturbance by cattle and agricultural practices (Figure 22.6), was developed. Half of the watercourse length has already been planted, and the remainder is expected to be concluded within one year.

Planting schemes were organized in units of 100 m long using tree and shrub species identified in the vegetation survey, and adapted to local moisture soil conditions with few maintenance demands. However, a one-year maintenance programme was included in the project contract, to assure irrigation and substitution of dead plants during the initial stage of development.

In the river channels the first priority was the development of cleaning and dredging procedures along the full lengths of all the watercourses. These procedures were undertaken with special care to eliminate litter, silt and other deposits, but avoiding the destruction of vegetation that protected banks from erosion. Woody vegetation obstructing the normal water flow was pollared. Either manual and mechanical processes were used as appropriate. The dense bankside vegetation sometimes overhanging the river channel in several stretches of the urban section of the River Xarrama was preserved as breeding bird habitat.

Public information was a special concern of this project. Its success will depend upon the involvement of residents and the public in general. Children were the main target of publicity. Several actions have been developed such as:

● Information panels concerning the project were erected during the dredging operations. Litter was made a particular target in the publicity.

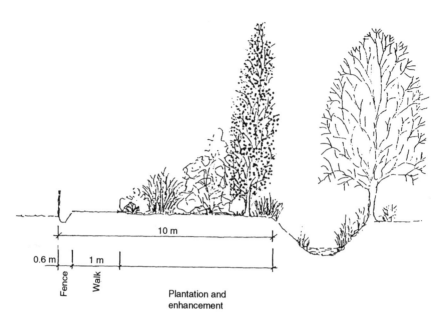

FIGURE 22.6.   Cross-section scheme of a planned buffer strip to implement in rural and transitional
reaches of Xarrama River

- A booklet was published and advertisments were placed in local newspapers and on radio.
- A video film was produced and has been shown in schools.
- Public awareness extended through conferences and technical meetings.

CONCLUSIONS

This project is a pilot experience on river restoration and reclamation projects in Portugal. These concerns are a new field of action and research in many countries in Europe and overseas. Traditionally, river management and maintenance involved "hard" techniques and the destruction of ecological and amenity values. The practices and techniques developed for this project represent a "starting point" in Portugal for more environmentally sound procedures for river management and enhancement based upon ecological information about the state of degradation and ecological targets for the restoration.

Good institutional collaboration at a regional and local levels and a multidisciplinary approach were achieved. However, some difficulties and conditions were encountered, such as the lack of a legal basis for river enhancement practices (e.g. planting on private land). Nevertheless, meetings with riparian owners to explain the project and obtain their agreement through an association of landowners to maintain the enhancement works has proved promising. Future monitoring to assess the impacts of this reclamation process, both its biological and social/amenity uses, is clearly necessary.

River management and conservation depends on the environmental concerns of responsible authorities and the public in general. In Portugal it is in its infancy. It is necessary to co-ordinate efforts, share experiences, develop research and spread public information and education to achieve these objectives. This project aims to employ a simple but comprehensive approach.

ACKNOWLEDGEMENTS

To Dr Rui Brandão for conducting the vegetation survey; to Prof. Manuel Mota for reviewing an earlier version of the manuscript and support on graphic solutions; to David Harper for useful suggestions; to Eng. Rui Bercmeyer for hydraulic research; and to Claudia Cruz, Maria João Silva, Sandro Nóbrega and Paula Simões for field work and laboratory support.

23

River Management and Mammal Populations

CHRISTOPHER F. MASON

University of Essex, Colchester, UK

INTRODUCTION

Mammal species may use rivers and river corridors in a variety of ways. Some species (e.g. river dolphins) spend their entire lives within rivers. Others (e.g. otters, desmans) are amphibious but obtain a large proportion of their food from within the river. A greater number of species (e.g. polecats, water voles) exploit the food resources of the riparian zone and may have their dens within river banks. Several bat species utilize the adult stages of aquatic insects. Some mammals may make particular use of a seasonal resource, such as bears congregating at salmon runs. Finally, many species visit the river casually, but importantly, for drinking and bathing.

The river corridor may serve as an important dispersal route, especially in extensively managed landscapes. Rivers may also prove barriers to dispersal, especially for small mammals. In tropical forest habitats it has been shown that many species of birds (and by implication other vertebrates) make intensive use of riverine strips, and the retention of such strips during logging results in the survival of the majority of species which are intolerant of conditions in recently logged forests (Johns, 1992). Although there is no direct evidence, intuitively one might expect the riparian zone of temperate rivers to support a greater number of mammalian species than the adjacent landscape — both those species from terrestrial habitats (woodland and grassland) and the riparian specialists.

Table 23.1 lists the European mammal species which are, more or less, associated with water (from habitat descriptions in Corbet and Ovenden, 1980; Corbet and Harris, 1991). Of the 23 species, seven are introduced, all but one due to the activities of the fur industry. Six species are bats. Of the 16 native species, eight obtain their food more or less exclusively from the aquatic or riparian zone, i.e. they are riparian specialists. Three of these (water shrew, northern water vole and otter) are resident in Britain.

There are four major aspects of river management which may impact mammal populations:

(1) modification of the riparian and within-river habitats;
(2) modification of food supply, especially fisheries management;
(3) management of water quantity (especially low flows and river regulation);
(4) modification of water quality.

The Ecological Basis for River Management. Edited by D.M. Harper and A.J.D. Ferguson. © 1995 John Wiley & Sons Ltd

TABLE 23.1.   European mammals associated with freshwater and riparian habitats

---

Pyrenean desman *Galemys pyrenaicus*: mountain streams and clear canals; localized distribution; feeds on aquatic invertebrates

Water shrew *Neomys fodiens:*[a] well-vegetated banks of a range of watercourses; feeds on aquatic and terrestrial invertebrates, small fish and frogs; may be affected by habitat destruction and pollution

Miller's water shrew *Neomys anomalus*: disjunct range; habitat, etc., as for *N. fodiens*

Pond bat *Myotis dasycneme*: wooded country near water

Long-fingered bat *Myotis capaccinii*: frequently near water

Natterer's bat *Myotis nattereri:*[a] open woodland, often near water

Daubenton's bat *Myotis daubentonii:*[a] often near water

Pipistrelle bat *Pipistrellus pipistrellus:*[a] often near water

Barbastelle bat *Barbastella barbastellus:*[a] woodland, often near water

European beaver *Castor fiber*: rivers and lakes with broadleaved woodland; vegetarian

Canadian beaver *Castor canadensis*: introduced to Finland, where more widespread than *C. fiber*

Coypu *Myocastor coypus*: introduced to Europe, extirpated Britain; marshes and canals with dense vegetation; vegetarian

Northern water vole *Arvicola terrestris:*[a] slow-flowing waterways, lakes, marshes; vegetarian on riparian vegetation; may have declined in Britain

Southwestern water vole *Arvicola sapidus*: as for *A. terrestris*

Muskrat *Ondatra zibethicus*: introduced; freshwater habitats; vegetarian on aquatic vegetation

Racoon-dog *Nyctereutes procyonoides*: introduced, expanding; woodland, especially by water; omnivore

European mink *Mustela lutreola*: marshes, lakes, river banks; carnivore on aquatic and terrestrial prey; severe decline

American mink *Mustela vison:*[a] introduced, expanding; habitat and diet as for *M. lutreola*

Polecat *Mustela putorius:*[a] woodland, marshes, river banks; carnivore, terrestrial feeder

Otter *Lutra lutra:*[a] freshwater habitats; carnivore on fish, amphibians, etc.; severe decline

Racoon *Procyon lotor*: introduced, spreading; woodland, especially near water; omnivore

Elk *Alces alces*: open woodland, especially river valleys and lakes; vegetarian

Chinese water deer *Hydropotes inermis*: introduced; open woodland and marshland; vegetarian

---

[a] Occurs in Great Britain.

These will be dealt with in turn below. In addition, I will briefly discuss what is known of the impact of introduced species on the native mammal fauna and on the management of rivers.

Ecologically unsound river management has been implicated in the declines of a number of native mammals associated with rivers in Europe: desman (Stone, 1991), water shrew (Churchfield, 1990), bats (Stebbings, 1988), water vole (Jefferies, Morris and Mulleneux, 1989), European mink (Saint Girons, 1991), polecat (Weber, 1988) and otter (Mason and Macdonald, 1986). However, in the majority of cases these implications consist of vague generalizations and speculation about habitat loss, water pollution, etc. Only in the case of the otter has there been any attempt to systematically investigate the effects of aspects of river management on populations.

HABITAT MODIFICATION

Adequate cover is essential to the well-being of most mammal species. Before the intervention of man, rivers were bordered by extensive forests or wetlands. The removal of bankside vegetation has been one of the major impacts on the river ecosystem. The density of bankside trees along rivers in Wales and eastern England has been compared (Figure 23.1). The majority of Welsh sites receive some management but tree removal has been extensive in eastern England. In Wales the majority of river banks support more than 200 trees (mature and saplings) per kilometre, whereas in eastern England most stretches have less than 50 trees per kilometre. Examples are shown in Figure 23.2.

There have been few studies which have attempted to determine the precise habitat requirements of riverside mammals or which have related river management to mammal distribution. Following channelization of a watershed in Vermont, USA, Possardt and Dodge (1978) found that populations of small mammals were almost half those in unmanaged areas, while two of seven species were not present in channelized stretches. However, none of these species could be described as riparian specialists. Channelization of streams in Oklahoma reduced the forest by 93%. A comparison of channelized and unchannelized, forested sites revealed reduced diversity and numbers of 11 species of small mammals and some larger mammals, but 16 other species of larger mammals (including beaver) were not affected or were favoured by channelization (Barclay, 1980, quoted in Brookes, 1988). In another study on the Alabama−Mississippi border, beavers and muskrats were more numerous in unmanaged reaches of a river compared with both old and recently channelized stretches (Arner et al., 1975).

In Canada, Racey and Euler (1983) conducted a detailed study of the effect of cottage development along lakeshores on mink populations. Removal of vegetation and road-building activity had a detrimental effect on the habitat of the mink, which required dense cover for their dens. Habitat changes also had an impact on food availability, which itself reduced the ability of the habitat to support dense mink populations.

All three of the native British riparian species (water shrew, water vole and otter) are said to be adversely affected by river corridor management. There is no evidence in the case of the water shrew (Churchfield, 1990). Jefferies, Morris and Mulleneux (1989) conducted a questionnaire survey into the status of water voles. Water voles were reported to have declined at 71% of sites where dredging was carried out whereas declines had occurred at only 32% of sites with no significant management. Water voles were considered absent at 36% of dredged sites but at only 3% of unmanaged sites. There is a great danger in such subjective analyses however. If people have formed the opinion that management is detrimental to water vole populations, then this will colour their observations and their approach to the questionnaire. In contrast, a field study by Weeks (1982) concluded that major river works had little effect on water vole burrows.

Rather more work has been done on the otter because of its threatened status (Foster-Turley, Macdonald and Mason, 1990). Macdonald and Mason (1983) investigated the distribution of otter spraints (faeces), as a measure of otter activity, in relation to a number of habitat and potential disturbance variables on fifty, 5 km stretches of rivers in Wales and adjacent counties. The number of signs of otters was significantly correlated with the density of mature ash (*Fraxinus excelsior*) and sycamore (*Acer pseudoplatanus*) trees and with the number of potential holts, 46% of which were in the root systems of these trees (Figure 23.3). Mature oak (*Quercus petraea*) made up a further 14% of potential holts. The frequency distribution of these trees on rivers with and without otters is shown in Figure 23.4. There were no other significant relationships between otter signs and other measured environmental variables.

Other studies have found relationships between the number of otter signs and bankside cover in Scotland (Jenkins and Burrows, 1980; Bas, Jenkins and Rothery, 1984), England

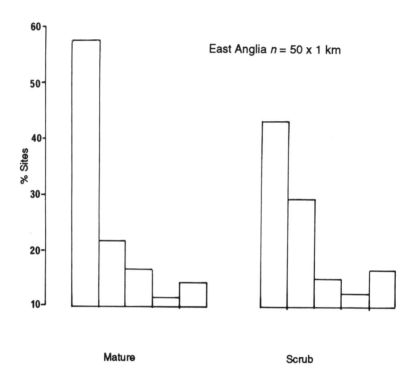

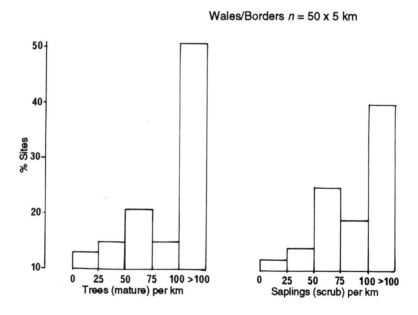

FIGURE 23.1. Frequency distribution of bankside tree and sapling (scrub) density in East Anglia and western Britain

(b)

(a)

FIGURE 23.2. Streams in (a) Wales and (b) East Anglia showing typical levels of management (January 1993)

FIGURE 23.3.   Characteristic otter den (holt) in the roots of a sycamore tree on a Welsh river

(Macdonald and Mason, 1988), Spain (Adrian, Wilden and Delibes, 1985; Delibes, Macdonald and Mason, 1991), Germany (Prauser, 1985) and Greece (Macdonald and Mason, 1982a; 1985). In Greece, high otter activity was associated with cover of *Phragmites* on irrigation canals, *Salix* on lowland rivers and *Salix* and *Rubus* on upland rivers (Macdonald and Mason, 1985).

A radio-tracking study in Scotland (Green et al., 1984) showed many above-ground resting sites of otters associated with scrub, while the most frequent holt sites were the roots of ash and sycamore trees. Otters radio-tracked as part of a re-introduction programme spent much of their time in riparian woodland (Jefferies et al., 1986), while a radio-tracking study of *Lutra canadensis* in North America (Melquist and Hornocker, 1983) showed that stretches of habitat with ample food were virtually unused by otters in the absence of sufficient cover and resting sites.

The study of Macdonald and Mason (1983) was conducted at a time when the otter population was at a low ebb and animals may have spent much of their time in the best habitat. Since then the population has expanded and some animals are forced into poorer quality habitat so that apparent relations between otter activity and habitat might be obscured. Indeed a study on the River Severn during 1986–88 (Delibes, Macdonald and Mason, 1991) found no relationship between otter activity and bankside cover. The study was, however, on a much smaller scale than the earlier investigation, only 20 stretches of 1 km being studied. In 1992 we undertook a survey of two catchments in south-west Scotland which held otters throughout (Macdonald, S. M. and Mason, C. F., unpublished). Of seventy-nine 1 km stretches of river surveyed (77 of which had evidence of otter activity) the cover was considered of poor quality for otters at 33%. Despite the ubiquitous

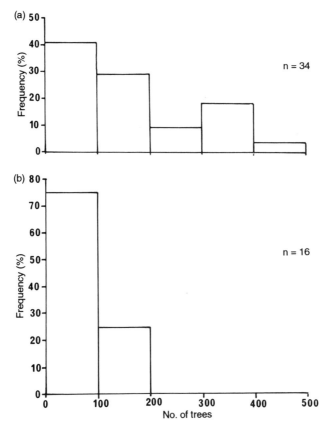

FIGURE 23.4.   Frequency distribution of mature bankside oak, ash and sycamore trees on rivers (a) with and (b) without otters

distribution of otters, there was still a significant relationship between otter activity and cover ($r = 0.34$, $p < 0.01$). Although otters will use stretches of river with poor cover, a range of studies has shown that cover is very important to them.

Pipistrelle bats are widely distributed over a number of habitats but in north-east Scotland they foraged extensively among riparian trees. Bats were not seen where the river passed through open fields with no riparian trees (Racey and Swift, 1985).

A single study has attempted to examine the effect of stream improvement on a riparian mammal (Burgess and Bider, 1980). In Canada, a 100 m channel was improved, leaving an adjacent channel as a control. Pools of up to 1 m deep were created by damming and cover was provided in the form of logs, stumps and rafts of alder. The activity of mink was higher over two years in the improved stretch than the control, though the improved zone was very small in relation to the normal range of mink.

Thus far I have considered the impact of habitat management on mammal populations. In contrast, the beaver acts as a keystone species to markedly alter stream structure and dynamics with a minimum of direct energy or nutrient transfer (Naiman, Melillo and Hobbie, 1986; Naimen, Johnston and Kelly, 1988). The beavers build dams and impound water, thus changing the discharge regime. The current velocity decreases and the area of flooded soil increases, with an increase in the retention of sediment and organic matter. A stream invertebrate fauna is replaced by a pond fauna, with collectors and predators becoming dominant over shredders and scrapers. The invertebrate biomass increases up to fivefold. By cutting wood, beavers virtually clearfell mature deciduous trees in the riparian zone; shrubby growth and eventually conifers becoming dominant. This in turn changes the pattern of inputs of organic matter into the water. Historically beavers have had a widespread influence on the landscape of river valleys in North America and they are now being used to rehabilitate streams with a long history of abuse (Johnston and Naiman, 1987; Naiman, Johnston and Kelly, 1988). The European beaver is less of a dam builder than its North American relative and its role in structuring stream communities has not been studied.

FOOD SUPPLY

Bankside and within-river management will influence the food supply of riparian mammals, as well as reducing cover, but there have been no studies. Otters are predominantly piscivorous and manipulation of fisheries could alter the carrying capacity of the habitat for the species. Food shortage was considered to be the ultimate cause of natural mortality in coastal-dwelling otters in Shetland, most deaths occurring in the spring when populations of intertidal fish were smallest (Kruuk and Conroy, 1991). Otters are resident in acidified streams in Scotland where fish numbers are low, with the animals diversifying their diet and taking a larger number of frogs (Mason and Macdonald, 1989). Nevertheless there was a relationship between otter activity and both pH and conductivity so it appeared that, although otter distribution was not affected, carrying capacity was. There is no information, however, on the minimum prey base required to support otters or on the prey biomass required to support a thriving population.

Fisheries affect otters in another way: many die in fyke nets (see review in Macdonald and Mason, 1994). Females and juveniles are most at risk and such losses are likely to be especially significant in depleted populations. In Denmark stop-grids must be fitted to the entrance of fyke nets (Madsen, 1991) and they are also compulsory in some parts of the UK. The Danish stop-grid is a 17 cm wire square divided into four smaller squares and tied at the entrance of the first funnel of the net. Eel catches are not affected by the stop grid.

WATER QUANTITY

Low flows in Britain are largely a feature of the lowland agricultural area of eastern and southern England. Nothing appears known of the effects of low flows on water shrews and water voles, while the otter is largely absent from this region for other reasons. Low flows may, however, have an impact on those otters being re-introduced to eastern England (Figure 23.5) and could become important should otters begin to recolonize former habitats in lowland England. More is known of the effect of low flows in continental Europe.

River flow is altered by the building of dams for hydroelectric schemes and by water abstraction largely for irrigation. In countries with mountainous regions, dam construction has proceeded rapidly reaching, for example, well over 2000 in Bulgaria alone (Spiridonov and Spassov, 1989). The dams themselves seldom provide suitable otter habitat, being too deep and steep-sided for successful foraging, and lacking, due to fluctuating water levels, adequate bankside cover. The negative effect of dams on otter distribution has been noted in, for example, Austria, Portugal, Spain and France (Macdonald and Mason, 1982b; Bouchardy, 1986; Ruiz-Olmo, 1991; Gutleb, 1992).

FIGURE 23.5.   Top of the River Stort, Hertfordshire, January 1992, dry due to overabstraction. Otters were re-introduced a few kilometres below here in December 1991 — an entirely unsuitable catchment for such a project (Mason, 1992, reproduced by permission)

In southern Europe and North Africa, rivers downstream of dams are subject to severely reduced flow or even desiccation, especially in summer. The resulting extirpation of the otter population on the Palancia River in Valencia was documented by Jimenez and Lacomba (1991). Water abstraction from downstream reaches for agricultural irrigation exacerbates the problem and is acute in east Spain where the current 51% of total natural water now abstracted is expected to rise to 93% (Jimenez and Lacomba, 1991). Over the last three decades otters have disappeared from Spanish rivers where flow has been reduced to 1 m^3 s^{-1}. Under conditions of low flow, relative concentrations of organic and industrial effluents increase.

Water abstraction is increasing in Mediterranean countries due to rising demands for irrigated land and to meet the requirements of growing tourist resorts. Wetlands such as the Coto Doñana marismas are drying as groundwater levels fall. Deforestation and overgrazing have also resulted in summer desiccation of rivers and hence the loss of fish populations and of local otters in southern Europe and in, for example, Morocco and Algeria (Macdonald and Mason, 1984; Macdonald, Mason and de Smet, 1985). In Tunisia the building of dams on the feeder rivers to Lake Ichkeul has reduced flow to such an extent that the lake is now hypersaline in summer and marshland is being lost. The effects on the food supply and habitat of the resident otters, a very important population in North Africa, may be severe.

WATER QUALITY

Pollution has been put forward as one factor in the suggested decline of the water vole (Jefferies, Morris and Mulleneux, 1989) but convincing evidence was not presented. Only for the otter has any attempt been made to link declines to water quality.

Some pollutants reduce the food supply of otters. The general deterioration of rivers due to farm effluents is of particular concern because it is taking place in otter strongholds in the west and north. However, there is no evidence that farm pollution has reduced the carrying capacity of affected rivers for otters. After a major fish kill on an East Anglian river, caused by piggery effluent, otters switched to a diet comprised largely of birds for several weeks. It was described above how similar dietary plasticity may allow otters to survive on acidified rivers with limited fish populations, though carrying capacity may be reduced (Mason and Macdonald, 1989). However, otters may not be able to live permanently in some acidified headwaters (Mason and Macdonald, 1987).

Otter populations and range are severely depleted over much of north-west and central Europe (Foster-Turley, Macdonald and Mason, 1990). Populations are thriving on the western seaboards and on the eastern periphery of Europe. Macdonald (1991) noted the plastics production in individual countries as an *index* of industrial output, and the direction of prevailing winds, and related these to otter distribution. Otters are extinct or threatened in those countries with high industrial output, or downwind of such countries. This indicates that the decline was caused by a contaminant that not only enters watercourses locally but is also widely dispersed by the winds. Because the decline was precipitate over a wide area it suggests a contaminant which reached critical levels during the late 1950s and 1960s.

Chanin and Jefferies (1978), through examining hunting records, concluded that dieldrin was the single most important cause of the decline in otters. The decline began more or less at the time that this pesticide was introduced into British agriculture. Other contaminants were discussed by Chanin and Jefferies, but dismissed. They ignored the fact that widespread industrialization in Britain and in much of western Europe in the 1950s, following recovery from the Second World War, would have led to increased contamination with compounds such as PCBs and mercury. Studies on lake sediments, for example, have shown an exponential increase in PCBs during the 1950s and 1960s (e.g. Sandars et al., 1992).

Heavy metals, especially mercury, are widely dispersed in the rivers of Britain and mercury occurs in eels often above the EC recommended level for human consumption (Barak and Mason, 1990; Mason and Barak, 1990). The majority of otters also contain heavy metals, especially mercury (Mason, Ford and Last, 1986; Kruuk and Conroy, 1991; Mason and Madsen, 1992; Mason and O'Sullivan, 1993a), but levels are such that it is unlikely that mercury was responsible for the widespread decline. Mercury does, however, act synergistically with PCBs in experiments with mink to reduce the survival of pups (Wren et al., 1987). PCBs are known to be more toxic to mammals than organochlorine pesticides and metals, and they have marked effects on reproduction, the endocrine system and the immune system of mammals. Perhaps they should be considered as a whole because there is little information on the physiological impact of suites of contaminants.

Figure 23.6 shows the concentrations of PCBs in otter tissues from various regions of Britain and Europe. The horizontal line at 50 ppm is the tissue concentration associated with reproductive failure in PCB-dosed laboratory mink (Jensen et al., 1977). Average concentrations of PCBs greater than 50 ppm were found in otters from south Sweden (population endangered), the Netherlands (extinct), East Anglia (wild population probably extinct) and Czechoslovakia (declining in many parts). The two otters with the highest levels from East Anglia showed pathological symptoms which included bleeding feet, deformed toes and claws, uterine tumours and skin lesions (Keymer et al., 1988). Symptoms were similar to those of Baltic seals considered to be suffering from PCB-induced adrenocortical hyperplasia, resulting in reproductive failure and dysfunction of the immune system (Bergman and Olsson, 1985; Olsson et al., 1992). Disorientation behaviour was observed in one animal from East Anglia. Such behaviour was also recorded, prior to death, in several otters from Ireland, from where blindness and pedal and integumentary lesions in otters were also recorded. Such symptoms are consistent with organochlorine poisoning and all such otters had elevated tissue levels of PCBs (Mason and O'Sullivan, 1992, 1993b).

A congener specific analysis of otter livers from Ireland, Britain and Denmark has shown similar patterns. Fourteen potentially toxic congeners were detected, comprising, on average, 62% of the total PCB concentration (Mason and Ratford, in press).

The otter is protected over most of its range and few bodies are received for analysis. The majority come from thriving populations where contaminant levels are likely to be low. To overcome this problem we have recently been analysing for organochlorine residues in spraints. The vast majority of the organochlorine residues measured in spraints are derived from the previous meal, i.e. that small proportion (some 10%) which is not assimilated. This can then be related back to intake and to likely accumulation in tissues (Mason et al., 1992). An example of the results obtained is given in Figure 23.7. Otters in East Anglia now belong largely, if not entirely, to a population that is being introduced from captive-bred stock. This region had the highest average level of organochlorine pesticide residues and PCBs in spraints. PCB levels in spraints (and presumably otters) increase rapidly following release (Figure 23.8).

Our studies have demonstrated the widespread presence of organochlorine pesticides and PCBs in aquatic ecosystems, although East Anglia showed the most consistently high level of contamination. There is a general increase in PCB levels from west to east in England and Wales (Mason and Macdonald, 1993a, b), northern Britain (Mason, 1993) and Ireland (O'Sullivan, Macdonald and Mason, 1993). Lowland stretches of rivers are more contaminated than upland stretches (Mason and Macdonald, 1993a). There are "hot-spots" of PCBs within this general distribution pattern (Mason and Macdonald, 1993c, 1994). Figure 23.9 relates the mean concentration of PCBs in spraints in regions to the percentage of positive sites for otters recorded in the most recent national survey. Clearly regions with a poor otter distribution are more contaminated with PCBs. Our study has indicated that, in some localities, PCB levels may be sufficiently high to pose a threat to otters.

IMPACT OF INTRODUCTIONS

Seven species of mammals which use river corridors extensively have been introduced into Europe (Table 23.1) and they can potentially interact adversely with the native fauna and

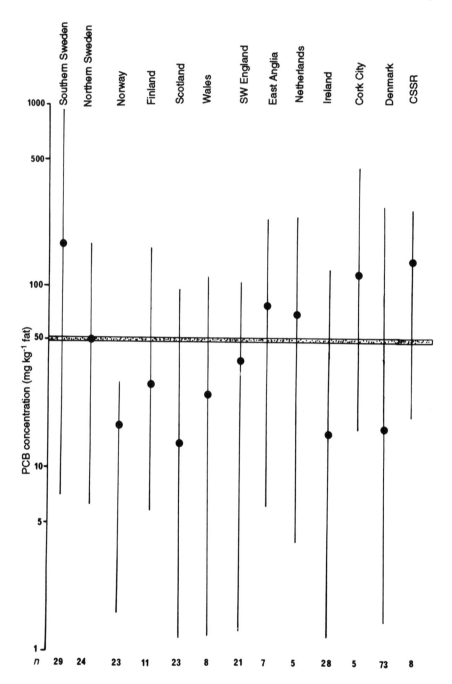

FIGURE 23.6.   Mean and ranges (mg kg^{-1} fat) of PCBs in otter tissues from various regions in Europe. The horizontal line at 50 mg kg^{-1} represents the concentration in tissues known to be associated with reproductive failure in mink. (Data from Olsson, Reutergårdh and Sandegren, 1981 (Norway, Sweden); Mason et al., 1986; Mason, 1988; Mason and Macdonald, 1994, unpublished (UK); Skaren, 1988 (Finland); Broekhuizen, 1989 (The Netherlands); Mason and O'Sullivan, 1992 (Ireland); Mason and Madsen, 1993 (Denmark); Hlavac, 1991 (Czechoslovakia))

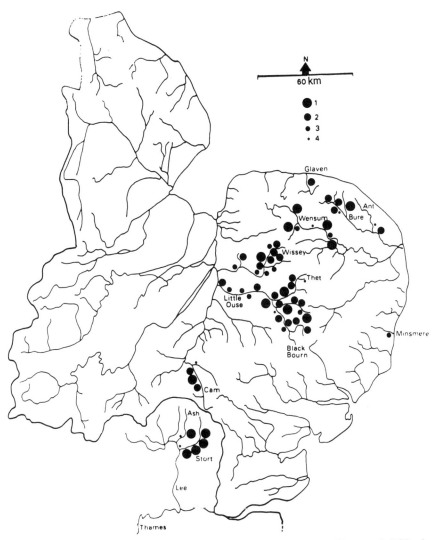

FIGURE 23.7. Mean total concentration of organochlorine pesticide residues and PCBs in otter spraints from sites in East Anglia, grouped in four levels. Contaminants in level 4 are likely to be associated with tissue contaminant levels causing adverse physiological effects on otters, while those in level 3 may cause effects. Samples in level 1 are considered to represent current background levels (Mason and Macdonald (1993b) provides more details)

flora or can damage man's resources. Particular concern has been directed at two introduced herbivores (coypu, muskrat) and one predator (American mink).

In Britain, the coypu had established a feral population by 1944 and quickly became widespread in East Anglia (Gosling, 1974), where it was accused of damage to river banks and crops, as well as to the indigenous flora. Systematic control began in 1962 and an eradication programme was started in 1981, which appears to have exterminated the coypu population (Gosling and Baker, 1991). Muskrats established populations in several counties

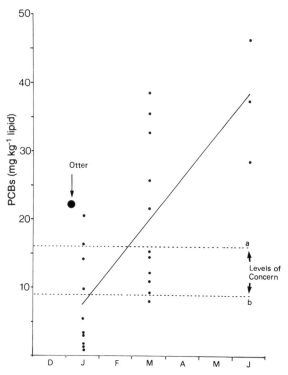

FIGURE 23.8.   Concentrations of PCBs (mg kg⁻¹ lipid) in otter spraints collected from the River Lea catchment from January 1992, following a re-introduction of otters in December 1991. Levels of concern a and b are considered likely or probably, respectively, to reflect tissue levels associated with adverse physiological effects on otters. Note the increase in PCB levels as otters expanded their range into the main river and downstream to the industrial town of Harlow. The otter was killed by traffic in the centre of a town, upstream of the release site, less than a month after release; it had already accumulated a substantial amount of PCB in its liver

of Britain in the 1920s but were eradicated by 1937 (Hills, 1991). Populations of coypu and muskrat persist on the continent of Europe.

The American mink was first recorded breeding in the wild in Britain in the 1950s and the species is now widely distributed in riparian habitats, as it is in a number of areas on the continent of Europe. Mink have been accused of depradations on livestock, on native wildlife (including competition with otters and European mink) and on fisheries. The occurrence of remains of domestic animals and reared game in the diet of mink is extremely rare (Dunstone and Ireland, 1989) and, at worst, the impact of mink on domestic livestock is no different from that of native carnivores; good husbandry is the best defence against livestock losses. Mink occur at low densities and, being small carnivores, have low energy requirements, so their impact on native prey, including fish, is likely to be slight (Dunstone and Ireland, 1989). Nevertheless, Woodroffe, Lawton and Davidson (1990) considered that in an area of Yorkshire where mink were spreading, they posed a serious threat to the survival of water vole populations. More work is required, especially on water vole populations in areas where mink have been long established.

Several lines of evidence suggest that mink are highly unlikely to be competitors of otters

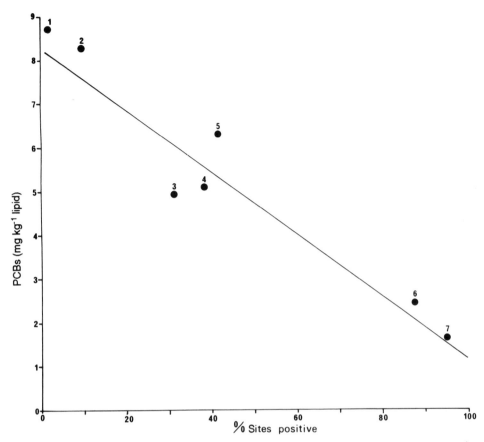

FIGURE 23.9. The mean concentration of PCBs (mg kg^{-1} lipid) in otter spraints from seven regions and the percentage of sites in the region positive for otters during the most recent national otter surveys carried out in the mid-1980s ($r = -0.96$, $p<0.001$). There were no significant correlations with individual pesticides. 1, south-west Ireland; 2, south-west Scotland; 3, Wales; 4, south-west England; 5, east-central Scotland; 6, northern England; 7, East Anglia

(Mason and Macdonald, 1986; Dunstone and Ireland, 1989). Competition between American mink and the rapidly declining European mink is more problematic. Saint Girons (1991) considers that there is "solid evidence" for competition and that it is "highly likely" but then states that competition "has not been formally proved by any serious study". Saint Girons suggests that the American mink is more prolific and more opportunistic than its European relative, making it a better competitor, and is also "more resistant to pollution". It is also suggested that the species can inter-mate. While this does not result in viable embryos, females will not come into heat again in the same year, leading to further declines in the population of the rarer species. Clearly the interactions between these two species in those continental areas where they co-occur should be an urgent research priority if any realistic conservation strategy for the endangered European mink is to be developed.

CONCLUSIONS

To effectively conserve mammal populations (and indeed any other animal group) in managed river corridors, it is essential to have detailed knowledge of their precise habitat and ecological requirements. Quite clearly this information is largely lacking for the majority of species, and conservation measures cannot be more than vague prescriptions. The otter provides an exception.

A number of studies have emphasized the importance of bankside cover to otters and several have identified specific features (e.g. tree species used as holts) in some areas. However, specific habitat features may differ in other regions, depending on availability, and we do not know the minimum amount of cover required to support a thriving otter population, or indeed how this may vary with other factors such as disturbance. For example, a river in northern Scotland, with poor cover but rarely visited by man, may have sufficient refuges for otters. A river in lowland England with equally poor cover but lined at weekends with anglers and with heavy pressure from walkers and their dogs, may prove totally unsuitable.

There have been a number of conservation initiatives for otters over the past decade. Project officers, often after only cursory surveys, invariably recommend tree planting schemes because, if nothing else, they are then seen to be doing something. Such planting is often quite unnecessary and indeed blanket planting of banks could be detrimental, for some stretches of bare bank clearly have major social significance to otters; they are heavily marked with spraints and other signs of activity, such as rolling places (Macdonald and Mason, 1987). Inadequate water quantity is likely to have a negative impact on otters where it affects food resources, especially fish numbers. Where rivers dry up completely and fish are absent, otters are likely to disappear. Nevertheless, where partial drying occurs, reducing the river to a series of pools where fish and amphibians may be concentrated, otter activity can be very high (e.g. in Portugal; Macdonald and Mason, 1982b). Otters certainly do live on rivers with a very poor food base but we do not know the minimum requirements necessary to support a viable population.

Bioaccumulating pollutants, and especially PCBs, are not only a major cause of the widespread and precipitate decline in otters over much of Europe but they are likely to be a major limiting factor preventing recovery in many areas. PCBs have been present in every otter, otter spraint and fish analysed in our laboratory and those PCB congeners known to exert adverse physiological effects make up a substantial proportion of the total. There is a strong negative relationship between the regional mean PCB content in spraints and the distribution of otters in the region determined in national surveys (Figure 23.9). Yet the statutory monitoring programme of river water carried out by the National Rivers Authority in England and Wales and the purification boards in Scotland rarely detect the presence of PCBs or other bioaccumulating contaminants. The statutory monitoring programme is therefore not presently adequate for protecting the wildlife resources of our rivers against the possible effects of these bioaccumulating contaminants.

We have derived some standards for PCBs (Table 23.2) from our recent detailed studies of organochlorines in otters and otter spraints which we believe would protect otter populations (Macdonald and Mason, 1994). The derivation of the standards can be criticized; for example, because the mink has been used as the experimental model to determine the adverse effcts of PCBs. We believe, however, that they are every bit as good as those used to protect human populations. We are, in effect, taking a precautionary approach to protecting

TABLE 23.2. PCB quality standards for protecting otter populations

In fish (whole body mince or flesh):	
PCBs	<0.026 mg kg^{-1} fresh weight is safe
PCBs	>0.05 mg kg^{-1} fresh weight *require action*
In otters (liver or muscle):	
PCBs	<10 mg kg^{-1} lipid weight is safe
PCBs	>30 mg kg^{-1} lipid weight *require action*
In otter spraints:	
PCBs	<4 mg kg^{-1} lipid weight is safe
PCBs	>9 mg kg^{-1} lipid weight *require action*

otter populations. Similar standards will also protect populations against organochlorine pesticides, which now occur at much lower concentrations in most rivers than do PCBs. Standards for this and PCBs should perhaps be considered in combination, however, because of previous concern over dieldrin.

It is clear that we still know very little about the factors influencing mammal distribution within river corridors and detailed research is urgently required. Most of the available resources for river conservation have been directed to river corridor surveys, especially of vegetation and birds. While the value of vegetation surveys is clear, for plants are static, those of animals are less so, because they vary in numbers over space and time. Their distribution in the year of the corridor survey may be very different from their distribution in the previous or following year — why they are where they are is not considered.

Many resources have been put recently into otter conservation projects but one cannot help feeling that many of these have been for publicity or cosmetic reasons and generate little conservation benefit. When funding is scarce it seems pointless to direct projects to areas where otters are thriving and no longer under threat, or to areas where otters are unlikely to colonize in the foreseeable future. Similarly re-introduction projects have little conservation merit for a country which still supports thriving otter populations, especially where preliminary studies prior to release are inadequate (Mason, 1992). Rather, conservation work on otters would be better targetted on those catchments at, or adjacent to, the leading edge of population expansion. Such targetted areas could be expanded if and when the population expands. Targetted action could allow a greater depth of study. For example, contaminant levels in fish are rarely examined in conservation projects even though contaminants were the main cause of the decline in otters. More research is also required to refine our understanding of the needs of the otter before conservation strategies can be effective — funds are rarely available for this!

24

The Management of Riverine Vegetation

MAX WADE

Loughborough University of Technology, Leicestershire, UK

INTRODUCTION

The club-tailed dragonfly (*Gomphus vulgatissimus*) has depended upon a range of vegetation associated with the river environment for as long as it has inhabited such European rivers as the Severn and Loire. Its life as a nymph is linked into the submerged weed beds which encourage sediment accretion and provide a refuge and source of food for its prey. Equally critical is the availability of marginal vegetation to enable the larva to crawl out of the water to undergo its transition to the adult state. Floodplain woodlands provide optimal habitat for the adult to hunt in, and meadows are used to seek out a mate before returning to the river for egg laying. Eggs are deposited in the mud whereas some other dragonfly species inject their eggs into the tissue of aquatic plants for protection. The management of riverine vegetation by those with responsibility for rivers increasingly reflects a need to consider vegetation outside of the confines of the channel not just for the conservation of rare dragonfly species, but in pursuance of better water quality, a complete river landscape, a healthier financial budget, and a more stable environment.

This chapter considers a variety of situations related to the management of the vegetation of river systems. It considers the range of communities associated with the river and its floodplain. These communities are considered in turn, starting with the truly aquatic plants and what has been traditionally known as aquatic weed control. The marginal and emergent zone is an important element of channel stability but equally critical is the transition from water to land. The bankside and riparian communities are a continuation of this ecotone, providing habitat such as riparian woodlands and establishing a buffering system between landward processes and the quality of the river water. The floodplain communities, where they remain, include a wide range of habitat types such as old channels, wet grasslands and, in recent decades, flooded gravel pits, all of which create a diversity of species ranging from aquatic plant communities to carr woodland. These components of the vegetation of the river system, although usefully considered separately, are part of a continuum and the chapter concludes by considering the integrated nature of the various communities and the need for management to recognize this interconnected nature. Boon (Chapter 19) emphasizes the importance of the management of rivers based on their plant communities and

The Ecological Basis for River Management. Edited by D.M. Harper and A.J.D. Ferguson. © 1995 John Wiley & Sons Ltd

the functioning of rivers through intimate contact between channel, riparian and floodplain habitats.

IN-CHANNEL COMMUNITIES AND AQUATIC WEED CONTROL

Aquatic vegetation — larger, macroscopic plants and microscopic algae — is of fundamental importance to life in rivers. All primary producers are an important food source, but additionally macrophytic plant structure adds both directly and indirectly to the physical diversity of the in-channel environment (see Chapter 18). In those rivers where there is little or no allochthonous organic material entering the water, the aquatic vegetation is the basis of the food chains. It is generally regarded that the more diverse the aquatic flora, the more diverse the fauna. This has been demonstrated for a range of freshwater habitats (Palmer, 1981) but further field investigation is needed to explore this relationship for rivers. The structure of the aquatic vegetation is important for fish in providing a substrate for egg laying and a refuge for fry. Certain species of waterfowl and fish feed on these plants and they can be important for nest construction.

Vegetation can also pose significant problems for river managers, ranging from creating resistance to flow to the impedance of navigation, leading to such management being referred to as weed control with an emphasis on weed cutting and the use of herbicides (e.g. Seagrave, 1988; Pieterse & Murphy, 1990). In many instances, however, the need to manage the vegetation arises from inappropriate use or management of the river as a system. Eutrophication by effluents from sewage treatment works, river regulation and the siting of human settlements on river floodplains, all result in changes to the river system which can induce the need to manage vegetation either because biomass has increased to a nuisance level or because hydrological criteria for the river have changed.

The first stage in developing a vegetation management strategy is to agree upon the target nutrient loading and flow characteristics for a given river in relation to the desired plant communities. Past data which exist for rivers can provide basic water-quality characteristics and a crude indication of plant species composition. Only very rarely do data exist for the size of standing crops and other such quantitative estimates. Nutrient concentration data are, in themselves, of limited value in that they need to be converted into mass flow and give little indication as to nutrient availability in sediments. This is of potential significance as roots have been identified as an important site of nutrient uptake for certain aquatic macrophyte species, at least in lakes (Barko & Smart, 1981; Smith & Adams, 1986). Biomass and tissue nutrient concentration are not always correlated with open-water nutrient concentrations for the aquatic macrophytes of rivers (Owens & Edwards, 1961; Ladle & Casey, 1971), which points to the importance of sediment nutrients to plant growth in flowing waters. In Danish streams, aquatic plant biomass and tissue concentrations were not correlated with either open-water or sediment concentrations, suggesting that some other factor or factors limited growth here (Kern-Hansen & Dawson, 1981).

In one attempt to implement aquatic plant control through nutrient limitation, results were unclear (Charlton & Bayne, 1986). Water quality and nuisance aquatic plant growth were monitored in the Bow River, Canada, after the installation in 1982 – 83 of phosphorus removal at Calgary's two sewage treatment works. The first two years, 1984 – 85, did not show a consistent reduction in plant biomass in spite of an 80% reduction in total phosphorus loading. An evaluation of nutrients and plant growth in the river from 1986 to 1988, however, revealed that a significant decrease in macrophyte biomass had occurred since

1983 at six of the 22 regularly sampled sites, with biomass also appearing to decline at some other sites (Sosiak, 1990). Most of the sites with a significant decline in biomass were in the mixing zone for Calgary's sewage treatment effluent.

Nutrient control is but one aspect of the river which could be managed to manipulate the submerged and floating vegetation. Flow (Nilsson, 1987; Chambers et al., 1991) and shading (Dawson & Haslam, 1983) are two other useful variables which can be used to modify plant growth.

EMERGENT AND MARGINAL COMMUNITIES AND CHANNEL STABILITY

An ecological approach to the management of river vegetation is not a new idea. For a long time the maintenance of ecological stability of the river bank through the manipulation of the emergent and marginal communities has been recognized. The submerged and floating aquatic plants can provide a degree of stabilization of wave action and general energy absorption, but it is the community of the reed-bank zone which has the most potential (Hemphill & Bramley, 1989). This zone comprises those plants which either grow up out of the water with emergent leaves and stems or grow along water margins with roots extending below the water table. Emergent plants grow in depths up to 2 m, but most species are confined to depths of less than 0.5 m.

Emergent aquatic plants can be important in protecting river banks and the floodplain from damage due to high flows and, in navigable rivers, boat traffic. Murphy, Bradshaw and Eaton (1980) assessed the protective potential of reed-swamp species on the lower slopes and stoloniferous grasses on upper slopes of canal banks in Britain: *Glyceria maxima* and *Sparganium erectum* grow in this marginal zone effecting good protection and *Agrostis stolonifera* is useful on the bankward part of this zone due to its binding root system and pliable upper parts. Studies on the River Thames, UK, have investigated the dissipation of energy waves produced by boats for various species. Sixty per cent of boat-wave energy is dissipated by a 2 m band of *Phragmites australis*, 70% by a 2−5 m band of *Scirpus lacustris*, 75% by a 2−4 m band of *Acorus calamus* and 60% by a 2−3 m band of *Typha angustifolia*. Such vegetation also encourages siltation by absorbing current flow energy, and thus reducing the sediment carrying capacity of the flow (Hemphill and Bramley, 1989).

Planting of such species can be achieved manually or by using machinery such as a dredger. The dredged material can be used either to return vegetation to sections of the river being managed, or for transplantation to another stretch of the same or a different river. Some species colonize more aggressively than others, thus selection of the most appropriate species to prevent problems arising through excessive growth and channel obstruction is necessary. This marginal zone is also prone to invasions by alien plants species such as *Impatiens glandulifera* and *Fallopia japonica* (also known as *Reynoutria japonica* or *Polygonum japonicum*) and care is needed to prevent the establishment of such invasive plants (de Waal et al., 1994; see Chapter 25). Planting programmes should use native, and preferably abundant, plants taken where possible from the same river or one of its tributaries. Some countries have legislation restricting the removal of plants; in the UK, for example, the Wildlife and Countryside Act 1981 prohibits the uprooting of any plants without the permission of the landowner or occupier.

The planting of a mixture of species rather than single stands leads more rapidly to a balanced community with a reduction in the likelihood of excessive growths developing of one species. Planting in early spring as the shoots first appear above ground level is recommended (Lewis and Williams, 1984) and it is preferable to transfer sediment with the plant material to aid establishment and protect the plants from any late frosts. Manuals such as that by Hemphill & Bramley (1989) provide specific advice on the water depth, soil type, water quality, flow velocity, planting method and growth for a range of emergent species commonly used in bank protection. Most aquatic plants are easy to propagate from cuttings, whole plants or rhizome fragments. *Phragmites australis*, a popular candidate for propagation, is an exception, noted for its unreliability as a species for planting. Propagation can be achieved from planting clumps, cuttings, by layering or using seed. Planting clumps and cuttings is usually successful but it is important to locate clumps above the mean water level (rhizomes planted below water level will usually rot), to avoid damage by trampling to the shoots in the donor stand and in the planted material, and, in the case of cuttings, to plant at water level and set them so that at least three leaf nodes are buried. Advice on the best time of year for planting varies, e.g. March to May (Hemphill and Bramley, 1989) or October to March (Lewis and Williams, 1984). Lewis and Williams (1984) describe the conditions necessary for germinating *Phragmites australis* seeds.

BANKSIDE AND RIPARIAN COMMUNITIES AND THE OPPORTUNITY ZONE

The focus on ecotones and their ecology in such programmes as UNESCO's "Man and the Biosphere" has encouraged studies and projects on the bankside and riparian habitats. One such habitat type is that of riparian woodland, and the creation of such woodlands or changing the management of existing woodland illustrates well the way that riparian communities can be integrated into the river environment. The value of riparian zones is recognized in the latest edition of the *Forests and Water Guidelines* (Forestry Commission, 1991) which emphasizes the need for sensitive woodland management and careful planning as the zone itself is fragile and water quality and habitat can both be easily damaged. In the UK, grants are available for planting trees and with recognition of the role of such woodland as buffer strips (see Chapter 19), there is strong potential and a need for riparian woodland creation.

An interesting example of such woodland management is the Kirk Burn, a tributary of the River Tweed, Scotland. The riparian vegetation for the stream has been coniferous forest for many years but a recent scheme has removed the coniferous trees from 20−40 m either side of the stream and replanted the area with deciduous trees. In a short period of time the ground flora has diversified, though for such schemes the vegetation which recolonizes may frequently be of less conservation value than the vegetation of neighbouring burns which have never been afforested. The process of revegetation is very slow, possibly because of nutrient status, depth of needle litter and depth of felling debris. Marginal vegetation growing down and over the water has improved habitat for trout. It is predicted that revegetation will be followed by increased bank stabilization, enabling the stream to become narrower and deeper with consequent improvements in the diversity and biomass of the invertebrate fauna, and in the biomass of the fish (Adamson, J., pers. comm.).

Studies undertaken by the Forestry Commission in conjunction with the Institute of Terrestrial Ecology in south-west England have also implicated shading by conifers as a factor which increases stream erosion, with revegetation having a beneficial effect in reducing erosion. Landscape and amenity advantages have been shown to encourage visitors to walk in the river valley. This and other such projects have been important in contributing to the *Forests and Water Guidelines* (Forestry Commission, 1991) in which three main points are emphasized with respect to riparian woodlands:

(1) The riparian zone should be properly identified and should include features such as headwater source areas, gully banks and terraces. The view of the marginal zone should be a broad one and include floodplains and adjacent wetlands.
(2) Management should ensure that either open or partially wooded conditions are maintained to stimulate a vigorous and thriving ground vegetation. Maintaining good stock-proof fences around woodland will help to prevent overgrazing of the riparian zone and favour the growth of bankside plants.
(3) Maintaining about half the length of the stream open to the sunlight, and the remainder under dapple shade from bankside trees and shrubs is desirable for a diverse aquatic ecosystem. It is particularly important to maintain open ground to the south of streams to allow more light to penetrate to the stream and increase the water temperature for better aquatic plant and invertebrate growth. The periodic cutting of natural regeneration to maintain sections will be required.

A number of factors need to be considered when deciding on which species of trees to plant. Shading intensity, for example, is important, with species such as oak casting a dense shade as compared with lighter foliaged trees such as birch, willow, hazel and rowan. Litter from the lighter foliaged trees decomposes in the water more rapidly than litter from oak, beech and most conifers, and can provide valuable organic material for the aquatic invertebrate community. Alder naturally occurs in the riparian zone but in addition to casting a heavy shade it can also contribute to acidification in some sites.

One of the problems with documents which provide guidance in the form of statements such as average distances, is that it is taken literally. For example, for riparian woodlands, streams should have unplanted strips either side of the channel not less than 5 m wide. Duncan Ray (pers. comm.) reports that too often the recommendation was taken literally, creating unplanted corridors with walls of conifers. These strips should include the riparian zone and extend into the adjacent land in order to qualify as buffer areas. Their function is to protect the stream from sediment and pollutants being widest at points of discharge from ditches.

The management of riparian woodlands might also necessitate felling in order to maintain habitat structure and diversity as well as ensuring the economic viability of such planting. This should be carefully planned as clear felling can lead to erosion and damage to the river, if not bank slip and damage to the riparian strip. Shading can be important in providing cool spots for fish to rest up in, or alternatively in suppressing aquatic macrophyte growth in relation to flood defence objectives. Selective felling would be preferable, allowing time for regeneration to occur.

In addition to the management and creation of woodlands, the riparian zone has been the focus of attention in developing and using buffer zones to enhance biodiversity and to protect water quality. In both instances the manipulation of the vegetation is fundamental to the success of the buffer zone and its sustainability. The opportunities associated with the riparian zone are considerable and further research is needed to understand more about the influence of community structure and successional processes in its functioning (Malanson, 1993).

FLOODPLAIN COMMUNITIES AND LOST OPPORTUNITIES

In many rivers the floodplain component of the system has been divorced from the river itself, often to the detriment of the ecology of both components. The vegetation of ox-bow lakes illustrates the value of this relationship. Studies in the Trent valley, UK, have shown how ox-bows have developed a particularly diverse aquatic flora; an ox-bow at Sawley, for example, has built up a community of between 50 and 60 species since its creation in about 1700. The dominant plants are characteristic of the flora of the River Trent itself but factors such as greater niche diversity, protection from pollution present in the main river and the absence of disturbance to the littoral zone have enabled other species to become established. This reservoir of species is important in providing a source of propagules for distribution along the river, with periods of high flow and flooding facilitating dispersal. Over an observation period of 12 years, species found to become distributed below the ox-bow included *Butomus umbellatus*, *Lemna polyrrhiza*, *Hottonia palustris* (in ponds adjacent to the river), and *Oenanthe fistulosa* (Wade, in prep.).

The vegetation of ox-bows provides a refuge for a range of riverine species. These include wildfowl which use such habitat for nesting and roosting, e.g. for herons and cormorants, using the river as a source of food or as a corridor. Gravel pits and other such water bodies provide a similar function. For fish, however, a direct connection between the river and ox-bow is important. The vegetation structure of the ox-bow allows spawning and also provides a refuge in which fry can grow up before moving out into the river (Pinder, C., pers. comm.). The submerged component of the aquatic plant community is of particular importance, including species such as *Ceratophyllum demersum* and the wintergreen *Callitriche* species. An ox-bow is also relatively buffered from pollution incidents in the main river and can act as a retreat for fish under such conditions.

An ox-bow, along with other floodplain habitat is not a static entity. Successional processes are in operation and the accumulation of sediment and organic detritus will over the decades cause the ox-bow to pass from an open water body to an alder—willow carr. Each seral stage has its own value in terms of species diversity and functional role within the river corridor, e.g. the provision of suitable habitat for different dragonfly species, otters or nesting birds. River management needs to take stock of the distribution and frequency of these various floodplain habitats and develop appropriate management. This might necessitate reconnecting or connecting certain water bodies to the river, such as ox-bows deliberately cut off from the river or recently created water-filled excavations. Additionally, management could be undertaken to slow down the rate of succession by removing accumulated sediment and detritus from a given ox-bow and cutting back or pollarding associated tree growth.

Ox-bows and other floodplain features are the types of sites which might be acquired by nature conservation agencies and it is important that integrated management plans are developed between such agencies and national or regional river management bodies.

CONCLUSION

The examples described above have consistently highlighted the need for an integrated approach to the management of the vegetation of rivers and their associated margins and floodplains. It is not only the dragonfly that exemplifies the interconnection of these plant communities; it is also seen in the habitat requirements of species such as the otter and the heron. Aquatic plants are not necessarily able to maintain themselves in the main channel, and may rely on recolonization from ox-bows and similar floodplain water bodies. There is an increasing body of information on which to base such integrated management — information which spans a range of traditional professions. For the establishment of buffer zones in the UK, guidelines are emerging from the Ministry of Agriculture, Fisheries and Food, English Nature and the National Rivers Authority to add to those already available from the Forestry Commission. Successful river vegetation management will require such agencies to work closely together to achieve agreed objectives.

Deriving these objectives will be one of the main challenges in river management over the next decade. These objectives need to take into account the whole spectrum of uses to which the river and its floodplain are put, ranging from flood defence through water quality and fisheries management to the agricultural and other land uses within not only the floodplain but the catchment itself. Fundamental to these objectives will be an agreement upon the nature and extent of the different plant communities within the river corridor. The vegetation can be mapped out at different scales for planning purposes. This could be at a general community scale, e.g. to show the extent of riparian woodland and different floodplain communities, or it could provide the necessary detail for a stretch of river, indicating the locations of individual trees and weed beds. Such descriptions of vegetation would also be valuable in laying down criteria by which a given management programme could be assessed. Assessment should be established on the basis of both species composition and the extent of a given community type. Such appraisals will be essential in learning how best to modify management schemes and in being able to justify the financial costs involved in such projects.

25

The Management of Three Alien Invasive Riparian Plants: Impatiens glandulifera (Himalayan balsam), Heracleum mantegazzianum (giant hogweed) and Fallopia japonica (Japanese knotweed)

LOUISE C. DE WAAL,

Wolverhampton University, Wolverhampton, UK

LOIS E. CHILD and MAX WADE

*Loughborough University of Technology,
Leicestershire, UK*

INTRODUCTION

The management and functioning of the riparian zone is of increasing significance to river authorities. This can be with bank stabilization in mind, for amenity purposes or for the creation and maintenance of buffer zones. Whilst the management of emergent vegetation is relatively well documented, little attention has been paid to the control of unwanted vegetation on the banks of rivers. The riparian habitat has been readily invaded by exotic species to the extent that almost half of all neophytes naturalized in Europe occur along waterways, including rivers and canals (Sukopp, 1972). This large number of exotic species in riparian communities is due to several factors. River banks are kept open by the action of water, resulting in the continuous presence of sites with little competition. Water provides an effective means of dispersal for many species and the considerable anthropogenic influence in these habitats, such as flood defence works, is also a significant factor. A comparative study of three such neophytes — significant weed species in Europe and in the case of *Fallopia japonica*, also in the USA (Locandro, 1973) — has been undertaken in order to define an ecological basis for the management of such invasive riparian species.

Introduced species achieve different measures of success in establishing themselves within the flora of a region. Four stages are distinguished in the process of naturalization of alien plants, each subsequent stage being more difficult to achieve than the previous one (Kornas, 1990):

The Ecological Basis for River Management. Edited by D.M. Harper and A.J.D. Ferguson. © 1995 John Wiley & Sons Ltd

(i)   introduction of propagules and appearance of first individuals;
(ii) establishment in heavily disturbed sites;
(iii) colonization of less disturbed sites;
(iv) invasion into undisturbed sites.

Impatiens glandulifera, *Heracleum mantegazzianum* and *F. japonica* have all reached stage (iii) and in certain instances they have demonstrated the ability to invade undisturbed sites. They have all become widely distributed, both throughout the British Isles and elsewhere in Europe, and have reached pest proportions, posing serious problems along river corridors (de Waal et al., 1994). An indication of the seriousness of these invasive species in the UK is the inclusion of two of them, *H. mantegazzianum* and *F. japonica*, under Schedule 9 (Part II), Section 14 of the Wildlife and Countryside Act 1981, such that it is an offence to plant or otherwise cause the species to grow in the wild.

These three tall and striking species were introduced into Europe in the 19th century as ornamental garden plants and have exhibited similar patterns in their spread and distribution, especially *I. glandulifera* and *F. japonica* which are particularly successful in invading and establishing themselves in the urban environment. There are many locations in which these latter two species can be found within the same community and examples have been noted in which all three of these neophytes occur within the riparian zone of stretches of a given river, e.g. on the River Lea, Greater London.

COMPARATIVE AUTECOLOGY

Impatiens glandulifera

I. glandulifera is native to the Western Himalayas and was introduced into the British Isles shortly after 1839 as an ornamental garden plant. It has rapidly become naturalized along canals and rivers, in waste places, open woodlands and shaded marshland (Beerling, 1990). *I. glandulifera* is an annual herb with seedlings developing rapidly into tall stems with foliage and flowers reaching up to 2.5 m. The flowers are slipper shaped, 25–40 mm long (Grime, Hodgson and Hunt, 1988) and are white through pink to purple in the British Isles (Dunn, 1970). Each year, from August to October, each plant produces up to 5000–6000 seeds (Perrins, Fitter and Williamson, 1990). Although *I. glandulifera* is reported as having no persistent seedbank, seeds kept at room temperature have remained viable for at least 3 years (Mumford, 1990) and in a field experiment it has been observed that under some circumstances *I. glandulifera* seeds persist as a viable seedbank for at least 18 months (Beerling and Perrins, 1993).

The dispersal of the seeds is dependent on the explosive dehiscence of the seed pods. This, plus the height of the capsule above the ground and the direction and strength of the prevailing wind, governs the distance the seeds can be spread. Using this mechanism, the plant has been recorded by Fitch (1976) to move by about 3 m year^{-1} and up to 5 m year^{-1} by Beerling and Perrins (1993). Such a dispersal distance is often sufficient for the seeds to reach water courses. Moist seeds are negatively buoyant and are capable of germination under water. The seeds may be redistributed in sediments by seasonal flooding, the natural means of long distance dispersal. Long distance dispersal is also achieved through human activities, both deliberately, with seeds still being sold by some seed companies, and inadvertently, by transport of topsoil containing viable seeds.

Virtually all seedlings appear over a period of about four weeks in February to March.

The synchronous germination of a large seedbank results in the establishment of a dense, evenly aged stand of *I. glandulifera*, competing with native species such as *Urtica dioica* (common nettle) (Beerling, 1990; Prach, 1994). The fleshy nature of *I. glandulifera* stems ensures that the plants rot away on senescence, leaving behind little or no evidence of the stand.

Heracleum mantegazzianum

H. mantegazzianum is native to the Caucasus and was introduced into the British Isles in the late 1800s as an ornamental garden plant. Since its introduction it has colonized not only riparian habitats but also agricultural land, road and railway verges, urban habitats and disused areas such as waste land, demolition sites and abandoned fields (Dodd et al., 1994).

    H. mantegazzianum is a monocarpic, biennial or perennial herb, growing from either a basal tap root or from seeds. The stem usually grows up to 5 m tall, and after $2-3$ years growth, produces white umbrella-like flower heads approximately 1 m across. The plant sets seed from July onwards, after which the parent plant usually dies back (Kees and Krumrey, 1983; Schuldes and Kübler, 1990; Schwabe and Kratochwil, 1991). One plant has the potential to produce approximately 120,000 seeds (Tiley and Philp, 1994) forming an extensive seedbank in the immediate area of the parent plant. Seeds, kept dry at room temperature, have been found to be viable after 7 years (Morton, 1978) but Lundström (1989) suggests a seed viability of possibly 15 years.

    The seeds are dispersed over short distances by wind, e.g. $2-10$ m in winds of $3-14$ m s^{-1} (Neiland, Proctor and Sexton, 1987), sufficient to reach a watercourse, another means of natural dispersal. The seeds are able to float for up to 3 days before they become waterlogged and sink. With a hypothetical river surface velocity of 0.1 m s^{-1} a seed could thus be transported up to 10 km by water (Clegg and Grace, 1974). Human activities have also contributed to the distribution of *H. mantegazzianum*. The seed heads are considered attractive decorations and are often transported over large distances, for example, on the top of vehicles. Seeds can be dropped during transport and the subsequent disposal of the seed heads can lead to the development of new colonies of *H. mantegazzianum* on rubbish dumps, and along road and railway verges (Lundström, 1984). Movement of topsoil contaminated with *H. mantegazzianum* seeds is another significant means of human dispersal.

    H. mantegazzianum seeds germinate in late March−April and the largest seedlings rapidly establish a leaf rosette but do not flower in the first year. During this first year a long fleshy tap root develops which persists for up to four years. Plants in their second, third or fourth year sprout vigorously in early spring from the tap root until the plant is ready to flower. The tall dead stems of *H. mantegazzianum* often remain after senescence. Given the size of the plant, these are at a relatively low density.

Fallopia japonica

F. japonica is native to Japan and was introduced into Europe in the early 19th century as an ornamental garden plant and as fodder for cattle (Conolly, 1977). Since its introduction it has become established in habitats such as waste places, along canals and railway embankments, river and stream sides, and road verges.

    F. japonica is a large rhizomatous perennial, which grows vigorously in spring from March onwards and develops into dense mono-specific stands with bamboo-like stems up to 3 m tall. The plant produces creamy-white flowers in clusters from August to October. Although this species is functionally dioecious, records of seed-set are rare and when observed, are most likely to be the result of hybridization, as only the female plant is recorded in the British Isles (Bailey, 1989). *F. japonica* only reproduces by vegetative means, primarily by rhizome fragments but also through sections of stem material. Recent studies have shown that as little as 0.7 g of rhizome material can support a new shoot, and 10% of stem material floating in water has been shown to produce healthy new shoots and adventitious roots (Brock and Wade, 1992; Welsh Development Agency, 1993; Brock et al., in press).

Natural dispersal is by means of fragments of rhizome and stem material, broken off the parent plant and washed downstream by high water flows. Human activities, such as the cartage of earth contaminated with fragments of *F. japonica*, are very important in the long distance transport and the invasion of new sites. *F. japonica* grows vigorously from an extensive rhizome system in early spring with growth rates as much as $5-10$ cm day^{-1} (Jennings, 1980). The stands die back in winter to leave dense thickets of erect, hollow, semi-woody stems which may persist for several years.

CONTROL AND MANAGEMENT

The successful control of *I. glandulifera*, *H. mantegazzianum* or *F. japonica* will largely be dependent upon incorporating three key components into a management programme:

(1) using the knowledge of the autecology of the particular species,
(2) developing a co-ordinated management programme,
(3) stopping further spread.

The autecology of *I. glandulifera* is characterized by a single regeneration strategy with the production of a large amount of seeds annually. *H. mantegazzianum* has two regeneration strategies; one vegetative, the formation of an over-wintering tap root, and the other relying on the annual production of a large amount of seeds developing into a persistent seedbank. If lasting control is to be achieved it is essential to prevent these two species from setting seed by controlling all plants before flowering. A continuing management programme over several years is necessary which aims to exhaust any remaining seedbank. Even if there is no persistent seed bed for a stand of *I. glandulifera*, re-invasion readily occurs from seeds produced by upstream stands which might not have been controlled. The autecology of *F. japonica* is characterized by a single vegetative reproductive strategy, so the success of a control programme and the prevention of further spread of this species is dependent on the containment and elimination of the large underground rhizome system. One of the major problems regarding the spread of *I. glandulifera*, *H. mantegazzianum* and *F. japonica* is the dispersal of these species *via* water. Therefore control programmes need to commence in the upper reaches of river catchments to limit the chance of re-infestation.

There are several methods available to control each of these species; chemical, manual (cutting) and grazing are practicable methods, while biological control is a future possibility.

Chemical control

Candidate herbicides need to be translocated to the rhizomes in order to be effective against *F. japonica* through destruction of these tissues. This is also a desirable aim for *H. mantegazzianum* but the herbicide must additionally prevent the plant from flowering and subsequently setting seed. The control of *I. glandulifera* is also dependent on killing the plant before flowering. One of the problems associated with these invasive plants is the exposure of river margins to erosion in winter when the plants have died back naturally. Herbicidal control could exacerbate this situation by killing other species, so it should be limited at all times. Any herbicides which could achieve these objectives must also be approved for use near water when managing riparian invasions (Department of the Environment, 1992).

There is only one herbicide in the UK which complies with the above requirements and which is effective against all three species — glyphosate. Glyphosate successfully controls both *I. glandulifera* and *H. mantegazzianum* when applied early in the growing season (March/April) to seedlings and newly forming rosettes respectively. A second slightly later application, in May, has been effective in controlling later germinating seedlings which may have been missed by the initial treatment. *F. japonica* has been shown to be susceptible to glyphosate, the addition of a wetting agent potentially improving uptake of the herbicide. Glyphosate should be applied twice in the year around May and July (de Waal, in press).

Glyphosate is a non-selective herbicide and therefore all vegetation is affected by spraying. However, it is non-persistent and replanting with native species can be carried out 10 days after treatment. Such re-seeding/re-planting, for example with grass, ensures that erosion of soil does not occur.

Manual control

Cutting is an effective method of control for *I. glandulifera*, either manually or mechanically, early in the season before flowers are produced. Once cut, the plant no longer produces flowers. However, cutting *H. mantegazzianum* has been less successful as satellite umbels of flowers may form after cutting and subsequently set seed. A further problem with cutting this species is the potential health hazard to the contractor by skin contact with the phyto-phototoxic sap. Cutting of *F. japonica* using either a brush cutter or a flail mower has been successful in controlling the growth of the plant whilst treatment is continued. It is necessary to carry out the cutting regime regularly throughout the growing season at two weekly intervals (Child et al., 1992). Baker (1988) reports control of *F. japonica* over a three-year period by hand-pulling all visible shoots.

Grazing

Cattle, sheep and goats will graze the young shoots of *F. japonica* (Beerling, 1990) and *H. mantegazzianum* (Vogt Andersen, 1994), though this method of control will only inhibit the growth of *F. japonica*, by removal of above-ground material, rather than eradicating the plant. No observations have been made on the grazing of *I. glandulifera*.

Biological control

Initial studies have been carried out to investigate the possibilities of finding a biological control agent for these species (Emery, 1983; Sampson, 1990; Holden, Fowler and Schroeder, 1992). Each of the three species, in common with other introduced exotic plants, has little associated invertebrate fauna feeding off it in the British Isles. None of the plants is known to be particularly susceptible to microbial pathogens. There is potential therefore for the introduction of an insect or microbial pathogen which would cause significant damage, thus reducing the competitive advantage of these plants.

The search for biological control agents is as yet in its infancy but biological control could provide a long-lasting and cost-effective control method for these species. In the meantime, both an integrated approach using a combination of the above methods and a co-ordinated strategy involving all agencies concerned in the control of the plants is the most appropriate.

CO-ORDINATED MANAGEMENT STRATEGY

A control programme for an invasive species is not only reliant on the treatment methods employed but also requires a co-ordinated approach by the various agencies involved in the treatment of the species (Child et al., 1992; Hill, 1994). As *I. glandulifera*, *H. mantegazzianum* and *F. japonica* are not restricted by artificial boundaries, successful treatment of the plants can only be achieved if the potential for re-infestation is removed. The dispersal mechanisms of each of these species pose a particular problem in riparian habitats where rhizome fragments, stem material and/or seeds can easily be transported by water, with the potential for the formation of new stands downstream, indicating the need for co-ordinated management strategies at a catchment level. In practice this means co-ordination and liaison between the different bodies involved in the treatment of the species at a local, regional and national level (i.e. those responsible for the interests of fisheries, amenity, recreation, flood defence and public safety). Effective containment and control necessitate that river management agencies liaise with other authorities, for example, those concerned with local administration, road and rail transport and private agencies. A working group made up of representatives from the National Rivers Authority, Swansea City Council and British Rail was established to tackle an infestation of *F. japonica* in Swansea, Wales, through which the River Tawe flows (Hill, 1994).

A clearly defined strategy of treatment targets and funding would not only improve the effectiveness of control measures undertaken in the long term but would also be cost-effective. Funds available for the treatment of these invasive species would not be wasted due to re-invasion but would contribute towards containment of the problem and eventually control of the species.

It is essential to avoid unnecessary movement and/or tipping of soil contaminated with either *F. japonica* rhizome and stem fragments or with *I. glandulifera* and *H. mantegazzianum* seeds. At present, there are no restrictions on the movement of soil within a region, which can obviously cause further spread of the species over a wide area. However, this problem can be tackled through a co-ordinated approach by limiting the free movement of contaminated soil. Where such soil needs to be moved it should be disposed of in areas already infested with either of the three species and subsequently treated. In the case of *F. japonica*, contaminated soil could be disposed of in a licensed landfill, burying the soil material to at least 5 m depth (Welsh Development Agency, 1993).

DISCUSSION

The management of alien weeds which have invaded the riparian zones of European rivers costs large sums of money: in the UK alone the cost is of the order of hundreds of thousands of pounds sterling annually. North America is not immune from such invasions and is currently experiencing serious problems from *Lythrum salicaria* (Minnesota Department of Natural Resources, 1992). This comparative study of *I. glandulifera*, *H. mantegazzianum* and *F. japonica* highlights the need to develop a management strategy for each species based on its known autecology. A uniform treatment of the riparian zone will at best contain the species; at worst it could lead to an extension of the infestation. Programmes of control will need to be planned over different time periods depending upon the species. Seed dormancy in *H. mantegazzianum* and *I. glandulifera* are key factors in their control whereas the persistence of *F. japonica* rhizomes will dictate the length of time needed to control this species.

Invasive riparian plants do not respect administrative boundaries and present a special weed control case, different in many respects to that of agricultural weeds. Effective containment and control must prevent the further spread of the target species (a difficult task where the propagules concerned are small seeds or small rhizome fragments) and should recognize the need for co-ordinated control programmes. Such integrated control programmes are difficult to initiate and few examples have been found. Swansea City Council has established a "Japanese knotweed Working Group" and the River Tweed Forum in Scotland is establishing a joint programme to achieve effective control of *H. mantegazzianum* in the catchment of the River Tweed.

The riparian zone is especially prone to invasion by alien species and this situation is unlikely to change. Despite the investigation of the three species compared in this chapter and investigations relating to other species, further research is needed to understand more fully why this habitat is so open to colonization by neophytes. Such research would undoubtedly indicate appropriate management for the riparian zone, not only in relation to invasive aliens but also concerning such aspects as conservation value and its role as a buffer between the land and water.

ACKNOWLEDGEMENT

This study was partly funded by the National Rivers Authority, under R&D Project 294, and the Welsh Development Agency.

26

The Ecological Basis for Management of Fish Stocks in Rivers

J. MALCOLM ELLIOTT

NERC Institute of Freshwater Ecology, Ambleside, Cumbria, UK

INTRODUCTION

There is a vast literature on the ecology of river fish and I will make no attempt to review it in this short contribution. Instead, this review discusses three aspects:

(i)   if there really is a need for management based on ecological principles;
(ii) to what extent ecological knowledge is integrated into management at present;
(iii) what is desirable and realistic for the future management of fish stocks.

The emphasis is placed on principles rather than detailed case studies. Although this review deals with the management of fish stocks, it is obvious that fish cannot be considered in isolation from other aspects of river ecology. For example, water quality is of obvious importance and is discussed elsewhere in this volume (see Chapter 11). Unlike most of the flora and fauna of rivers, many fish species provide food for humans and therefore their accumulation of toxic materials is of increased importance. Chandler (1990) presented convincing arguments to support the view that systems of fisheries management cannot exist in isolation and should be considered as part of an integrated river basin management. Such a view is reflected by the integrated nature of this volume and must surely be the ultimate goal for the future.

IS THERE A NEED FOR MANAGEMENT BASED ON ECOLOGICAL PRINCIPLES?

As an animal ecologist, I used to assume it was self-evident that if a species was a valuable resource, then ecological knowledge would be essential for its conservation and management. I have since realized that this was a naive assumption! One extreme, but not uncommon, view is that the best way to manage a freshwater fishery is to put in the fish the anglers want and ensure that most are caught. The argument is therefore that most effort should be placed on the science of rearing fish in hatcheries and breeding fish that will please most anglers; the fish should be large and easy to catch. Another view is that all you need to

The Ecological Basis for River Management. Edited by D.M. Harper and A.J.D. Ferguson. © 1995 John Wiley & Sons Ltd

do is ensure a high water quality and the fish will look after themselves. It may surprise some readers to learn that I have some sympathy with the latter view. Cummins (1992) stated: "The concept of 'management' is one of the more arrogant human notions."

Man's attempts to manage nature are most frequently seen in agriculture and forestry. Although there have been many successes, there have also been some disasters — some of these long after supposed "successes". The best management of wild species is probably to minimize human interference and rely on natural processes, a strategy followed to a certain extent in many of the world's national parks. Perhaps the ideal strategy is to manage the human populations and their activities, rather than natural communities of animals and plants.

Unfortunately we do not live in an ideal world and man often upsets natural balances with the innocence of ignorance. The following summary of the situation by Thomas (1974) is apt:

> "We are now the dominant feature of our environment. Humans, large terrestrial metazoans, fired by energy from microbial symbionts lodged in our cells, instructed by tapes of nucleic acid stretching back to the earliest live membranes, informed by neurons essentially the same as all the other neurons on earth, sharing structures with mastodons and lichens, living off the sun, are now in charge, running the place for better or worse."

Although benign management is always desirable, it is not always feasible in a world in which ecosystems are threatened and conservation is accepted as essential. The reasons for this were summarized in the three aims of the World Conservation Strategy: to maintain essential ecological processes and life support systems; to preserve genetic diversity; and to ensure sustainable utilization of species and ecosystems (International Union for Conservation of Nature and Natural Resources, 1980). All these are applicable to river ecosystems and their fish populations.

Many fish species provide valuable commercial and sports fisheries and thereby qualify as naturally sustainable resources. Some populations may be useful as sources of genetic material for aquaculture, and for providing fish for restoration in rivers in which fish populations have been destroyed by human activities. Fish are also an important and integral part of the river ecosystem with some species serving as the top predators or keystone species within the community.

Boon (1992) has reviewed the many pressures on rivers and their flora and fauna. He suggested five options for river conservation: preservation or conservation of natural or semi-natural systems; limitation of catchment development in rivers of high quality; mitigation for rivers of lower quality (essentially a salvage operation!); restoration where a river has been severely damaged by human activities; and dereliction which is self-explanatory. All five options are also applicable to fish populations and all except the last require a progressive increase in the role of management. Additional conservation options for managing freshwater fish populations are stock transfer to new sites, captive breeding and cryopreservation (see review by Maitland and Lyle, 1992).

An example of restoration is provided by the radical improvement of water quality in the River Thames, southern England, and the restoration of Atlantic salmon, *Salmo salar*, to the river (Banks, 1990). A similar success story is the recovery of salmon stocks in the River Tyne, northern England (Champion, 1992). A good example of the practical problems of managing fisheries in two contrasting British rivers is provided by a summary of management practices for the River Severn, a relatively clear-water salmon river that is the longest in Britain (354 km), and for the River Trent, a shorter river (280 km) flowing through several industrial regions and facing all the associated problems (Templeton and Churchward, 1990).

Such management requires a detailed knowledge of the ecological requirements of the fish. Unfortunately such knowledge is often lacking, especially at the quantitative level. It is clear, however, that there is a need for management of fish populations in rivers and that it should have an ecological basis. The degree of management depends upon the extent to which the fish are exploited or stressed by human activities. It also depends on both financial and intellectual resources, the latter being the amount of ecological knowledge available for each species.

ECOLOGICAL REQUIREMENTS OF FISH POPULATIONS

Requirements relevant to river fish management will be considered under four headings:

(i)   habitat requirements
(ii) intra-specific differences between stocks and the role of density-dependent processes
(iii) growth
(iv) fish communities

In each section, the basic ecological knowledge is briefly reviewed, the present use of this knowledge in management is considered and some suggestions for future developments are summarized. Although these four aspects are considered separately, they are interlinked, often in a complex way. They therefore must all be considered in defining the ecological requirements for the management of a fish population.

Habitat requirements

Simple qualitative descriptions of habitat requirements can be found in many popular guides to freshwater fish. More detailed qualitative descriptions have been used in the visual assessment of habitat evaluation, this being by far the commonest method of assessing the habitat requirements of river fish (for an example of this approach in the context of UK Atlantic salmon enhancement programmes, see Kennedy, 1984). A slight improvement is so-called semi-quantitative assessment in which a visual assessment is combined with the measurement of a few quantitative variables, the most frequent being width, mean and maximum water depth, water velocity (or stream gradient), stream-bed substratum, and cover. Examples of semi-quantitative methods are provided by Herrington and Dunham (1967), Duff and Cooper (1976), Platts, Megahan and Minshall (1983), and Milner, Hemsworth and Jones (1985).

Finally, there is a fully quantitative assessment, using a wide range of variables. In their review of models that predict standing crop of stream fish from habitat variables, Fausch, Hawkes and Parsons (1988) list 20 such variables, of which five are associated with the drainage basin, and eight with channel morphometry and flow; the remaining seven are cover, stream-bank stability, depth and velocity preference, abundance of invertebrate drift food, substratum type, temperature and water chemistry.

Various models for habitat evaluation have been developed, based on different combinations of these variables. An example is the "Habitat Quality Index" (HQI) based on nine variables that could account for 90% of the variation in salmonid biomass in Wyoming

(USA) streams (Binns and Eiserman, 1979). The "Instream Flow Incremental Methodology" (IFIM) is designed to quantify available habitat for different species of fish and was developed by the US Fish and Wildlife Service to provide a standard analytical technique for recommending suitable flow in a stream (Bovee, 1978, 1982). The US Fish and Wildlife Service has also developed a "Habitat Suitability Index" (HSI) for river salmonids, using up to 18 habitat variables for four life stages; embryo, fry, parr and adult (Raleigh, 1982; Hickman and Raleigh, 1982; Raleigh, Zuckerman and Nelson, 1986). Several variations of these models have been developed for salmonid streams elsewhere in North America (e.g. Scarnecchia and Bergersen, 1987; Wesche et al., 1987).

In the UK, Milner, Hemsworth and Jones (1985) compared three habitat evaluation techniques for brown trout in Welsh streams: quantitative (multiple regression of fish abundance on 13 habitat variables), qualitative (linear functional regression of fish abundance on scores from a visual fish cover index), semi-quantitative (linear functional regression of fish abundance on scores from a "HABSCORE" questionnaire). The quantitative analyses worked well for medium and large, but not small, trout at hard-water sites, but were not very successful at soft-water sites (< 25 mg litre^{-1} $CaCO_3$). The qualitative method worked well only for medium-sized fish in hard water but the semi-quantitative method produced better results, especially in hard waters.

Many of these methods assume a linear or log−linear relationship between fish abundance and habitat variables. This is often erroneous. The IFIM uses only four variables (water depth, velocity, substratum and cover) but does utilize a variable probability of preference for different species rather than assuming a linear relationship. This method can, however, be criticized on both statistical and biological grounds (Mathur et al., 1985). One frequent assumption is that fish select each habitat variable independently of the others. As Heggenes (1988) has indicated, this basic assumption has rarely been tested. Violation of this assumption has been demonstrated for some warmwater species (Orth and Maughan, 1982; Mathur et al., 1985).

It has been proposed that water depth, velocity and stream-bed substratum cannot be considered as independent additive variables, or be used alone to assess the suitability of a physical habitat for fish (Moyle and Balz, 1985; Mathur et al., 1985). In his review of physical habitat selection by brown trout in rivers, Heggenes (1988) concluded that interactions between habitat variables may explain some of the different ranking of variables between streams and may also explain the poor performance of the different models of trout habitat quality arrived at in the literature. He also emphasized that it is important to consider total available habitat data and not habitat use data alone (see also DeGraaf and Bain, 1986; Gatz, Sale and Loar, 1987; Heggenes and Saltveit, 1990).

Quantitative habitat evaluation is often used as a tool for fisheries management in North America, but it is seldom used elsewhere and the most common method appears to be based on subjective, intuitive judgements. There is clearly a need for such a management tool, but it is clear from this brief review that models currently available are based on false assumptions. As the habitat requirements of a fish species change through the life cycle and the abundance of a fish population changes with many factors other than habitat variables, it is perhaps an impossible goal to expect to predict fish abundance simply by measuring a suite of habitat variables.

Instead of repeating errors in previous models, it might be best to return to basics and consider ideas elsewhere in ecology. Central to most is the concept of a multi-dimensional niche, originally formalized by Hutchinson (1957, 1978). He considered the niche to be defined by the complete range of variables to which a species must be adapted for it to survive. The axes of a multi-dimensional niche are of two kinds. Bionomic axes relate to variables directly involved in the lives of organisms and are usually resources, such as food and space, for which there may be competition. Scenopoetic axes represent the tolerance limits to certain physical and chemical variables such as temperature and oxygen. Although animals can rarely be said to compete for scenopoetic resources, the latter may differ between species or between different life stages of the same species. The animals are therefore partially or wholly partitioned and this facilitates resource partitioning with a reduction in competition.

Many niche components interact so that as more axes are added, the hypervolume-niche changes (e.g. there are three axes in Figure 26.1(a) — a fourth variable would lead to visualization problems!). A real example for brown trout illustrates the interaction between one scenopoetic axis (water temperature) and one bionomic axis (daily energy intake). As the latter decreases, the optimum temperature for the dependent variable (growth) decreases (Figure 26.1(b)). This example shows why it is difficult to define even a relatively simple variable such as temperature in trying to describe the habitat requirements of a fish species. The distribution of a species was once thought to be controlled by the habitat variable for which the organism has the narrowest range of adaptability or control (e.g. Bartholomew, 1958), but this simplistic limiting-factors approach is rarely tenable. As shown above, one variable may influence another. Also, habitat variables are measured in different units, so in no absolute sense can one variable be declared broader or narrower than another. More recent developments have seen the transformation of Hutchinson's niche into the concept of the utilization niche which incorporates the fractional use of the resources along main niche axes, the most obvious being food, space and time (see review by Schoener, 1989).

It is clearly an impossible task to measure all components of the multi-dimensional niche and therefore only a few of the niche axes are used. The selection of these axes clearly requires a great deal of quantitative knowledge about a species. Finally, it should be remembered that even when it is possible to define the major components of a niche, this provides a description of only the fundamental niche, i.e. the suite of conditions in which a species can exist in the absence of competitors. Few fish species enjoy this freedom, but some of the salmonids often occur as the only dominant river species of fish. Many species exist in a fish community and therefore there is niche overlap with each species occupying a realized niche, i.e. a suite of conditions that is less than the fundamental niche. The definition of habitat requirements in fish communities is therefore even more complex than that for single species, and will be discussed later.

Intra-specific differences between stocks and the role of density-dependent processes

Unlike many terrestrial species that can move freely from one habitat to the next, each species of freshwater fish is usually represented by many geographically isolated populations. Even migratory species, such as many salmonids, are usually reproductively isolated in discrete populations. It is therefore not surprising that populations of the same species often differ phenotypically in their morphology, physiology, ecology and behaviour. There is insufficient space in this short review for a comprehensive survey of the literature. Instead, some important general conclusions will be discussed and illustrated wherever possible by work on one of the most frequently studied species, the brown trout. In his comprehensive review of the salmonid literature, Taylor (1991) provided numerous

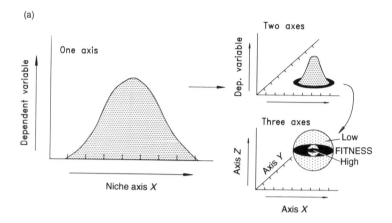

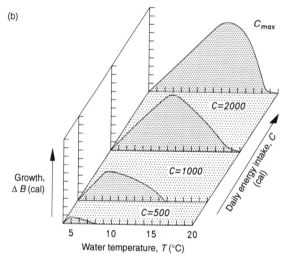

FIGURE 26.1. (a) The relationship between a dependent variable (e.g. survivor density, growth, reproductive success) and one niche axis or two or three niche axes. (b) Relationship between growth (ΔB, cal day^{-1}) and water temperature ($T°C$) for 50 g trout on maximum rations (C_{max} cal day^{-1}) and ration levels of 2000, 1000 and 500 cal day^{-1} (adapted from Elliott, 1979)

examples of ecological differences between populations. Some of these differences have been described as local adaptations but few can be linked to genetic differences. Such phenotypic differences cannot be considered as local adaptations unless they are shown to have a genetic basis. It is also necessary to detail the nature of the selective mechanism that has favoured such adaptive traits. Few studies meet these exacting requirements and therefore the evidence for local adaptation must be regarded as largely circumstantial and inconclusive for many species. Successful introductions of salmonids to non-native habitats are cited as evidence that they are flexible species without narrow local adaptations (Larkin, 1981). However, the same evidence is used to support the view that the apparent flexibility of such populations could itself result from adaptation in highly variable local environments through selection for phenotypic plasticity (Via and Lande, 1985; Taylor, 1991). Even when direct evidence for genetic variability is obtained, it is often difficult to show that the variability is of functional significance, i.e. is it the result of natural selection rather than genetic drift (Ferguson, 1989)?

There are, however, some excellent examples of divergence between neighbouring trout populations. The study of three sympatric morphotypes of brown trout in Lough Melvin, north-west Ireland, is exceptional because it provides morphological, meristic, ecological and genetical evidence for divergence between the three types. Reproductive isolation is the key factor that ensures the three populations are distinct (references in Ferguson, 1986, 1989). Evidence is accumulating for several salmonid species, including brown trout, that isolated populations living above impassable waterfalls often show marked differences from those living below them. There are well-documented differences in their ecology, behaviour and genetics (review by Northcote and Hartman, 1988). Many of the isolated brown trout populations possess a genetic marker for the ancestral type of this species (Ferguson, 1989; Hamilton et al., 1989; Marshall et al., 1992).

Information on fish genetics has increased markedly in the last 20 years but, again, the emphasis has been on salmonids probably because of their value in aquaculture. It is important to realize that fish are phenotypically more variable than other vertebrates (Allendorf, Ryman and Utter, 1987). For example, the intra-specific range of weight differences in Arctic charr is over 4000%, which is more than 15 times as great as the corresponding differences between all species of Darwin's finches! It is therefore not surprising that many fish species exhibit a large phenotypic variation. In contrast, heritability estimates (within populations) for different characters in various vertebrate species are usually much lower in fish populations than within populations of other vertebrates (Allendorf, Ryman and Utter, 1987). Higher levels of phenotypic variation together with lower heritabilities indicate greater susceptibility to environmental factors, and this explains many of the ecological differences found between populations of the same species.

There has been a rapid development in techniques in the last 20 years from classic cytogenetics to electrophoretic methods and, more recently, analyses of mitochondrial DNA. These methods have elucidated relationships between species, e.g. rainbow trout have been transferred from the Atlantic salmon and brown trout genus *Salmo* to the Pacific salmon genus *Oncorhynchus* (Smith and Stearley, 1989). Excellent reviews of the application of electrophoretic methods to fish are available (Ferguson, 1980; Allendorf and Phelps, 1981; Utter, Aebersold and Winans, 1987; Cross, 1989). The literature on mitochondrial DNA in fish has been reviewed by Billington and Hebert (1991). The advantages (especially enhanced resolution) and disadvantages (mainly cost and technical complexity) of DNA analysis in fishery management have been assessed by Hallerman and Beckmann (1988).

The most important conclusion for fisheries management from all this work is that many species, especially salmonids, occur as reproductively isolated populations as well as geographic races (reviews by Cross, 1989; Davidson, Birt and Green, 1989; Verspoor and Jordan, 1989; Ferguson, 1989; Hershberger, 1992). These genetic differences can occur at a very small geographical scale, for example, some of the variant alleles in brown trout are found in populations restricted to a single drainage system or even part of a drainage system (Ferguson, 1989). The maintenance of such differences provides strong evidence for the precise natal homing of trout before spawning commences. It is perhaps surprising that there appears to be no genetic dichotomy between resident and migratory brown trout spawning at the same time in the same locality (Ryman, 1983; Ferguson, 1989; Hindar et al., 1991; Cross et al., 1992).

The available evidence therefore suggests that for many fish species, especially the salmonids, there are distinct populations or "stocks". The general concept of isolated stocks has important implications for fisheries management and is not therefore without its critics, especially in the marine field (e.g. Gauldie, 1991). Most fish biologists agree that genetically unique stocks should be identified and appropriate measures taken for their conservation, but the basic philosophy and practice remain unclear (Meffe, 1987). These "gene pools" are important sources of material for both aquaculture and restocking restored rivers that once supported fish populations. Restocking to augment existing populations should be performed with fish reared from the indigenous population because the latter should always contain the optimum genotypes for a particular locality. Hatchery strains will rarely be suitable because their genetic diversity is usually greatly reduced. For example, Swedish hatchery stocks of brown trout have retained, on average, only 25% of the mt DNA variability of wild populations (Gyllensten and Wilson, 1987). Many of those responsible for the management of river fish now understand these principles and are more careful about the source of fish used for stocking.

Finally, I will consider the role of density-dependent processes in populations living in contrasting habitats. Two hypotheses of Haldane (1953, 1956) are important for the argument. The first emphasizes the importance of density-dependent processes in natural selection and the second predicts that changes in population density will be largely due to density-dependent processes in high-density populations living in favourable areas, and to chiefly density-independent processes in low-density populations living in unfavourable areas. Haldane added that natural selection will be for different genotypes in the two types of habitat, but only if the two types of population are isolated from one another. Examples of terrestrial populations that agree with these hypotheses were summarized in Elliott (1987).

Both hypotheses have been shown to be apt for brown trout in a comparative study that contrasts a migratory-trout population (Black Brows Beck) with a population of resident-trout living above a waterfall in a neighbouring stream (Wilfin Beck), both streams being part of the Windermere catchment in north-west England (Elliott, 1987, 1988, 1989a,b). Population density in Black Brows was much higher than that in Wilfin Beck and was regulated by density-dependent survival in the early life stages. There was no evidence for similar density-dependent regulation in Wilfin Beck; simple proportionate survival occurred with fairly constant survival rates. Black Brows trout were always larger than Wilfin Beck trout of similar age; fry size at the start of the growth period being chiefly responsible for these differences. Variations in water temperature were chiefly responsible for differences in growth rates between year-classes within each population. There was no evidence for genotypic differences in mean growth rates between the two populations. The relative variation in the size of individual trout was not a stable characteristic of either population; it was density-dependent in Black Brows and age-dependent in Wilfin Beck. Experimental studies in the laboratory showed that these changes were due to natural (stabilizing) selection. There is some evidence for genotypic differences between the two populations (see summary in Table 26.1) and therefore support for Haldane's hypothesis that natural selection will be for different genotypes in the two types of habitat.

These differences between populations have important implications for the management of

TABLE 26.1. Summary of possible genotypic differences between the two populations

Black Brows Beck	Wilfin Beck
Favourable habitat with high population density	Less-favourable habitat with low population density
Migratory behaviour	Non-migratory behaviour
Females usually breed once	Females usually breed two or more times
30–37% of females's energy content invested in good production	16–17% of females's energy content invested in egg production
Maximum CV for live weight *c.* 50%	Maximum CV for live weight *c.* 59%
Genotypes adapted chiefly to density-dependent factors	Genotypes adapted chiefly to density-independent factors

fisheries. In theory, stocking should do little harm to low-density populations living in unfavourable habitats and most of the stocked fish will have a low probability of survival. The situation is more complex in the more valuable high-density populations in favourable habitats. Even when juveniles reared from the same population are used for stocking, the effect on the population can be harmful. For example, the optimum egg density for maximum survival of juveniles is 40–50 eggs m^{-2} in Black Brows Beck. If egg density is similar to or exceeds this value, stocking with eggs or juveniles will decrease survival and hence the number of recruits to the population. It will also reduce the variation in the size of individual fish because high initial densities produce a more uniform fish size (see also Elliott, 1989a). Therefore overstocking of a valuable commercial or sports fishery can lead to lower catches and the absence of the rare, but large, specimen fish beloved by anglers. There is no evidence in the literature that these relatively new concepts are being integrated into management at present.

Growth

The literature on fish energetics and growth has increased markedly in the last 20 years and has been summarized in several reviews (e.g. Brett and Groves, 1979; Elliott, 1979, 1982; Ricker, 1979; Pitcher and Hart, 1982; Weatherley and Gill, 1987; Wootton, 1990). This brief account is therefore restricted to some aspects relevant to the management of river fish.

It is generally agreed that water temperature, fish size and the level of energy intake are the three most important variables of the many that can influence the growth of freshwater fish. Temperature requirements vary considerably between species. For example, a comparison of brown trout and carp shows that thermal stress, or even death, in each species occurs at temperatures that are optimal for feeding and growth in the other species, but both species have a relatively narrow range for egg development (Elliott, 1981). Within a species, optimum temperatures can change for different functions and are affected by other variables such as energy intake (e.g. Figure 26.1(b)). All the components of an energy budget, including growth, are affected by fish weight. Extensive reviews of the literature show that the relationship between specific growth rate and fish weight is described by a negative power function with an exponent of *c.* 0.4 (Brett, 1979; Jobling, 1983). Similar relationships exist between fish weight and energy intake or metabolic rate when the latter are expressed in terms of unit weight (e.g. Elliott, 1979, 1981, 1982). It is therefore misleading to express absolute growth or food consumed as g g^{-1} or cal g^{-1} fish weight.

The estimation of energy intake of fish in the wild is difficult and remains a weak link in studies of energetics and growth. A critical review of direct methods of estimating food consumption in the field showed that most of these methods ignore the assumption of an exponential rate of gastric evacuation and often underestimate the daily food consumption (Elliott and Persson, 1978). Two new methods were proposed in this review. Both methods require detailed field studies and are therefore best used to check the adequacy of less accurate methods.

An alternative approach is to develop a growth model for a particular species and to use it to determine if the growth of fish in the field is at a maximum or is retarded by shortage of food or some other factor. Most of the reviews listed at the start of this section include work on growth models. Early empirical models based on age and maximum size lacked a physiological basis and some of their assumptions, such as asymptotic growth, can be criticized. These models have been used frequently in the management of fish stocks, especially in North America, but are best reserved for impoverished data sets that are unsuitable for models based on metabolism and energetics.

Early versions of the latter models were based on differences between anabolism and catabolism, but later models included rate of food consumption whilst others incorporated the concept that growth was related to size or growth rate already achieved. An example of a model incorporating all these aspects is provided by one developed for growth of brown trout (Elliott, 1975a,b). Apart from reservoir populations, this model has now been used to investigate growth potential in at least 40 stream populations of brown trout (e.g. Crisp, 1977; Edwards, Densem and Russell, 1979; Crisp, Mann and Cubby, 1983; Elliott, 1984, 1985; Allen, 1985; Mortensen, Geertz-Hensen and Marcus, 1988; Jensen, 1990).

Such a model can be used to compare growth in different populations and different year-classes of the same population. For example, the 1970 year-class in Black Brows Beck was typical of the faster growing year-classes whereas the 1975 year-class exhibited the poor growth that was typical of year-classes affected by severe droughts. The model predicts correctly the marked difference in growth rates (Figure 26.2). It can also be used to predict the effects on growth of marked changes in temperature due to natural and human activities. An example is a rise in temperature due to climate change (Figure 23.3). The "normal curve" for the 1967 year-class in Black Brows Beck can be compared with growth curves for rises of 2°C and 4°C. Smolt weight would be slightly below normal for +2°C but markedly less for +4°C. The smaller size of the trout at the end of the first summer will almost certainly affect their survival. The increases in temperature could therefore lead to a reduction in numbers as well as growth rates.

There is little evidence up to the present that these more complex growth models have been used in the management of river fish stocks. This brief review has shown that such models provide some insight into the mechanisms regulating growth, and can be used to detect periods when growth is restricted and to predict growth changes due to natural and human activities. There is clearly a need to develop similar models for different fish species so that they can be used as tools for the conservation and management of the fish stocks.

Fish communities

Few rivers contain one dominant species, but some salmonid rivers are fairly close to this

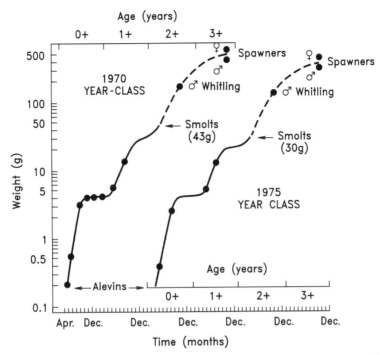

FIGURE 26.2. Growth pattern for the 1970 and 1975 year-classes; solid lines estimated from growth model in increments of 15 days, broken lines represent possible growth curves for post-migratory trout; actual mean weights (+) are given to show adequacy of the estimated values

with only one or two dominant species. The problems of managing these relatively simple systems increase in rivers containing mixed fisheries, often with several important species. Management is obviously improved by taking the interactions between species, and between fisheries, into account. To a certain extent, this is already happening at the qualitative level but so little is known about the ecological requirements of many species that management at the quantitative level in the near future is unlikely for most multi-species communities of river fish. A good example of the complex problems was provided by Schiemer and Waidbacher (1992) in their synthesis of strategies for the conservation and management of a large community of 52 species in the Austrian Danube.

At the qualitative level, it has long been recognized that fish species can be used for river classification. Hawkes (1975) has argued that they best reflect the general ecological conditions of rivers because they are at the top of the aquatic food chain. He provided a useful review of early ideas on river zonation and classification using fish species and, for many years, it was common to refer to a trout zone, a grayling zone, a barbel zone and a bream zone in considering a large river from source to mouth. More complex systems of classification now exist and have been reviewed by Naiman et al. (1992) who also summarized the advantages and disadvantages of using fish communities for classification purposes (Table 26.2).

In North America, the "Index of Biotic Integrity" (IBI) is used to classify rivers, and incorporates both structural and functional components of fish communities to assess the environmental quality of stream sections (Karr, 1981; Fausch, Karr and Yant, 1984;

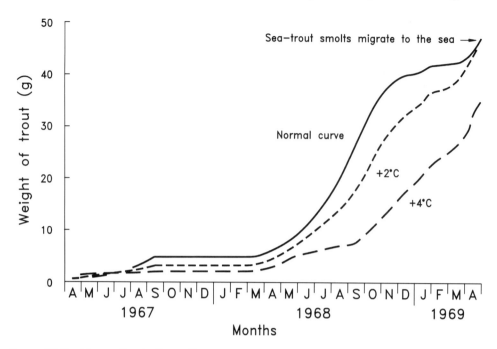

FIGURE 26.3.   The predicted effects of increased temperature (+2°C, +4°C) on the growth of brown
trout

TABLE 26.2.   Advantages and disadvantages of using fish communities for stream classification (from
Naiman et al., 1992)

Advantages
   Community function should be related to habitat quality
   There should be a predictable response to habitat change
   Community attributes integrate local and upstream habitat quality
   Species are indicators of stream function
   Biotic resources are coupled to physical habitat

Disadvantages
   Species composition varies across zoogeographic regions
   Structure can vary between drainages due to differences in biotic and abiotic controls
   Intensive quantitative sampling is often required (difficult in large rivers)

Leonard and Orth, 1986; Steedman, 1988). Work on the ecology of river fish communities
presents a rather equivocal picture with support for both a stochastic and deterministic
structure, and both strong and weak species–habitat relationships (Naiman et al., 1992).
Habitat requirements become more complicated because of frequent overlap between
species. However, such problems are now being addressed in ecology and Hutchinson's
multi-dimensional niche has now been extended to a community perspective by Litvak and
Hansell (1990). These authors have used these procedures to investigate the niche in a

community of cyprinids in a Canadian lake but they could equally be applied to river communities.

The stream hierarchy model of Meffe and Vrijenhoek (1988) describes how fish populations within rivers are partially isolated from one another with some gene flow between them. Higher frequencies of gene flow will cause greater genetic similarity among local populations and lower frequencies of gene flow will obviously cause the opposite. When rivers are impounded or river habitats are degraded, populations previously in contact may be fragmented and isolated, leading to a decline in genetic diversity and even extinction in extreme cases of habitat fragmentation (see also Sheldon, 1988; Nyman, 1991).

In managing fish communities, there is often an implicit assumption of balance, with the community persisting despite variation in the level of exploitation. This assumes that predators and their prey remain at an appropriate level of dynamic equilibrium. This view has been criticized by Ryder and Kerr (1990) on the grounds that it is difficult, if not impossible, to define a management goal. They suggest as an alternative the concept of harmonic communities but recognize that their proposal is neither complete nor sufficient. A harmonic community is persistent over time, functions as an integrated unit and is a moderately stable, hierarchic level of an ecosystem. According to these authors, harmonic communities have evolved over time through niche complementarity or niche packing, and an optimization of individual niche scope either through genetic or phenotypic diversification and adaptation. The appropriate management strategy is to regulate the keystone top predator species in the community as well as the first three or four harmonic species. Weighted mortality coefficients may be appropriate targets for regulation. A strong case is argued for this approach but the authors are aware that their views may well prove to be controversial. The usefulness of this approach for river fish communities remains to be tested.

FORTUNE TELLING

Examples of ecological knowledge being applied in fish management have been given already in the appropriate sections and some suggestions for future development have been provided. This final section does not attempt to catalogue all possible developments in the future but concentrates instead on a few aspects that may lead to an improvement in the management of fish stocks in rivers.

It is a truism for most fish species that the probability of a newly hatched larval fish reaching maturity is very low. Mortality rates are high in most species, especially in the early life stages. This important point should be remembered by those who wish to kill predators on fish, such as cormorants, mergansers and herons. Most fish taken by these predators are probably the sick, the lame and the lazy. An interesting analysis was made of the percentage mortality of brook trout and brown trout in a North American river, the major predators being mergansers, herons, kingfishers, brown trout, mink, otters and anglers (Alexander, 1979). Brown trout were the chief predators of eggs and juveniles, whilst anglers were by far the chief predators of older fish. Although the removal of predatory birds is advocated by some (e.g. Elson, 1962), this can lead to an increase in the rate of predation by other predators or an increase in intra- and inter-specific competition (Wootton, 1990).

Although mortality is high between birth and maturity in most fish species, the mortality rate can vary considerably throughout the life cycle and between populations of the same species. There are critical periods in which many fish die, usually because of limited resources. For most species, both habitat and dietary requirements change throughout the life cycle and therefore the availability of resources will vary for different life stages. Absolute population size will be limited by the resource, or combination of resources, in shortest supply. This resource limitation will affect the life stage most dependent on that resource and there will be a critical period for survival.

Such resource "bottlenecks" usually occur either early or late in the life cycle (see review by Sinclair, 1989). A good example of the former is provided by the population of brown trout in Black Brows Beck. There is a critical period for survival in the first few weeks after fry emergence; survivor density, mortality rates and selection intensity are strongly density-dependent during this period but proportionate survival with stable loss rates occurs after it (Elliott, 1989c, 1990a,b). Bottlenecks can also occur much later in the life cycle when there is a marked change in diet (e.g. plankton to benthic invertebrates to fish), or after sexual maturity (Weatherley and Gill, 1987; Shuter, 1990). In the latter case, it is the resources exploited by the adults that ultimately limit population abundance and these resources include suitable spawning habitat. Little quantitative information exists on population bottlenecks for most species of river fish but clearly such information would be of great value to managers. If the effects of such a bottleneck could be reduced, then population size should increase. It is also important to know when such critical periods occur so that the fish are not subjected to more stress at this time. This is already acknowledged to a certain extent in the operation of a close season for many species (see Chapter 32).

As stressed already, the precise habitat requirements of most fish species in rivers remain poorly defined in quantitative terms. This remains one of the major weaknesses in our knowledge of the ecological requirements relevant to fish management. A closely related problem is what determines the carrying capacity of a stream. This question remains largely unanswered but for some species, more information on territorial behaviour would be useful. For example, the quantitative study of brown trout in Black Brows Beck has shown that the spatial heterogeneity and territorial behaviour of the juveniles affect their density (Elliott, 1990b). Surprisingly, estimates of the total area occupied by the juveniles are only a small percentage of the total area of stream bed available to the fish: 3.4−6.8% about one month after fry emergence, 9.1−14.7% towards the end of the first summer of the life cycle. It is not obvious why a large part of the stream is an unsuitable habitat but the preferred places are probably those that provide energetically profitable feeding stations. Territorial behaviour is therefore one of the fundamental mechanisms affecting the carrying capacity of a trout stream but the precise habitat requirements of the territorial fish remain largely unknown. If such information is lacking for brown trout, one of the most frequently studied species, it is not surprising that so little is known about carrying capacity and habitat requirements for other river species.

The brief review of intra-specific differences showed that populations or stocks of the same species may require different management plans. There is therefore a need to classify populations of the same species according to their ecological requirements. The hypotheses of Haldane have already been mentioned and were supported by the comparative study of two trout populations. This work has led to the hypothesis that brown trout populations can be classified between two extremes. At one end are high-density populations living in favourable habitats. Such populations are regulated chiefly by density-dependent processes in the juvenile stage and are inherently stable with a small relative variation in their density (measured, for example, by the coefficient of variation). At the opposite extreme, there are low-density populations living in unfavourable habitats and showing little or no density-dependent regulation. Such populations are unstable with a large relative variation in their density.

An analysis of annual commercial and rod catches of adult sea-trout in 67 rivers in England and Wales provides some support for this hypothesis (Elliott, 1992). Rivers with high catches were nearly all those with a small relative variation in catches from year to year. In marked contrast, rivers with a large relative variation in catches were chiefly those with low annual catches that varied considerably between years of "boom or bust". There were, of course, rivers intermediate between these two extremes, but the analyses generally support the hypothesis and suggest that such a classification may be feasible. It would be useful to try some analyses on catch data for other species.

Finally, the importance of model development must be mentioned. Some examples of the use of mathematical models have been given in this account and it is obvious that some quantitative predictions would have been impossible without them. The development and improvement of such models requires large sets of data. For testing models of population dynamics, long-term quantitative studies are essential. There has to be a change in the common held view that environmental problems affecting fish populations can be solved rapidly by short-term, low-budget research projects. A major objective for future work on the more important fish species in rivers should be the development of models that can be used to predict the optimum density, the maximum attainable growth rate and the size variation of the fish in different populations of the same species. Such models should also facilitate the prediction of the effects on fish populations of environmental changes due to natural causes (e.g. droughts, spates and climate change) or human activities (e.g. fish stocking, fish farming, flow regulation, water storage, land drainage, forestry, changes in land management and, of course, pollution). The ultimate goal in the future should be therefore the development, testing and continuous improvement of realistic predictive models that can be used as tools for the conservation and management of the fish stocks.

27

Natural Factors Influencing Recruitment Success in Coarse Fish Populations

RICHARD H. K. MANN

NERC Institute of Freshwater Ecology, Huntingdon, UK

INTRODUCTION

Most European rivers are subject to perturbations, including land drainage, water abstraction, flood control, navigation requirements and the input of a variety of domestic and industrial waste products. The restoration and management of such disturbed systems in order to improve the status of the fish populations has had varying degrees of success. To be effective in the future, such operations must be based on a knowledge of the natural factors that regulate fish populations and which underlie subsequent anthropogenic influences. The domination of many coarse fish populations by progeny from relatively few years is well documented (Mills and Mann, 1985). An example is the chub (*Leuciscus cephalus*) population in the River Stour, Dorset, which is dominated by fish from only three or four year-classes although individuals may live up to 22 years (Mann, 1976, 1979). It is also recognized that the strength of a particular year-class is determined early in the life cycle, usually in the first few weeks of life. Once determined, the relative strength of the year-class persists throughout its life span.

What are the key factors that determine the year-class strength (YCS) of a particular species under natural conditions ? Unlike the salmonids, most coarse fish populations are not regulated by density-dependent factors. It is also evident from several studies (Mills and Mann, 1985) that both biotic and abiotic factors are involved. It has been argued that in unpredictable habitats (e.g. upland streams with sudden variations in discharge) fish populations are controlled principally by the abiotic characteristics of the environment. In contrast, it is suggested that biotic features play a more important role in relatively stable habitats such as lowland rivers (Zalewski and Naiman, 1985). Although this division is broadly true, other studies (Mills and Mann, 1985; Harvey, 1991a) have shown it to be an oversimplification and that biotic and abiotic factors often act simultaneously. Moreover, some biotic factors may be subject themselves to abiotic influence. For example, water temperature and river discharge can affect phytoplankton and zooplankton cycles and, hence, the initial food sources of newly-hatched fish (Mann and Mills, 1986).

The Ecological Basis for River Management. Edited by D.M. Harper and A.J.D. Ferguson. © 1995 John Wiley & Sons Ltd

Very high levels of mortality between the egg and adult stages have been estimated; Braum (1978) quotes a range of 99.87 – 99.99%, with most mortality occurring in the very early stages. To ensure that, despite this loss, some progeny attain sexual maturity, fish are generally highly fecund. Hence, even a small percentage decrease in early mortality can lead to a large increase in the absolute number of recruits to the spawning population.

In this review, the role of the most influential environmental variables affecting the egg and larval stages of coarse fish is examined. Reference is made to the way these variables may be disrupted by anthropogenic disturbances, and attention is drawn to those areas of knowledge that require further investigation to improve the design of river management policies.

EGG STAGE

Spawning requirements

Coarse fish species utilize a range of spawning substrata and, consequently, they can be grouped into spawning categories or guilds according to their preference (Balon, 1975, 1981; Mills, 1991). The eggs of most species are initially sticky but, once they are attached to the substratum, they lose this property and do not gather particles except by passive deposition (see later). The lithophils are those that spawn on clean gravel in flowing water; they include the dace (*Leuciscus leuciscus*), chub (*L. cephalus*) and barbel (*Barbus barbus*). The phytophils are dependent upon aquatic macrophytes on which to deposit their eggs, often in backwater areas of the river. They include the tench (*Tinca tinca*), rudd (*Scardinius erythrophthalmus*), silver bream (*Blicca bjoerkna*) and pike (*Esox lucius*). However, the most ubiquitous species in Europe are the phytolithophils (non-obligate plant spawners), which can utilize a range of spawning substrata. For example, the roach (*Rutilus rutilus*) is known to spawn on the moss (*Fontinalis*) growing on vertical metal and wooden pilings (Mills, 1981a; author, pers. obs.), on *Salix* roots (Diamond, 1985; author, pers. obs), on *Elodea* (Diamond, 1985) and on gravel (Holcik and Hruska, 1966). Other phytolithophils are the common bream (*Abramis brama*), bleak (*Alburnus alburnus*) and perch (*Perca fluviatilis*).

The lithophils and phytophils are the most susceptible of the three groups to environmental changes. For example, small side channels of the River Frome, Dorset, which are used by *E. lucius* as spawning areas, dried out in the hot summer of 1976 and spawning was carried out in the main river channel (Mann, 1980). Bank erosion by floods can lead to the de-oxygenation of gravel spawning areas through the deposition of silt (Mills, 1981b), and this can result in high egg mortality. Many rivers have lost a substantial part of the original backwater areas that are favoured by the phytophils, with a consequent decreased annual recruitment of these species and an overall reduction in their population numbers. For example, the loss of spawning habitat has been implicated in the reduced populations of *Tinca tinca* and other species in the River Great Ouse (Copp, 1990a). Some phytolithophils (e.g. *R. rutilus*) tolerate quite marked habitat changes, but others (e.g. *A. brama*) apparently have more specific habitat requirements, although these are not fully understood.

Many spawning sites are used year after year, and adult fish often make substantial spawning migrations to them (Kennedy and Fitzmaurice, 1968; Goldspink, 1977; Diamond, 1985). Thus, Vollestad and L'Abee-Lund (1987) observed a 90.2% precision in the return of spawning roach from a lake to their home tributaries. However, from the recapture of marked (fin-clipped) *L. leuciscus*, Mann and Mills (1986) found that, though a gravel spawning bed was used for over 10 successive years, some of the adults had not spawned there in previous years. Hence, details of the extent of homing is little understood, though it has clear implications for fisheries management. The provision of suitable spawning sites within a river system is clearly crucial to the resident fish species. However, the spawning requirements of many fish species are not known precisely. For example, in the River Great Ouse, *A. brama* spawns at only a few sites, compared with *R. rutilus*, which is ubiquitous (Copp, 1990a). Our knowledge is insufficient to explain why *A. brama* do not utilize other sites that appear to be suitable.

Restriction on the number of sites, for whatever reason, can lead to reduced egg deposition or the use of inferior sites, especially for those species with very specific requirements. Loss of spawning habitat has been connected with the increase in hybridization in some rivers (Wheeler and Easton, 1978; Mann, 1991) through the enforced sharing of spawning substrata by more than one species. However, this claim has yet to be substantiated.

Predation

The eggs of most coarse fish species are not protected in any way and, therefore, they are vulnerable to predation by fish and invertebrates. A notable exception is the bitterling (*Rhodeus sericeus*), which deposits its eggs in the mantle cavity of freshwater mussels. There has been little research on the mortality of coarse fish eggs on spawning substrata, although eel (*Anguilla anguilla*) predation on *R. rutilus* eggs has been recorded (Diamond and Brown, 1984). Mills (1981b,c) found neither fish nor invertebrates heavily predated *L. leuciscus* eggs in a chalk stream. Instead, egg mortality was attributed largely to oxygen deficiency through the eggs being laid in silty areas, or through the gravel becoming silted after the eggs had been deposited. Mortality in one year from this cause was estimated to be between 78.4 and 91.4%, compared with a 5% loss from invertebrate predation.

Water temperature

Egg incubation periods for all fish species are controlled largely by temperature. For example, Mills (1991) for *L. leuciscus* and Diamond (1985) for *R. rutilus* demonstrated a negative exponential relationship between temperature (T) and the egg incubation period in days (D). This took the form:

$$\log D = a + bT$$

Thus there is an approximately constant incubation period in terms of number of degree days for each species. Data for *R. rutilus* (Diamond, 1985) showed that the incubation period was 15 days at 12°C and 12 days at 14°C. Consequently, the period of incubation can vary from year to year (or site to site) for a particular species. Such variations in egg incubation periods could result in differences in the mortality rate. However, the *L. leuciscus* studies (Mills, 1981b,c, 1982) showed that the annual variations in egg mortality rate were insufficient to account for the observed level of variation in YCS. Incubation temperatures outside the optimum range (which will vary between species) can increase egg mortality rates. Thus, *P. fluviatilis* showed increased mortality rates at the eyed egg stage at temperatures above and below the optimum range of 8°C−12°C (Guma'a, 1978).

River discharge

Eggs of *R. rutilus* are often laid just below the water surface on various substrata. Mills (1981a) observed sudden decreases in water level in a chalk stream following weed-cutting operations, which could potentially destroy many eggs before they hatched. *R. rutilus* eggs have been seen stranded above the water surface in the River Great Ouse, following changes in the river level as a result of the operation of locks and sluices (author, pers. obs.).

NEWLY HATCHED FISH LARVAE

Water velocity

Newly hatched coarse fish larvae are weak swimmers because of their small size (usually < 10 mm) and their lack of fully-formed fins. Some species initially attach themselves to vegetation by adhesive glands (e.g. *R. rutilus*, Copp, 1990b), but this period is very short although there are inter-specific differences (see later). Hence the larvae are soon susceptible to displacement by the river current unless suitable refugia exist close to the spawning sites. There is little information on the fate of such displaced larvae, though it is often supposed that their survival rates are low. Many rivers have structures that may restrict later, compensatory, upstream movements should displacement occur. In the USA, Harvey (1991b) recorded very high mortalities of fish larvae < 10 mm in length in floods. He considered that biotic factors only influenced larval survival and recruitment success within the framework allowed by flooding. Pearsons, Hiram and Lamberti (1992) observed that proportionately fewer fish were lost in floods in hydraulically complex stream sections than in more simple channels. The former habitats also contained a higher fish species diversity. They also noted that the timing, frequency and intensity of floods had a strong influence on fish community structure through effects on recruitment success and age structure. Thus, recruitment success of spring spawning cyprinids was severely impaired by late spring/early summer floods.

To illustrate the relatively low current velocities required to displace coarse fish larvae, Figure 27.1 shows the results of experiments carried out in a small artificial channel at the IFE River Laboratory (Mann, R. H. K., unpublished). They show the maximum water current velocities that can be withstood by *R. rutilus* and *L. leuciscus* larvae of different sizes. However, it should be noted that swimming performance may be influenced by the rate of energy intake rather than the availability of stored energy (Ware, 1975). These critical current velocities are not the same as the velocities selected by the same-sized fish under natural river conditions. Lightfoot and Jones (1979) found that *R. rutilus* larvae 7.5 mm in length selected areas of river, usually in macrophyte beds, where the current speed did not exceed 20 mm s^{-1}, even though their fatigue velocity (50% of larvae displaced after one hour) was 69 mm s^{-1}. Mann and Mills (1986) noted that areas with flow velocities < 20 mm s^{-1} were concentrated along the margins of the River Frome and represented only 2−3% of the river's surface area. This narrow riparian zone is vital as a refuge for newly hatched fish larvae, and any modification of it can have a major impact on local fish populations. Surveys of river bank morphology along the Austrian section of the River Danube showed that only c. 10% was suitable as refugia for larval fish. Schiemer et al. (1991) considered this to be the cause of a decline in the stocks of previously common fish species.

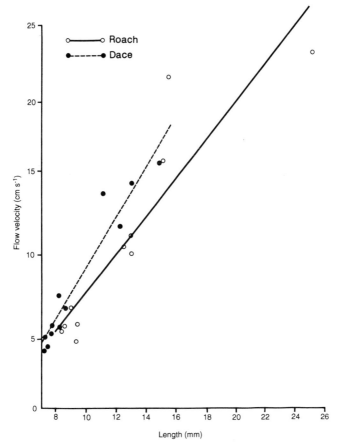

FIGURE 27.1.   Critical flow velocities of roach *R. rutilus* larvae in an experimental channel. Each point is the mean velocity, based on five replicates of 10 larvae, at which 50% of the larvae were washed onto the downstream screen of the channel

Figure 27.2(b) shows that river current speeds in a side channel (Lees Brook) in the Great Ouse system over a short (80 min) time period ranged from -21 mm s^{-1} to +135 mm s^{-1} during a period of relatively stable flows. Discharge and water velocity in Lees Brook can alter rapidly when control sluices and boat locks operate, particularly during periods of low flow in the summer (Figure 27.2(a)). From Figure 27.1 it can be seen that the maximum value (135 mm s^{-1}) would displace *L. leuciscus* and *R. rutilus* larvae up to *c*. 12 mm length. The changes in available low-velocity habitat at different discharge levels in Lees Brook is demonstrated in Figure 27.3. Thus, suitable velocity patches are not fixed and fish must frequently seek new areas as flow conditions change.

Temperature

The causal relationship between water temperature and fish growth rate is well established, and higher growth rates of coarse fish have been observed in summers with above-average water temperatures than in cooler summers (Mann, 1991; Mills and Mann, 1985). Mills and

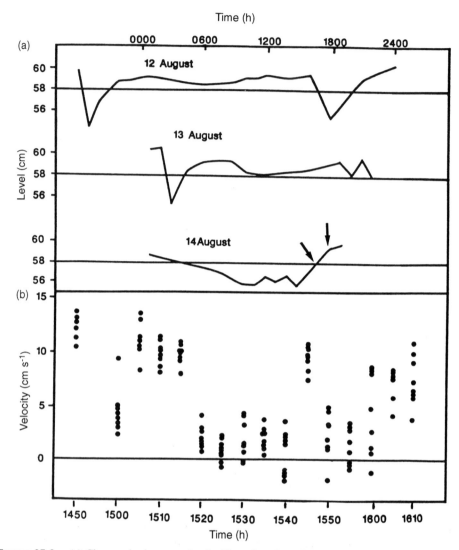

FIGURE 27.2. (a) Changes in the water level of Lees Brook, a side channel of the River Great Ouse.
(b) Changes in the flow velocity at a single point in Lees Brook, taken 10 cm below the water surface
and 100 cm from the bank, during the period indicated by arrows in (a). Each point is the mean
velocity over 10 5s; each series for a particular time was taken within a *c.* 1 min period

Mann (1985) demonstrated a strong correlation between the larval growth rates and
subsequent YCS for several coarse fish species. Also, Cerny (1975) recorded a poor
development and a high mortality rate of *R. rutilus* larvae reared at 8.9°C, and good
development with low mortality at 22.1°C. It is argued that the variation in larval growth
rates between years is the prime cause of YCS variation because faster-growing larvae are
vulnerable to predation by aquatic invertebrates for a shorter time period. Also, faster-
growing larvae are vulnerable to water currents for a shorter period. Few data exist to
support the claim regarding predation, but Mann and Mills (1986) found that, in aquaria,

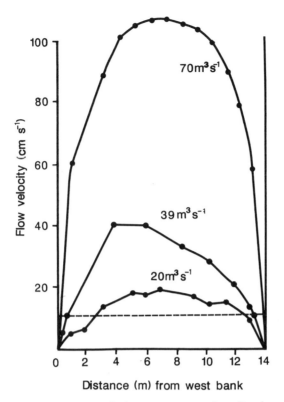

FIGURE 27.3.   The flow profiles along a single transect across Lees Brook at three discharges. The distance along the broken line from each bank to the intercept with a particular flow profile indicates the width of riparian habitat with a flow velocity < 10 cm s^{-1}

three Odonata species ate significantly more small *L. leuciscus* larvae than larger larvae. Indirect evidence for high predation rates on fish larvae is shown from the initially high survival rate ($> 95\%$) of *L. leuciscus* larvae stocked in predator-free cages suspended in a river before density-dependent effects were observed (Mills, 1982).

It is equally clear, however, that water temperature does not account for all the variation in fish growth rates. Thus, temperature accounted for only 65.5% of growth variation of 0 group *L. leuciscus* in the River Frome (Mills and Mann, 1985). Water temperature also accounts for only part of the variance in recruitment success (YCS). Reported values include 44.3% (*L. cephalus*), 62.2% (gudgeon *Gobio gobio*) and 69% (*L. leuciscus*) (Mills and Mann, 1985).

Recent studies of *R. rutilus* larvae in the River Great Ouse suggest that the availability of suitable exogenous food particles at the time of yolk sac absorption can affect fish growth (Mann, R. H. K., unpublished). Mooij and van Tongeren (1990) calculated the maximum growth rate of 0 group *R. rutilus* in relation to initial size and water temperature, based on laboratory feeding studies in which the fish were fed *ad libitum*. Figure 27.4 shows the observed growth of 0 group *R. rutilus* in the River Great Ouse in 1989 and 1991, and the theoretical growth trajectories for the same two years based on Ouse water temperatures and the Mooij and van Tongeren (1990) growth model. In both years the observed growth rates

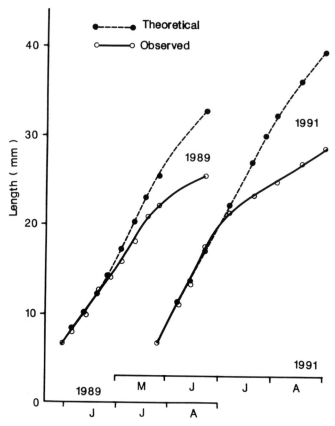

FIGURE 27.4.   Observed mean lengths (mm) of 0 group roach *R. rutilus* in the River Great Ouse in 1989 and 1991, and theoretical growth curves based on observed water temperatures and the growth model of Mooij and van Tongeren (1990)

initially agree closely with the theoretical rates, which suggests that food is not a limiting factor. However, from late June onwards, the observed rates fall behind the predicted rates. The onset of this change coincides with a switch in 0 group *R. rutilus* diet from invertebrates (Cladocera, chironomid larvae, etc.) to one dominated by detrital aufwuchs from plant (*Nuphar*) leaf surfaces (Mann, R. H. K., unpublished).

Results from the Great Ouse indicate that the presence of sufficient suitable prey (diatoms, rotifers) at the time of hatching depends on previous climatic conditions (which affect river flows). Also, some of the fish species have peak spawning and hatching times that differ from year to year. For example, peak hatching times of *R. rutilus* from 1988 to 1992 ranged between 10 May (1990) and 31 May (1988). Sometimes they were synchronous with the peak abundance of diatoms (*Stephanodiscus*) and planktonic rotifers, sometimes they were not. Many authors (e.g. Balon, 1956; Tesch, 1962) regard the transition from yolk sac dependence to exogenous feeding to be a critical phase in the early life of fish. Lack of suitable prey organisms at this time can lead to starvation, increased vulnerability to predation and, hence, to low larval survival and poor recruitment. Vladimirov (1975) observed a critical period for *A. brama* 4–6 days after yolk sac absorption, with those fish dying being less well developed and in poorer condition than their contemporaries. However, Cerny (1975) concluded from an examination of published data and from his experiments with fed and starved *R. rutilus* larvae, that the increased mortality rate of several fish species before and after yolk sac absorption could not be explained by a shortage of convenient food organisms. Evidence of inherent differences in early development rates have also been demonstrated. El-Fiky and Wieser (1988) showed that the larvae of some species (*R. rutilus*, *L. cephalus* and *L. leuciscus*), which typically are able to swim and feed within three days of hatching, show a more rapid differentiation of their gill structures than do species such as bleak (*Alburnus alburnus*), or Danube bleak (*Chalcalburnus chalcoides*), which rely for a longer period (7 – 10 days) on gas exchange across the whole body surface. Other species (*A. brama*) occupy an intermediate position.

Thus, the growth rates of newly hatched coarse fish larvae can be affected by many factors, which can show substantial year to year variation and can lead to marked between-year variations in larval survival and recruitment success. Inherent inter-species differences in larval growth rates are affected by water temperature and food availability. These show within-season variations and, hence, larvae hatching at different times of the year may vary in their recruitment success. For example, Mills and Mann (1985) found no significant correlation between *R. rutilus* and *L. cephalus* YCS in the River Stour, Dorset, over a 12 year period. In this river, *R. rutilus* usually spawned in mid to late May, whereas *L. cephalus* spawned from mid June onwards.

Microhabitat

It has been already shown that fish larvae are constrained in their choice of habitat by river current velocities. However, there are other environmental attributes that determine microhabitat preference and, hence, subsequent larval survival rates and recruitment success. These attributes can differ between fish species. Copp (1990b) showed that, after the initial stage of attachment to vegetation in the spawning area, *R. rutilus* larvae became positively associated with shallow depths, aquatic vegetation, woody debris and lentic areas. Later, as they entered the L3 stage of larval development (fin-fold differentiating but fins not fully formed), they swam more actively and selected more open areas with water depths of 0.2 – 0.5 m. Further studies (Copp, 1992) showed that coarse fish larvae could be grouped according to their microhabitat preferences, with water depth, channel topography, substratum and water temperature being the most important variables. Water current was important, but appeared to be less influential (providing that it did not displace the larvae), though it is clear that this variable must be affected by water depth and channel depth and width. Rincon, Barrachina and Bernat (1992) took this a stage further in proposing that the non-random use of most habitat traits resulted from their dependence on water velocity.

Ontogenetic shifts in habitat use by fish larvae have been recorded for the River Danube in relation to substratum, water velocity and food availability (Schiemer, 1985, 1991; Schiemer and Spindler, 1992). The larvae of all species were at first located only in shallow bays where current velocities usually remained low despite changes in water level. As the larvae grew, the rheophilic species moved to adjacent shallow gravel banks and, by the end of their first summer, were found in areas with current speeds <50 cm s^{-1}. Eurytopic species remained in the sheltered riparian areas. Though these studies have considerably advanced our knowledge of the habitat requirements of coarse fish larvae, the priorities of different environmental attributes on recruitment success are still not known precisely for the various fish species. In general, there are three sets of environmental attributes that are essential for recruitment success:

(i)   availability of spawning locations that are close to the areas that are suitable as refugia for the newly hatched larvae;
(ii) sufficient microhabitat diversity to satisfy the changing needs of developing larvae, and the needs of different species;
(iii) backwater areas away from the main river, both to act as refugia during floods and to provide the specialized habitats for phytophilous species.

In addition, water temperatures during the early larval period need to be sufficiently high to produce fast growth rates and, hence, shorter periods of vulnerability to predation and displacement by river currents. However, little is known about the precise relationships between all these factors and the year-class strengths of coarse fish species. The collection of this essential information, based on long-term studies (minimum 20 years) to allow causal relationships to be established, is imperative in order to underpin the short-term management objectives that are current in much of present-day river management. Without such fundamental knowledge, inefficient empirical management procedures will persist.

ACKNOWLEDGEMENTS

The assistance of Miss Emma Jerman in carrying out the critical velocity experiments is gratefully acknowledged. The manuscript benefitted greatly from the comments of my colleagues, Jon Bass and Paul Garner. The work was funded by the Natural Environment Research Council.

28

Factors Affecting Recruitment Success in Salmonids

GERSHAM J. A. KENNEDY and WALTER W. CROZIER

River Bush Salmon Station, Bushmills, Northern Ireland, UK

INTRODUCTION

This chapter focuses attention mainly on stock and recruitment in salmon, *Salmo salar* L., on which the bulk of our studies have been concentrated. Also, since this volume is concerned with river management, we limit our discussion to those aspects affecting stock and recruitment in the freshwater phase of the salmon life cycle.

Our work on salmon ecology and survival has been carried out on the River Bush in Northern Ireland, which is recognized as one of the North Atlantic's ICES "index" rivers (Anon., 1985). The project has now been under way for 20 years, and overviews of the work are available in Kennedy and Johnston (1986) and Crozier and Kennedy (1991). A comprehensive analysis of the freshwater production data on the River Bush in relation to other index rivers on both sides of the Atlantic has also been undertaken (Kennedy and Crozier, 1993), and many of the conclusions from this will be drawn into the present chapter.

We have been asked to consider the state of ecological knowledge required for the effective management of salmonid stocks. In relation to salmon in freshwater this can be approached on two levels, i.e. in terms of the whole river system or at the localized habitat level. Distinctly different concepts and research approaches are required for management of the salmon resource at each level. What is common to both is a wide range of variability in published survival figures — both temporally and geographically.

RANGES OF PRODUCTION

Smolt count data series are now available for a number of river systems on both sides of the Atlantic, e.g. Ireland: River Bush (Crozier and Kennedy, 1991), River Corrib (Browne, 1990), River Burrishoole (Anon., 1990a); Scotland: N Esk (Shearer, 1986), Girnock Burn

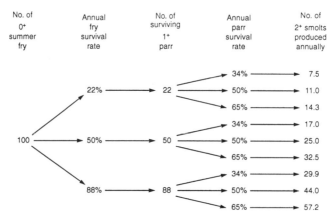

FIGURE 28.1.   Potential 2+ smolt production from the range of published annual survival rates per 100 0+ summer fry

(Buck and Hay, 1984); Atlantic Canada: Western Arm Brook, Little Codroy River, Sandhill River, Big Salmon River, Black River (Chadwick, 1985a, 1988, 1991). Annual within-river fluctuations of up to about sixfold in absolute smolt numbers have been reported. The literature also indicates that there is considerable variability in both survival rates (at all juvenile stages) and in overall production in different river systems (for a review, see Mills, 1989). Kennedy (in prep.) has produced a model using the published annual survival values for 0+ and 1+ parr in the British Isles to indicate that the range in potential 2+ smolt production from summer fry is over sevenfold (see Figure 28.1), i.e. after the initial high fry mortalities following emergence have already occurred. Symons (1979) concluded that the likely range of smolt production for Atlantic salmon covered an order of magnitude, from about 1 to 10 smolts 100 m^{-2}, and that this was related to the age of the smolts produced.

On the River Bush, annual smolt counts have varied by over fourfold, from 10 779 to 43 958, and ova to smolt survival from each cohort has varied by over fivefold, from 0.40% to 2.13% (in Figure 28.2). These data have been incorporated with marine survival and exploitation information into a production model (Kennedy and Crozier, 1991), which indicates that the potential for variation in grilse returns to the River Bush from one ova deposition is in the region of 20-fold (Figure 28.3). An understanding of the biotic and abiotic mechanisms resulting in these levels of variability is clearly a prerequisite for the management of the salmon resource and its environment.

THE STOCK – RECRUITMENT RELATIONSHIP

The precise shape of the stock – recruitment curve is still the subject of debate. Solomon (1985) reviewed the historical data and concluded that the evidence favoured the flat-topped version for salmon. However, Solomon also noted that most of the data on salmon stocks come from the lower end of the ova deposition range in the stock – recruitment model, well away from any possible right-hand decending limb. In his long-term study on Black Brows Beck, Elliott (1984b, 1989a,b) has conclusively shown that, for sea trout, a dome-shaped

Ricker curve is the best fit to the data for fry surviving in May and June from a wide range of ova depositions. By August/September the relationship becomes more flat topped, with a slightly domed curve continuing to give the best statistical fit.

In the River Bush, salmon ova depositions from 1973 to the present have varied from 1.06 million to 4.79 million. Total counts of $1+$, $2+$ and $3+$ smolts are available up to the brood year 1988, and a Ricker curve provides a marginally significant fit to the data (Figure 28.4). The results to date therefore suggest a dome-shaped density-dependent stock − recruitment relationship, with density-independent variation producing a wide scatter of points, particularly at low ova depositions (Figure 28.5). Elliott (1984) drew a similar conclusion for sea trout, noting that density-dependent mortality predominated over density independent mortality only in year-classes with an initial egg density greater than the optimum value. However, in the Girnock Burn, Hay (1989, 1991) found high-density independent variability, resulting in large annual fluctuations in smolt production, at both low and high ova depositions. This led him to conclude that an "envelope" bounded by diverging upper and lower lines defining the limits of smolt production best represents the stock − recruitment relationship (Figure 28.6). Hay considers that in years when adult salmon penetration of spawning streams is good, their eggs are spread uniformly over a wide area, and smolt production tends towards the upper edge of the envelope. Conversely in years when spawning is restricted to smaller areas and initial juvenile densities are localized, smolt production is reduced. Ongoing research into the environmental variables involved suggests that this effect is regulated by rainfall and

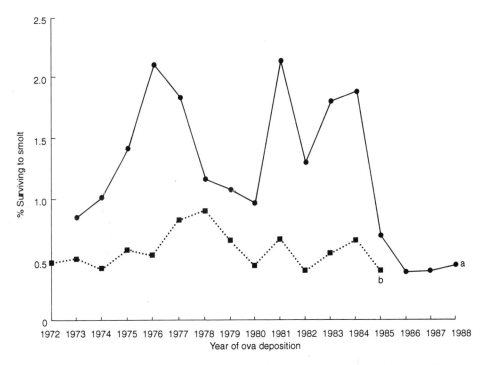

FIGURE 28.2.   Percentage survival from ova to smolt for each ova deposition (a) on the River Bush 1973−88 (● ____ ●), and (b) on the River Burrishoole 1970−85 (■ · · · · ■). The River Burrishoole data are taken from the means of the published ranges of ova to smolt survival for this river (see Anon., 1990). (Reproduced with the kind permission of Dr K. Whelan, Salmon Research Agency of Ireland)

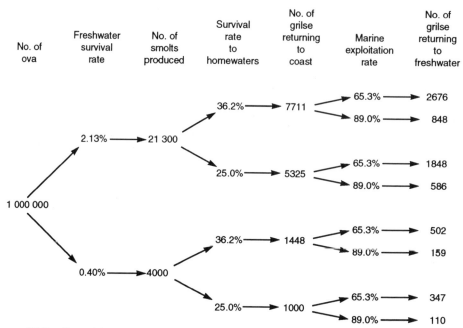

FIGURE 28.3. Potential grilse returns to the River Bush from an ova deposition of one million, as derived from the ranges of observed freshwater and marine survival rates, and the range of marine exploitation around the Irish coast on microtagged wild fish

river flows, with "windows of opportunity" for spawning in different stream areas being related to certain optimum water levels at critical periods (Hay, 1989).

Preliminary analyses of environmental variables on the River Bush similarly indicate that variation in smolt production at lower ova depositions is correlated to rainfall and river flows, with warm, dry spring and early summer periods resulting in the best survivals (Kennedy, G. J. A. and Crozier, W. W., unpublished). It is hypothesized that this is related to the critical time of emergence and setting up territories (Elliott, 1989a,b), when high flows may cause high mortalities in small swim-up fry by disruption of territory formation. High flows can also cause washout and siltation of redds, and low flows result in the stranding of redds (Harris, 1978), or the limited dispersal of emerging fry (Kennedy, 1982). However, temperature may also play a part, and in the River Bush, 1+ smolt production is positively correlated to water temperature during the period June−September ($r = 0.61$; $p<0.05$). In the less temperate climate of Newfoundland, Chadwick (1982) reported significant inverse correlations of egg to fry survival with both river discharge and winter temperature. Apparently extreme winter temperatures caused freezing of redds at low river levels, resulting in high egg mortalities. Chadwick excluded a very low smolt production figure for the 1972 year-class on Western Arm Brook from his stock−recruitment analysis on the basis that it was an "outlier" resulting from such extreme weather conditions. Elliott (1989) similarly excluded the 1983 and 1984 year-classes from his analyses on the grounds that their mortality rates were exceptionally high due to summer droughts in these two years. Drought is also considered to be the main factor regulating recruitment in the sea trout population of a small Swedish stream (Titus and Mosegaard, 1992). Annual fluctuations of

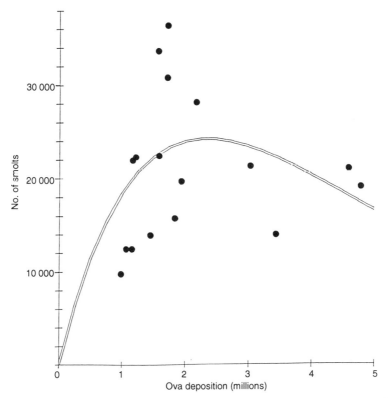

FIGURE 28.4. Total smolt counts on the River Bush from varying ova depositions 1973−88. A Ricker curve has been fitted to the data ($r^2 = 0.442$; $p < 0.10$)

up to 17-fold were attributed both to the effects of low flows restricting access for adult spawners and to drought-induced mortality on summer fry.

Bird predation has also been implicated as a major factor influencing density-independent mortality fluctuations in some rivers, e.g. the River Bush, where cormorant (*Phalacrocorax carbo L.*) predation has been found to be significant — particularly on older parr and smolts (Kennedy and Greer, 1988; Kennedy and Crozier, 1993; Warke and Day, in press). The effects of sawbill duck predation on smolt production in Canadian streams has been documented for many years (e.g. Elson, 1962), but only on a limited basis in the British Isles, where these birds are increasing in both numbers and range (Mills, 1989; Shearer et al., 1987).

In general, the effects of predation and environmental variables as causes of density-independent variation are not well studied or understood. Indeed in some areas the effects of density-independent mortality in freshwater are considered to be insignificant compared to other variables. For example, in Newfoundland, Chadwick (1991) reported a 20-fold variation in the marine survival rates of smolts from Western Arm Brook, but only a 1.9-fold variation in smolts per spawner at ova depositions below the asymptote of the freshwater stock−recruitment curve. This led him to conclude that "further research on factors other than egg deposition will not necessarily improve our ability to predict returns of adults". However, in the River Bush stock, where variation in marine survival is much less than in

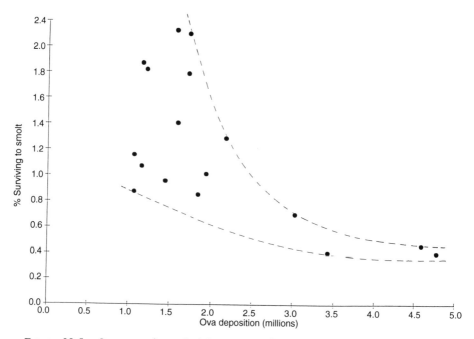

FIGURE 28.5. Ova to smolt survival for a range of ova depositions on the River Bush

Newfoundland (<1.5-fold to the coast and 3.1-fold to the river (Crozier and Kennedy, 1993, 1994; Kennedy and Crozier, 1993)), freshwater density-independent variation is a major influence on production. It equates to a variation in smolt production from about 14 000 to 35 000 from a low ova deposition. For management purposes this is not an acceptable level of unpredictable variability and further research on the causes of density-independent fluctuations in freshwater mortality need to be undertaken.

INDEX RIVER DATA

Long-term data series on a range of parameters, including stock and recruitment relationships are being accumulated on a number of so-called "index" or "monitored" rivers on both sides of the North Atlantic. These data sets are improving our knowledge of the factors controlling survival and production within these river systems, but are they providing models which can be applied more widely, and on which broader stock management can be based? How comparable are the models from different rivers? Can they be used to estimate national smolt yields and assess overall production? A detailed evaluation of these questions has been undertaken in Kennedy and Crozier (1993), and the main conclusions are summarized below.

Smolt count data

In Newfoundland, Chadwick (1985b, 1988) reported significant correlations between three smolt count data sets and adult returns outside the river systems providing the counts:

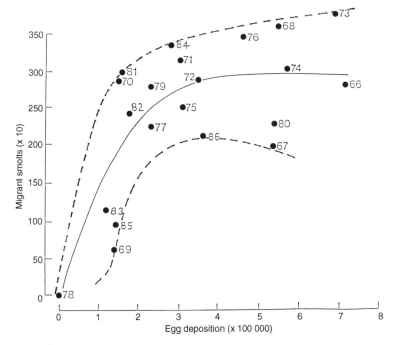

FIGURE 28.6. Total smolt counts on the Girnock Burn from varying ova depositions 1966–86. The dotted lines define the envelope of production proposed by Hay (1991). (Redrawn with the kind permission of D. Hay, Scottish Office, Agriculture & Fisheries Department)

(1) Counts of grilse in the Torrent River were correlated to counts of Western Arm Brook smolts in the previous year.

(2) Western Arm Brook smolt counts could also be used to predict total adult returns to local home waters, where they only represent about 10% of the stock.

(3) Returns of adults to the neighbouring Grand Codroy and Little Robinson Rivers could be predicted from Little Codroy smolt counts.

No such correlations have been reported for river systems in the British Isles, and no information at all is available on the geographical extent of such parallel stock fluctuations. Further investigation is required if the growing numbers of data sets on smolt numbers are to fulfil wider management functions outside the monitored rivers.

Ova to smolt survival data

Correlations in survival

Comparison of the two Irish index rivers, the Bush and the Burrishoole (see Figure 28.2) indicated that ova to smolt survival rates were significantly correlated for 12 out of the 13 years from 1973 to 1985 ($r = 0.61$; $p < 0.05$). It was therefore concluded that some sort of overlying regulating mechanism influenced ova to smolt survival similarly in both rivers in most years. It was hypothesized that the mechanism is environmental, and possibly related to similar rainfall patterns and the consequential effects of river flow. The lack of a complete

TABLE 28.1. Comparison of smolt production with catchment characteristics

River	Catchment size (ha)	Maximum smolt production per cohort	Maximum smolt production per hectare	Range of ova to smolt survivals (%)	Area of standing water in catchment (ha)
Bush	33 700	36 360	1.08	0.40−2.13	0
Burrishoole	8 550	>15 000	>1.75	0.39−0.95	410

correlation in freshwater survival rates between the two rivers in all years was taken to indicate that some of the climatic factors influencing annual production can be very local in their effects (the two catchments are separated by about 250 km).

No other similar analyses appear to have been undertaken between monitored rivers, and as with smolt count data, further such investigations are required if index river data are to perform a useful predictive management function — even within a limited geographical area.

Comparative productivity

It was also noted that the mean and maximum survivals from ova to smolt on the River Burrishoole were less than half those on the River Bush. This apparently reflects the differing land use and productivities of the two catchments. The River Bush drains an intensively farmed catchment with associated enrichment from fertilizer and organic leachate, whereas the Burrishoole catchment is comprised of much poorer land supporting only rough grazing, mainly for sheep. However, inspection of the smolt production figures for each catchment revealed an apparant anomaly. Although the Burrishoole has a lower productivity than the Bush, as measured by ova to smolt survival, the maximum smolt output per hectare of catchment is higher on the Burrishoole (1.75 smolts ha^{-1}) than on the Bush (1.08 smolts ha^{-1}) (see Table 28.1). The difference appears to be the result of production arising from lake-dwelling parr in Lake Feeagh (410 ha) in the Burrishoole catchment. The Bush catchment has no comparable standing water.

The considerable capacity of lakes for rearing smolts has also been noted in Iceland (Einarsson, Mills and Johannsson, 1990) and in Newfoundland (Chadwick, 1985; Chadwick and Green, 1985), where it can apparently account for up to 70% of production in some systems. However, although various fry densities have been investigated for the artificial stocking of standing waters in Canada (Pepper, Oliver and Blunden, 1985) there is a paucity of information on the natural production potential of lakes, the ova depositions required to optimize this potential, and the recruitment mechanisms operating between standing waters and their feeder streams.

It is therefore clear that for management purposes there are considerable dangers in extrapolating measures of productivity from one catchment to another. In addition to considerable variation in the holding capacity of streams of differing productivity (Symons, 1979; Mills, 1989), neither ova to smolt survival nor the type and quantity of rearing habitat are necessarily comparable between catchments. Detailed field surveys are a prerequisite for management estimates of smolt production or target ova deposition on all systems.

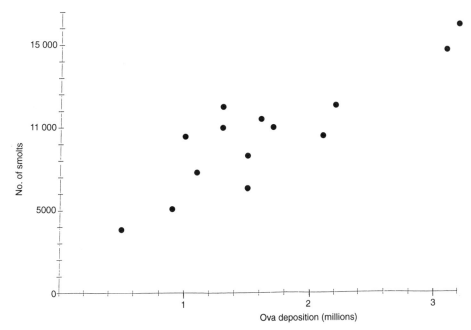

FIGURE 28.7. Total smolt counts on the River Burrishoole from varying ova depositions 1970–85. (The ova depositions are taken from the mean of the published range (see Anon., 1990)). (Reproduced with the kind permission of Dr K. Whelan, Salmon Research Agency of Ireland)

However, the information available on the contribution of standing waters to overall salmon production is inadequate at present for management purposes, and further research in this area is required.

Relative variability

The final point of difference between the Bush and the Burrishoole survival data was in the range of variation from year to year. Ova to smolt survival on the Burrishoole from 1973 to 1985 varied from 0.39% to 0.95%, i.e. by about a factor of 2.4, whereas the comparable data from the River Bush varied by a factor of 5.3, from 0.40% to 2.13%. Comparison of the stock–recruitment curves for the two systems indicated that while the River Bush data included high ova depositions beyond the asymptote of the curve (see Figure 28.4), ova depositions on the River Burrishoole were all below the asymptote of the curve (see Figure 28.7). This observation explains the apparent discrepancy between the two systems, since density-dependent mortality increased above the asymptote on the River Bush. On the analagous ascending limb of the stock–recruitment relationship for the Bush, ova to smolt survival ranged from 0.85% to 2.13%, i.e. by a factor of 2.5.

It is therefore clear that for effective management of a salmon river, an understanding of both the shape of the stock–recruitment curve and the status of each spawning escapement/ova deposition in relation to the asymptote is an important factor, e.g. for the setting of a target ova deposition and for the setting and the application of regulations either for commercial or sport fisheries in relation to achieving the target ova deposition.

An example of this approach is on the River Foyle, where Elson (1975a) made recommendations for achieving target spawning escapements in relation to a stock—recruitment analysis. These recommendations are now implemented by controlling commercial fishing operations in response to target adult numbers counted at a fish counter (Crawford, W. G., pers. comm.).

Given the growing evidence for dome-shaped stock—recruitment curves for salmon production in freshwater, excessive curtailment of either commercial fishing or sport fishing may not be an optimum management strategy in some rivers. However, the dangers of exceeding the asymptote should not be overstated, and Elson's warnings in this respect have now been largely discounted due to a general lack of acceptance of steeply domed stock—recruitment curves (Chadwick, 1982, 1985b; Solomon, 1985). From the evidence now available it seems certain that, due to limited data at high ova depostitions, Elson misintepreted density-independent variability in River Foyle stocks for a steep downward trend in numbers beyond the asymptote of the curve.

SETTING TARGET OVA DEPOSITIONS

Whatever the precise shape of the salmon freshwater stock—recruitment curve, the smolt production capacity of a river system only becomes apparent at the asymptote of the curve. This area of inflexion indicates the minimum number of eggs that have the potential to produce the maximum number of smolts from a river system, given optimum density-independent survival conditions. This is known as the target ova deposition, and for management purposes it is important to realize that this is not synomymous with the potential egg deposition of the adult freshwater escapement.

Elson (1975b) suggested that a figure of 25% freshwater mortality should be applied to the freshwater escapement to cover angling catches, non-catch fishing mortality and natural mortality — and in the frequent absence of hard data this has become something of a standard. For example, this correction factor has been incorporated into the recommended egg deposition rate of 2.4 m^{-2} now used as a basis for calculating target spawning requirements in Atlantic Canada. Chadwick (1985b) has questioned the applicability of this rate to all rivers without taking account of unique local factors. Certainly in the British Isles, freshwater exploitation rates from sport fisheries vary greatly between systems (Mills, 1991) or on different stocks within one system (Gee and Milner, 1980), and it is not appropriate for managers to apply a standard correction factor when setting target ova depositions. Further research is also required in the difficult areas of non-catch fishing mortality and natural mortality, which may vary greatly — both between systems and between years.

We have already seen that it is inappropriate for managers to extrapolate from one river system to another on the basis of catchment size. It is evident that one recommended egg density is also inappropriate for all habitat types within any one system, and that any recommended target ova deposition is highly dependent on the proportion of different types of habitat in different river systems. If a fishery manager diligently carries out a full catchment and habitat survey, the next question is — can the carrying capacity and production potential be calculated and a meaningful target ova deposition for the catchment be set?

In a previous paper (Kennedy and Crozier, 1993) we calculated that the target ova deposition for the River Bush probably lay within the range 1.6 million to 2.4 million, but even with a 20 year data set it is not possible to be more precise due to high density-

TABLE 28.2. Options for expressing target ova depositions in the range $1.6 \times 10^6 - 2.4 \times 10^6$ for the River Bush

Habitat type		Area in River Bush catchment	Target ova deposition
(i)	Total river catchment	33 700 ha	$47.5 - 71.2$ ha^{-1}
(ii)	Total wetted surface of river	84.55 ha	$1.9 - 2.8$ m^{-2}
(iii)	Total usable salmonid nursery habitat	41.06 ha	$3.9 - 5.8$ m^{-2}
(iv)	Total usable grade A salmonid nursery habitat	23.38 ha	$6.6 - 9.8$ m^{-2}
(v)	Total area of grade A salmonid nursery habitat normally used	16.91 ha	$9.5 - 14.2$ m^{-2}

independent variability. Habitat assessments on the River Bush allow the transformation of this target ova deposition to ranges of ova deposition which vary according to the habitat definition (Table 28.2). This permits comparisons to be made with other catchments where the proportions of different habitat types are not directly analogous to the River Bush. For example, the Girnock Burn is an upland tributary with a catchment of 2800 ha and 5.8 ha of usable salmonid habitat (Hay, D., pers. comm.). If the asymptote of the stock–recruitment curve on this system is taken to lie within the range of about 200 000–300 000 ova (see Figure 28.6), this is equivalent to an ova deposition of $71.4 - 107.1$ ha^{-1} of catchment, or $3.5 - 5.2$ m^{-2} of usable salmonid habitat.

Symons (1979) related the range of smolt production levels in streams to differing productivities, and consequently recommended different overall target ova depositions, varying from 0.8 m^{-2} to 2.2 m^{-2}. However, he also recognized that smolt production and target ova depositions could be higher for restricted areas of productive nursery habitat. The most comprehensive recommendations made for artificial stocking densities in relation to habitat type and productivity have been produced for Scottish streams by Egglishaw et al. (1984). These range from two to ten ova or unfed fry per square metre, depending on the altitude and size of the streams. In early studies on Lake District streams, Le Cren (1973) found that the asymptote for brown trout fry was achieved at a stocking density of about 12 m^{-2}. However, the high carrying capacity of 8.7 fry m^{-2} was obtained in screened stream sections which did not contain any other fish, and work on the Altnahinch tributaries of the River Bush by Kennedy and Strange (1980, 1982, 1986a,b) has clearly shown that both inter- and intra-specific competition affect the survival, growth and distribution of stocked salmonids. Elliott's (1984, 1985, 1989a,b) work on Black Brow's Beck has shown that, in a small spawning tributary, the asymptotes of the dome-shaped stock–recruitment curves are achieved at ova depositions of 40.0 m^{-2} and 62.5 m^{-2} for the May/June and August/September sampling periods respectively. However, there is no indication what these may represent in terms of ova per square metre for all the nursery habitat available on a catchment basis.

A similar very high optimum density for stocked salmon is indicated by recent stocking experiments in the Altnahinch streams. Trials of various "swim-up" salmon fry stocking densities of up to 30 m^{-2} in the presence of other age classes of salmon and trout have apparently not achieved stream carrying capacity (Figure 28.8(a)). However, transformation of the data to percentage survival plotted against ova deposition (Figure 28.8(b)) suggests (a) that survivals are high but variable at low stocking densities (presumably in response to density-independent variables in an analogous way to those on the main river; see Figure 28.5), and (b) that percentage survivals are consistently low at high stocking densities. A cost−benefit analysis of the optimum usage of fry for stocking suggests that a stocking density of about 5 m^{-2} gives the optimum return from the stocked fish in this stream (Crozier, W. W., and Kennedy, G. J. A., unpublished), i.e. stocking higher densities than this may increase the numbers of fish in the river, but this is not a cost-effective use of a limited resource.

There is also evidence that natural ova depositions should not be equated with those calculated from artificial stocking (Kennedy, 1988). Apparently the careful dispersal of eggs and fry during stocking operations can improve survival by over threefold (O'Connell, Davis and Scott, 1983; O'Connell and Bourgeois, 1987) compared to natural mortality rates induced by the clumping of adult spawners (as described by Hay (1989) in the Girnock Burn) and the subsequent limited dispersal of fry (Kennedy, 1982). However, the parameters involved in regulating survival from stocking operations have never been satisfactorily quantified, and recommended artificial stocking densities and natural ova deposition rates are frequently cited in the literature as if they were synonymous.

The answer to the question posed above must therefore be a qualified yes — fishery managers can set a target ova deposition based on catchment surveys, but at present these must take the form of the "most likely range". Before this can be more tightly defined, further work is needed to identify the nature of the stock−recruitment relationship in a range of rivers, and more fundamental research is required to define habitat types more closely and to determine the egg depositions which are appropriate for these in both running and standing waters. For optimum management, target spawning escapements cannot continue to be set on the basis of one historical density applied across continents, in adjacent catchments or even in different portions of any one catchment. Inappropriate target ova depositions may well be masked in the short term by the high annual variation in smolt output from rivers as a result of high levels of density-independent mortality. There may also be changes in smolt size (Chadwick, 1991) or age structure (Kennedy and Crozier, 1993) resulting from fluctuations in annual ova deposition, and these in turn can affect management strategy by impacting on marine survival and exploitation patterns (Issakson and Bergman, 1978; Crozier and Kennedy, 1993, 1994). However, there is no concensus on these aspects in the literature and the regulating mechanisms are not well studied or understood.

MANAGEMENT PLANS

In the absence of data on freshwater escapement and likely ova depositions, river management must, of necessity, take the form of a staged approach aimed at optimizing each individual aspect of the resource. This compromise strategy involves acquiring the best possible local knowledge, and integrating this with the best available "state of the art" information from index river and other studies on similar catchments. This need not necessarily be a "piecemeal" approach, but an appropriate management plan spelling out the priorities and the action required to fulfil defined objectives and policies is a prerequisite. Such an integrated fishery managment plan has been developed for the River Tweed (Anon.,

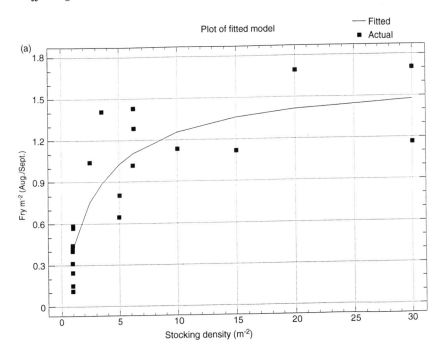

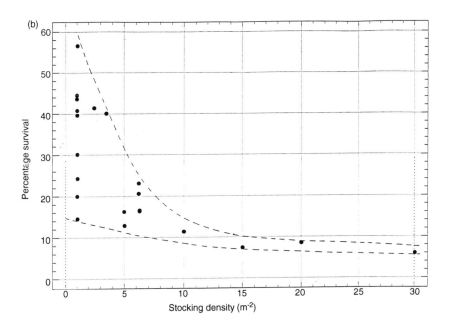

FIGURE 28.8. (a) Density (with fitted Beverton and Holt curve; $r^2 = 0.760$, $p < 0.001$) and (b) percentage survival of summer 0+ salmon fry from varying stocking densities in the Altnahinch stream 1984–92. (dashed lines drawn by eye)

1990b). This identifies the objectives, the actions necessary to meet the objectives and the appropriate agents who have responsibility for carrying out the work. It covers both the management of the fish stocks and their environment (through assessment, protection and improvement) and the management of angling (through advisory and liaison work).

These aspects of catchment management are dealt with in more detail later in this volume, and will continue to form the core of most fisheries management work for the foreseeable future. What we have tried to indicate in the present chapter is that, despite considerable effort over the last two decades on stock — recruitment modelling, we are still a long way from understanding all the parameters regulating recruitment success in salmonids — or from a full assessment of the wider applicability of index river models for use in stock management.

ACKNOWLEDGEMENTS

We wish to thank Messrs I. Moffett, L. Brownlee, B. Hart and B. Jones for their dedication and assistance with many aspects of this chapter relating to the River Bush research. We also gratefully acknowledge Mr D. Hay of the Department of Agriculture and Fisheries for Scotland, and Dr K. Whelan of the Salmon Research Agency of Ireland for their permission to cite unpublished material and reproduce their data in Figures 28.2(b), 28.6 and 28.7.

29

Management of Physical Habitat for Fish Stocks

DIEGO GARCÍA DE JALÓN

Universidad Politecnica de Madrid, Madrid, Spain

INTRODUCTION

Fisheries managers have been promoting physical habitat improvements for the past 60 years (Hubbs, Greeley and Tarzwell, 1932) although the theoretical considerations on which the improvements are based evolved much later, from the relatively modern science of fluvial ecology developed during the last three decades. This chapter deals with physical habitat improvement (PHI) techniques used in stream fisheries management. The ecological knowledge on which they are based is analysed, and planning steps and the main improvement measures are described.

Effective management of physical habitat for fish stocks requires an understanding of fish behaviour and its relationship with the physical conditions of the streams in which the fish are living. Fish stocks can be limited by food, refuge and spawning habitat, flow fluctuations and water-quality constraints, and also by factors other than habitat such as biological interactions (competition, predation and parasitism). Improving physical conditions for fisheries should take into account natural recovery processes (Cairns, Dickson and Herricks, 1977; Gore, 1985; Reice, Wissmar and Naiman, 1990) and the biogenic capacity of the reach, in order to act with nature (Heede and Rinne, 1990) rather than against it, allowing nature to improve the stream herself (White and Brynildson, 1967).

ALTERNATIVES TO FISHERIES MANAGEMENT

In many rivers, fisheries management is subjected to improvement demands made primarily by fishermen. The most usual answer to this has been fish stocking, not only because it directly satisfies the "fish" demand, but also because aquaculture is a more comprehensible technique for most fisheries managers.

Fish stocking risks in the short term the introduction of non-native species, the disappearance of threatened species, and the propagation of fish diseases; and in the long term risks genetic contamination of native populations. García Marín (1992) has shown how

The Ecological Basis for River Management. Edited by D.M. Harper and A.J.D. Ferguson. © 1995 John Wiley & Sons Ltd

stocking with northern European brown trout has made native brown trout disappear from several streams all over Spain. The replacement of original populations with stocked genetic material seems to be a world-wide problem for the management of salmonid fisheries (Hindar et al., 1991).

Fish stocking should be restricted to stream reaches which have suffered from, or are occasionally subjected to, any sort of human impact causing the disappearance of fish; and are disconnected from pristine reaches which ensure the integrity of native stocks. In these reaches stocking can be complemented with physical habitat improvements, as Jutila (1992) has shown in Finland for the restoration of salmonid populations in previously channelized rivers, obtaining promising results.

Physical habitat improvement represents an alternative to fish stocking, not as much as a mean of speeding up the recovery of fish population, or compensating for excesive fishing pressure; rather as a procedure to mitigate unfavourable effects on the stream from afforestation, flow regulation and channelization. Hale (1965) has shown in two Minnesota streams that PHI increased the carrying capacity for trout populations, and was a more cost-effective practice than stocking.

It is first necessary to understand what is meant by stream habitat in PHI. Gordon, McMahon and Finlayson, (1992) define the physical habitat as those factors that form the "structure" within which fish make their home. It must include the physical conditions of the stream channel (substratum, aquatic vegetation, hydraulic conditions, water quality, etc.) and the riparian zone. From a fisheries management point of view, stream habitat is seen as the space where fishes develop, including refuge and cover areas, spawning and egg-incubating zones, alevin developing areas and food supply areas.

PHI PLANNING

PHI is best implemented through a formally planned project, with multiple objectives (Gardiner, 1991) and directed by multi-disciplinary teams using a bioengineering assessment (Orsborn and Anderson, 1986). The steps identified in planning PHI, which are suggested by White and Brynildson (1967) and Wesche (1985), can be grouped as follows:

(1) stream selection
(2) evaluation of fish populations and their habitat
(3) diagnosis of habitat problems
(4) design of a habitat improvement plan
(5) implementation of planned measures
(6) monitoring and evaluation of results

The selection of rivers or reaches to be rehabilitated should be based on previous stream surveys of the region. Although aspects of cultural ecology or social interest have to be considered (Kern, 1992), the criteria for stream selection must rely on the natural properties of the stream, based upon the PHI objectives. These properties must be evaluated through a fisheries population inventory. For increasing fish production, the priority should be given to the reaches with highest difference between actual fish carrying capacity (low) and potential fish carrying capacity (high), and with a high capacity for natural recovery processes.

Conservation goals should give priority to reaches containing either habitats of special interest because of their rarity or their typicality or representativeness; or fish species that should be protected because they are rare or endangered. Reeves et al. (1991) have reviewed most of the evaluated PHI projects carried out in North America, and concluded, not surprisingly, that the best results from habitat enhancement were obtained when prior conditions were worst. In this case, not only the type and the intensity of the perturbation have to be considered, but also its duration (longer disturbed reaches will take longer time to be recolonized by the biota). Obviously, the greater the potential biotic capacity of the stream, the greater effectiveness. Stream channel recovery depends on flow energy and sediment disposability (Brookes, 1992). Biotic community recovery rates greatly depend on their resilience, natural streams frequently disturbed showing faster recovery (Reice, Wissmar and Naiman, 1990). This is the case of upstream communities, which usually exhibit a more rapid recolonization rate than downstream ones (Schlosser, 1990; Poff and Ward, 1990; Zwick, 1992).

EVALUATION OF FISHERIES HABITAT

Evaluation of fisheries habitat needs biological criteria. Traditionally, managers have focused these criteria on the habitat requirements of "indicator species", considered as those of social, economic or ecological interest. Their habitat needs are thought also to represent those of co-existing species. Theoretically, evaluation of habitat should take into account the requirements of all life history stages of the species included in the stream community (even species other than fish), but limitations of knowledge, personnel and money will make this choice unpractical. There is an extended bibliography on the habitat needs of salmonids (Wesch, 1985; Raleigh, Zuckerman and Nelson, 1986; García de Jalón, 1992) and on certain well-appreciated angling species like the pike, but for most coarse fisheries the knowledge of their habitat demands are still scarce (but see Chapter 27).

The instream flow incremental methodology (IFIM) has been developed to evaluate the physical habitat of fisheries at different flow intensities, taking into account the habitat needs of different life cycle stages of some species (Bovee, 1982). The IFIM has become accepted in USA as a "standard" method and, even though it has been criticized because it only reflects potential habitat, is an effective tool as it incorporates not only biological aspects, but hydrological and geomorphological aspects also.

In stream fisheries habitat analysis, five main components can be recognized. These are spawning areas, food-production areas, refuge zones, flow regimes and water quality.

Fish require good spawning habitat to reproduce and incubate their larvae. Frequently, salmonid population densities depend on the quality and quantity of their spawning-beds (Bear and Cardine, 1991). A review of rehabilitation techniques for spawning gravels can be found in White and Brynildson (1967) and Reeves et al. (1991).

Often, benthic invertebrates represent the most important food resources for river fish. Macroinvertebrate density, biomass and diversity are usually higher in riffles than in pools (Logan and Brooker, 1983; Brown and Brussock, 1991), and because of this, combined with drift, riffles are the main food-supply areas. Reviews of techniques for restoration of food-supply areas are contained in White and Brynildson (1967) and Gore (1985).

Refuge areas are those that provide protection to the fishes from swift currents and from predators. This protection is given by vegetation cover, boulders or undercut banks, or by depth or water turbulence. In small streams refuge is often a limiting factor for larger sized fishes, which are the main interest in angling, and therefore cover enhancement has been an important objective in PHI.

Stream flow is important to fisheries in the provision of an adequate space for fish. Flow requirements greatly vary with species composition and between seasons, depending on development stages. Life histories are adapted to natural flow regimes, but stream flow regulation may prevent complete developement. Often, instream minimum flows have to be specified for the maintenance of fluvial populations (see Chapter 1). Techniques for minimum ecological flows have been discussed by Demars (1985), Gàrcía de Jalón (1990) and Gordon, McMahon and Finlayson (1992).

Water quality is mainly outside the scope of physical habitat, but considerations of temperature regimes may be important. In Mediterranean rivers, small changes in summer temperatures may cause significant shifts in fisheries composition. This is especially true in the transitional zones of the longitudinal river zonation from rithron to potamon.

One of the most important question facing PHI concerns the suitability of habitat to fish populations. At the heart of this question is the degree to which habitat structure can be related to fish population size and structure. Physical habitat factors are generally more predictable, less variable and more easily measured than biological ones, and are thus preferable descriptors of streams. Also each fish species has a different range of tolerances to any given factor, with some factors being more critical than others. Therefore, physical factors can be used only with caution to make predictions about the abundances of fish populations (see Chapter 28).

North American rivers have been a focal point for the development and testing of several habitat models for predicting salmonid standing stocks (Binns and Eiserman, 1979; Wesche, 1980; Wesche, Goertler and Frye, 1987a; Bowlby and Roff, 1986; Kozel and Hubert, 1989; Platts and Nelson, 1989). In these studies, cover has proved to be remarkably successful at predicting the standing stocks of trout in a large number of streams. This result is similar to that obtained by Nielsen (1986) in Danish streams, and Baglinière and Arribe-Moutounet (1985) in a French river, where cover, represented by undercut banks and riparian vegetation, along with depth proved to be the best predictor of trout population density.

Fausch et al. (1989) reviewed 99 models predicting the standing crop of stream fish from habitat variables. They concluded that relatively precise models often lacked generality and that the assumption that fish population was limited by habitat rather than by competition or predation was not adequately addressed. Other authors have not found any single habitat variable predicting overall population abundances, but the some have successfully related densities of certain developmental stages, such as the length of stream edge and area of lateral habitat for the prediction of young-of-the-year cutthroat trout (Moore and Gregory, 1988); or pools as habitat for juvenile coho salmon (Nickelson, Beidler and Mitchell,1979).

Some kinds of habitat features are thus better at predicting fish abundances than others. When they are, we can reasonably assume that "habitat" is acting as a limiting factor and thus fish production is constrained by discrete habitat factors. Special attention has to be paid to identifying these physical aspects that are acting to limit populations. Examples of these habitat "bottle-necks" along life-history events of salmonids are provided by Mason (1976) and by Murphy and Meehan (1991).

Therefore, for effective planning of PHI we should know the characteristics of each stream's fish population (composition, structure and dynamics). The evaluation of the physical habitat allows the identification of controlling factors, in order to diagnose specific

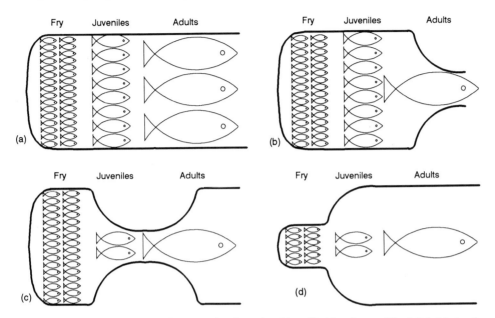

FIGURE 29.1. Fisheries limiting factors as bottle necks: (a) no limiting factor; (b) adult habitat acting as limiting factor; (c) juveniles habitat as limiting factor; (d) spawning habitat as limiting factor

problems. When a deficiency in a population is found, it is necessary to first confirm that it is caused by a habitat feature. Then, PHI may be applied (Figure 29.1).

MEASURES FOR HABITAT IMPROVEMENT

The term PHI has often been used to describe man-made structural and non-structural instream channel modifications, but could also include riparian restoration, flow regulation, sediment control or extreme water-temperature mitigation. PHI must be based on a more comprehensive and integrated approach than the merely biological point of view, taking into account geomorphological, hydraulic and engineering constraints. Once the physical habitat needs have been identified, the possible improvement measures can be evaluated from a geomorphological and engineering point of view.

Geomorphological considerations for PHI

Three important geomorphological factors must be considered in the implementation of PHI measures: channel stability, stream energy and channel morphology.

Channel stability

The first precaution necessary when improvement measures are being planned is to be certain the channel is in a stable condition. Although there is still insufficient knowledge on streams' dynamic equilibria (Heede and Rinne, 1990), instability can be detected where there is imbalance between sediment transport capacity and sediment supply. A useful tool in the interpretation of this imbalance is the equation proposed by Lane (1955):

$$Q.J = k. \ Qs \ . \ \phi s$$

where Q is the flow, J is the slope, k is a constant, Qs is the sediment load and ϕs is the mean substrate size. Even if the channel is initially stable, the PHI may create instability. An increase in flow and/or in slope must be compensated by increases in sediment load and/or substrate size, and *vice versa*. These sediments will either be taken from the channel (erosion), or deposited on it (siltation).

The instability of the stream bed depends on the substrate particle sizes and their distribution pattern. The ranges of critical water velocity values for the movement of particles of different sizes are shown in Figure 29.2.

Stream energy

Each stream has energy available for modifying its channel morphology by reworking stream banks and bed materials. This potential energy is defined by the stream power per unit of stream-bed area (Ω), or as the work done per unit of time by the shear stress at the bed. It evaluates the potential energy that can be used to transport bed load, and is usually quantified by:

$$\Omega = \delta.g \ . \ Q.J \ (\text{W m}^2)$$

where δ is the density of water, g is gravity, Q is the discharge and J is the bed slope.

There is a stream power threshold for the maintenance of general stream-bed equilibrium, discriminating those channel substrata that are eroded from those that are not. However, stream-bed substrata is usually composed of different patches, each with characteristic particle sizes. This mosaic of patches is a key factor influencing the fluvial biodiversity (see Chapter 18). The instability of certain stream patches of biological importance, such as spawning gravels, can be reached at lower values of stream power than the general threshold. Reice, Wissmar and Naiman (1990) proposed a relationship between stream patch instability and stream power (Figure 29.3). In the ascending part of the curve, gravels will

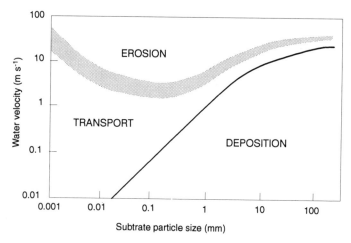

FIGURE 29.2.   Critical water velocities for the movement of different substrate particle sizes, showing their thresholds for erosion, deposition and maintenance at transport (after Hjulström, 1935)

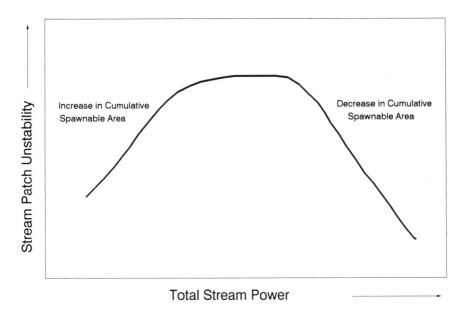

FIGURE 29.3.   Stream power and stream bottom patch unstability as proposed by Reice, Wissmar and Naiman (1990). At low stream power values, increasing patch unstability eliminates fines and, thus increases spawnable gravels, while at high stream power gravel-erosion produces reduction on spawning area

be deposited and the spawning area will increase with increasing stream power; in the descending part of the curve, gravels will be eroded, so there will be a reduction of spawning area and a stabilization of stream patches.

The recovery of 60 channels subjected to restoration projects in England and Wales was evaluated by Brookes (1992), who concluded that those with stream power close to 35 W m² were most successful. In these streams, excessive erosion or deposition was not a problem for the stability of rehabilitation measures. Brookes proposed guidelines for channel restoration taking into account the stream power and the sediment supply from upstream or channel erosion:

(1) In streams with high power (>35 W m²), structures to improve the channel will always have a significant risk of failure through erosion. If sediments are available, channel adjustments will take place, especially in cross-sections with flow convergences, and the pool−riffle sequence can be destabilized. Where the sediment is in limited supply, channel erosion and a reduced recovery can be expected.

(2) In low power streams (<35 W m²), improvement measures will have a low risk of failure. If sediments are available, there will be the formation of vegetation-stabilized benches, while pools will be poorly developed and instream structures (i.e. deflectors) may be buried by sediments. If sediments are limited, there will be no natural recovery and non-structural measures (pool dredging or gravel dumping) will be effective.

Channel morphology

Pools are important as refuges (especially at low flows) and riffles are important for fish food production, so the proportion of both must be maintained not only for geomorphological reasons, but also for biological ones. Leopold, Wolman and Miller (1964) established a pool/riffle spacing optimum of around six times the channel width, and this has been recommended for stream restoration (Besctha and Platts, 1986).

The main hydraulic mechanisms of pool formation in rivers are as follows (Beschta and Platts, 1986):

(1) scour resulting from accelerated flows due to convergence, common downstream from riffles in the centre of the stream if the channel narrows;
(2) flow deflection by the opposite bank, or in meanders by point-bar deposits;
(3) channel obstructions may provoke scouring by flow over, under or at the base of boulders, logs, or debris dams.

Riffles are formed at higher flows and often are major storage locations of bed material (Figure 29.4). They are located on the elevated reaches of the thalweg. In meandering streams riffles are found between consecutive meanders in the straight reach of the channel. The main mechanism for the formation of these riffles is the diversion of spate flows from the centre to the sides of the channel with consequent deposition.

The ratio of riffles to pools in stream length has been suggested as a criterion for habitat fitness for certain fisheries (Platts, Megahan and Minshall, 1983). For rainbow trout, the optimum ratio value is next to 0.5 and for brown trout, 0.4; lower values are given for charr.

Meandering is a stream mechanism for dissipating excess stream energy and thus any PHI that modified meanders in alluvial rivers may cause channel instabilities (Hasfurther, 1985). On natural rivers relationships have been derived (Leopold, Wolman and Miller, 1964) between meander length (M) and bankfull width (W), where $M = 11.03W^{1.01}$ (metres).

ENGINEERING MEASURES

The different structures and enhancement measures of PHI can be listed as follows: current deflectors, low dams, boulder placement, and riparian improvement. Each of these may have different applications for habitat enhancement. Other instream treatments, such as half-logs, trash catchers, bank overhangs, ripraps and submerged brushes are also described in the literature but are not considered here.

Current deflectors

Also called wing-deflectors (Figure 29.5(a)), these are the most communly used structures (Wesche, 1985). Their purpose is to change flow direction in order to protect banks, or to scour pools, to concentrate low flows, or to create riffles. Deflectors are easily built from a variety of materials, but must be carefully installed to prevent erosion and removal by floods. Wesche (1985) gives details on their technical construction. Generally they are designed with a triangular shape, anchoring the largest side in a stream bank.

Small dams or weirs

These are low-profile dams (Figure 29.5(b)), generally used to create or deepen pools and to collect and hold spawning gravels in steep-gradient small streams. The dams are built of stones, logs or gabions with a "spillway" to enable migrating fish to pass during low flows.

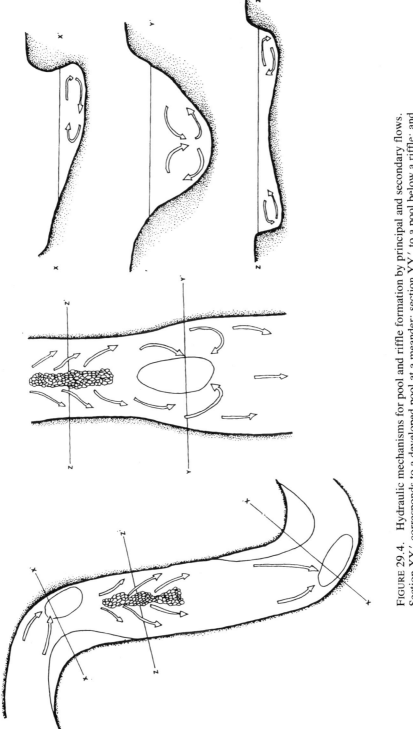

FIGURE 29.4. Hydraulic mechanisms for pool and riffle formation by principal and secondary flows. Section XX' corresponds to a developed pool at a meander; section YY' to a pool below a riffle; and section ZZ' to a riffle

(a)

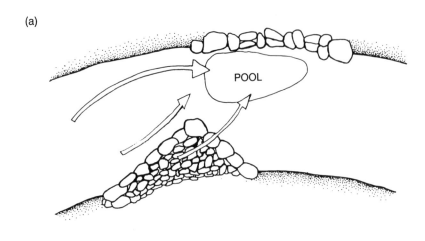

POOL

(b)

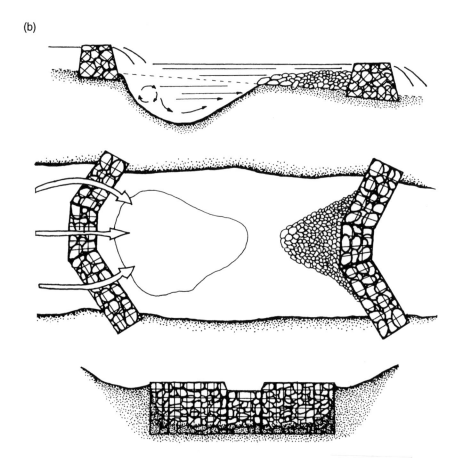

FIGURE 29.5.   Types of engineering habitat improvement measures: (a) a wing deflector, (b) a low-
profile dam in longitudinal view, plan view and transverse view

Recent modifications have found that the V-shape dam concentrates flow at the centre and protects the anchoring sides (Reeves et al., 1991), although this can create bank failure at high flows (Hey, 1990). Best results have been obtained with consecutive dams when the bottom of the upstream weir is level with the top of the next downstream dam. Other similar types of structures with the same functions are the Hewitt ramp, log sills and cedar board structures, which are basically submerged vanes with gaps to avoid material retention. In lowland streams, small dams may be also used to create riffles below them in order to enhance macroinvertebrate benthos production.

Boulder placement

Large boulders can withstand high floods, provide fish cover and may increase rearing habitat. Thus, instream boulder placement, dispersed or in groups, is a simple technique for streams of any sizes.

Riparian improvement

Protected and controlled riparian vegetation at the stream edges plays a major role in increasing stream fish stocks (White and Brynildson, 1967). It provides fish cover, bank stability, terrestrial food-supply, cool summer water-temperatures and overwinter protection. Also, it controls cross-sectional shape, favouring deeper stream channels that have higher refuge capacity. The improvement of fisheries as a consequence of fencing overgrazed stream banks is well known (Platts, 1991). Therefore, protecting and promoting riparian vegetation is an effective PHI treatment, inexpensive and "natural" in appearance. However, the ecological succession from a overgrazed bank to an overforested one (see White and Brynildson, 1967) provides in the intermediate stages (dominated by grasses, low brushes and small willow (*Salix*) species) the most beneficial conditions for fisheries (Figure 29.6). Thus, it is recommended that these stages are maintained, with controlled burning, large tree pollarding or willow basal pruning (coppicing). In addition, the prevention of excessive channel shading enables adequate macrophyte development which itself provides refuge and enhances macrobenthos production.

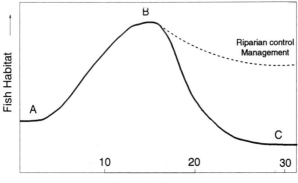

FIGURE 29.6. Relation between the riparian ecological succession that takes place from an overgrazed situation to an overforested one, and fisheries habitat quality

In larger streams, bank-side trees such as alders (*Alnus*) provide submerged cover through their roots and lower branches (as well as other ecological benefits; see Chapter 18). In banks deficient in riparian cover, re-vegetation techniques are an essential component of PHI. These techniques are reviewed by Risser and Harris (1989) and by Sainz de los Terreros, Garcia de Jalon and Mayo, (1991).

MONITORING

The final step, after PHI measures have been implemented, is to monitor the effectiveness and the maintenance of the structures. Most improvements lack quantitative evidence to evaluate their success, although some have been monitored (Hunt, 1988; Reeves et al., 1991).

Fish population response to habitat changes may have a delay of several seasons, so at least four years should be allowed before a clear picture of the effect of reclamation works on fisheries can be achieved (Wesche, 1985). Reeves et al.(1991) have suggested that the assessment of the effectiveness of stream-habitat enhancement projects must include both a quantitative evaluation of habitat change and an evaluation of change in populations. Both these evaluations must be done by comparison of before and after PHI implementation. In order to make this comparison, the same methodologies used to identify limiting factors must be applied for both evaluations.

FURTHER AHEAD

The basic idea of "the stream in its valley" (Hynes, 1975) should be applied to PHI. A catchment-scale perspective is needed to detect causes of physical habitat deterioration. On eroding basins, it is better to solve the problems on slopes (in-site erosion) than in the channel (see Chapter 4). The same principle should be applied if the fisheries problem is pollution or flow regulation. Since these problems cannot be solved at stream-reach level it is useless to apply habitat improvement. PHI must never be viewed as a substitute for habitat protection, not only for aesthetic reasons, but also because prevention of initial degradation of the stream channel is more effective than repairing it once it is degraded.

There is still insufficient knowledge to eliminate uncertainty from the creation of dynamic equilibrium in streams (Heede and Rinne, 1990) and thus, measures for PHI should be flexible. However, some structures for fisheries enhancement are designed to be permanent and rigid, so they may prevent geomorphological adjustments. Therefore, more emphasis must be given to projects that encourage and maintain riparian vegetation and stable channels and less to those which alter them.

Techniques applied successfully in one stream may not be directly applicable in others. A practical approach is to analyse the principles underlying these techniques with an open mind, and search for their adaptation to the conditions of a particular stream based on experiments.

Indeed, the current state of PHI is deficient in appropiate expertise in biological, geomorphological and hydraulic engineering linkages. Frissell and Nawa (1992), in a recent analysis of 161 fish habitat structures after bank-full floods, concluded that commonly prescribed structures were often inappropiate in streams with elevated sediment loads, high peak flows, or erodible banks. Thus, the research needed must be focused on building up a database of cases (Brookes, 1992) and increasing the quantitative and holistic descriptions of pre- and post-treatment conditions.

30

Fish Stock Assessment—A Biological Basis for Sound Ecological Management

Ian G. Cowx

University of Hull, Hull, UK

INTRODUCTION

Inland fisheries need to be managed for a variety of reasons. The objectives are usually associated with the types of use as well as with socio-economic factors connected with the associated community. A typical, but not exhaustive, list of such objectives is:

(1) production of food;
(2) maintenance of stocks for sports fishing;
(3) maintenance of stocks for recreational fishing;
(4) supply of seed or broodstock for aquaculture production;
(6) supply of ornamental fish;
(7) control of unwanted organisms (disease vectors, vegetation, rice borers);
(8) maintenance of employment within a community;
(9) conservation or other aesthetic values.

All these objectives would normally demand management action aimed at the maintenance of the fish stock. The current approach to managing inland fisheries therefore relies heavily on having some measure of stock abundance or associated population parameter(s) on which to formulate the decision-making process or evaluate the impact of a particular management activity. For example, stock assessment is usually deemed necessary to gauge the effects of activities such as the impact of overfishing, a pollution event, land drainage improvement works, river regulation, habitat restoration, stocking, introduction of a new species, or merely to respond to requests for advice on the management of waters. However, in many cases, particularly in large water bodies, the results obtained are disappointing because of the problems of estimating absolute abundance of the fish populations by conventional scientific methods. Often the root cause is poor fish sampling efficiency and subsequent failure to meet the assumptions underlying the assessment methodology/models. Many practitioners, having knowingly violated the assumptions to some (unmeasured) extent, still proceed to interpret the results as if no possibility of error existed.

The Ecological Basis for River Management. Edited by D.M. Harper and A.J.D. Ferguson. © 1995 John Wiley & Sons Ltd

Consequently management of inland fisheries is far from an exact science. Decisions are often based on poor-quality information, supplemented by the experience of the manager. However, this need not always be the case. New developments in sampling methodology (e.g. hydroacoustic and multiple anode electric fishing gear) and strategic planning of the stock assessment exercise could overcome many of the limitations. This chapter considers a strategic approach to stock assessment which meets the demands of both the manager and the scientist.

OBJECTIVES OF STOCK ASSESSMENT IN RIVER FISHERIES MANAGEMENT

To manage fisheries successfully there is a prerequisite for "adequate" information on the status of the fish stocks. However, the information that needs to be collected is, in part at least, dependent on the management objectives for the fishery. The objectives of fisheries management tend to fall into a number of categories:

(1) evaluation of the status of the fish stocks for conservation and enhancement purpose;
(2) monitoring long-term population changes as a result of natural or anthropogenic activities;
(3) evaluation of the response of management activities directly targetted at the river system or its fisheries, e.g. restocking or introductions, habitat improvements, water-quality improvements, flow regulation;
(4) assessment of environmental damage, e.g. post-impact appraisal of a pollution incident, fish kill or natural catastrophe such as a drought or flood;
(5) as a function of an environmental impact assessment exercise to predict the effects of development activities on the fisheries.

The data requirements for each of these objectives vary according to the desired precision necessary to support the management decision-making process. Sometimes the aim is to assess total population size, e.g. when comparing populations with respect to yield, or population density (number per unit area), e.g. when studying population regulation as a function of stocking. Both parameters are absolute parameters and require considerable resources to estimate them. Often, however, the aim is to assess temporal or spatial changes and trends, such as in environmental monitoring or the impact of management regulatory measures. In these cases it may be sufficient to make an estimation of relative parameters (presence/absence or catch per unit effort), allowing comparisons but not absolute determination. Relative estimation of population parameters is usually less costly than absolute estimation, and in many cases may be more than adequate.

PRECISION LEVEL FOR STOCK ASSESSMENT

Before identifying possible mechanisms for undertaking studies with different objectives, it is important to consider the desired information with respect to individual fish or populations, and the accuracy and precision that must be achieved.

Accuracy is associated with the type of error or bias in the data. Poor accuracy tends to lead to assessments that considerably, but consistently, over- or under-estimate. Precision is associated with the "noise" (usually expressed as the variance or coefficient of variation, CV, of the estimate) generated by the sampling procedure, and is usually reduced by larger sample sizes or repetitive surveys (Southwood, 1978). A highly reliable estimate will have a low coefficient of variation. The precision of the stock estimate dictates the change in stock parameters that can be detected.

If population parameters are being determined, the required precision of the estimated abundance or magnitude of change (spatial or temporal) that needs to be detected, must be determined in relation to the objectives. This minimizes the risk of obtaining a precision too low or high for the purpose. As the choice of precision level will strongly affect the resource input, it is worth considering this question in relation to the objectives at the planning stage.

Bohlin and co-workers (Bohlin et al., 1989; Bohlin, Heggberget and Strange, 1990) suggested a rough guide for establishing precision levels for fisheries surveys. This is based on three categories.

- *Class 1*: Studies in this class require a high level of precision; a population change, in time or space by a factor as small as 1.2 (e.g. 83 ← 100 → 120) has to be detected with about 80% probability when using a 5% significance level. In the case of an independent estimation, this level of precision corresponds approximately to a coefficient of variation not larger than about 0.05.
- *Class 2*: Studies in this class require an intermediate level of precision; a population change, in time or space by a factor as small as 1.5 (e.g. 67 ← 100 → 150) has to be detected with about 80% probability when using a 5% significance level. In the case of an independent estimation, this level of precision corresponds approximately to a coefficient of variation not larger than about 0.10.
- *Class 3*: Studies in this class require a low level of precision; a population change, in time or space by a factor as small as 2.0 (e.g. 50 ← 100 → 200) has to be detected with about 80% probability when using a 5% significance level. In the case of an independent estimation, this level of precision corresponds approximately to a coefficient of variation not larger than about 0.16.

Although the choice of precision level is ultimately the decision of the manager or scientist, there are certain standards that need to be achieved to meet the objective of the project/ survey. The minimum acceptable precision level for each of the objectives of stock assessment outlined earlier is given in Table 30.1. This illustrates that absolute abundance is not a prerequisite of all stock assessment exercises, and this should be taken into consideration when planning surveys.

In addition to collecting data of the desired precision to formulate the management response, there are two other reasons for considering precision level:

(1) Data collection exercises tend to be expensive and high cost has to be paid for in time, manpower and financial resources.

TABLE 30.1.   Minimum desirable stock assessment data for various management activities

Activity	Abundance estimate	Precision level
Evaluation of the status of fish stocks	Relative/absolute	2−3
Monitoring long-term population changes	Relative	3
Evaluation of the response of management activities	Absolute/relative	2−3
Assessment of environmental damage	Absolute/relative	1−2
Enviornmental impact assessment	Absolute/relative	1−2

(2) It is often difficult to obtain precise information on the status of the fish populations because the sampling technology is not available.

METHODS OF ESTIMATING FISH POPULATION ABUNDANCE

Fish abundance in river systems can be estimated by a number of methods, the most frequently used of which are listed below.

(1) successive removal or depletion sampling (Leslie and Davis, 1939; DeLury, 1947; Zippin, 1956; Seber and LeCren, 1967; Carle and Strube, 1978; reviewed by Seber, 1973; Otis et al., 1978; White et al., 1982; Cowx, 1983);
(2) capture−recapture methods (Cormack, 1968; Seber, 1973; Ricker, 1975; Youngs and Robson, 1978; White et al., 1982);
(3) gear calibration methods (Bayley, 1985, 1993; Serns, 1982, 1983);
(4) catch−effort sampling strategies, such as creel censuses, log books, catch statistics (Cowx, 1991);
(5) remote sensing using hydroacoustic gear (Thorne, 1983; MacLennan and Simmonds, 1992; Kubecka, Duncan and Butterworth, 1992);
(6) high efficiency methods used in backwaters or restricted channels such as poisoning (Davies, 1983; Bayley and Austen, 1990) or explosives (Layher and Maughan, 1984; Bayley and Austen, 1988).

Successive removal methods

Successive removal methods have become a standard approach for abundance or biomass estimates in the UK. They provide discrete estimates in space and time, but are subject to a number of underlying assumptions.

(1) Individual fish must have equal probabilities of being caught during successive fishings;
(2) No emigration of immigration occurs during the total sampling period;
(3) All members of the population must be equally vulnerable to the gear;
(4) A significant proportion of the population must be removed during each fishing.

Several of these assumptions can be violated when sampling some species or habitats (Mahon, 1980). For example, Lelek (1966), Cross and Stott (1975), Mahon (1980) and Bohlin and Cowx (1990) noted changes in efficiency of electric fishing after successive runs. Bohlin and Cowx (1990) also estimated that a proportion of the individuals (between 10 and 30% of the population) were apparently not vulnerable to capture. This tends to produce a significant, but unmeasurable, negative bias in the estimates.

Unfortunately these methods are often used indiscriminately because the catch data appear to fit the models, and they produce "apparently" reasonable confidence limits. The methods should be subject to greater critical appraisal to assess whether the assumptions are violated and the results are valid or, indeed, accurate. This can only be achieved by calibration against fish populations of known size (Bohlin and Cowx, 1990).

Another limitation of these methods is that they only provide information on discrete sites at a specific time, both of which may not reflect the true status of the fishery in a target river. For example, it may not accommodate clumped distribution or migratory patterns of fish. It is possible to reduce the potential error by sampling a large number of sites or by sampling several times a year (Bohlin et al., 1989; Bohlin, Heggberget and Strange, 1990), although this may be constrained by resource and time restrictions.

Capture – recapture methods

Capture – recapture methods are most valuable when it is difficult to isolate a section of river or the efficiency of the sampling method is unlikely to deplete the population size during successive samples. The methods are also subject to a number of assumptions.

(1) A random sample is marked and/or marked individuals mix randomly with unmarked fish;
(2) The marking must not influence the catch probability of the marked individuals. Marked fish must therefore be equally vulnerable to capture as unmarked fish;
(3) The marking must not influence the mortality of individuals.

Capture – recapture methods are subject to the same problems of differential vulnerability of individuals, gear efficiency, and spatial and temporal distribution patterns as depletion sampling. The approach also requires a significant rate of recapture to avoid large confidence limits. The approach usually requires a minimum of two sampling occasions which may impinge on resource availability.

Gear calibration

Gear calibration methods have received comparatively little attention yet are possibly the most cost-effective mechanism of assessing fish abundance. They are particularly useful in large rivers where depletion sampling is not possible because the reach cannot be isolated or gear efficiency is too low.

There are two approaches to gear calibration methods (Bayley, 1985): the whole system approach (e.g. Serns, 1982, 1983) and the point estimate approach (Bayley, 1983; Lazauski, 1984). In either case an independent estimate of the vulnerable population size and efficiency of the gear is required.

In the whole system approach the efficiency of the gear (probability of capture) is determined for an isolated population of known size. This can be an independent water body where the size is known because it can be directly counted (stocked numbers of fish or dewatering) or it has been assessed by an alternative method (e.g. capture – recapture). The probability based on the independent assessment is then transposed to the system to be evaluated. Often the greatest source of error is the low precision of the methods available to estimate the vulnerable population. Also the probability of capture in the test system may not be representative of the probability in the target water.

The point estimate is based on blocking off a small area of the habitat to be evaluated and assessing the vulnerable population size by a high efficiency method. For example, in a large river a length of one margin is isolated by a block net and a successive removal estimate carried out. The probability of capture is determined as the proportion of the first catch against the estimated population size. Evasion from the area when it is being enclosed is unimportant, since only the population vulnerable to capture by the gear being calibrated needs to be estimated. Problems arise when the high efficiency alternative method is not 100% efficient or subject to other biases (see removal methods).

In both cases long lengths of river are fished once using the calibrated gear and the population abundance (N) determined by $N = C/P$, where C is the catch and P the probability of capture.

The advantages of using gear calibration methods are numerous:

(1) The problem of unknown changes in catchability during repetitive sampling is removed.
(2) Problems with contiguous or non-random distribution in a river are reduced by surveying long stretches.
(3) The non-migration and escapement assumptions with respect to blocking procedure do not apply, and thus the method can be used in habitats that cannot be isolated.
(4) The effort expended in setting block nets and repeat sampling can be spent surveying greater lengths of river, thus increasing the overall precision.
(5) Situations where part of the population is not vulnerable can be identified and effective catchability adjusted.
(6) Estimates of abundance, biomass or production can be obtained from previously calibrated data.

Catch–effort methods

Monitoring changes in fish populations by catch–effort methods is an extremely cost-effective approach which has been used with considerable success on both exploited (commercial and sport) and unexploited (recreational) fisheries (see Cowx, 1991, for an overview of the methods and limitations). Creel surveys provide good information about relative changes in the population structure and abundance against which many management activities can be formulated. In exploited fisheries, where long time series of catch and effort data are available, assessments of surplus yield have proved possible (Alimoso, 1991) and the methods are adaptable for large river systems where alternative scientific sampling methods are impractical (Meredith and Malvestuto, 1991).

These methods suffer from many biases associated with data recording and recollection (Malvestuto, 1991), but are an invaluable tool in large river systems.

Hydroacoustics

Hydroacoustic methods have received little attention in fresh water because of the large investment in the gear. The gear also needs calibrating against an independent method, is more or less restricted to mid-water fish and cannot be used in shallow water or where physical barriers (e.g. weeds) obstruct the operation.

Although estimates of abundance are possible, particularly as new dual beam equipment comes onstream, it provides little information on benthic species or species composition and allows no opportunity for further data capture on individual fish parameters (e.g. growth rates).

High efficiency methods

These methods provide an assessment of absolute abundance but they are usually based on destructive sampling methods. Consequently, they are unacceptable in fisheries where the resource is valuable in recreational, commercial or conservation terms. The methods are also impractical in large rivers with a reasonable flow because the fish are difficult to collect.

SAMPLING METHODS AND STRATEGIES USED ON RIVERS IN THE UK

In the UK rivers, fish populations are sampled by a variety of methods and strategies depending primarily on the size of the system, but to a lesser extent on the efficiency of the gear (Table 30.2).

Small streams tend to be sampled by electric fishing with three or four personnel wading with one or two hand-held electrodes. Current output is usually pulsed DC at 50 or 100 Hz from a small generator, although backpack gear is employed occasionally. In this type of habitat, depletion sampling a section of river (varying between 20 and 300 m) tends to be

TABLE 30.2.   Summary of fish sampling methods used in the UK to evaluate the status of fish stocks

Water body	Sampling gear[a]	Sampling strategy
Small streams <5 m wide	1 hand-held electrode, DC, 50/100 Hz PDC electric fishing, generator supply, wading (3)	Depletion sampling between stop-netted sections
Small rivers 5—15 m wide, pool—pool/riffle topography	1 or 2 hand-held electrode(s), DC, 50/100 Hz PDC electric fishing, generator supply wading (3) or boat-based in deeper sections (3—6)	Depletion sampling between stop-netted sections, if possible
Small rivers 5—15 m wide, pool topography, >1 m deep	Boat-based, 2+ hand-held electrode(s), DC, 50/100 Hz PDC electric fishing, generator supply (3—6) Multiple anode boom array (4)	As above One-catch relative assessment Calibrated sampling
Large rivers and canals >15 m wide >1 m deep	Boat-based, 2 boats (7—8), 2+ hand-held electrodes, 50/100 Hz PDC generator supply. Multiple anode boom arrays (4) 4—7.5 kVa generators	Depletion sampling Calibrated sampling Relative assessment
	Seine netting (wrap around technique) (6+)	Depletion or calibrated sampling
	Catch statistics/ licence returns	Catch effort and trend analysis
	Creel census	Catch effort and trend analysis
	Hydroacoustics (2)	Calibrated biomass and density estimates
Large still waters	As for large rivers, electric fishing and netting in margings only	As for large rivers

[a]Number in brackets indicates the minimum number of personnel used for survey

reasonably successful, although the population size is probably underestimated due to the variable susceptibility of individual fish to electric fishing (Bohlin and Cowx, 1990).

In small rivers the same equipment and procedure are used but fishing is usually based on a boat, except in shallow riffle areas where operators wade and pull the boat housing the generator. Two electrodes tend to be standard but more manpower is usually used. The population estimates suffer from the same biases as in small streams but accuracy can be further affected by reduced gear efficiency.

In large rivers and canals a variety of gears is used. Active methods are based on either boat-mounted, multi-anode boom array electric fishing (Cowx, Wheatley and Hickley, 1988; Cowx et al., 1990; Cowx and Harvey, 1993) or seine netting (Coles, Wortley and Noble, 1985; Pygott et al., 1990). Isolating sections of most rivers with stop nets is generally difficult because of the current, thus depletion methods using electric fishing as the capture technique are rarely appropriate, although such estimates are often derived. The accuracy of such results is subject to large errors and should be treated with extreme caution. In slow-flowing rivers and canals it can be possible to isolate a section of water and carry out a depletion sampling. Results from electric fishing tend to be variable and probably do not provide accurate estimates of population size because of poor sampling efficiency. Conversely, where seine netting is possible, reasonable estimates of fish biomass and density are feasible (Coles, Wortley and Noble, 1985; Cowx et al., 1990), although the method is manpower-intensive. Occasionally, when resources have been available, capture − recapture exercises have been carried out on large rivers with varying degrees of success (Hunt and Jones, 1974; Bowles, Frake and Mann, 1990). This approach is generally applicable for estimating population size of fish in large rivers with electric fishing if the resources are available, if sufficient is known about the behaviour and migratory patterns of the target fish species, and if marked and unmarked fish are equally susceptible to capture.

The difficulties of efficiently sampling large rivers by active methods have resulted in a range of alternative strategies being adopted. These include, hydroacoustic methods (Kubecka, Duncan and Butterworth, 1992), catch statistics (Bunt, 1991; Churchward and Hinckley, 1991) and angling creel census methods (Malvestuto, 1983, 1991; Cowx, 1990, 1991). These methods have their limitations, but are all cost-effective methods of obtaining valuable data on the fisheries. The alternative strategy which has received little attention is gear calibration methods (Bayley, 1993).

Large, still water bodies, such as lakes and reservoirs, are sampled by seine netting, although electric fishing using hand-held or boom-mounted electrodes is utilized in the margins. Population estimates are impractical because of the problems of sampling efficiency, although capture − recapture methods have been attempted.

STRATEGIC APPROACH TO STOCK ASSESSMENT

Although estimates of fish population abundance are routinely made, critical evaluation of the validity of the results is rare. This does not suggest that stock assessment work should be discontinued, merely that greater consideration should be given to the assessment procedure. Selection of the appropriate methodology depends on a number of criteria, including efficiency of fish capture, habitat structure and availability of resources (mainly manpower). However, the whole procedure can be improved by strategic planning of the assessment exercise. Selection of the appropriate methodology for assessment relies on having knowledge of:

(1) the limitations of the various gears in different sampling situations;
(2) limitations of the various stock assessment methods;
(3) the biology and behaviour of the target species and populations.

It is extremely doubtful whether sufficient detailed information is available on riverine species. Indeed the information that is available suggests that the assumptions, such as equal catchability, are likely to be violated, because of differential movements within stocks. Thus before embarking on a stock assessment exercise, particularly on large rivers, a critical appraisal of the procedures should be carried out. Flow charts, such as that shown in Figure 30.1, can be drawn up to guide the operator towards the most appropriate method.

In conjunction with this process there is a need to strategically plan the survey to meet the objectives. This is an aspect which is frequently neglected by scientists and managers, often to the detriment of the project's outcome. All assessments should be treated as individual projects and planned to achieve the desired outputs of the exercise. Goals should be set and activities should be orientated towards meeting the aims. In the planning process, a number of phases should be considered (Table 30.3, Figure 30.2).

Identification phase

As a preliminary to any stock assessment exercise, the overall management objectives must be conceived and formulated with the intention of maintaining, improving or developing the fishery. Within this identification phase gross activities to meet the objectives, from doing nothing to complete rehabilitation of the fishery, need to be perceived. It is only after this stage has been formulated that stock assessment requirements can be considered in context.

Preparation phase

The preparation phase should review the overall management objectives and determine the fish stock information required to meet these aims. The data to be collected must be defined in terms of precision and type of parameters. This must propose realistic data outputs which can be achieved with the resources available (i.e. manpower, equipment, time, finances). It is pointless expecting to get an absolute population estimate of population size on a large river based on depletion sampling with a single anode electric fishing gear. Consideration must therefore be given to the precision level of the data (Table 30.1) needed, and whether the results obtained using the resources available will be valid.

Equally important at this stage is a mechanism to measure the applicability of the stock assessment procedure as it evolves. A commonly used technique in development projects is the logical framework (Anon., 1982). The technique was developed by the US AID organization in the late 1960s and is useful in setting out the design of a project in a clear and logical way so that any weaknesses that exist can be brought to the attention of the planners or managers. The identified deficiencies may then be remedied at an early stage, or if insuperable, the project may be aborted. The logical framework can be easily adapted to fit the planning of stock assessment exercises.

Table 30.4 illustrates a logical framework format that might be adopted at the onset of a development project. The logical framework technique clearly defines the objectives of the proposed exercise and focuses attention on the desired outputs in terms of data required to meet these objectives. Indicators (second column) are used to determine the extent to which the objectives are achieved and can be measured at different times, notably in the monitoring

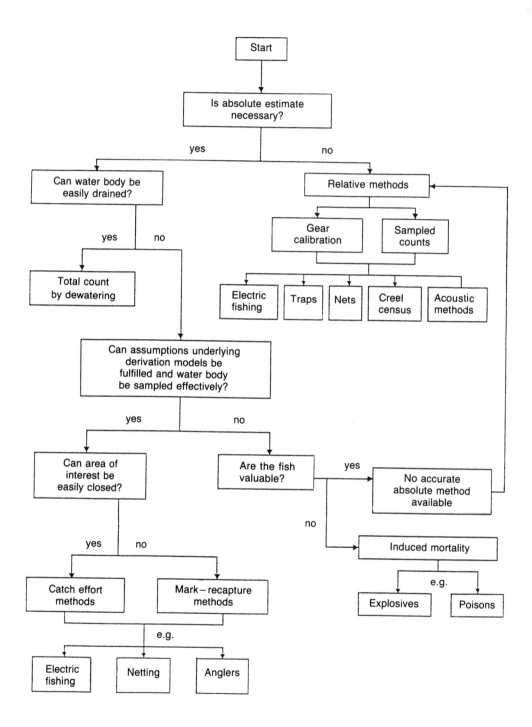

FIGURE 30.1. Schematic approach to selecting the most appropriate gear and stock abundance
methodology (modified from O'Hara, 1981)

TABLE 30.3. The project approach to stock assessment

Phase	Characteristics
Identification	Establishing relevance of overall project to management policy Generation of the management options Determining the suitability and feasibility of project
Preparation	Assessment of the feasibility of the various project options established during the identification stage Review objectives of stock assessment in relation to the overall project Establish design of stock assessment survey to meet objectives Assessment of technical, financial and resource requirements Preparation of logical framework
Appraisal	Review feasibility of carrying out the assessment in relation to the technical, financial and resource requirements Review feasibility of assessment meeting desired accuracy and precision level
Implementation	Detailing of work plans and financial arrangements Translation of logical framework into activity schedules Carry out stock assessment surveys
Evaluation	Evaluation of the success of the assessment in relation to defined objectives Post-project surveys assessing the overall project Feedback of evaluation for future project preparation

of stock assessment performance. Where possible, the indicators should define the expected data output and the level of precision and accuracy. The section devoted to the risks and conditions of the logical framework (third column), is concerned with establishing realistic output parameters for the environment in which the assessment is to be carried out. As the project develops so the logical framework can be modified to take account of new information likely to affect the project elements.

The theoretical example of the logical framework (Table 30.4) typifies many of the problems facing fishery managers. A stretch of river is deemed by the fishermen to have deteriorated in productivity as a recreational fishery and their concerns have been transmitted to the river management authority who are asked to implement a restocking programme. The chances of sustained improvement in the fishery by merely stocking are likely to be minimal, and it would probably be of marginal benefit to commit scarce resources which would only result in short-term, easily dissipated benefits. The example is thus orientated towards addressing the problem of the perceived poor quality of the recreational fishery stretch in the river. The stock assessment programme is embodied in the overall development plan and the manager/scientist must decide what information on the status of the fish stocks is required, and whether it can be obtained, to formulate the action plan.

Similar logical frameworks should be drawn up for each management exercise on each new river. It is not possible to transpose the framework from one river system to another, or between activities, because each has its own intrinsic problems. The benefits accrued from drawing up the logical framework, to focus attention on the quality and quantity of data required, far outweigh the time and resources spent trying to collect stock assessment data in which no confidence can be placed.

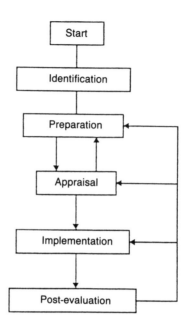

FIGURE 30.2.   Phases in the project approach to stock assessment showing the feedback mechanisms

Appraisal phase

In the appraisal phase the feasibility of carrying out the assessment to the desired standards is evaluated, preferably by an independent observer. At this stage the technical and resource implications of the proposal are integrated into the assessment scheme and the potential validity of the output data is reviewed. If the conclusions suggest that the stock assessment strategy is impractical or will not provide data of sufficient quality, the procedure will have to be reconsidered with a view to restructuring the overall project, or, if insuperable, aborted.

Implementation phase

Once the stock assessment strategy has been ratified, detailed work plans and financial/ resource arrangements can be drawn up. The stock assessment activity programme is then carried out. During implementation it is important to constantly evaluate the outputs in relation to the established indicators. If the desired standards and outputs are not achieved the strategy may have to be adjusted or reconsidered. It is imperative that the operators do not allude themselves that they are achieving the desired results when it is quite clear they are not.

Evaluation phase

The final aspect that should always be carried out is an evaluation of the stock assessment exercise in relation to the defined objectives. Without active feedback on the success of assessment exercises it will be difficult to improve the methods and strategies. Both

TABLE 30.4. Example of the logical framework approach in project development indicating the relevance of stock assessment in the overall scheme (from Crean, 1993)

Project structure	Indicators/means of verification	Assumption/risks
Overall development aim: Assess the factors which are contributing to poor sports fishing quality of River X	Listing of factors and weighting of key determinants	That there is a real basis to the complaint That the factors contributing to the sports fishery quality of the river can be identified and separated
Specific objectives: Assess the status of current fish populations Determine whether the sports fishers has declined compared to previous years Suggest and implement remedial action to improve the river fishery (where appropriate) Liaise closely with sports fishery users	Assessment of the fish biomass and species composition of the river fishery Monitoring of sport fishing performance over 30 year period Identification of stock deficiencies and and indication of absolute value of shortfall	That the river can be effectively sampled That the methods of population assessment are appropriate and the results reliable That historical records are available and contribute a non-ambiguous body of information to the study That causes can be clearly identified and that remedial action is feasible given resource constraints
Outputs: Generation of clear picture of current status of fish populations Identification of key factors affecting quality of river fishery Identification/implementation of a stock management plant where appropriate	Monitoring of time series data describing fish population (cycles of) abundance Monitoring of changes in water quality, fish population sizes, introductions, climate, etc. over the period of the study Monitoring of size/composition of anglers catches over the period Monitoring change in angling technology (e.g. rod weight, bait, line thickness, etc.) over time	That good-quality time series data are available to be evaluated and assessed That the resources are available to carry out the study That the managers of various data sources collaborate with the study That the data sources indicate unambiguously which factors (if any) are the cause of the perceived problem
Inputs: A 3 man-month survey of the current status of fish population in 10 km stretch of River X Survey of research undertaken on river over a 30-year period Analysis of supporting environmental data over a similar period of time Organized angler opinion survey Assessment of all other sources of coverage relating to the river fishery over recent history, e.g. historical records, press reports, personal views, etc.		

successes and failures, problems encountered and the way they were tackled, and limitations (relating to manpower, gear, methods of estimating abundance and finance) of the procedures should all be reported.

CONCLUDING REMARKS

The current practices for carrying out stock assessments are based on well tried and tested sampling methods and data analysis, e.g. depletion sampling with electric fishing. However, in the majority of cases the output is known to be inaccurate, having often violated the assumptions on which the models are based, to some (unmeasured) extent. For obvious reasons this mechanism needs addressing!

The project approach has proved to be a valuable tool in the development of inland fisheries, mainly in situations where finance has been provided by international development banks. It has yet to be accepted as a technique that could be used more widely.

The project approach offers a sequential and comprehensive methodology for assessing the viability of management-orientated goals and the feasibility and/or suitability of undertaking stock assessment exercises which have a well-defined output. At its most effective, the project approach can indicate a stock assessment procedure which, if adopted, will give the greatest opportunity for, and make the best use of, resources, and provide the most accurate information possible. The technique is not limited in application but may be applied as readily to the comparison of any research or development option. The logical framework is a technique which links the various stages of the project approach. It is to be recommended, and ultimately provides the indicators and analysis framework to determine the stock assessment's success or failure. The project approach shifts the emphasis away from the basic data collection exercises and encourages wider thinking to cover the scientific, technical, resource and financial aspects. Finally, the project approach should be viewed as a long-term system to aid development of stock assessment procedures, and, as such, deserves careful consideration by those who manage the inland fisheries environment.

31

Variability in Growth, Density and Age Structure of Brown Trout Populations under Contrasting Environmental and Managerial Conditions

FELIPE G. REYES-GAVILÁN, RAMÓN GARRIDO,
ALFREDO G. NICIEZA, M. MAR TOLEDO and FLORENTINO BRAÑA

Universidad de Oviedo, Oviedo, Spain

INTRODUCTION

Brown trout (*Salmo trutta* L.) is a widely distributed species (MacCrimmon and Marshal, 1968; Baglinière, 1991) whose populations exhibit a broad range of life-history variation, mainly associated with plasticity for growth and reproduction (Jonsson, 1977; L'Abée-Lund, 1990; Champigneulle et al., 1991). Changes in habitat volume and structure of the channel throughout the river influence distribution and demography of brown trout populations (Jonsson and Sandlund, 1979; Baglinière and Arribe-Moutounet, 1985; Kennedy and Strange, 1986a) as well as fish community structure and species richness (Gorman and Karr, 1978; Schlosser, 1982; Angermeier and Karr, 1983; Zalewski et al., 1990; Penczak et al., 1991). Furthermore, growth intensity varies along environment gradients and has a strong influence on several life-history traits, such as age at first reproduction, longevity and reproductive investment (Jonsson, 1977; García and Braña, 1988). Latitudinal variation in demographic and life-history traits has been reported for brown trout in connection with climatic factors (Fahy, 1978; L'Abée-Lund et al., 1989). Altitudinal zonation incorporates the potential superimposed effects on growth and population structure of climatic gradient and downstream increasing habitat complexity. Lastly, elevation axis parallels the abiotic – biotic regulatory continuum throughout the river (Vannote et al., 1980; Zalewski et al., 1990), and has been identified as a major factor accounting for variation in biomass and production of some brown trout populations (Scarnecchia, 1983; Lanka, Hubert and Wesche, 1987; Kozel and Hubert, 1989a).

Brown trout populations in central northern Spain are continuous from sea level to an elevation above 1500 m, thus experiencing a wide range of environmental variation within a geographically small area. This is a particularly favourable situation for identifying some of

The Ecological Basis for River Management. Edited by D.M. Harper and A.J.D. Ferguson. © 1995 John Wiley & Sons Ltd

the factors affecting growth, density and age structure of these populations in connection with climatic and stream habitat gradients. Assuming that physical habitat structure is a key factor for species distribution and for within-species ontogenetic niche sifts (Schlosser, 1987; Schiemer and Zalewski, 1992), we assayed, in addition to altitude, distance to sea, depth (mean and coefficient of variation), width and substrate diversity as potential predictors for trout distribution. The influence of additional parameters as stream order, canopy, basin substrate and fishing regime was also studied.

METHODS

Study area and fish community

The study was conducted in first to fourth order streams (5 to 1380 m above sea level and 2.7 to 45.0 m mean width) at the northern end of the Iberian Peninsula (Figure 31.1). These are short (usually no longer than 100 km), steep, coldwater streams located at the meridional limit of the anadromous salmonid range in Europe (MacCrimmon and Marshall, 1968; MacCrimmon and Gots, 1979, Doadrio, Elvira and Bernat, 1991). Few fish species, most of them exhibiting diadromous cycles, inhabit these streams. Brown trout (*Salmo trutta*, both anadromous and nonanadromous) is the prevailing species throughout the study area, and the only one at altitudes above 500 m. Atlantic salmon (*Salmo salar*), European eel (*Anguilla anguilla*), and European minnow (*Phoxinus phoxinus*) also frequently occur in low and medium reaches. Three additional species were captured in five or less sampling sites: sea lamprey larvae (*Petromyzon marinus*), flounder (*Platichtys flesus*) and thick-lipped mullet (*Chelon labrosus*).

Fish sampling and population parameters

We electro-fished 76 stream reaches, from mid June to mid December, during 1986−93 (most of the samples were taken in two intensive sampling periods in 1989 and 1991). Sampling stretches (surfaces ranging from 106 to 720 m^2) were selected to represent available microhabitat heterogeneity (raceways, riffles, pools). Each sampling consisted of two to four electro-fishing operations without replacement, from which trout number and density were estimated using Zippin's (1956) method. In some sections where the river was too wide to apply this method (only 8% of sites), we obtained an estimate of capture per unit effort (CPUE) in a single electro-fishing operation, and then a value of density was estimated from the least-squares regression of density (DEN; individuals per square metre) on CPUE (fish per minute in the first electro-fishing operation) obtained for most of the sections that supplied good Zippin's estimates. The following equation was applied:

$$DEN = 0.177 \; CPUE + 0.04$$
$$(r^2 = 0.62; \; N = 61; \; p < 0.0001)$$

Fork length (*FL*, mm) of all trout were measured and scales of most individuals were removed for age determination. Age was assigned by inspection of length−frequency distributions (Petersen's method) in the few samples for which scales were not available.

Weights (*W*, g) were estimated on the basis of the regression on fork length (mm) obtained from a subsample of individuals from several populations covering the entire size range of the whole sample, according to the following equation:

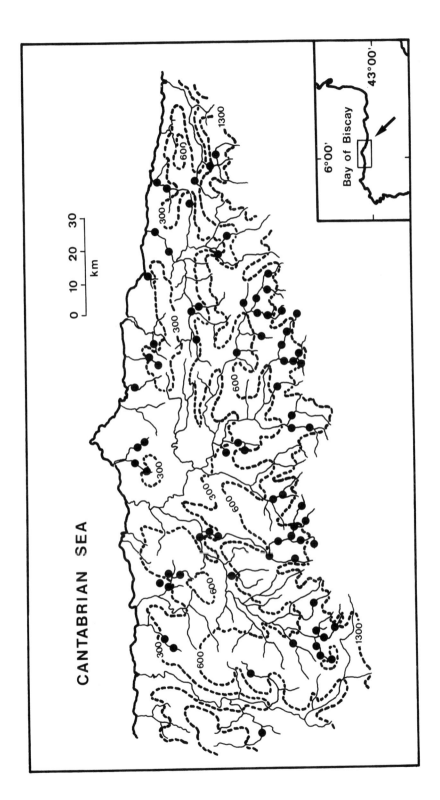

FIGURE 31.1. Map of the study area showing the location of the sampling sections, with contour lines at 300, 600 and 1300 m above sea

$$\log W = 3.02 \log FL - 4.98$$
$$(r^2 = 0.995; N = 150; p < 0.0001)$$

Estimated weights were then employed to obtain values of biomass per square metre and per individual. Except when otherwise stated, fish of age 0+ were not used to obtain density, biomass or variables describing age structure of the population, as this age-class could not be fully recruited in some of the earliest samples of the year. In order to prevent inconsistent data variation in performing analyses on age structure parameters, we have not considered values from samples of less than 10 fish of age 1+ and older.

Growth intensity was estimated by means of fork length at the end of the first winter (FL_1). For this purpose, we considered directly the average length of 0+ trout for samples taken from October to December, as growth has virtually ceased at this period, and back-calculated length at the first winter of 1+ individuals for samples obtained before October. Although FL_1 could incorporate some differences among sections associated with asynchrony in hatching time, we preferred this measure because it is the only age-specific length free of possible size-selective mortality or migration associated with reproduction or smolting in any section (García and Braña, 1988; Toledo et al., 1993). Back-calculations were made according to the body proportional hypothesis (Francis, 1990) with parameters based on the regression of fish length (FL, mm) on scale radius (S, mm):

$$FL = 16.65 + 148.41\, S$$
$$(r^2 = 0.892; N = 171; p < 0.0001)$$

Therefore, back-calculation was performed according to the following equation:

$$FL_1 = [(148.41 + 16.65\, S_1)/(148.41 + 16.65\, S_c)]\, FL_c$$

where FL_c and S_c are measures at catch and FL_1 and S_1 are corresponding values at the time of formation of the first annulus.

Habitat structure and fishing regimes

Habitat was measured in width, depth, bottom type and canopy. Width was measured every 10 m along the stream and depths were taken in the centre and near the edges in transects across the channel. Proportions of bottom surface occupied by the six following substrate categories were recorded: mud, particles less than 10 cm diameter, small boulders (10–20 cm), large boulders (20–50 cm), small blocks (50–100 cm) and large blocks (>100 cm). Substrate diversity (H; Shannon-Weaver) was calculated from these proportions. A value of overhanging canopy was assigned to each section according to the proportion of stream bed shaded, in vertical projection, by trees and brushes, according to the following scale: (1) <10%, (2) 10–25%, and (3) to (5) in 25% increments till 100%. Sampling sites were roughly classified as calcareous or siliceous, according to the stream bed prevailing substratum as reported in geological charts. Altitude (m.a.s.l.) and distance from the sea were also recorded.

Fishing regimes, ranked in order of decreasing exploitation pressure, are:

(1) free fished sections (fishing season from March to August), with limitations concerning minimum takeable size (18 cm) and daily bag limit (16 trout per angler and day);

(2) mountain sections, where fishing begins one month later;
(3) sections where there are additional limitations in number of angling days and number of anglers/day;
(4) sections where fishing activities are forbidden during the whole year.

Statistical analysis

Raw data were transformed, when necessary, to achieve statistical assumptions of linearity, normality (Kolmogorov-Smirnov test) and homogeneity of variances (Cochran's test) required in most analitycal techniques (Sokal and Rohlf, 1981). In order to fulfill the above conditions we customarily used natural logarithmic and square root transformations, and arcsine $\sqrt{x}$ for percentage data.

Principal component analyses (PCAs) were used to explore variation in several population parameters and to relate biomass per individual and density values to a reduced-axes space of habitat information. Variables selected from PCAs were further analysed with simple and stepwise multiple regression, correlation and analyses of variance and covariance. Homogeneity of slopes was checked prior to analysis in ANCOVA. Group means were compared with the Tukey test in ANOVA and with multiple t-test in ANCOVA, setting in this case the significance level of a single test adequately to control the probability of committing a type I error in the multiple comparison (Zar, 1984). The nonparametric Kruskal-Wallis test was applied instead of one-way ANOVA when large-scale heterogeneity of variances (Cochran's test) was detected. Multiple comparisons were then carried out using Dunn's procedure (Zar, 1984). Pearson's correlation coefficients were employed to describe the relationship between pairs of variables whenever they were continuous and normally distributed, and Spearman's rank correlations otherwise.

RESULTS

Growth

Length at the end of the first growth period (FL_1) showed a huge variation among sections, ranging from mean values lower than 50 mm in some mountain sections to about 120 mm in low elevation, third or fourth order ones. In order to assess the influence on growth intensity of several habitat attributes we performed a stepwise multiple regression analysis with FL_1 as the dependent variable (Table 31.1). Variables significantly contributing to the regression model were distance from sea, mean width, and altitude, which accounted for 61% of the variation in FL_1. Growth increased with river width and declined with distance from sea and altitude, this last variable alone accounting for 46% of variation in FL_1 (Figure 31.2(a)).

Fish in low-order streams grew more slowly than did those in high-order ones (ANOVA, $F_{3,66} = 10.72$, $p<0.0001$; particular significant differences were found between orders $1-4$, $1-3$ and $2-4$; Tukey test, $p<0.05$; see Figure 31.3(a)), but this tendency was not significant when the effect of altitude, stream width and distance to sea were removed (ANCOVA; $F_{3,62} = 1.71$, $p>0.1$).

Brown trout showed significantly faster growth in scarcely covered or uncovered sections than in more shaded ones (ANCOVA, altitude as covariate, $F_{4,62} = 6.82$, $p<0.0001$; particular differences existed among the following canopy categories: $1-3$, $1-4$, $1-5$ and

TABLE 31.1. Results of stepwise multiple regressions of trout population variables with habitat attributes. To achieve normality, mean width and mean biomass per individual were *Ln* transformed and density square root transformed

Dependent	Variable entered	Coefficient	r^2	F	p
Fork length at age 1	Elevation	−0.002	0.493	59.27	<0.001
	Mean width	0.972	0.553	37.15	<0.001
	Distance from sea	−0.021	0.611	30.94	<0.001
	Constant	8.185			
Density	Distance from sea	0.025	0.329	31.43	<0.001
	% Large boulders	2.822	0.389	20.07	<0.001
	Mean width	−0.991	0.462	17.78	<0.001
	Constant	3.757			
Biomass per m²	% Large boulders	14.613	0.100	7.13	<0.05
	Constant	7.463			
Biomass per individual	Elevation (*Ln*)	−0.550	0.569	80.70	<0.001
	Mean width	0.511	0.675	62.30	<0.001
	Constant	6.299			
Mean age (4th quartil)	% Large boulders	1.374	0.099	5.91	<0.05
	CV of depth	2.094	0.217	7.34	<0.01
	Constant	2.308			

2−5; t-test, $p < 0.05$; Figure 31.3(b)). Growth was also faster in streams with a prevailing calcareous substratum than in those flowing over predominantly siliceous materials (ANCOVA, altitude as covariate, $F_{1,67} = 4.09$, $p < 0.05$).

Trout density and biomass

Brown trout density, all age-classes included, varied from 0.02 to 1.07 fish m^{-2} in the whole sample, with more than 75% of values in the range from 0.10 to 0.60 fish m². The range for total biomass was $1 - 45$ g m^{-2}, with values lower than 20 g m^{-2} amounting to 72% of cases. Differences in canopy, stream order or substratum nature (calcareous − siliceous) did not significantly influence brown trout density (residuals on altitude) nor biomass m^{-2} (ANOVAS, $p > 0.05$ in all cases; Figure 31.3).

Habitat attributes, several of them correlated with elevation, were synthesized in the first two axes of a PCA (39.9% variance explained; Table 31.2). When density at each site was plotted in the two axes space (Figure 31.4(a)) the highest values appeared concentrated in the lower right quadrant, corresponding to distance from sea, elevated reaches, with intermediate width and depth values. Low density values corresponded, on the other hand, to deep, wide, close to sea reaches with a high proportion of mud substrate. Stream order was negatively correlated with both axes I ($r_s = -0.502$; $p < 0.001$) and II ($r_s = -0.360$; $p < 0.01$).

Variation in density and biomass per square metre in relation to habitat attributes was further analysed by means of regression analyses. For density, parabolic regression with altitude gave the best-fitting equation to a single variable, as density values were greater in mid-elevation reaches and declined at both high and low altitudes (Figure 31.2(b)). However, a stepwise regression involving distance from sea, proportion of large boulders and width provided the best fit (46% of variance explained; Table 31.1). For biomass per square metre, the only variable to enter the multiple regression at the 0.05 step level was the proportion of large boulders, though the percentage of variance accounted for by the regression was small (10%).

A scatterplot of mean biomass per individual in the space defined by the two first axes of the PCA (Figure 31.4(b)) showed a noticeable segregation around the first factor, with the highest values in low altitude, close to sea reaches, rather deep reaches with a high proportion of mud substrate. This distribution reveals the inverse link of density with mean biomass per individual ($r = -0.456$; $p < 0.001$). However, this relationship did not persist when residuals on altitude of both variables were computed ($r = 0.136$; $p > 0.1$).

Mean biomass per fish significantly decreased with elevation (Figure 31.2(c)) and increased with stream width. These two variables explained 67% of the variability in mean biomass per fish in a stepwise multiple regression model (Table 31.1). Mean biomass per individual differed in relation to overhanging canopy (ANCOVA, $F_{2,62} = 2.94$, $p < 0.05$; Figure 31.3(b)), though we have only found it to be significantly larger in the least covered sections than in the most shaded ones (t-test, $p < 0.005$). The occurrence of a calcareous stream bed was a marginally positive factor on mean fish weight (ANCOVA, $F_{1,70} = 3.36$, $p = 0.07$).

Higher order reaches showed greater values of biomass per individual (ANOVA, $F_{3,69} = 15.88$, $p < 0.05$; significant differences existed between orders $4-1$, $4-2$ and $3-1$, $3-2$; Tukey test, $p < 0.05$; Figure 31.3(a)). Unlike growth, the stream order effect was not totally removed when elevation, width and distance from the sea were introduced as covariates (ANCOVA, $F_{3,63} = 3.21$, $p < 0.05$; significant differences remained between orders $3-1$ and $3-2$; t-test $p < 0.05$).

Age structure relationships

Mean age was significantly correlated with the proportion of large boulders ($r = 0.33$, $p < 0.05$), and also exhibited a negative significant correlation with growth (FL1; $r = -0.25$, $p < 0.05$). The proportion of large boulders and the coefficient of variation of depth (CVD), both of them with positive coefficients, were involved in the stepwise multiple regression model with mean age of fish in the upper quartile as the dependent variable (Table 31.1), but the regression accounted for only 22% of variation in this variable. Age diversity was positively correlated with depth variability (CVD residuals on distance from sea; $r = 0.29$, $p = < 0.05$; Figure 31.2(d)).

The first two axes of a PCA performed on eight relevant trout population variables explained 66.7% of the variance brought by that parameters (Table 31.3). On the first axis, mainly related to age structure of the sample, have high positive scores populations diverse, longevous, with low proportion of 1+ individuals. Low density, fast growing ones with large biomass per fish exhibited high positive scores on the second axis. Both altitude and distance from sea were negatively correlated with the second axis ($r_s = -0.660$ and $r_s = -0.613$, respectively; $p < 0.001$ in both cases), but uncorrelated with the first one, thus indicating a decrease in density and increase in growth rates proceeding downstream and close to the sea.

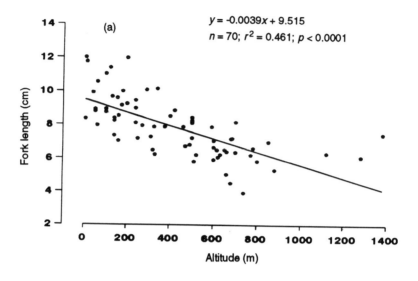

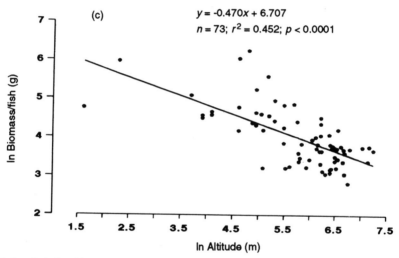

FIGURE 31.2.   Relationships between (a) fork length, (b) density and (c) biomass per individual with altitude; and (d) age diversity with depth variability

Effect of fishing regime

Populations differed both in mean age (ANOVA, $F_{3,63} = 4.85$, $p < 0.005$; Figure 31.3(c)) and age diversity (ANOVA, $F_{3,63} = 5.40$, $p < 0.005$) according to fishing regime. These differences tended to weaken when the effect of the main correlates were removed (percentage large boulders for mean age, ANCOVA $F_{3,53} = 3.49$, $p < 0.05$; and CVD residuals for age diversity, ANCOVA $F_{3,53} = 3.10$, $p = 0.05$). The most heavily exploited populations were younger and showed lower age diversity, but the only significant differences were found in both cases between the two extreme situations: free-to-fishing sections versus the absolutely closed ones (Tukey test, $p < 0.05$). The proportion of trout

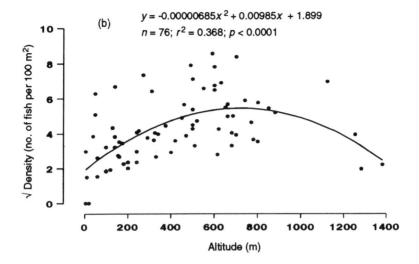

$y = -0.00000685x^2 + 0.00985x + 1.899$
$n = 76; r^2 = 0.368; p < 0.0001$

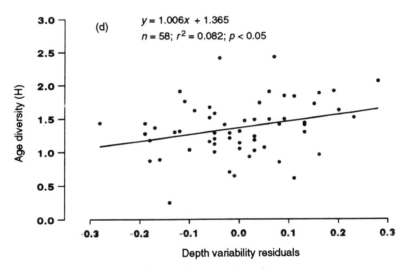

$y = 1.006x + 1.365$
$n = 58; r^2 = 0.082; p < 0.05$

FIGURE 31.2. *continued*

larger than 18 cm (the legal fishing limit) was higher in less exploited sections than in most fished ones (ANCOVA, altitude as covariate; $F_{3,63} = 3.35, p < 0.05$; a particular difference was evident between closed-to-fishing and mountain-fished sections; t-test, $p < 0.05$). We have also found a consistent difference according to fishing status in mean biomass per individual (Figure 31.3(c)), with higher values in reaches subjected to controlled angling exploitation than in free-fished and mountain ones (ANCOVA, with altitude as the covariate; $F_{3,68} = 4.96, p < 0.01$; t-test, $p < 0.05$).

Densities of legal-sized trout in non-fishing reaches were superior to those occurring in free-fished and mountain ones (Kruskal-Wallis, $H = 17.50, p < 0.001$; Dunn, $p < 0.05$; Fig. 31.3(c)), but there were no differences in overall (excluding 0+) density in relation to

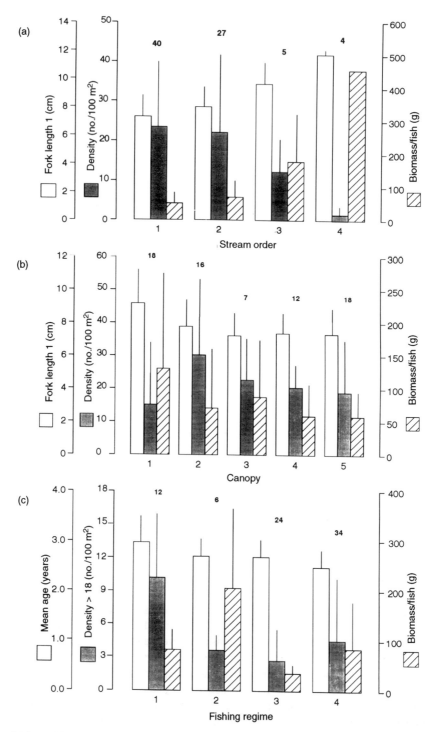

FIGURE 31.3.   Effects of (a) stream order, (b) canopy and (c) fishing regime on several brown trout population parameters. Mean (+1SE) and sample size are indicated

TABLE 31.2. Loadings of habitat variables and variance explained by the first three factors of a Principal Component Analysis

	1	2	3
Distance from sea	0.818	−0.232	−0.041
Altitude	0.767	−0.071	−0.016
Mean width	−0.428	−0.453	0.070
Mean depth	−0.600	−0.522	0.132
CV of depth	−0.384	0.315	0.606
Substrate diversity	−0.247	−0.094	0.743
% Mud	−0.659	−0.161	−0.343
% Sand and gravel	−0.479	0.094	−0.409
% Small boulders	0.046	0.855	0.126
% Large boulders	0.448	−0.421	0.132
% Small blocks	0.268	−0.274	0.309
% Large blocks	0.009	−0.421	0.037
Variance explained (%)	24.56	15.34	11.32

fishing regime (ANOVA of residuals from the regression on altitude; $F_{3,72} = 1.53, p > 0.1$). Biomass per square metre was larger in closed-to-fishing sections than in mountain-exploited reaches (Kruskal-Wallis, $H = 13.00, p < 0.005$; Dunn, $p < 0.05$).

DISCUSSION

In our study area brown trout showed extreme variability in growth intensity, covering most of the range of fork length at the end of the first growth period reported for European populations (e.g. Mortensen, 1977; Jonsson, 1977; Lobón-Cerviá, Montañés and Sostoa, 1986; Mann, Blackburn and Beaumont, 1989; Maisse and Baglinière, 1991). The main single factor accounting for such a high variation was elevation, whose negative relationship with growth has been reported for brown trout elsewhere (Scarnecchia, 1983; Zalewski, Frankiewicz and Brewińska-Zaras, 1986). This negative relationship presumably operates through the associated decrease in water temperature as elevation increases, demonstrating the importance of temperature on trout growth (Weatherley, 1972; Elliot, 1988; Baglinière and Maisse, 1990). Nevertheless, the relationship between growth intensity (as displayed by length at the end of the first winter) and elevation might be reinforced by linked correlations, e.g. lower temperatures result in longer incubation periods (Jensen, Johnsen and Heggberget, 1991; Kamler, 1992) and longer time from hatching to initial feeding in salmonids (Jensen, Johnsen and Saksgard, 1989). In addition, spawning is somewhat delayed in high altitudes, and the growing season shortened (authors' unpublished data). In fact, the elevation axis integrated a great amount of the variability in physical habitat attributes, as correlated with river width, river depth, size of substrate, gradient and stream order. Consequently, we have found the influence of altitude, either directly or mediated by variables with which it is linked, on almost every relevant population parameter.

Brown trout densities and biomass per individual were not independent of habitat features and both strongly varied along the longitudinal axis. Densities were greatest at intermediate elevations and declined both at higher and lower altitudes. Biomass per individual, on the

(a)

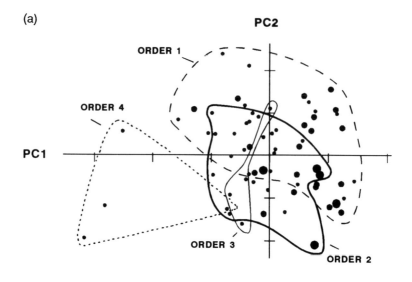

(b)

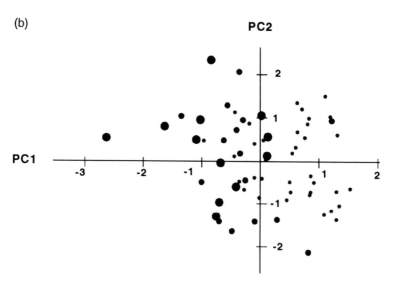

FIGURE 31.4. Brown trout (a) density and (b) biomass per individual at each sampling location, plotted on principal components axes I and II. Size of dots (small, medium and large) represents increasing levels of density (<20, 20–50 and >50 individuals per 100 m²) and biomass per individual (<50, 50–100 and >100 g per individual)

other hand, was low in the higher sections and increased as elevation decreased. This pattern is similar to that reported by Schlosser (1987) for small warmwater streams, in relation to a gradient of increasing habitat heterogeneity and pool development. Higher reaches, supporting low densities of small-sized individuals, were narrow and shallow, lacking deep pools that provide more protective and stable habitats suitable for larger/older individuals (Schlosser, 1982, 1987; Harvey and Stewart, 1991). Furthermore, high altitudes correspond to low-order reaches, in which total production and food availability for fish could decline

TABLE 31.3. Loadings of trout population variables and variance explained by the first three factors of a Principal Component Analysis

	1	2	3
Maximum age	0.694	− 0.261	− 0.142
Age diversity	0.891	0.198	− 0.117
Mean age	0.934	− 0.065	− 0.243
Fork length at age 1	− 0.137	0.844	0.381
Density	0.072	− 0.757	0.613
Biomass per individual	0.572	0.642	0.395
Biomass per m²	0.408	− 0.173	0.882
% Age 1⁺	− 0.890	0.068	0.166
Variance explained (%)	43.22	23.05	19.69

(Zalewski et al., 1990). In fact, some authors have reported faster growth for brown trout as stream order increases (Zalewski, Frankiewicz and Brewińska-Zaras, 1986). In our study, on the contrary, among-order differences in FL_1 were almost entirely explained by a set of simple habitat parameters: elevation, mean width and distance from sea. However, mean biomass per individual increased with stream order, and these differences were only partially removed by the above covariates. Low reaches often support populations with a high proportion of anadromous individuals (sea trout) in our study area, which implies high biomass per individual and pronounced fluctuations in density and age structure in relation to migratory and reproductive cycles (Baglinière et al., 1989; Toledo et al., 1993).

Although downstream increase in food and space availability would allow higher densities at lower altitudes, increased within- and between-population interactions could play a major role in limiting fish abundance. In lower sections several species join the fish community, so that inter-specific competition for habitat and food might emerge and constrain brown trout densities. Previous studies have shown that competition for food and space could exist between young brown trout and juvenile Atlantic salmon (Kennedy and Strange, 1980, 1986b; Gibson and Cunjak, 1986; see, for our study area, Suárez, Reiriz and Anadón, 1988), and to a lesser extent with eel (Mann and Blackburn, 1991). Predation of large silver eels upon juvenile trout could occur (Mann and Blackburn, 1991), but the main predator for young brown trout in our study area are the larger/older individuals of the same species (Suárez, Reiriz and Anadón, 1988; Toledo et al., 1993). Under pressures of predation or predatory risk, small trout would select shallow waters and tend to move towards deeper sections as growth progresses (Solomon and Templeton, 1976; Maisse and Baglinière, 1990), and then, as larger pools are available, a change towards fewer and bigger individuals could take place (Schlosser, 1982, 1987; Harvey and Stewart, 1991). This tendency may be reflected, to some extent, by the complementary distribution we have found between density and biomass per individual on the PCA space. Furthermore, we have found older fish and higher age diversity associated with high depth variability, which agrees with the expected patterns according to the age-specific distribution in the channel.

Trout distribution was not only influenced by variation in habitat volume or biotic interactions, but also by the physical structure of the channel. Mean width of the channel was negatively related to trout density, although its influence could partially reflect covariation with other parameters, such as elevation or mean depth. Nevertheless, the increase in stream width could produce a depressing effect on density, as this usually implies a decrease in the bank length/channel surface ratio and so a reduction in the availability of the bank microhabitat, advantageous for brown trout (Baglinière and Arribe-Moutounet, 1985). We also found that density, biomass per square metre and mean age of trout were positively related to the proportion of large boulders. The increase in the proportion of large-sized substrates may enhance habitat quality for trout by improving shelter availability and providing more energetically profitable feeding positions (Bachman, 1984).

As reported for other salmonids, effects of riparian canopy on population parameters are somewhat confusing (Hawkins et al., 1983). In our study, shading negatively influenced growth but did not affect trout density or biomass. Shading can negatively affect fish growth by decreasing light availability for autochthonous primary production, thus diminishing benthos abundance (Angermeier and Karr, 1983; Behmer and Hawkins, 1986); by decreasing water temperature (Brown and Krygier, 1970; Scarnecchia, 1983); or by impairing the efficiency of prey capture (Wilzbach and Cummins, 1986) and reducing the effective time for food searching, given that salmonids are visual feeders (Blaxter, 1970; Wankowski and Thorpe, 1979). However, local production of benthic macroinvertebrates could not be an appropriate measure of food availability, as brown trout in rivers of this study region mainly feed on drift from upper reaches (Suárez, Reiriz and Anadón, 1988; Rincón and Lobón-Cerviá, 1993). A link between shading and decreased water temperature is also difficult to find, as this probably requires shading over extensive upstream areas (Hawkins et al., 1983). These possibilities, however, cannot be absolutely rejected, as we could assume some link to exist between the shading status of the sampled sections and the condition of immediate upstream sections.

In addition to water temperature and light availability, calcium concentration may strongly influence fish growth and production (Le Cren, 1969; Scarnecchia, 1983; Beaudou and Cuinat, 1990; Zalewski et al., 1990) by means of its effect on primary production and, consequently, on zoobenthos abundance (Schiemer and Zalewski, 1992). In our study, trout in sections overlying calcareous beds showed faster growth and a slightly larger biomass per individual than those in sections over siliceous substrates, but did not show differences in density or biomass per square metre. This could be indicative of weak effects only detectable in variables that directly reflect individual food status (fish length and weight), but not in population parameters subjected to multiple interactions and sources of variation.

We found consistent decreases in mean age, age diversity, abundance of trout over the legal fishing limit and mean biomass per individual as angling pressure increased. These are the usual outcomes of sport angling exploitation on brown trout stocks, as a direct consequence of removing the largest/oldest fish from the population (Shetter, 1969; Avery and Hunt, 1981; Büttiker, 1989; Braña, Nicieza and Toledo, 1992). Despite the clear shift in age structure, we did not detect effects of fishing regime on overall trout density or biomass per surface unit, so we can assume that the present harvest rate does not endanger population renewal. This might be partially due to the existence of within-population regulatory mechanisms with potential compensatory effects on density and standing stock, as several of them, such as increased juvenile growth and recruitment (Healey, 1980; Donald and Alger, 1989) or fecundity (Healey, 1978), have been reported in relation to fishing exploitation in salmonids.

ACKNOWLEDGEMENTS

We thank Luis Reiriz, América García for their help in field work, and Carmen Alvarez for improving the English. Electrofishing, with subsequent release of fish, were made under the authority of a scientific sampling permit issued by the Consejería de Medio Ambiente y Urbanismo, Principado de Asturias. This work was partially supported by FICYT and CICYT-PB92-0093 grants.

32

Ecological Basis for the Management of Recreation and Amenity: The Norfolk Broads

TIMOTHY O'RIORDAN

University of East Anglia, Norwich, UK

INTRODUCTION

There is a statutory duty on both the National Rivers Authority (NRA) and the private water utilities positively to further the interests of wildlife habitat, fisheries, recreation and amenity (see Chapter 19). This means:

(i) that the interests of ecological welfare are an integral part of all future water management, and
(ii) that economic justification for such management effort should be legitimately part of any future cost − benefit analysis.

It will take time before the NRA can fully adjust to this wide-ranging statutory duty and it certainly places rather more of a burden on good ecological understanding of the effects of management and mismanagement on waterways. Because of this, it is likely that the principle of precaution will be applied — playing safe in the face of scientific uncertainty — which may result in the setting aside of vulnerable areas even where there may be a degree of tolerance for some intervention, and in the over-commitment to ecological protection in areas where scope for manipulation is not well known. This in turn will require more flexibility from the UK Government Treasury than has happened up until now in incorporating the social valuation of environmental improvement in river management schemes. It will also mean a change in the legal status of minimum flows to ensure that adequate water is maintained in rivers and groundwater feeding sources to maintain healthy fluvial ecosystems (see Chapter 1). To date, this particular issue has hardly been addressed as a statutory duty on both the private water abstractors and the NRA. The arrangements so far are primarily voluntary.

This chapter looks at some of these issues as they apply to the forthcoming management plan of the Norfolk and Suffolk Broads, in eastern England. The main points of that exercise are:

(i) a commitment to maintaining the ecological integrity of the Broads as the prime objective of management;

The Ecological Basis for River Management. Edited by D.M. Harper and A.J.D. Ferguson. © 1995 John Wiley & Sons Ltd

(ii) joint policy measures, backed by collateral funding between the Broads Authority and the NRA;

(iii) intensive experimental research to reduce ecological ignorance and narrow the zones in which the precautionary principle has to be involved;

(iv) extensive consultation with all stake-holders *via* a series of partnership deals and consultative panels which act as mediators of disputes and not just as advisers.

Despite this, the Broads' experience highlights the enormous expense of overcoming ecological ignorance and the real dangers of trying to become too certain *via* modelling and analytical guesswork. Precaution probably has a greater part to play than is normally assumed, but it will only be successfully applied when the economics of all-round river management are fully accepted. Right now, we are still in an early phase of transition.

THE ECOLOGY OF THE BROADS

The Norfolk and Suffolk Broads are equivalent to a UK National Park, but are designated as a special statutory authority under the Norfolk and Suffolk Broads Act 1988. It is best to describe the Broads as "part of the national park family but not a national park". This is to ensure that the general principles of management that apply to the 10 designated parks in England and Wales also apply to the Broads, but that the Broads Authority is free to extend its management practices in ways that the other County Council controlled parks are less willing to explore.

The structure of the Broads Authority is unusual in British local government. It is comprised of 35 members, of which 18 are drawn from the six District Councils (two each) and the two County Councils (four from Norfolk, two from Suffolk). The remaining 17 members are appointees by the Environment Secretary drawn from the interests of navigation, angling, farming, tourism and conservation. In general there is no great difference of view between the non-local authority members and their elected counterparts. The main division of opinion lies between the conservation interests and those of navigation. This is not so much because of the personalities involved, as an outcome of the 1988 Act. This laid down a number of statutory safeguards for navigation, safeguards that the boating interests have sought to exploit. These provisions are:

(i)   an executive navigation committee with powers to raise tolls on all craft and spend the revenue on navigation-related purposes;

(ii) co-opted navigation interests to that committee to ensure that it always has a majority navigation bias;

(iii) a formal statutory procedure should any navigable waterway be closed, even temporarily, for the purposes of ecological recovery in the broads and rivers;

(iv) ring-fenced protection of the navigation toll revenue for use only with regard to navigation purposes.

These points are emphasized because the broads management plan seeks to straddle the conflicting interests between ecological welfare and rights of navigation. Legally, the right of navigation is hazily defined. If a waterway is regularly used for boating, with at least 20 years of proven use, or if the waterway is tidal, then there is a presumption in favour of navigation. The limit of tidal influence is always a matter of conjecture. In any case, a number of broads and waterways have been closed by landowners for over 20 years. The precise legal right of navigation over such areas is so uncertain that frankly it is best to enter into voluntary management agreements. The basis for such agreements must be mutually accepted between the management agency and landowners. This is a prime reason for advancing ecological well-being as a key ingredient of management strategy. But where the legal status is in dispute, or at least unenforceable, then agreement can only be achieved through partnership *via* voluntary negotiation. This is why the institutional structure of the management authority is so important. It is also critical why it is so necessary to get the ecological understanding on a reasonably secure footing.

The Broads cover an area of 30 292 ha, essentially the river valleys of the Bure, the Yare and the Waveney, in eastern Norfolk and Suffolk. No land in the Broads Authority executive area is over 25 m above sea level. It is the ultimate lowlands park. The area is visited by over one million people per year, 500 000 of whom take to the water in hire craft or private vessels. The boat industry is worth over £30 million to the local economy, adds £0.30 to every £1.00 of tourist spending in the region and employs over 700 people. In relative terms, apart from agriculture, which on the unimproved marshes is fairly unproductive these days, the boat industry is the main source of economic activity outside of tourism. Only 10% of the area is not privately owned, so virtually all the ecological research requires the co-operation of landowners. Luckily, in terms of securing agreements, the most interesting fen communities are owned by relatively few people, the vast majority of whom are supportive of conservation. Some 70% of the marshes are farmed, 10% is woodland (mostly unmanaged), 8% is fen (mostly managed), and 6.5% is open water. The area is rich in conservation designations with 21 Sites of Special Scientific Interest (though 80% of the fen is of SSSI quality), two "Ramsar" sites with at least one more about to be listed, and 25 local trust nature reserves. Over the period 1985–92 the Authority spent £345 320 on conservation and restoration of the broads and fen with current annual expenditure of £150 000. From 1993 to 1996, under the EC LIFE arrangement, the Authority will spend over £712 000 on ecological research alone, aimed at understanding more precisely the mechanisms through which freshwater ecosystems can be restored to a pre-degraded state. This will be a jointly funded programme under the auspices of the NRA regionally, the NRA nationally and the Broads Authority, with over half of the funds coming from the European Commission and additional contributions from the Soap and Detergent Industry Association and English Nature.

Since 1985, parts of the Broads grazing marshes have been specially managed to ensure that grazing continues, even against the economic tide, and that the drainage systems (dykes) are manipulated to retain or enhance the diversity of aquatic plant life that used to be found throughout the Broads, but is now virtually confined to spring-fed margins. Nowadays the whole of the Broads grazing marsh is designated as an Environmentally Sensitive Area (ESA) for which the Ministry of Agriculture is spending £2.5 million annually to ensure suitable water levels and to maximize the potential for birdlife and invertebrates.

PRINCIPLES OF BROADS MANAGEMENT

The 1988 Act gave the Broads Authority three somewhat incompatible statutory duties:

(i) to preserve, protect and enhance natural beauty and amenity,
(ii) to provide for and extend public enjoyment, and
(iii) to protect and promote the interests of navigation.

In the debates that preceded the passage of the Act there was much dispute over the possibility of "conservation primacy", namely that the three duties were listed in order of priority. In fact, the UK Government in 1976 conceded the principle of conservation primacy in its response to the Sanford Report on national park management. Where there is an irreconcilable conflict between the integrity and viability of ecosystems that sustain the very tourist and scenic interest of a park, then the conservation need is the greater. This provision does not explicitly extend to the Broads. Most parks have subsequently adopted strategies that steer people and economic activity generally away from sensitive zones, notably where the ecological case can be justified.

In the Broads case that ecological case is based on the following arrangements:

(i)   an intensive and long time-scale programme of linked ecological research co-funded by the Authority, English Nature, the NRA and the European Commission and supported by landowners;
(ii) a detailed structure of ecological monitoring and assessment of research results through joint study teams, regular workshops and a scientific advisory panel;
(iii) a regular review of experimental research, coupled to an international scientific exchange arrangement with French, Dutch and Romanian national parks aimed mainly at learning from different management trials;
(iv) a wardening service backed up by voluntary conservation bodies to manage the fen *via* grant aid and payments in kind;
(v)   the introduction of ecological − economic cost − benefit studies based in part on the contingent valuation method. This involves asking visitors to estimate the amount they would be willing to pay to see the Broads, or parts of the Broads where conservation management is maintained or enhanced;
(vi) study period for staff in applied management training, often in international field workshops.

This is a major commitment. It absorbs over 40% of the full Authority budget, it takes up over a third of the staff, and it dominates the overall Broads management strategy. Such an approach pays off. Until now, the Broads Authority budget allocation has never been refused by local authority members, even though some of its local authorities are rate capped and must divert each from other needy programmes to pay for the Broads restoration.

But by far the most significant development is not the research or the cash allocation; it is the philosophical underpinning of the Broads management plan, which forms the fundamental basis for all strategic policy decisions and management agreements throughout the area. The following principles have been agreed regarding that Plan, which is due to be published in early 1995.

(i)   The natural processes which sustain the Broads ecosystems must be maintained without impairment.
(ii) Such processes have both a *functional value* in broads management terms in supplying services of cleansing absorbing and buffering, as well as an *intrinsic value* in biodiversity terms. This is the so-called *in situ* value of ecosystems which have a right to exist on their own account, but whose existence can be evaluated *via* social accounting measures and subsequently justified as part of management.
(iii) All management decisions regarding the statutory objectives must be subservient to the guaranteed maintenance of the ecological integrity of the Broads region, and that where the integrity cannot be guaranteed then the precautionary principle must be invoked.

These are powerful commitments. They put a value on ecological maintenance and restoration that reinforces the statutory duties of furthering conservation held by the NRA and the Anglia Water Services plc. They ensure that navigation rights are not absolute, but subject to the best management principles of ecosystem stability. They enable voluntary agreements to be signed that limit boating in space and time in areas where bird breeding is endangered, or where bankside restoration techniques are being experimented. They allow for the manipulation of bird feeding habits where reed margins or subsurface macrophytes are being damaged during a period of broads restoration. In short, they are both pro-active and very interventionist. But to be successful, they must be justified by the very best ecological research, full consultation, and carefully programmed experimentation undertaken step by step from the small scale to the modestly large. There is no quick route to the restoration of ecological integrity.

ECOLOGICAL RESEARCH AND MANAGEMENT UNCERTAINTIES

Despite 12 years of intensive experimentation and research in broads restoration, it is reluctantly the case that we still do not know for certain how to recreate a freshwater ecosystem. As is well established, the Broads suffer from nutrient enrichment, two-thirds of which is due to the discharge of phosphate-rich effluent from sewage works and the remainder from septic tanks, industrial wastes and farm runoff. Most of the latter is difficult to control without expensive surveillance, so the main effort has been directed at the sewage works. At present, seven plants are subject to phosphate removal *via* the addition of ferric sulphate to the final effluent. This programme was originally jointly funded by the Broads Authority and Anglian Water. But since privatization, Anglian Water Services plc has paid the bill. It should be stressed that the main reason why they have do so is because of the joint policy framework on water-quality standards and consent conditions established between the Broads Authority and the NRA. Without that joint action framework it is unlikely that AWS would have volunteered to reduce phosphate discharges to 3 mg litre^{-1} as requested by the Authority.

Even so, the politics and economics of actually achieving these standards are very tortuous. Anglian Water Services remains unwilling to invest in the very high standards of phosphate removal demanded by the Broads Authority and brokered by the NRA. In practice, Anglian Water Services requires authority from the water rate tariff regulation OFWAT to raise prices to meet these high standards of effluent treatment. OFWAT is under political pressure to keep prices down. The Broads are not designated so that low phosphate levels would be a mandatory requirement. NRA has to negotiate a deal of transitional lower standards with the aim of phasing in higher standards (i.e. to 2 mg litre^{-1}) over three to five years. This is unsatisfactory but may be politically expedient.

The trouble is we have no idea if 2 mg litre^{-1} is the correct level for revival. The Authority is working on the assumption that this effluent level will result in an ambient water phosphorus level low enough over the most important stretch of designated sites to result in sufficient conditions for restoration. But there are two stumbling blocks:

(i)   phosphate-rich sediment releases phosphates even under oxic conditions, so has to be removed, or at least precipitated from the surface layers;
(ii) the actual sediment surface conditions appear to be critical in determining the physical basis for the reproduction of submerged macrophytes. What these conditions are is not fully known, but they could be an absence of physical disturbance and the stabilization

of flocculent sediment. It is most likely to be the exclusion of light by phytoplanktonic algae, and lack of control of the algae by zooplankton which are themselves supressed by planktivorous fish. It is possible that conditions for macrophyte regeneration can only be created by a barrier of sorts, an exclosure to fish which facilitates healthy zooplankton population development, leading to clear water and thus macrophyte growth. To implement the sequential exclosure of all key broads would be a tall order, certainly beyond the capacity of the Authority's funding programme at present.

Nevertheless the first part of this restoration exercise is well under way with four broads mud-pumped to 1 m to remove the nutrient-rich sediment. This proved to be a costly programme, running at around £10 000 ha^{-1}. Obviously it is a costly solution for a large broad of say 100 ha. For example, to suction-dredge Barton Broad, on this scale, is estimated to cost £1.5 million over five years.

So the Authority is embarking on an experimental programme of ferric chloride dosing of the surface sediment to see if this treatment will precipitate the surface phosphate to harmless low levels. To date, tank experiments have shown that the ferric chloride technique has not proven entirely satisfactory, possibly due to the absence of organic detritus and the uncertain role of chironomid larvae. A small field trial was ruined by wind-assisted sediment redistribution and freak salt incursion. It will take two years of detailed research on an isolated small broad before all the various relationships are fully understood. Yet the cost of these modest experiments alone will be £175 000 over the period 1993–95.

The re-establishment of submerged and surface plants in treated broads is also proving to be a major problem in ecological research. In three of the treated broads, despite clear water conditions, seed germination did not take off, even though it did take place in its early stages. One of the factors may be selective grazing by coot, which are observed feeding on submerged vegetation. Bird exclosures have revealed a luxuriant growth of macrophytes, even in moderately turbid water. In addition, fish have been removed to allow for increased populations of zooplankton, notably the cladoceran *Daphnia*, principal grazers of phytoplankton. For the next three years, coot-feeding habits and the conditions that render exclosures so apparently successful will be monitored. But the fact remains that the fundamental causes may lie in the characteristics of the seedbank, the quality of germination and the instability of sediment. The cost of all this work will be £126 000 over the period 1992–95 with again no guarantee of a successful outcome.

Both fish population manipulation and sediment chemistry alteration will have an influence on the surface conditions, favouring or discouraging processes that could be counter-productive to broads restoration. It is possible that chironomid larvae influence phosphate release through their feeding activities, but that their densities are influenced by populations of benthivorous fish. So fish numbers manipulation may have to be fine tuned. In a large broad, that will be a very difficult and expensive task. Under stable conditions, filamentous mats of algae can grow, dampening seed germination and altering the physical and chemical state of the sediment surface. So field trials will require extensive replication if they are to prove anything (although replication may have to be sacrificed in favour of the opportunity of scale).

All this goes to show that ecological research is still in its relative infancy when it comes to the practical restoration of living watercourses used for boating and public enjoyment. What is impressive is the commitment by the Broads Authority and the NRA to finance this work, now backed by the EC but not originally so, and the international recognition that this work is receiving. But it has to be said that the state of ecological knowledge is rudimentary. So there is nothing for it but to invoke a variant of the precautionary principle and invest in pro-active research along the lines proposed. Meanwhile if bird culling is necessary, if boat movement has to be temporarily halted to ensure the correct reliable conditions, so be it. This is the essence of precaution.

Politically speaking, that point has not been accepted in the Broads. Precaution is a challenging concept, but it requires the full-hearted support of affected interest groups before it can be successfully put into effect. This requires regular workshops and a ground variant of shuttle diplomacy. This is also why senior officer time is so dedicated to communication and bridge building.

ECOLOGICAL ECONOMICS AND COST–BENEFIT ANALYSIS

The Broads Authority is working closely with the sea defence arm of the NRA to devise a comprehensive flood alleviation strategy for the region. Salt intrusion is a serious pollutant because the salty water passes through the floodwalls long before they are overtopped. The evidence is still patchy because data trends are short, and the fieldwork is time consuming and specialized. But it seems as if salt pollution, coupled to rapid runoff from upland catchments, is having a devastating effect on the marginal plant-rich dykes of broadland. If true, this would be a very serious problem, given the value of these ecosystems as plant reservoirs for broads restoration. Also the broads in the process of restoration are extremely vulnerable to salt intrusion. So it is vital that salt water is kept out of the middle reaches of the Bure system at least, and also the marginal locations of the Waveney marshes.

Saltwater enters the Broads under two conditions: *via* a tidal surge when North Sea water is piled up in a north-westerly wind, and when freshwater river flows are very low so that normal neap tides can creep up the rivers on successive floods. The problem in the second case is where the rivers are either over-abstracted, even though the amounts taken are within statutory consents, or when the groundwater feeder flow is reduced due to drought.

The tidal surge can be contained by improved floodwall protection for the moderate surges, for the lower reach floodwalls are sinking all the time, and by the construction of a barrier for the very severe events of over 1 in 10 year probability. The purpose of this chapter is not to discuss either the merits or the politics of the two major proposals still being reviewed, namely a Yare barrier at the Haven bridge with a 1 in 200 level of protection, or a Bure barrier plus large washlands on Haddiscoe Island, the latter with a 1 in 20 level of protection for the lower Waveney and Yare marshes (Figure 32.1).

At present the Local Flood Defence Committee has opted for the Yare barrier, but the opposition of all the main conservation bodies, including the Broads Authority, remains. The most likely outcome is a prolonged phase of improving the river walls on a cost-effective basis, followed by a Bure barrier when the various parties have put their case to a public inquiry. Tidal washlands are unlikely to prove cost–effective, so controlled overspill may be a necessary option for the lower Yare and Waveney marshlands. It will be at least two years before a public inquiry is convened.

The issue for this discussion is the role of ecological economics in all this (see Chapter 40). I am broadly in agreement with observations on the mental contortions expected of a respondent when asked to valuate an ecological or scenic resource. My concern is the economics of freshwater flow assimilation and the biological diversity created by washland complexes. The difficulty here is threefold:

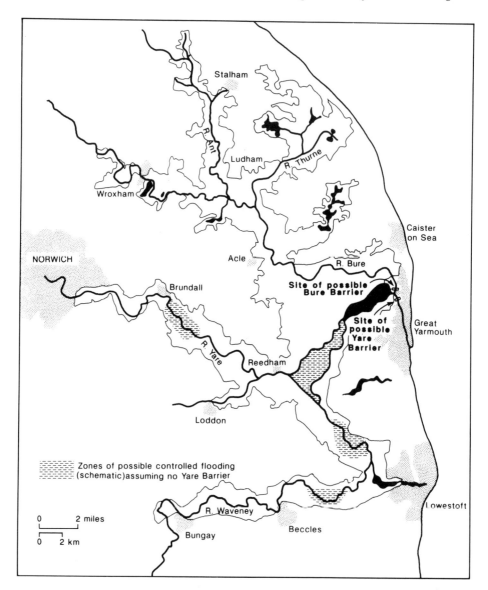

FIGURE 32.1.   The floodplain of the lower Yare catchment which forms the boundary of the Broads Authority. All of this area is subject to tidal and fluvial flooding unless protected by river walls or tidal barriers. The upper reaches are deliberately left unprotected as fen. The schemes designed to safeguard the area from tidal incursion are improving the river floodwalls, the construction of a Yar or Bure barrier, or a series of controlled washlands. The most likely combination is floodwall improvement to 1:10 standard, a Bure barrier and a number of controlled tidal spillways on the lower Yare and Waveney marshes

(i)   Ecological knowledge is limited as to the likelihood of the ecological gains of any given set of measures to improve habitat biodiversity in advance of investment taking place.

(ii) Contingent valuation is a problematic technique at the best of times. It is even more awkward if applied to hypothetical future ecological states where the uncertainties of actual delivery are very great.

(iii) The precise ecological functioning of improved habitat, for example in sequestering nutrients, is still poorly understood. So the economic significance of the utilization functions are at best guesswork.

Yet biodiversity improvement measures should be economically justified. One should at least try to measure the likely possible gains for various units of investment. Otherwise the whole process is based either on whim or on the pushiness of key players in the political chess game known as resource management.

We need more trial and error case studies. The Broads washlands scheme, which should result in extended bird habitat and a return of both higher and lower saltmarsh with its scope for nutrient sequestration, can be modelled on the basis of sediment chemistry and ecological models. Various values can be imputed either *via* expert groups or through community panels invited to bid against different levels of habitat diversity. Images of how the washlands might look under various flood regimes could be used to show people more clearly what future landscapes might come about. We have to be imaginative because this is a new area of research and the old techniques will no longer do.

On the matter of low flow augmentation even more is at stake, given the somewhat confused legal status of abstraction rights and statutory duties. The utilities cannot simply meet their abstraction licences because these were authorized in another era when environmental well-being was not an issue. Both the NRA and the water supply companies will have to give way, in deference to their responsibilities. The actual point of consensus depends critically on the ecological economics of habitat restoration and hence all the issues already alluded to in this chapter. A number of trial schemes will be necessary to see just what advantages are to be had by maintaining reasonable flows, as for example in the middle Bure under salt intrusion conditions, by means of groundwater injection, abstraction variation, water charging and special management agreements. This will prove to be a lovely piece of ecology, economics and politics wrapped up in groundwater hydrology and ethics. To train people to form complementary teams for such a challenging task will be one of the more significant roles of adventurous environmental science in the late 1990s (O'Riordan, 1994).

THE NEXT STEPS

This is no substitute for representative decision-making and co-ordinated consultation amongst interests who operate within a climate of trust and good fellowship. Ecological research cannot, on its own, provide the answers needed for purposive management. That research is not readily funded even when there is a need for it. Universities and research councils rarely take an interest in applied management research even when there are innovative schemes in prospect. Local authorities are usually too strapped for resources to tinker with ecological experimentation unless there is clear payoff. So it is desirable to seek pragmatic coalitions of co-funders, where NRA, English Nature, local authorities and private institutions and landowners can join forces to produce a catalogue of research that forms an essential background to consultation and involvement by key players.

Precaution cannot be followed purely on the basis of ecological research. By definition, precaution forges ahead of models and predictive equations to secure a sense of directional purpose amongst competing interests. This can only be achieved in an atmosphere of trust and commitment to common goals. This therefore requires imaginative use of ecological economics, innovative experimentation, excellent monitoring and regular appraisal of results. The NRA needs to be ultra-sensitive to this line of approach and ready to work with various resource management partners to make it work. All the signs are that the NRA is ready for this task.

33

Ecological Management of Angling

PHIL HICKLEY, CHRIS MARSH and RICK NORTH

National Rivers Authority, Solihull, UK

INTRODUCTION

Recreational fishing is the ritual pursuit of pleasure associated with the experience (Carlton, 1975). A model presented by Hudgins (1984) identified two major components: an angling factor, including the number of strikes and the number and size of fish caught; and a recreational factor, including non-catch aspects of the trip such as social group involvement and personal satisfaction. Fishing trip satisfaction was defined by Holland and Ditton (1992) as the fulfillment of various psychological outcomes. Important aspects were a sense of freedom, excitement, catching a fish, relaxation, enjoying the natural setting and thinking about past fishing experiences, with only 6% of those questioned rating catch more important than any other factor. Good management must, therefore, facilitate the enjoyment of angling as a sport, but this must be achieved in a way that is not detrimental to the fishery being exploited. However, any measures introduced to protect fish stocks must be such that they in turn do not deprive anglers of their pleasure.

Successful recreational fisheries have two associated requirements running in parallel; management of the fish population and management of angling activity. The first is described in Chapters 23–27; the second can be achieved by the introduction and enforcement of various regulatory contraints.

This chapter describes the regulation of river-based angling in England and Wales, particularly within the Severn–Trent region of the National Rivers Authority (NRA), and discusses both the purpose and the ecological basis for the different controls. Implementation and monitoring are then considered, along with recommendations for the future.

REGULATION

The Bledisloe Report (1961) stated that there are many arguments for leaving angling free from statutory regulation. It suggests that a man's pleasure is hardly suitable to be regulated by law and public opinion in sporting matters ought to be a more powerful sanction against malpractice than resort to the courts. Howarth (1987), however, pointed out that angling has features not shared by other sports; in particular, the need to conserve stocks from over-

The Ecological Basis for River Management. Edited by D.M. Harper and A.J.D. Ferguson. © 1995 John Wiley & Sons Ltd

exploitation and to protect fish from illegitimate fishing practices by force of law. Angling, therefore, has a legal character distinct from other sports. Regulation thus appears necessary and can be applied at various levels from primary legislation through to voluntary codes of practice using a wide range of techniques.

The principal motive behind the regulation of angling activity is to control the amount of fishing effort in order to reduce the pressure of angling on fish stocks. In commercial fisheries the management target is often a maximum sustainable yield. This can also be the case for recreational fisheries where all fish caught are removed, but not for any catch—release situation. It is better to adopt a concept of maximum permissible activity, with regulatory actions aiming to protect the welfare of both populations and individual fish.

History of regulation

The need for regulation of freshwater fishing has resulted in many and varied Acts of Parliament (Association of River Authorities, 1974; Howarth, 1987) (Table 33.1). Legislation was introduced as early as 1295 when a Statute of Westminster provided a close season for salmon. Further close season restrictions plus protection for immature fish of all kinds came in 1538 and 1558. In 1860, such was the decline in status of salmon that commissioners were appointed to investigate and report. The Salmon Fishery Acts of 1861, 1865 and 1873 followed. A few years later, specific reference to freshwater (coarse) fish was made in a Bill aimed at "Putting a stop to the wanton and mischievous waste of piscine resources of our freshwater lakes, rivers, ponds, canals and streams". This led to the passing of the Freshwater Fisheries Act, 1878. It was this Act which first laid down the statutory close season for freshwater fish as 15 March to 15 June inclusive.

Coarse fish and game fish were brought into line with the Salmon and Freshwater Fisheries Acts of 1907 and 1923. More minor reforms were followed by another significant Salmon and Freshwater Fisheries Act in 1972. A major consolidation took place with the passing of the Salmon and Freshwater Fisheries Act 1975, which although modified slightly by the Salmon Act 1986, Water Act 1989 and their consolidation under the Water Resources Act 1991, continues to provide the framework for regulation of fishing and fisheries. A list of present-day legislation which can impinge to varying degrees on angling activity is given in Table 33.1.

TABLE 33.1. List of current legislation affecting anglers in England and Wales

The Diseases of Fish Act, 1937
The Theft Act 1968
The Salmon and Freshwater Fisheries Act 1975
The Land Drainage Act 1976
The Wildlife and Countryside Act 1981
The Diseases of Fish Act 1983
The Salmon Act 1986
The Public Order Act 1986
The Water Act 1989
The Environmental Protection Act 1990
The Water Resources Act 1991

Levels of regulation

Parry (1978) identifies three levels of agency for the regulation of fishing pressure on fish stocks — the law, byelaws and owner or club rules. In recent years voluntary codes adopted by individuals provide a fourth. Regulation at all levels can be general or species-specific but, as Parry (1978) warns, only regulations which are sound and comprehensible will be followed in the main by fishermen and taken seriously by the law courts.

The law in England and Wales limits fishing to licensed instruments and bans a range of methods of taking fish absolutely or conditionally. It therefore provides a national framework within which byelaws can address the detail. The Salmon and Freshwater Fisheries Act 1975 gives the purposes for which byelaws can be made. These are altering close seasons, setting size limits and the fishing distance from dams, requiring catch returns and regulating lures and baits. There is also a "catch-all" purpose of better execution of the Act and better protection, preservation and improvement of fisheries. Under rules set by fishery owners and angling clubs it is quite common for an additional range of restrictions to be imposed. Voluntary codes usually relate to good angling practice and are often proposed by umbrella organizations such as the National Federation of Anglers.

Techniques of regulation

Gulland (1971) suggests that six techniques exist whereby fishing effort within commercial fisheries can be regulated. They are:

(i)   closed areas
(ii) closed seasons
(iii) limitation of total catch
(iv) limitation of the total amount of fishing
(v)   restrictions on the type of gear used
(vi) restrictions on the sizes of fish that may be landed

In general terms, the regulatory measures used for angling can be classified in the same way. Table 33.2 summarizes which of the legislative levels can be used to apply each technique and Table 33.3 shows the relationship of the techniques to the ecological requirements of the fish stock.

TABLE 33.2.   Techniques of regulating angling effort and the level of control used

Regulatory technique	Primary legislation	Regional byelaws	Fishery rules	Voluntary codes
Closed areas		*	*	
Close season	*	*		
Catch limit		*	*	*
Amount of fishing		*	*	
Type of gear	*	*	*	*
Size of fish		*	*	

TABLE 33.3. Techniques of regulating angling effort and the ecological requirements of fish stock which are addressed

Regulatory technique	Population no.	Protect broodstock	Undisturbed spawning	Free passage	Fish welfare
Closed areas	*	*		*	
Close season		*	*	*	*
Catch limit	*				
Amount of fishing	*	*			
Type of gear					*
Size of fish	*	*			*

Closed areas

Closed areas are designed to protect stock directly by denying access to the angler. For resident populations this can provide refuge for both broodstock and young fish. For migratory fish there is the opportunity for free passage through potentially risky areas. Establishment of closed areas should be based on knowledge of the species habitat, lifestyle and spawning grounds. Closed areas enforced by law are generally for the benefit of migratory salmonids in estuaries and also where the fish congregate near to weirs (Parry, 1978). The efficacy of a sanctuary area within a fishery has been described by Hill and Shell (1975) who compared a sanctuary lake with non-sanctuary lakes to show the better spread of exploitation.

The NRA Severn–Trent regional byelaws prohibit fishing for salmon for fixed distances above and below certain weirs on the River Severn. The apparent decline in salmon catches in recent years (Churchward and Hickley, 1991) provided the scientific base for extending the number of closed areas from just Shrewsbury weir to eight major weirs when the byelaws were revised in 1991.

Closed seasons

The ecological basis for the imposition of closed seasons is that of allowing the uninterrupted reproduction of fish including, for migratory fish, free passage to spawning grounds. In theory, decisions affecting the timing and duration of closed seasons should rely on a thorough knowledge of the spawning habits of the species concerned and retain a degree of flexibility in their application.

The Salmon and Freshwater Fisheries Act 1975 imposes a duty to make byelaws fixing seasons for salmon and trout, other than rainbow trout, and minimum durations are prescribed. For freshwater fish, the Salmon and Freshwater Fisheries Act 1975 grants the power to dispense with the close season altogether but, if in place, it must be 93 days. Also, unless modified by byelaw the period is 15 March to 15 June inclusive. Eel fishing is prohibited under the Salmon and Freshwater Fisheries Act 1975 unless specifically authorized by byelaw. Table 33.4 shows the number of NRA legislative regions (ten in total although for administrative purpose there are 8) which have close seasons in place for the various species.

TABLE 33.4. Number (out of 10) of legislative NRA regions with close seasons of various kinds

Fish type	River	Canal	Pool
Salmon	10	N/A	N/A
Migratory trout	10	N/A	N/A
Brown trout	10	N/A	9
Rainbow trout	10	N/A	2
Freshwater fish	9	8	5
Eels	7	5	3

Parry (1979) argued that in fisheries where the catch is removed there is probably no real justification for taking fish in one part of a season as against another. If a certain number of fish are going to be caught in a year, it does not really matter from the point of view of the fish stocks when this occurs. He stated that the effect of a close season for species that are taken for human consumption, however, is to ensure that the fish are exploited at a time of year when they are most fit to eat and flesh quality has not suffered as a consequence of gonad development.

In coarse fishing in the UK the catch is almost invariably returned to the water. There is a need, however, to consider whether fish are harmed by capture and handling at a time when they could be already stressed by rising water temperatures and the onset of spawning activity. In addition, the close season for freshwater fish coincides with breeding of waterside birds and mammals and this should, perhaps, be deemed important (Tydeman, 1977). On the need for a freshwater fish close season, the Association of River Authorities (1974) suggested that varying the duration might provide a satisfactory compromise in certain cases. The Association also stated that, where hook and line eel fishing is allowed, it is difficult to ensure that freshwater fish are not caught just as readily. The National Federation of Anglers (1989) reported the results of 213 member clubs answering a detailed questionnaire. Irrespective of any scientific evidence appertaining to spawning or the ultimate survival of coarse fish there was an almost unanimous view in favour of retaining some form of close season, 75% of replies wishing to retain the present close season of 93 days from 15 March to 15 June. Whilst protection of fish and maintenance of the overall fish population was obviously the main reason for the close season, 90% put the conservation of wild riverside plants, mammals, birds and aquatic animals and plants as a major reason for setting a close season.

It is often argued by critics of the close season that the timing is wrong, especially in the case of that for freshwater fish. Table 33.5 gives details of the spawning times for selected coarse fish in the NRA Severn–Trent region as determined from collections of brood fish for Calverton fish farm (Nottinghamshire). All dates listed fall within the specified close season period. With regard to salmon in the River Severn, times of migration through the fish counter at Shrewsbury were used as justification for extending the salmon fishing season by one week when byelaws were revised in 1991.

Limitation of total catch

Limits are placed on total catch to prevent over-exploitation of stock and thus maintain population numbers, the concept being similar to maximum sustainable yield of commercial

TABLE 33.5.   Dates on which different species of coarse fish were taken from rivers in the Severn-Trent region to Calverton fish farm for spawning. Fish were either ovulating or within a week of doing so

Year	Dace	Chub	Barbel	Roach
1988	3 April	10 May	19 May	—
1989	31 March	17 May	10 May	—
1990	15 March	4 May	2 May	—
1991	14 March	9 May	9 May	26 April
1992	18 March	6 May	14 May	30 April
1993	18 March	10 May	4 May	2 May

fisheries. In addition, such regulation allows for the sharing of catch of migrant fish or where there is otherwise restricted stock. In order to set sensible limits, details of population size, mortality rates and exploitation levels should be known.

Catch limits are not necessary for coarse fish in the UK as it is customary for all fish caught to be released. However, should this tradition not be maintained, restrictions on take would be essential. As an example of how fast coarse fish populations could be denuded, Hickley (1980) demonstrated that up to 60% (29.5% average) of the standing crop of takeable size fish would be landed during an angling contest on the River Eden, Kent.

Bag limits are commonly applied on a daily or visit basis and mainly to game fish which either breed locally or are stocked artificially (Parry, 1978). However, the disadvantage for anglers is that on what might be their one good day of the season they have to cease fishing. Mills (1979) claimed that if there were to be no restrictions of this sort there is no doubt that stocks of game fish could be severely reduced. Catch limits are usually set and enforced by angling clubs and fishery owners but NRA South West Region imposes a bag limit on migratory salmonids.

An alternative to a bag limit is to return all fish alive. In an independent review entitled "Angling" (Anon., 1991) several conclusions were drawn. Whilst it is common for trout fisheries to set limits on the daily catch, either to conserve stocks of wild fish or to balance the input to stocked water, catch limits are increasingly being applied by fishery proprietors in an effort to conserve spawning stocks of migratory fish. In addition, catch and release has become more common in game fishing, although the spread is slower than in North America or mainland Europe. Noteworthy, however, is a statement that, for catch and release to be successful, it is important that proper procedures are followed with regard to playing, handling and release.

Limitation on amount of fishing

The control of the amount of angling that takes placed is based on similar ecological criteria to catch limits. The technique aims to protect stocks of fish by reducing the opportunity for fish to be caught, thus enhancing escapement of broodstock and reducing pressure on catch−release fisheries. Two aspects of the latter are important. First, there is the potential for frequently-angled fish to learn a degree of hook avoidance which disrupts sport (Raat, 1985) and secondly, mortality can be induced by repeated angling damage (Cooper and Wheatley, 1981).

Restrictions on the total number of anglers or their spacing along the bank are often used at owner or angling club level, sometimes for stock or habitat conservation and sometimes to preserve an enjoyable freedom from interference (Parry, 1978). In addition, byelaws within the NRA Severn–Trent Region prohibit the use of more than two fishing rods at the same time.

On occasion it would be of significant management benefit to place a limit on the overall number of anglers exploiting a stock of migratory fish. However, whilst the Salmon and Freshwater Fisheries Act 1975 permits the number of netting licences to be limited for a period of 10 years, this facility does not exist for rods. Licence price can offer some degree of indirect limitation, but this control disappeared with the recent issue in 1992 of a single national rod licence covering all species (although a separate salmon and migratory trout licence has now been re-introduced for 1994).

Restrictions on type of gear used

Gear specification is used to reduce exploitation of populations by influencing both the efficiency of angling and the species selectivity of different fishing methods. There is also the requirement to protect individual fish by reducing any stress and physical damage associated with being caught and retained if the intention is to return them alive. It is necessary to possess sound information on fish lifestyle and the effectiveness of different angling techniques in order to develop appropriate constraints.

There is absolute prohibition of a number of ways of catching fish contained in the Salmon and Freshwater Fisheries Act 1975. Thus, the law is capable of directing people away from catching fish in certain ways and by implication, towards catching them in other ways (Parry, 1979). For example, it is lawful to hook a fish in the mouth but not to snatch it by foul hooking (Parry, 1978). The objective of restrictions on baits and lures is to confine capture to recognized "fair" means as being the favoured way of exploiting the resource. Gear may be banned either because it is too efficient for the target species or because it may endanger others.

The Salmon and Freshwater Fisheries Act 1975 calls for the NRA to regulate fishing by the issue of licences for specific instruments and specific fish. In the anglers' case this is a rod and line for salmon, trout, freshwater fish and eels. The validity of the licence is necessarily dependent upon compliance with the Act and any regional byelaws.

Restrictions imposed by the Salmon and Freshwater Fisheries Act 1975 itself include the prohibition on the use of barbed gaff, otter board, stroke haul and snatch; in other words, methods that could not be deemed fair angling. Section 30 of the Act prevents introduction of fish without consent and this is especially important with regard to livebaiting and the associated risk of disease transfer.

Under NRA Severn–Trent byelaws, various lure, bait and gear restrictions give important support to close season regulations in aiding the protection of certain species. For example, during the close season for freshwater fish, persons angling for salmon and trout are restricted to conventional baits; persons fishing for eels cannot use a hook less than 12.5 mm gape and the use of floats and keepnets is banned. Gear specification under byelaw can also aid the welfare of individual fish. An example is the improvement in keepnet design. The NRA Severn–Trent Region banned the use of knotted mesh keepnets after December 1981 and at the same time specified a minimum requirement based on "state of the art" manufacturing. None the less, further consideration might be required as the

National Federation of Anglers has sponsored a scientific investigation to assess whether retention in keepnets of modern design is harmful to fish due to stress.

At either club level or on a voluntary basis, anglers can do much toward the protection of populations and individual fish. The "fly fishing only" rule, which is often imposed on trout rivers, not only reduces exploitation and fulfils any purist desires but also protects against accidental capture of under-sized trout and salmon parr. For catch — release fisheries the use of barbless hooks is becoming more common. Also, fewer pleasure anglers are retaining coarse fish in keepnets; instead releasing them immediately.

Restrictions on size of fish taken

The Association of River Authorities (1974) stated that the primary aim of a size limit control is to ensure that a particular species of fish is not removed permanently from the water until individuals have attained a length at which they are considered to have matured and thus to have spawned and reproduced themselves at least once. The Association also noted that individual size limits may have been perpetuated through successive byelaw revisions without particular attention to their scientific and practical relevance. For size limits to have a proper basis, detailed information should be available on population size structure, fish length at different ages, fish length at sexual maturity and natural mortality rate.

Under the Salmon and Freshwater Fisheries Act 1975 it is an offence to knowingly take, kill or injure (or attempt to do so) any fish which is immature. Therefore minimum size limits serve a practical purpose in aiding the angler in the interpretation of what is meant by immature. In coarse fisheries, size limits for retention of fish in keepnets can reduce damage and improve post-release survival.

One problem with a minimum size limit is that, in preventing removal of non-breeding fish and increasing survival of the protected size, there is increased exploitation of broodstock. A second may be that the size limit can result in the slowest-growing fish being the ones that breed, passing on slower growth characteristics such that in time size limits have to be readjusted downwards (Parry, 1979). Worthy of attention, therefore, is the combined use of minimum and maximum size limits which would allow exploitation of a mature, abundant size group whilst protecting the broodstock population.

NRA Severn — Trent byelaws specify takeable sizes for brown trout and rainbow trout where they spawn naturally. Even with a good scientific base it is remarkably difficult to set an appropriate takeable size. None the less, with knowledge gained from field surveys the size limits were adjusted when the byelaws were reviewed in 1991.

IMPLEMENTATION

Regulations required for angling activity need to be implemented once established. Where the law has a part to play, the role of enforcement is critical. Education of individuals and user groups is very important also however, particularly but not exclusively, where legal powers are lacking. The more that regulatory controls can be seen to be for genuine protection of stocks and the stronger the ecological basis of these controls, the greater the likelihood of successful implementation.

Enforcement

Until 1827 the use of spring guns and man traps was permitted and if a poacher walked into such a device the owner was not held responsible (Parry, 1979). Although this facility has now been lost, the Salmon and Freshwater Fisheries Act 1975 does give bailiffs considerable powers. They can examine instruments and baits, require production of licences and initiate the prosecution of offences. Anti-poaching, spawning protection and close season patrols are carried out. Licence-checking activity facilitates the additional benefits of byelaw enforcement and liaison. The Salmon Act 1986 requires the licensing of salmon dealers and provides for investigation of the handling of salmon in suspicious circumstances. Under the Theft Act 1968 riparian owners can take action against anglers fishing without consent, as under a special clause of this Act it is unlawful to take or destroy, or attempt to take or destroy, any fish in water which is private property. Behrendt (1977) suggested that the possibility of expulsion from a club or fishing organization can be a useful deterrent and that the onus is on the fishery manager to explain that rules are there to protect the fishery, which in the end makes for better fishing.

Education

Education should aim for a meeting of minds between scientists, fishery managers and anglers. The NRA, charged with the duty to "maintain, improve and develop fisheries", should work with anglers to balance the needs of fish and sport.

Educational methods can vary. Most NRA regions will offer fisheries management advice, covering topics such as stocking, weed control and habitat improvement. As individuals, anglers can improve their understanding of fish ecology by turning to appropriate books (e.g. Pratt, 1975). Angling is becoming more of a target for anti-fieldsports campaigning, increasing the need for angling organizations to issue Codes of Practice, especially in relation to the welfare of fish and wildlife.

In many cases, there is a need to change anglers' perceptions which are often entrenched on the grounds of heresay rather than science. For example, restocking remains a popular cure-all for coarse fisheries without clear scientific support, rather than, for example, habitat improvement. Scientific data can be collected and used to win arguments, such as the popular notion held by game anglers that on waters inhabited by salmon parr the use of float-fished maggots by coarse anglers should be banned. Research by North (1983) showed that there was no biological justification for this and that the worm fishing used by the game fishermen caused greater mortality.

Behrendt (1977) described the type of fishery manager who, finding a loophole in the law, extends their fishing season merely to make extra money. A good example is the willingness of fishery owners in the Midlands to promote any method of trout fishing so as to tempt anglers to keep fishing during the freshwater fish close season. Dill (1978) identified an increasing pattern of greed, with more anglers competing for trophies or money together with the promotion of angling as a basis for business. The introduction of zander into the Midlands (Hickley, 1986) was illegally carried out by a few anglers looking after their own desires without regard for others. These types of selfish behaviour have to be countered at every opportunity.

MONITORING

The better that anglers and their fishing success are monitored, the easier it becomes to tailor regulation to real needs. The least productive method is the receipt of unsolicited complaints. Wherever possible, anglers should be encouraged to keep records. Malvestuto (1991) detailed the principles behind the establishment of creel surveys, which include such techniques as telephone interviews, questionnaires, use of log books and riverside inspections of catches. Major surveys of angler catches have been carried out on the River Trent by Cowx and Broughton (1986) and on the River Severn by North and Hickley (1989). For River Severn salmon, catch returns are required by NRA Severn−Trent byelaw (Churchward and Hickley, 1991).

ALLIED ISSUES

Management of the angling environment to create the right conditions to satisfy the angler, beyond merely providing the fish population, is a necessary ingredient for a successful recreational fishery. The requirement for good access, car parking, provision of fishing platforms and so forth, should be addressed with regard to ecological considerations above and beyond the river channel. As experience of spiritual or existence values is one of the most prized conditions of being human (Kellert, 1984), a conservationist's overview of protection of the landscape is bound to enhance the long-term enjoyment of angling as a recreation.

One blight on the environment is litter, but it need not be so. Concern over discarded fishing line resulted in the National Federation of Anglers and the fish tackle trade producing a "nylon line and litter" code and labelling packaging with a "protect wildlife, take line home" logo (Anon., 1991).

A conservation issue of major importance in recent years was the poisoning of swans by anglers' lead weights. In the end the power to specify fishing gear was used, albeit for the protection of swans and not fish, when all NRA regions adopted a byelaw prohibiting the use of certain sizes of lead weights for fishing.

THE FUTURE

The fisheries resource of England and Wales has to be managed to achieve a balance between the sustained development of recreational and commercial fishing on the one hand and conservation of fish populations on the other. Parry (1978) regarded it as optimistic to expect that regulatory measures could be a complete force in controlling the stock level of any particular species, but if none were applied to our sport fisheries there is no doubt that they would be in jeopardy. He concluded that the objective for the future should be to use regulations in as sensitive manner as possible for the application of scientific knowledge, and to be as sparing in their imposition as is compatible with preserving the ethic of conservation of the stock for wise and acceptable use. Such a sentiment is still valid today.

Wherever possible the introduction or variance of regulatory measures should be based on sound ecological principles. However, there is often a lack of reliable quantitative information available for the purpose. There are many recently commissioned NRA national R & D projects on such subjects as fisheries classification, survey methodology, catch statistics, sea trout and coarse fish biology, the outputs of which will assist the regulators in future.

Guiver, writing in 1973, asked whether, with the advent of Regional Water Authorities on 1 April 1974, there was a good case for reviewing existing legislation to determine whether some rationalization was desirable. Severn Trent Water Authority went on to combine byelaws for Severn and Trent and now the NRA has identified a need for reviewing byelaws in all its regions, but already it is 20 years after Guiver's (1973) suggestion. The most urgent requirements are to standardize the coarse fish close season (currently being addressed), the specifications for keepnet design and the number of rods that can be used at any one time. In the long term, the general requirement for close seasons would be better served by their setting becoming a legislative power but not a duty, with a removal of restrictions on duration. Such a facility would be more in keeping with a philosophy that regulation should be based around ecological fact.

Education and liaison with fishermen, fishery owners and the general public should be improved so they can more closely identify with the management of the fishery resource. Efforts should be made to promote good angling practice through regulation at all levels including byelaw reforms and user group voluntary codes. More education, concentrating on the sustained enjoyment of the sport and the desire for conservation, should reduce the tendency for a minority to be motivated by greed. Regulations could be reduced if a few people did not set out to break the spirit of the law. For example, all restrictions relating to bait and gear during the freshwater fish close season could be reduced if anglers were willing to comply properly; a simple "don't fish for coarse fish" should be enough.

In general, there is scope for the continued maintenance of an appropriate balance between exploitation and protection of stocks. The NRA for its part, however, must continue recent efforts to re-establish the organization's reputation with the angling community, which has been previously eroded by poor handling of national licensing and the potential "Section 142" income scheme (NRA, 1992h). On the anglers' side, a speedy return to the caring attitudes which give anglers a good reputation as conservationists (National Federation of Anglers, 1989) would be welcomed.

ACKNOWLEDGEMENTS

The authors thank their colleagues for assistance in gathering information. The views expressed are those of the authors and not necessarily those of the NRA.

34

Ecological Impacts and Management of Boat Traffic on Navigable Inland Waterways

KEVIN MURPHY

University of Glasgow, Glasgow, UK

NIGEL J. WILLBY and JOHN W. EATON

University of Liverpool, Liverpool, UK

INTRODUCTION

Man has used and adapted rivers for navigation for thousands of years. Inland water transport was a well-established feature of ancient oriental civilizations, both on rivers and on artificial connecting channels dug mainly for irrigation and land drainage. In Europe, the larger rivers, such as the Danube, have been important trade routes for many centuries. About 1000 km of non-tidal British rivers were navigable in their natural state and from the 14th century onwards river engineering gradually extended this total, reaching 2100 km by the start of the canal building era in 1760 (Skempton, 1984). With the stimulus of the Industrial Revolution the system then expanded rapidly, mainly by the construction of artificial channels, to reach its zenith in about 1840, when over 6400 km of interconnecting canals and river navigations were in use.

Le Cren (1972) suggested that the major effects of navigation on river ecosystems are associated with canalization, dredging and flow regulation. In these respects river management for navigation uses techniques already in place to support water supply, land drainage or flood control functions, so serving multiple objectives. Until the 19th century, boats were small and moved by human, horse or wind power, so their passage probably caused little disturbance to the rivers. Only with the advent of larger, motorized craft in the last 150 years has the actual movement of the boats become a major additional effect.

To keep down construction costs, artificial channels were dug as small as possible in water cross-sectional area. In these confined dimensions, the effects of boat traffic are substantially magnified, even though the size of craft using the channel may be small. Artificially

---

The Ecological Basis for River Management. Edited by D.M. Harper and A.J.D. Ferguson. © 1995 John Wiley & Sons Ltd

constructed navigation canals originally designed for horse-drawn traffic, together with some drainage channels, are now used extensively by motorized boats in Britain and elsewhere. These waterways differ from rivers in a number of important features. Their uniform cross-sections offer lower habitat diversity than all but the most rigorously engineered lowland rivers. Periodic scouring floods are absent and flows are very slow, especially in canals built specifically for navigation, where the main aim is to retain rather than convey the water. Most lengths are uniformly depositional habitats and, despite their outward appearance, are much more like linear ponds than rivers. Furthermore, canals extend over watersheds and hence have opened up water connections between river catchments, thereby increasing opportunities for dispersal of native and introduced species of waterplants (Willby and Eaton, 1993; Simpson, 1984) and aquatic invertebrates (e.g. Malacostraca: Holland, 1976; Pygott and Douglas, 1989; molluscs: McMillan, 1990; oligochaetes: Kennedy, 1965), as well as creating a habitat type new to some of the areas in which canals were constructed (Hanbury, 1986).

In Britain the original freight transport function of inland navigation had disappeared on all but a few of the largest waterways by the mid-20th century, as rail and later road transport deprived them of traffic. Many former river navigations were re-engineered for land drainage only. Some canals were filled in and lost, while others retained water but became derelict. By 1957, usable river navigations had decreased to 1696 km and canals to 2397 km (Economic Commission for Europe, 1958). Surviving navigations did, however, start to be used by increasing numbers of recreational boats, developing a strongly seasonal type of traffic previously mainly confined to the Norfolk Broads, and a few rivers and lakes (Thames Water Authority, 1980; Price, 1977; Broads Authority, 1982; see Figure 34.1).

Recreational traffic continued to increase until the late 1980s, in some cases bringing traffic densities on inland waterways, back to or even beyond those experienced during the heyday of waterborne freight transport in the 19th century (Stabler and Ash, 1978; Murphy, Eaton and Hyde, 1982; Adams et al., 1992). Demand for additional facilities led to the restoration of boat traffic on about 160 km of rivers and 480 km of canals and drainage channels during the period 1955–92 (Murphy, 1992; Inland Waterways Association, 1993), and further reconstruction schemes are currently in progress or planned.

Recreational boating has therefore greatly increased its effects on waterways in Britain, in both intensity and geographical extent. If the present momentum of restoration schemes is maintained, effects are likely to be further extended in the future.

A large international technical literature has developed on the direct and indirect effects of navigation on channel ecology. Direct effects of boat movement have been reviewed by Jackivicz and Kuzminski (1973), Liddle and Scorgie (1980), Allen and Hardy (1980), Nelson (1982) and Wright (1982). Adams (1993) assessed environmentally sensitive predictors of boat traffic loading on inland waters. Both the direct effects and indirect influences through channel and flow modifications were included in an annotated bibliography of 452 papers compiled by Pearce and Eaton (1983); and more recently, Brookes and Hanbury (1990) have reviewed the subject specifically in Britain, with special reference to bank protection techniques.

In this chapter we consider the nature of boat traffic effects on channel habitats and consequent influences upon components of the aquatic ecosystem, concentrating only on representative, mainly recent literature. We then identify likely future trends, management options and research needs in Britain.

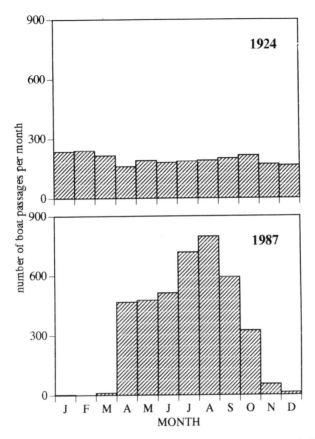

FIGURE 34.1. Annual patterns of boat movements in 1924 and 1987 at Bank Newton Top Lock, Leeds and Liverpool Canal

EFFECTS OF NAVIGATION ON HABITAT

River engineering for navigation

Adequate depth and controlled flow are the main aims of navigation engineering. These are achieved by a combination of dredging, construction of weirs and locks and, where the natural course of the river is tortuous, the insertion of artificial cut-off channels (Skempton, 1984). The riffle−pool sequence, if previously present, is replaced by a predominantly depositional state, in all except high flood, often requiring regular local dredging to maintain the desired depth. Formerly large seasonal variations in water level are replaced by a new, fairly high, mean operating level, exceeded in flood periods but otherwise exhibiting only small, cyclical changes when locks are used. Prolonged dry weather may result in very slow flows verging temporarily on lacustrine conditions, but only in extreme drought will water levels fall.

Overall, habitat diversity is reduced both spatially and temporally, though local increases in channel complexity are created by cut-offs, weirs, lock bye-channels and tail waters.

Effects of boat traffic on habitat

Physical effects

Boats may pollute freshwaters through their inputs of physical energy as well as by releasing chemical substances.

Kinetic energy dissipated *via* boat wash and/or propellor churning may cause bank erosion to rise above rates attributable to natural causes and wind-induced wave action (Zabawa and Ostron, 1980; Garrad and Hey, 1988a); especially when channels of small cross-sectional area are navigated by powered boats. In some heavily trafficked Broadland rivers, bank erosion rates may reach 0.3 m per year (Broads Authority, 1987). Fine, unconsolidated sediment is thereby redistributed across the channel bed. As this accumulates and is supplemented by allochthonous sources of alluvium, it becomes increasingly susceptible to resuspension within the water column each time a boat passes, making the channel bed habitat very unstable (Karaka and van Hofen, 1974; Williams and Skove, 1981; Smart et al., 1985). Increased suspended solids loading results in reduced water clarity due to light scattering and absorption by the suspended particles (Kirk, 1985). The periodic resuspension increases transport rates of sediment both downstream in the mass flow and laterally into backwaters, where present, causing the latter to infill with unnatural speed (Smart et al., 1985).

Most inland navigations are hydraulically "constraint" channels, i.e. small enough for the size and shape of the waterway to be a significant influence on the progress of the craft, as compared with deep lakes or the open sea where, apart from weather conditions, craft features alone determine movement characteristics. The amount of disturbance caused by an individual boat passage therefore depends on a range of variables, principally the size and shape of the hull, the speed of the craft and the cross-sectional dimensions and flow in the channel. Research, mostly on large craft in ship canals, has shown that the hydraulics of such movements are very complex and currently defy detailed quantitative analysis. Nevertheless, it is clear that resistance and hence disturbance rise steeply with increasing blockage factor (the ratio of submerged cross-sectional area of hull to water cross-sectional area of channel) and with craft speed, especially once those speeds are reached at which either a breaking wave is generated or the hull has tilted back sufficiently for the stern to be running near or on the channel bed (squat effect). It is also clear that the shape and total displacement of the hull and the speed and depth of the propellor are further major influences on the magnitude of the actual disturbance created (e.g. British Transport Docks Board, 1972; Dand and White, 1977).

In navigable systems a small part of the energy used to propel a boat is converted to sound energy *via* movement and vibration of engine components leading to noise pollution, the effects of which may be amplified underwater (Boussard, 1981).

Chemical effects

Powered boating produces direct chemical changes to water quality by adding fuel combustion products and a variety of other pollutants, but also causes indirect changes as a consequence of physical disturbance of water and sediment.

There is an extensive literature on the effects of pollution by fuel leakage and boat-engine emissions (hydrocarbons and combustion gases) which may themselves contain significant amounts of unburned fuel (e.g. English, McDermott and Surber, 1963; Kempf, Ludemann and Pflaum, 1967; Jones et al., 1980). Individual bioassay studies have shown that outboard motor emissions may be quite harmful to fish and invertebrates (Surber, 1971; Brenniman et al., 1979; Swanberg and Tarkpea, 1982). However, the extensive assessment of the technical literature by Pearce and Eaton (1983) suggested that while hydrocarbon-related pollution from boat traffic may be of local importance, usually in enclosed marinas, the overall risk which it poses to waterway ecosystem functioning is usually quite low. Less tangible effects include the visual detraction of thin oil films on the water surface. These appear to originate more often from road drainage entering at bridges than from boats and may be of little ecological consequence. Prior to the introduction of lead-free fuel, there was concern that boat engines might also be a source of lead pollution in navigated waters (e.g. Kuzminski and Mulcahy, 1974; Byrd and Perona, 1980).

Discharge of sewage from boats is a potential source of organic loading and bacterial contamination of the waterways system (e.g. Lear, Marks and Schminke, 1966; Faust, 1982). In Britain, however, sewage discharge from boats using inland waterways has been prohibited for some years under local byelaws and most boats are now fitted with sewage-holding tanks which are pumped out to land-based receivers for disposal via public sewerage systems.

Concern has mounted in recent years over the environmental effects of anti-fouling paints applied to the hulls of sea-going boats (Simmonds, 1986; Laughlin and Linden, 1987). In particular, one common active ingredient, tributyl-tin (TBT), is a membrane toxicant which produces well-documented lethal and mutagenic effects at extremely low doses in freshwater as well as marine organisms (Sarojini, Indira and Nagabhushanam, 1990), and may undergo trophic accumulation during chronic low level exposure, especially in filter-feeding shellfish (Ebdon, Evans and Hill, 1989). Normally perceived mainly as a marine issue, the danger to freshwater ecosystems has not been properly assessed, despite the potential local risks on ship canals and lower reaches of river navigations used by sea-going vessels. Marine studies suggest greatest risk where craft painted with TBT-containing substances are left moored for long periods, or dry-docked for scraping and repainting (Bellinger and Belham, 1978), especially in enclosed harbours or marinas from which contaminants are slow to disperse (Langston, Burt and Zhon, 1987). These threats may now be diminishing due to improved paint formulations, the wider dissemination of guidance on good working practices, and legislation which, since 1987, has banned the use of TBT-based paints on craft under 25 m. However, monitoring of potential bioaccumulators, such as unionid bivalves, from TBT-susceptible freshwaters, remains desirable to determine the extent of residual impacts.

There does not yet appear to have been any environmental impact assessment of the possibility that galvanized sheet steel trenching, used extensively for bank protection, may leach zinc, with potentially toxic effects on sessile organisms and linked higher trophic levels. There is a need for work to determine the risk associated with zinc leading into freshwater systems.

Disturbance of the channel bed by boat passages may produce lesser known, indirect chemical changes. These may be beneficial to invertebrate fauna, as for example, if sediment oxygen demand is reduced by aeration and mechanical comminution and dispersal of particulate organic matter, or they may create stresses through, for example, increased nutrient cycling (e.g. phosphate: Yousef, McLellon and Zebuth, 1980) or remobilization of industrial contaminants such as heavy metals, pesticides and PCBs (Munawar, Norwood and McCarthy, 1991) adsorbed onto particulate organic matter (Weber et al., 1983). The ecological impact of resuspended contaminants is, however, unclear. Mixed effects have been reported, including both stimulation and inhibition of phytoplankton production in the field (Munawar, Norwood and McCarthy, 1991) and variable toxicity in bioassays (Santiago et al., 1993), suggesting that resuspended contaminants do not always enter the trophic web. Clearly further research is needed to define overall effects.

ECOLOGICAL CONSEQUENCES

Aquatic vegetation

Boats affect aquatic vegetation in four main ways.

(i)   Direct physical damage is caused by propellers and contact with moving hulls (Cragg et al., 1980; Liddle and Scorgie, 1980). Zieman (1976) described the damage to seagrass beds in Florida Bay due mainly to the severing of rhizomes by propellors, but in general physical damage to macrophytes by boats is rarely quantified. Nymphaeids and tall-growing elodeids appear to be most vulnerable (Murphy and Eaton, 1983) although recovery may be rapid when boats are excluded (Eaton, 1986). Direct damage to submerged and floating-leaved plants is probably most important at low traffic densities, above which turbidity increases become the main influence (Murphy and Eaton, 1983).

(ii) Boat-generated waves and currents cause physical damage and uprooting (Schloesser and Manny, 1989; Vermaat and de Bruyne, 1993). Intermittent pulsed disturbance due to boat passage and lockages in otherwise sluggish waters may lead to washout of free-floating plants such as duckweeds (Lemnaceae), uprooting of lightly anchored non-rhizomatous species, and mechanical damage by drag and tearing on more firmly-rooted species, such as tall-growing, fragile elodeids and nymphaeids which lack adaptations to periodic high flows. Marginal reed-beds have a limited resilience to boat-wash, beyond which they are damaged by stem breakage and erosion and fragmentation of the root-mat (Bonham, 1980; Garrad and Hey, 1988a).

(iii) Eroded and resuspended sediment shades submerged plants (Westlake, 1966; Tanner, Clayton and Wells, 1993) and at very high levels may cause abrasion damage to plant tissues (Edwards, 1969). The importance of shading due to settlement of particles on exposed leaf surfaces is uncertain. Turbulence which accompanies resuspension or continuous flow in rivers may prevent stable deposition at points within the water column. Shading effects caused by boats moored for long periods in shallow marginal areas may also restrict the development of aquatic vegetation (Vermaat and de Bruyne, 1993).

(iv) Macrophyte establishment and spread are likely to be inhibited by soft, accreting, unstable and periodically resuspended layers of sediment. Jones (1943) attributed the absence of plants from the River Rheidol in Wales at least partly to the soft, unstable nature of the bed although mine wastes, not boats, were the source of the problem in that instance.

In the canal system controlled by the British Waterways Board, Murphy and Eaton (1981, 1983) demonstrated a significant negative relationship between boat traffic and quantity of vegetation, which included a major loss of vegetation from canals with annual traffic of 2000 movements ha^{-1} m depth^{-1} [my] due to the mechanisms outlined above. Above this traffic density there was also a sharp increase in suspended solid loading, providing a significant positive relationship between water suspended sediment load and boat traffic density. These findings therefore suggested an interaction between boats, bank erosion problems and aquatic vegetation, mirroring the results of studies elsewhere (Anderson, 1975; Liou and Herbich, 1976; Gucinski, 1981).

At low traffic densities, the reduction in biomass in canals is due mainly to the suppression of a few fast-growing, competitively dominant elodeids and reedswamp species such as *Glyceria maxima* by an intermediate level of disturbance. Consequently aquatic plant communities in lightly trafficked waterways are often diverse, sometimes including scarce species such as *Potamogeton compressus* or *Luronium natans* (Willby and Eaton, 1993) and can therefore be of considerable conservation value, despite their artificial character (Hanbury, 1986). As traffic increases further, rising turbidity compresses the euphotic zone into the upper $0.1 - 0.2$ m of water. Combined with physical damage and bed instability, this reduces the maximum rooting depths of fringing vegetation, causing the whole hydrosere to retreat towards the margins. The most trafficked canals are often completely devoid of aquatic plant cover and problems of erosion may then be so severe that expensive sheet steel or concrete bank hardening is introduced to protect banks from boat wash erosion.

In some large navigable systems with eutrophic water and long retention times, phytoplankton blooms are a separate cause of turbidity which may contribute little to the suspended solids load. The Norfolk Broads are an example where algae are the largest source of turbidity (Moss, 1977; Hilton and Phillips, 1982), yet at least in the connecting rivers, the usual daily cycle of boat-induced suspended solids still occurs (Garrad and Hey, 1988b). The loss of macrophytes is attributed by Moss (1986) to interference with the growth of submerged plants by phytoplankton shading, as well as the decline of reed fringes through coypu grazing and failure of regrowth to withstand natural or boat-related erosion due to an unstable morphology induced by high nitrate concentrations in the water.

There is no evidence that aquatic macrophytes are directly affected by chemical pollution from boats (Jackivicz and Kuzminski, 1973; Dietrich, 1974).

Invertebrates

The above- and below-ground parts of aquatic macrophytes, depending on their architecture, provide invertebrates with an effective anti-predator refuge (Crowder and Cooper, 1982), protection from water movement (Harrod, 1964), emergence and oviposition sites (Rooke, 1984), an indirect food source by providing attachment sites for epiphytic algae (Cattaneo, 1983), and ultimately a source of detritus for benthic organisms during senescence (Engel, 1985). The progressive reduction in macrophyte cover with increasing boat traffic therefore removes an important resource for invertebrates (Murphy and Eaton, 1981) and increases the reliance of benthic detritivore food chains on material of allochthonous origin (Hynes, 1960). High suspended solids loading may also reduce epilithon food quality and biomass by shading (van Nieuwenhuyse and La Perriere, 1986; Davies-Colley et al., 1992), leading to a lowered phototrophic component, and reduction in invertebrates which feed by scraping algal films or grazing benthic algae.

At high traffic densities invertebrates probably respond directly to increased turbidity and water movement. There is an extensive literature on the effects of suspended solids on aquatic organisms, dealing mainly with the effects of solids from mining, deforestation or soil erosion, but very little on the effects of turbidity caused by boating. It is of limited value to draw parallels with other studies reporting suspended solids concentrations typical of the range found in navigable canals and rivers, unless the size and nature of the particles making up the suspended load are known to be comparable. Whenever suspended solids concentrations are raised above background levels as a result of boat-induced resuspension, measurements based on samples collected at or near the surface will greatly underestimate the concentrations to which benthic invertebrates and bottom-dwelling fish are subject. The characteristics of suspended solid loading in canals and rivers, namely chronic exposure to low amounts, increasing to moderate amounts throughout the day, interspersed with brief very high pulses that decline rapidly, may also be in complete contrast to the conditions under which other studies have been made, especially those based on laboratory trials.

Where there are no direct studies of boat-induced increases in suspended solids loads, cautious extrapolation from work in other fields suggests several direct effects (Ward, 1992). Respiratory structures may be clogged or damaged by abrasion although some stream insects may be tolerant of suspended material loads of up to 2000 mg litre^{-1} for 48 h without adverse effect, despite heavy accumulation on the gills (Gerisch and Brusven, 1982). Feeding rates and efficiency may decrease due to nutritional dilution of suspended organic matter by inert solids and saturation of filter-feeding structures in benthic detritivores (Gray and Ward, 1982) and planktivores (McCabe and O'Brien, 1983; Hart, 1986; Kirk, 1991). Aldridge, Payne and Miller, (1987) exposed three species of filter-feeding unionid mussels to frequent intermittent disturbance of silt to simulate the effects of sediment resuspension during boat passages and found reduced food clearance rates and, due to starvation, a lowered metabolic rate and switch to stored body reserves. The rate of macroinvertebrate drift may increase (e.g. Ciborowski, Pointing and Corkum, 1977), either as a behavioural mechanism for avoidance of the previous stresses (White and Gammon, 1977), as an intrinsic response to darkening of the bed (Brittain and Eikeland, 1988), or due to substrate mobility (Culp, Wrona and Davies, 1986) as caused by boatwash (Seagle and Zumwalt, 1981).

Despite these potential effects, several authors have concluded that invertebrate communities are fairly resilient to turbidity per se and that the greatest damage occurs during subsequent siltation (e.g. Cline, Short and Ward, 1982; Ward, 1992), or as a result of lowered bed permeability due to infilling of interstices (e.g. Quinn et al., 1992). Other deleterious effects may be due to accumulation of shifting, unstable deposits (e.g. Nuttall, 1972) which may smother the hard surfaces required by sedentary filter feeders, or cause reduced epilithon food quality for invertebrate scrapers due to settlement or entrapment of fine silts and clays in the biofilm and detritus that collects on stone surfaces (Graham, 1990; Davies-Colley et al., 1992). Even small increases in turbidity therefore generally result in decreased invertebrate densities and species richness (Quinn et al., 1992), although in areas subject to navigation, the exact nature of these changes and their effects on other trophic levels have yet to be elucidated.

Fish

Direct physical damage to fish, for example by propellor scarring (Rosen and Hales, 1980), appears to be rare. Noise and disturbance, especially from high-speed boats, may adversely affect fish behaviour and survival (Boussard, 1984) in proportion to craft speed and intensity of usage (Lagler et al., 1950; Mueller, 1980). Lagler et al. (1950), however, found that outboard motor exhausts had no significant impact on production or various behavioural

TABLE 34.1. Bluegill (*Lepomis macrochirus*) reproduction in two ponds (1.47 and 1.87 ha) in Michigan, USA, respectively with no use of outboard motors, and with 194 h of outboard motor use between May and August 1949 (adapted from Lagler et al., 1950). Data are for young-of-the year fish

Pond	Mean size (cm) (August)	Density of young fish (per ha)	Net density (per ha)
No outboard use	2.44	179 500	76
Outboard use	3.12	178 800	67

features of bluegills or largemouth bass in ponds (Table 34.1).

The ecological effects of increased suspended solids on fish have been thoroughly documented, but most studies have concentrated on sport fish and salmonids and, as with invertebrates, are concerned mainly with turbidity from sources other than boat traffic (e.g. Newcombe and MacDonald, 1991).

Suspended solids may affect fish in several ways. Clogging of gill rakers and gill filaments may interfere with respiration, especially in juvenile fish (Wilber, 1983), but it is highly unlikely that boat activity alone can raise concentrations of suspended solids sufficiently high to kill adult fish ($> 20\ 000$ mg litre^{-1}; Wallen, 1951). Boat-related turbidity may disrupt courtship displays and spawning behaviour dependent on visual cues (Wilber, 1983). The loss of vegetation as an anti-predator refuge for small fish may, however, be more than compensated for in such waterways by the visual refuge provided by increased turbidity. Turbidity may reduce the visual feeding efficiency of fish, leading to shifts in prey selection on the basis of body size, colouration, orientation or movement (e.g. Barrett, Grossman and Rosenfeld, 1992). Nevertheless, in heavily trafficked canals in Britain, most fish appear to recruit well, despite high turbidity (Pygott, O'Hara and Eaton, 1990).

As with invertebrates, the greatest effect of turbidity appears to be via silt deposition rather than the presence of the suspensoids in the water column. Egg and larval stages may be more susceptible to direct effects, including physical damage through abrasion or sedimentation of suspensoids over eggs, leading to anoxia, accumulation of waste products, or exposure to pathogens (Wilber, 1983). In Belgian rivers, Boussard and Falter (1982) showed that the redistribution of sediments produced by motor boats could significantly reduce the hatching rate of roach eggs (*Rutilus rutilus*). Siltation also reduces breeding success by siltation and smothering of gravel spawning areas (Hansen, Alexander and Dunn, 1983) and the loss of macrophytes as spawning substrate or fry nurseries (Reynolds and Eaton, 1983).

Fish in heavily trafficked waterways are also influenced indirectly by effects on their prey organisms, due to loss of substrate stability and heterogeneity caused by siltation (Sparks, 1975; Stern and Stickle, 1978), organic impoverishment of sediments due to reduced primary production, and loss of macrophyte cover (Hansen, Alexander and Dunn, 1983). The loss of macrophytes removes a staple foodbase of some fish (Prejs and Jackowska, 1978), but the attendant reduction in populations of large plant-associated macroinvertebrates (Murphy and Eaton, 1981) may be especially important, since these help sustain the growth rate of larger fish (Mittelbach, 1981).

Community structure also changes, favouring opportunist or turbidity-tolerant bottom feeders, for which elevated turbidity offers a visual refuge from predators (Sorenson et al., 1977; Ewing, 1991). Hence, in the British canal system, increasing boat traffic results in decreasing fish biomass, and a shift from a community characterized by weed-associated tench (*Tinca tinca*) and sight-hunting pike (*Esox lucius*), to one dominated by gudgeon (*Gobio gobio*), which are equipped with barbels for tactile bottom-feeding, and by small opportunist roach (Linfield, 1985; Pygott et al., 1990). Stocked carp (*Cyprinus carpio*), which favour waters of high turbidity, perform well at high traffic densities and reach large sizes (Pygott et al., 1990).

Birds and mammals

Short-term effects of boats on water birds are well documented, mainly in relation to sailing and powerboating on open inland waters such as gravel pits and reservoirs, which may be important sites for wintering wildfowl (see Batten, 1977; Ward, 1990). Hume (1976), for example, recorded flight responses by goldeneye (*Bucephala clangula*) as a result of the approach of a powerboat within some 550−700 m of a flock on Cannock Reservoir, Staffordshire, England.

Both noise and visual stimuli may cause behavioural change in less-heavily navigated waterways, with reduction in bird occupancy, or even complete avoidance of the water as craft usage increases (Reichholf, 1974; Marchant and Hyde, 1980; Tuite, 1982). Species sensitive to disturbance may react to traffic densities well below those which cause outright deterioration of their habitat. All water-related human activities probably cause some psychological stress to water mammals and birds, and in a multi-recreational resource it may be difficult to distinguish between the stresses attributable to boating and those from land-based sources, such as angling and walking. Conversely exposure over a long period to a uniform intensity of slow-moving boat traffic might lead to habituation.

Table 34.2 illustrates the effects of increasing boat traffic on the breeding densities of three waterway birds which nest at or near water-level. The reductions in breeding populations are probably due to destruction or deterioration of suitable nest sites by unsympathetic management practices (Taylor, 1984) or boat wash. The retreat and fragmentation of reed-beds, removal of overhanging branches which provide material for nest construction, anchorage and concealment and swamping of nests by boat waves (Sharrock, 1976; Batten, 1977) may all be involved. Dense fringing vegetation is also a general habitat requirement for skulking waterbirds, such as the little grebe (*Tachybaptus ruficollis*). Species which dive for submerged aquatic vegetation (e.g. coot, *Fulica atra*), or for small fish and large weed-associated macroinvertebrates (e.g. little grebe), may also be affected indirectly by reduced food availability or increased water turbidity which disrupts visual feeding.

Much less is known of how waterway mammals react to boat traffic. In one of the few studies undertaken, Clark (1981) suggested that a population of muskrat (*Ondatra zibethicus*) in a navigable stretch of the Upper Mississippi showed relatively little sign of adverse impact directly connected with boat traffic. Water voles (*Arvicola terrestris*) also persist on less intensively trafficked waterways in Britain, where soft banks allow burrowing and there is an adequate supply of emergent vegetation as food, but heavily trafficked, turbid canals or highly engineered river navigations are unsuitable habitats due to their steep, reinforced banks (in a few cases a deliberate defence *against* water voles) and corresponding

TABLE 34.2. Breeding densities (pairs/10 km channel) of three widespread species of waterbirds in canals of varying intensities of boat use

Species	Disused canal[a]	Basingstoke Canal (1978)[b]	Kennet & Avon Canal (1983/4)[c]	Used canal[a]	S. Oxford Canal (1983)[d]
Little Grebe (*Tachybaptus ruficollis*)	5.1	14.7	10.5	0.2	0
Coot (*Fulica atra*)	4.7	9.9	21.8	2.5	0.3
Moorhen (*Gallinula chloropus*)	37.8	48.1	47.3	22.5	19

[a] From Marchant and Hyde (1980) based on BTO Waterway Birds Survey data
[b] From Clarke (1978)
[c] From Brown (1988)
[d] From Edwards et al. (1978)

lack of emergent vegetation and general bankside cover, combined with frequent disturbance by boats and other recreational users (Jefferies, Morris and Mulleneux, 1989). Boat wash may also flood burrow systems. Expansion of navigation on to quieter rivers may have implications for the long-term recovery of otter (*Lutra lutra*) populations in lowland England, through a combination of increased human disturbance and habitat modification (Chanin and Jefferies, 1978).

MANAGEMENT FOR CONSERVATION AND FISHERIES

Limited progress has been made towards controlling boat traffic effects in the management of waterways for wildlife conservation and fisheries. Two approaches have been developed, namely accommodation of effects by such responses as bank protection and reduction of those effects by regulation of craft design and usage.

Bank protection

The traditional solution to bank erosion problems is the installation of vertical sheet steel trenching or concrete piles. Although durable and undoubtedly effective, this approach is costly, visually intrusive and tends to retain wave energy and direct it into scouring of the bed. Alternative forms of bank stabilization which combine cost-effectiveness and energy dissipation with visual and ecological sensitivity are therefore desirable (Jorga and Weise, 1981).

Bonham (1980) studied the influence of stands of four species of emergent aquatic plants on the propagation, height and breaking of waves generated by passing boats at a site on the River Thames. By comparing maximum wave heights recorded at the inner and outer face of a stand of emergent plants with those on an unvegetated length of bank, he demonstrated that a reed fringe 1.7–3.0 m in width would dissipate 60–80% of wave energy (see

TABLE 34.3.   Boat wash energy reduction, measured as the rato of maximum wave heights between two probes located on the inner and outer sides of stands of emergent vegetation ($h_\mathrm{I}^\mathrm{max}/h_\mathrm{O}^\mathrm{max}$), at five locations on the River Thames, England (adapted from Bonham, 1980)

Species	Stem density (m^{-2})	Probe spacing (m)	$h_\mathrm{I}^\mathrm{max}/h_\mathrm{O}^\mathrm{max}$
Control: no plants	0	2.90	1.15
Phragmites australis	216	2.15	0.38
Scirpus lacustris	408	1.65	0.48
Typha angustifolia	80	2.97	0.32
Acorus calamus	248	2.60	0.24

Table 34.3) and provide complete protection against bank erosion. This endorses the widespread, successful use of managed emergent vegetation as natural protection against bank erosion, reviewed by Brookes and Hanbury (1990). This "living protection" is self-sustaining, and management needs and costs are minimal, because the tendency for encroachment of the reeds into the navigation channel is curbed beyond a certain threshold by the physical effects of boat wash (Oksiyuk, Merezhko and Volkova, 1978). Thus there is a range of benefits in re-establishing vegetation along denuded channel margins, where harsh conventional engineering techniques would normally be applied, and utilizing this as natural bank protection.

Some low-key management is often needed to protect the vegetation during its establishment phase (Broads Authority, 1993) and the following examples illustrate how this may be achieved.

Use of plants protected by semi-permanent solid structures

In the Thames, gravel or sand/clay parent material tends to erode by bank collapse as the protecting vegetation is lost (Bonham, 1980). Replanting can be achieved using material imported from a nearby convenient source by simply dumping sediment containing rhizomes onto the bank and leaving them to grow. Alternatively, individual clumps of plants with foliage intact can be planted into the backfill material. A simple structure of chestnut paling and old tyres is sufficient to protect the replanted vegetation. In rivers on peaty substrates, underwater erosion tends to undermine the reed-beds, which then float away and cannot regenerate naturally. A more substantial underwater tyre barrier, tied and anchored to the bank may then be needed to support the replanted reed-beds. Using these techniques, Bonham (1980) found that introduced reed-beds produced a useful density of plants within two years of planting, although even in the first year they afforded some protection.

Reinforced vegetative bank protection

At higher traffic densities, protection is needed to establish and maintain reed fringes. Emergent vegetation incorporated into a woven geotextile fabric has proved durable in some of the most heavily trafficked canals in Britain as a means of protecting against bank erosion (Hanbury, 1982; Brookes and Hanbury, 1990). Using this technique, clumps of a non-encroaching, robust, stand-forming plant (e.g. *Carex acutiformis*) are inserted into pockets in the fabric mat which is laid over the reformed bank. The reed fringe which subsequently develops offers maintenance-free protection of the bank at less than one-third of the cost of galvanized trench sheeting, as well as providing a more aesthetically pleasing margin, beneficial to invertebrates and waterside birds.

Dredging

Since enlarging the channel decreases the blockage factor, dredging is beneficial in that it reduces the intensity of disturbance caused by subsequent boat traffic (British Transport Docks Board, 1972). It can also be advantageous when large accumulations of soft, unstable silt are removed and a firmer bed for plant and invertebrate colonization is exposed. Additionally, the increased water volume in the channel reduces fluctuations in level due to lockage, especially on lengths of canal which have frequent locks or restricted water supplies. In rivers, these gains have to be weighed against the potential loss of habitat diversity due to dredging, noted earlier.

The act of dredging is always ecologically disruptive in the short term, but removal of strongly competitive plants, such as dense reed stands, and the opening up of new habitat to primary colonization may encourage the establishment of species and communities of conservation interest (Eaton, Murphy and Hyde, 1981; Murphy and Eaton, 1981), which can be subsequently stabilized in a state of arrested succession by resumption of boat traffic at low density (Willby and Eaton, 1993).

Fisheries

In British navigations, the change from a fishery dominated by tench, large roach, perch and pike in low-traffic, clear, weedy waters to dominance by numerous small roach and gudgeon in turbid, high-traffic conditions can be accommodated by developing match angling, where evenness of fish distribution and frequency of catch are valued characteristics (Pygott et al., 1990). However, the recent spread of zander (*Stizostedion lucioperca*) may further alter the fish community by introducing a predation pressure which is largely absent from these turbid navigations.

Craft design

The review in IWAAC (1983) found scope for reducing the disturbance caused by recreational craft to aquatic ecosystems by alterations to both hull shape and mode of propulsion. Since then, research to develop low wash hulls has resulted in the introduction of production models by at least one hire company, Alvechurch Boats (Billingham, 1993). Most recreational craft have a long life, so replacement of traditional hulls by the new ones would, at best, take many years. Additional difficulties are the higher production costs of the new designs, as compared with the traditional ones and the considerable capital expense involved in changing boat building procedures. There may be a tendency for users to travel faster while generating the same level of wash. This problem could be addressed in part by the use of less powerful engines in the more efficient hulls — the solution adopted by Alvechurch Boats. Increasing environmental concern amongst boat purchasers (British Marine Industries Federation, 1992) may go some way towards making higher-priced hulls more acceptable, but new design regulations or licence fee incentives may be needed to ensure their widespread introduction. Deeper-set, more slowly revolving propellors may also reduce wash, and electric propulsion is a further development which reduces noise nuisance (Wagstaffe, 1993).

Boat traffic regulation

The observation and enforcement of speed limits low enough to prevent the occurrence of breaking wave wash are widely practised on inland navigations. The speed limits applied to the various parts of the Norfolk Broads and rivers have recently been lowered, mainly to reduce bank erosion (Broads Authority, 1993).

Temporal zoning of traffic densities has only limited potential, due to the strongly seasonal demand for boating. On some reservoirs and lakes used by overwintering wildfowl, there is scope for confining recreational boating to the summer months, thereby avoiding disturbance during the critical October to March period (Ward and Andrews, 1993). On some British waterways, however, the peak demand for boating (April to September) coincides with the main period of growth and reproduction of most aquatic and waterside biota, so any time zoning controls during these months would be highly restrictive of recreational use.

Space zoning is feasible in extensive or branched navigation systems such as the Norfolk Broads (Broads Authority, 1993) and some large reservoirs and lakes (e.g. Ward and Andrews, 1993), but is likely to be very unpopular on linear waterways, where it will interfere with through navigation. Sidewaters connected to a navigation but carrying little or no traffic can function as wildlife and fisheries reserves. Their effectiveness is greatly reduced, however, if they receive substantial turbidity and depositing silt from the main channel, which can inhibit submerged plant colonization and accelerate seral succession to reedswamp and terrestrial vegetation (Bhowmik and Adams, 1989).

Cross-channel zoning attempts to limit boats to a defined track, and exclude them from sensitive marginal waters. This is feasible on some wide rivers, for example to protect lily-beds and fish-fry nursery areas (Eaton, 1986). Restriction of mooring to designated sites with deep water and hardened banks will safeguard other unprotected lengths where manoeuvring and securing of craft, and landing of crews would damage marginal vegetation (British Waterways Board, 1993a).

Purpose-built off-channel mooring basins can be promoted in preference to on-channel linear moorings (British Waterways Board, 1993b). They reduce the length of main channel subject to mooring manoeuvres and also concentrate pollution risks associated with boat servicing and maintenance into a limited area, which can be isolated quickly in the event of accidental spills of toxic materials.

Direct regulation of traffic densities to achieve particular wildlife conservation or fisheries objectives may be necessary in some sensitive areas. Licensing and lockage restrictions are feasible, but unpopular with navigation interests. Often, however, it is possible to manipulate traffic densities by judicious control over the siting and sizes of private moorings and hire-boat bases, since boat movements from a mooring are generally predictable and, in the case of private boats, mainly short distance.

Selective marketing and the controlled development of tourist attractions at particular sites are other ways in which recreational traffic patterns can be influenced without resort to compulsion (British Waterways Board, 1993b).

FUTURE TRENDS AND RESEARCH NEEDS

The extension of recreational navigation seems likely to continue in Britain. Restoration works were in progress in 1993 on 160 km of rivers and 1014 km of canals; plans for another 136 km of rivers and 143 km of canals are being developed and there is a substantial further list of waterways for which proposals have yet to be evaluated (Inland Waterways Association, 1993). Recent proposals include a particular emphasis on projects which reconnect, or in some cases connect for the first time, existing navigations, to form larger networks.

Long-term increase in boat traffic is expected in Britain's waterways (Leisure Consultants, 1989), despite a decline in some areas in the late 1980s due to economic recession. It seems likely that as economic conditions improve, growth in boat traffic will resume, partly in response to the new cruising opportunities offered by the expanding network and partly by increasing usage of existing waterways, only a minority of which are near their physical carrying capacity. This expansion of both the network and its usage poses questions of environmental impact and management.

For canals, with their hydraulic uniformity and well-researched traffic characteristics, impacts of navigation on the aquatic ecosystem can be defined and the effects of any proposed change or development can be predicted. Conservation of the substantial wildlife interest of lightly trafficked lengths requires appropriate regulation of both channel maintenance and traffic generation. For this, the procedures may be defined in general terms and can be defined in detail as management plans for specific lengths, where the conservation interest is great enough to warrant this level of effort (e.g. Basingstoke Canal Authority, 1993). Likewise fisheries requirements can be included in the management of both the channel and its traffic.

Canal restoration usually proceeds slowly, so there is often time for newly-created channel habitats to be colonized before traffic commences. Better understanding of colonization processes is nevertheless needed, so that any necessary planting procedures and traffic controls can be operated until the desired ecosystem and fisheries are established (e.g. Devon County Council, 1993).

In rivers, where navigation is only one of a range of hydraulic influences, and an integrated management strategy is essential (National Rivers Authority, 1993b), there are much greater problems than in canals in the discernment and management of specifically navigational impacts. Whereas canals, with their tendency to seral succession to reedswamp and to infill with sediment (Twigg, 1959), clearly depend upon channel maintenance and light boat traffic to sustain their conservation and fisheries values, rivers differ markedly in their self-scouring nature, and more complex channel morphology and flow characteristics. There appears to be no systematic knowledge of the separate impacts of navigation engineering and traffic movement on river ecosystems. Also it is only rarely possible (Jeffray, 1980) to draw valid comparisons between navigable and unnavigable lengths on a single river, since the latter are usually upstream and differ in such basic features as channel size, gradient and bed structure.

The large-scale national surveys of comparable sites with a range of navigation impacts from zero upwards which are available for canals (e.g. Murphy and Eaton, 1983) have not been attempted for rivers. Therefore there is at present no objective basis on which to evaluate the environmental impact of proposals to extend recreational navigation on British rivers, against the increasingly important conservation aspect of the management of these waterways (Boon, Calow and Petts, 1992).

ACKNOWLEDGEMENTS

Thanks are due to the following for providing information or commenting on drafts of the peper: Michael Handford, Inland Waterways Association; Roger Hanbury and Glen Millar, British Waterways; Brian Moss, University of Liverpool; Paul Wagstaffe, British Marine Industries Federation.

35

Ecological Impact of Angling

Peter S. Maitland

Fish Conservation Centre, Stirling, UK

INTRODUCTION

The relationship between angling and wildlife conservation is a complex and often ambiguous one. Many anglers consider themselves to be also naturalists and major guardians of water quality and freshwater habitats. In contrast, some people believe that angling involves substantial cruelty to fish and considerable damage to aquatic ecosystems. Most conservationists probably have views somewhere between these two extremes.

There is no doubt that many human factors other than anglers have a much greater impact on the aquatic environment — some of them totally destructive. These include domestic and industrial pollution, harmful runoff of fertilizers and pesticides from various forms of land use (notably farming and forestry), ditching and canalization of streams and rivers, and filling in of ponds. Recently, new threats have come from fish farming, acid deposition and global warming.

However, the objective of this chapter is to explore angling and river ecology and how they relate to each other. Both share many similar objectives, but it is important to examine areas of conflict to see if problems can be solved or compromises reached. A recent example of a conflict solved gives some cause for optimism. Waterfowl in many countries were being poisoned by lead deposited in the water by anglers. In particular, mute swans (*Cygnus olor*) in England were in serious trouble because of the large amounts of angler's lead (from discarded or lost weights) which they were ingesting. A voluntary code followed by legal measures has meant that most anglers have now replaced lead by substitute compounds and there has already been a marked decrease in swan deaths from lead poisoning.

Game fishing

Game fishing is the traditional form of angling in many parts of the British Isles. Exact definitions vary, but game fishing is usually understood to mean angling for salmonid fish, especially Atlantic salmon *Salmo salar*, trout *Salmo trutta* (both sea trout and brown trout) and rainbow trout *Oncorhynchus mykiss*. By extension, angling for other Salmonidae (arctic charr *Salvelinus alpinus* and brook charr *Salvelinus fontinalis*) is also included in the term,

The Ecological Basis for River Management. Edited by D.M. Harper and A.J.D. Ferguson. © 1995 John Wiley & Sons Ltd

but there is usually some argument as to whether grayling *Thymallus thymallus* (Thymallidae) are game fish or not.

Coarse fishing

There are also various interpretations of what exactly the term "coarse fishing" means. Commonly it would certainly include fishing for the larger Cyprinidae: common carp *Cyprinus carpio*, crucian carp *Carassius carassius*, tench *Tinca tinca*, bream *Abramis brama*, rudd *Scardinius erythrophthalmus*, roach *Rutilus rutilus*, chub *Leuciscus cephalus* and dace *Leuciscus leuciscus*. As well as cyprinids, most other large non-salmonid species are generally regarded as "coarse" fish, e.g. pike *Esox lucius*, eel *Anguilla anguilla* and perch *Perca fluviatilis*.

CURRENT MANAGEMENT PRACTICES

Angling and fishery management practices in the British Isles include many activities which are actually harmful to the aquatic environment and some indeed to the fisheries themselves (Maitland and Turner, 1987).

Stocking practices

As well as stocking waters which are fishless as a result of poisoning, various other types of stocking are practised by fishery managers. Commonly, the wild stock is supplemented by additional fish in an attempt to improve fishing. Often, very large numbers are introduced; these may be young fish which need to grow before reaching a catchable size, but more and more waters are receiving fish of a catchable size and in such "put and take" systems fish may only be in the water a few hours before being caught again.

Such intensive management practices can create many problems. There may be damage to the original native stock through competition, predation, genetic dilution or the introduction of diseases or parasites. Sometimes the original stock may be completely wiped out and the fishery is forced into a permanent "put and take" situation. In addition, it is often forgotten that such fisheries rely on fish farms for their stock and thus they exacerbate the problems created by fish farms (e.g. abstraction, pollution, disease, parasites, escape of stock, attraction of predators).

Stocking practices have varied widely in the British Isles in the past and have rarely had any scientific basis (Maitland, 1892). Commonly, stocking is carried out after the removal of unwanted species, but it is also practised at other times — even when there is no indication of a lack of stock in the water concerned. Sometimes the objective has been the introduction of "new blood". For over 100 years the principal species involved have been salmonids, but in recent years, many waters have been stocked with cyprinid species. Fish are stocked at all stages of their life cycle — as eggs, fry, juveniles or adults — depending on the objectives and the finance available. The price per fish increases greatly because of rearing costs from egg to adult.

Stocking with adult fish (often rainbow trout) is generally regarded as creating a "put and take" fishery. Often such fish may be caught within hours, usually days, of release. The fisheries involved are often intensive ones and the number of fish stocked and caught per year may be very high relatively to the size of the water and its natural fish production.

There is considerable concern, but relatively little information, about the effect of various stocking practices. Obviously, the introduction of diseases or parasites is an important issue, discussed below. If the species being stocked is a new one to the system concerned ("introduced"; see below) or is being released in very high numbers, then changes may be expected in the ecosystem concerned via the food web or in some other, probably indirect, way. Vulnerable fish or invertebrate species could be eliminated through predation. Often the stocking practice is intended to "enhance" the native (wild) stock in a water but concern is that it may do exactly the opposite. In particular, if the numbers stocked are large compared to the wild population, the latter may be reduced in number or eliminated through competition for food, spawning grounds and other resources. The loss of the genetic integrity of the local native stock which is assumed to have adapted to local conditions over thousands of years is a major issue, discussed more fully elsewhere (Maitland, 1989).

Introductions

All stockings are are effectively "introductions" but the latter term is usually reserved for fish which are introduced in the following ways:

(i)   intentionally, to create a new population, e.g. of a desired species (Maitland, 1974), or to develop safeguard stocks of rare species (Maitland and Lyle, 1992);

(ii) casually, by the release of excess livebait species at the end of a day's fishing or the dumping of unwanted aquarium or pond fish (Maitland, 1971);

(iii) through accidental escapes from fish farms or garden ponds.

Some species (e.g. the grayling) owe their entire Scottish distribution to introductions for angling (Gardiner, 1991). Certainly the status of fish stocks in the British Isles today owes much to introductions in the past (Wheeler and Maitland, 1973).

The concerns over introductions are similar to those for stocking. However, by definition, new species are involved so, as well as the threat of new diseases and parasites, there is a very real danger of a major impact on the ecosystem concerned (Raat, 1990). An outstanding example of this is at Loch Lomond, where ruffe *Gymnocephalus cernuus* (previously unknown in Scotland) are believed to have been introduced by pike fishermen from England about 1980. They were first detected in the loch in 1982 (Maitland, East and Morris, 1983) and have since become one of the most abundant species, posing a threat to native fish. Three other species new to Loch Lomond (gudgeon *Gobio gobio*, *L. cephalus* and *L. leuciscus*) have also been introduced by coarse fishermen over the last 10 years and all have become established.

One of these species, *G. gobio*, was not introduced casually but brought into the area with the specific intention of establishing a new population. It was first introduced to a small pond, but quickly found its way into the River Endrick. There are many other instances of anglers moving coarse fish into new waters with the objective of establishing a fishery there regardless of the nature and importance of the water involved.

In Scandinavia, there has been a tradition of introductions of various new species (both invertebrates and fish) to lakes with the objective of "improving" the fisheries. With hindsight, several of these experiments have clearly been unwise. Introductions of invertebrates (and disease) may also result from the use of equipment (including angling tackle) in different waters (Reynoldson, Smith and Maitland, 1981).

Diseases and parasites

There is no doubt that various diseases and parasites have been moved around from continent to continent, from country to country and from catchment to catchment. The means of transfer and the outcome have varied greatly from one situation to another (Secombes, 1991).

Maitland and Price (1969) found that a population of the North American largemouth bass *Micropterus salmoides*, naturalized in a pond in Dorset, was host to the monogenetic trematode parasite *Urocleidus principalis*. This parasite, which is specific to the genus *Micropterus*, had never before been recorded in Europe but was known as a common parasite of *M. salmoides* in North America. It was assumed that the parasite came to Great Britain with its host and became established here with it.

The outstanding example of parasite transfer in Europe in recent years has been the outbreak of the parasitic fluke *Gyrodactylus salaris* in Norway which has been spread by introductions from farmed salmonids to wild populations, and several native stocks of salmon have been virtually wiped out (Dolmen, 1987). Over the last decade it has proved necessary to use poison to eradicate entire fish stocks and communities in some rivers as the only means of eliminating the parasite. The original infection appears to have arisen from the import of parasitized stock from fish farms in Sweden, and the parasite is now known from 28 rivers and 11 hatcheries in Norway.

One of the surprising features of the incidence of this parasite in Norway is its virulence in the wild populations, whereas it does not seem to be a problem to salmon in its native Swedish rivers. It has been suggested that this may be due to the fact that Norwegian wild stocks were unadapted to it and therefore had no resistance, but it is also suggested that the resistance of the wild stocks in Norway had been lowered in genetic terms by the introduction of alien stocks from fish farms over the years.

There are several other examples of problems created by the introduction of diseases and parasites. For example, Norway also created enormous problems for its salmonid fish when it imported furunculosis-infected stock from Denmark in 1966. This caused substantial losses of farmed fish but the effect on the wild fish was not monitored. Thus there must always be potential dangers from the annual introduction of non-native stocks of fish into lochs via cage systems, however stringent the checks for disease on these fish are.

Removal of unwanted species

Because of the tendency for specialism in game fish, coarse fish or even a single species, to many anglers other species may be undesirable. For this reason, there has been a regular removal of unwanted species from waters in different parts of the country. Various methods have been adopted to remove fish, including poisoning, netting, trapping, electro-fishing and drainage.

A number of lakes (Morrison and Struthers, 1975) and some streams (Morrison, 1977, 1979) in Scotland have been poisoned, in most cases to eliminate unwanted populations of *E. lucius* and *P. fluviatilis*, or both, usually with the objective of introducing *S. trutta* or *O. mykiss*. In a few cases it has been used as a recruitment control measure (Walker, 1975). Occasionally, other species (e.g. *L. leuciscus* or *A. anguilla*) have been the targets of poisoning. The main poison used has been Rotenone (a derivative of Derris), for the use of which a licence is required from the government agriculture and fisheries office, SOAFD. Usually, relatively small waters have been involved (e.g. Fincastle Loch burn) but occasionally larger water bodies have been poisoned; perhaps notable among these was Loch Choin (Munro, 1957) which was poisoned to eliminate the population of *E. lucius* there.

Netting to remove unwanted fish has involved the use of both seine nets and gill nets. Seine nets are rarely used directly for this purpose because of the substantial manpower required, but when seine netting is carried out for other reasons, "unwanted" species of fish are often removed. For example, at Loch Lomond over a period of many years during regular seine netting (net and coble) for *S. salar* and *S. trutta*, it was common practice to destroy all the *P. fluviatilis* and powan *Coregonus lavaretus* taken — sometimes hundreds of fish in a day. Gill netting is the more usual method of removing unwanted species (especially *E. lucius*) and has been carried out at many lochs either as a "one off" management policy or else on a regular, usually annual, basis. Normally the nets used have a large mesh size in order to avoid taking too many, usually smaller, game fish.

Trapping is carried out less frequently for predator control. Sometimes fyke nets are allowed to operate for *A. anguilla* on the basis that a predator and competitor with game fish is being removed and some income is generated from the catch. In a few waters, *P. fluviatilis* traps (Le Cren, Kipling and McCormack, 1967)) have been used to remove large numbers (e.g. at Windermere). In the past, live *P. fluviatilis* trapped in Loch Leven were sold to English coarse fisheries.

Electro-fishing has been used on a number of salmonid nursery streams mainly to remove *S. trutta*, assumed to be a competitor and predator of young *S. salar*, thus leaving the habitat free for the latter. The trout thus removed may be used to stock hill lochs or other angling waters. This practice has operated on the River Thurso for many years.

The removal of unwanted species can create various ecological problems, depending on what method is used. Clearly, drainage of the whole system is enormously destructive, especially if the basin is left empty for a long period. Poisoning too can cause substantial damage, not only of course to indigenous fish, but also to various other aquatic taxa, including Amphibia (Morrison, 1987). The impact of gill netting varies; with coarse mesh only, the target species (usually *E. lucius*) may be the only fish directly affected, but with indiscriminate mesh sizes many fish (and some birds) may be killed. The use of seine nets may damage macrophytes and affect the substrates where used regularly, but few fish are killed (i.e. most could be released if wanted), and the same is true of trapping methods and electro-fishing.

Predator control

Several birds and mammals and even some fish (e.g. *E. lucius*) compete with anglers for fish and such piscivores are especially attracted to highly managed waters with unnaturally large populations of stocked, easily available, farmed fish. In the British Isles, heron *Ardea cinerea*, cormorant *Phalacrocorax carbo*, merganser *Mergus serrator*, goosander *M. merganser*, otters *Lutra lutra*, mink *Mustela vison* and seals e.g. *Halichoerus grypus* are often regarded as a threat to fisheries and are normally discouraged by shooting. This may have a short-term effect but normally the void created is simply filled by another of the same species.

The loss, or imagined loss, of fish to such predators has long been of concern to anglers (Mills, 1987) and in the past many predatory fish, birds and mammals have been persecuted as a result (Draulans, 1987; Carss and Marquiss, 1991). Scientific information on this topic is often fragmentary and sometimes equivocal, leading to heated debate between anglers and conservationists. Usually anglers would like to see some or all of the predators removed but occasionally the reverse is true, e.g. where *E. lucius* (or other predatory fish species) are introduced to populations of stunted fish with the objective of reducing the number and increasing the average size.

The basic argument of anglers is that any fish (of a valued species) eaten by a predator is a loss to them. Arguments countering this attitude point out that

(i) often only young fish, with a high natural mortality, are eaten,

(ii) it is frequently diseased or damaged fish which are vulnerable to predators,

(iii) it is often stocked fish (from farmed sources) which are most vulnerable because of their reduced capacity to cope in the wild,

(iv) predators may actually benefit a fishery by reducing dense populations of small fish,

(v) the operation of the "sump effect", where individual predators destroyed are simply replaced by others from neighbouring areas, and

(vi) most predators (e.g. *A. cinerea*, kingfisher *Alcedo atthis*, *L. lutra*) have a high intrinsic value to the public and merit protection.

The main conservation problems caused by predator control are the direct effects on the populations of the predator species concerned and the indirect effects on the ecosystem resulting from its disappearance. The results are often unpredictable.

Groundbaiting

Of considerable importance in England and Wales, until recently, the practice of groundbaiting (using cereals, maggots and various other baits) has been of little importance in Scotland because of the lack of interest in coarse fishing. However, it is now used regularly in grayling fishing, and with interest in coarse fishing expanding rapidly, there is now justification for paying attention to possible problems created by groundbaiting in all parts of the British Isles.

Several potential problems are created by groundbaiting where it is carried out over a long period. Although banned on many water supply reservoirs in England, on the grounds of protecting water quality, its contribution to nutrient budgets seems low, except under very high application rates and angler densities (Edwards, 1990). In five out of six reservoirs studied, Edwards and Fouracre (1983) estimated that groundbait contributed less than 0.3% of the phosphorus load. However, the contribution in the sixth reservoir was about 6%. In experimental work on a shallow coarse fishing reservoir, Cryer and Edwards (1987) found substantial reductions in the benthic fauna where groundbait was distributed, with the exception of tubificid worms which increased. In addition, the oxygen consumption in these areas increased 100-fold and caused local deoxygenation under warm calm conditions.

HABITAT MANAGEMENT

The riparian zone is frequently modified by fishery managers to avoid damage or aid angling access to the water. In some cases, as well as making physical alterations, attempts are made to alter water quality with the objective of improving the fishery. Materials may be added direct to the water or its tributaries or spread more diffusely within the catchment. Recently, a number of fisheries have installed equipment to pump oxygen into the water at critical periods.

Habitat management is widely practised by fishery managers (Murphy and Pearce, 1987; Swales and O'Hara, 1983) and can take many forms, some of them extremely damaging to the integrity of a site, others less so. Management may include any of the following practices: construction of groynes and fishing jetties, creation of fishing pools and loose gravelled salmonid spawning areas, raising of the water level using small dams, cutting off access to spawning streams to control recruitment (Campbell, 1967a,b), blocking of outflows (usually with a grid of some kind) to prevent stocked fish escaping, removal of natural barriers to fish migration (e.g. waterfalls), provision of fish ladders, addition of fertilizers, liming, clearance of aquatic weeds by herbicides (Brooker and Edwards, 1975), the use of grass carp, cutting of bankside vegetation (including trees), and introduction of fish "food" species (usually invertebrates but sometimes small fish species). Several of the other practices described elsewhere (removal of unwanted species, introductions and predator control) are also part of habitat management.

This topic involves such a wide range of practices which can be carried out in varying degrees, that it is difficult to review all impacts briefly. Presumably a basic start to any consideration is to assume that if the habitat concerned is a pristine one (a condition normally assumed to be desirable by most conservation criteria), then any management could be detrimental. Thus, the opening up of a new spawning tributary to a loch or river (by removing, say, a barrier such as a waterfall) may be regarded as highly desirable by local fishery managers but may be detrimental to conservation interests if the fish (or invertebrates) in the loch, river or tributary are of scientific importance.

One of the few biological methods of habitat management developed in recent years has been the use of grass carp *Ctenopharyngodon idella* which has been introduced to a substantial number of waters in Great Britain as a method of biological control of water weeds and to provide sport fishing. The initial introductions were carried out on an experimental basis (Buckley and Stott, 1977; Stott, 1977) and research has shown that this fish is efficient at controlling weed, is popular with anglers and is never likely to breed naturally here. On this basis its transfer to a number of waters has been encouraged by the Ministry of Agriculture Fisheries & Food for England and Wales.

HABITAT PROBLEMS

Physical access to fishings and other types of recreation (Sukopp, 1971; Liddle and Scorgie, 1980) can involve a number of issues which relate to the environment. Trampling of sensitive littoral or riparian vegetation is a common problem, as is disturbance to nesting birds and other wildlife. Lighting of fires, digging turf to find worms for bait, deposition of litter (including nylon line and old hooks) can all create environmental problems.

Damage

Apart from some (usually unintentional) damage to fences and gates, the main direct damage created by anglers is to the riparian vegetation. This may be caused by trampling across sensitive marsh or bog communities or actively destroying these to gain better access to the water. Often tree branches are removed or even entire trees felled — sometimes along a whole stretch of bank. Where worms are sought as bait it is common to find patches of turf dug out. Often, fires are built near the water's edge and surrounding vegetation may be further damaged.

Disturbance

Angling is normally a quiet and often solitary occupation. Nevertheless, in sensitive areas, substantial disturbance to wildlife can take place. Commonly waterfowl can be disturbed. Sometimes the disturbance is minimal, and especially with tolerant species like mallard *Anas platyrhynchos* there is probably little effect. However with more sensitive species, such as divers *Gavia* spp., the impact can be serious. Most damage is done at nesting time when birds are disturbed on the nest or prevented from gaining access to their nests. Much of this disturbance is done unwittingly by the anglers concerned.

At one time it was thought that some of the decline of the *L. lutra* population in southern Britain could be due to disturbance but research has shown that the species is less sensitive than previously thought and is rarely affected by human activity at the water's edge. The only exception is with breeding females where it is important that quiet and secure places are available for mothers to rear young in seclusion (Jeffries, 1987).

Litter

Many groups of humans deposit litter in the countryside and anglers are no exception. Angling litter is commonplace around popular angling sites and much of it includes discarded everyday items such as drink cans, polythene bags, paper cartons, etc. It is characterized, however, by the inclusion of angling items such as discarded bait containers, makeshift rod rests, nylon line and fish hooks.

Monofilament nylon line and fish hooks, as well as being discarded intentionally, are also frequently lost inadvertently during angling and are commonly found attached to riparian vegetation and stones in the water. Birds frequently become entangled in this and the RSPCA deals with hundreds of entangled birds each year. Fish too can be found with nylon streaming from them — attached to hooks which have been swallowed or are embedded in the body. Such fish are normally in very poor condition.

Outwith the national and local close seasons for fish, there is no legal restriction of access to fishings, provided: (i) permission has been obtained from the owner or lessee of the fishery, (ii) permission has been obtained from the riparian owner (not necessarily the same as the owner of the fishery), and (iii) the proposed method of fishing is legal.

Legal complications over access rarely cause conservation problems unless ownership is unknown or in dispute, when uncontrolled access could create difficulties. However, physical access may create significant problems for local conservation interests, especially through damage to vegetation and disturbance to wildlife (Cooke, 1987; Cryer, Corbett and Winterbotham, 1987; Jeffries, 1987). Both of these may be difficult to avoid if angling pressure is heavy. However, damage from fires, litter (Bell, Odin and Torres, 1985; Cryer et al., 1987; Edwards and Cryer, 1987) and bait digging should be perfectly avoidable with proper angler management and good local codes of practice (Mackay, 1987).

THE FUTURE

Some issues involve both ethical considerations and damage to the environment. In general, game fishing involves the least handling and pain to fish which, other than when they are actually being "played", are reeled in and either released at once if they are too small or despatched and taken to be eaten. In coarse fishing, on the other hand, fish are rarely taken for eating. Regardless of size they are normally removed from the hook and kept (possibly for hours) in a keep net (in shallow, often dirty and warm water) until they are released at the end of the day — usually after a rather lengthy weighing process. Substantial mortalities can occur during this period and after release; mortalities of up to 98% have been recorded after fishing competitions (Bylander, 1990) in North America.

In addition to fish which are the quarry, other smaller species are caught by hand net or rod and line and used subsequently as live bait. This means that they are kept alive in a small container, transported to the fishing site, impaled on one or more hooks and cast into the water in the hope of attracting large predatory species such as *E. lucius*, or *P. fluviatilis*. Fortunately such bait fish usually die a short time after impalement. Many also die in the container; those which do not are either destroyed, released at the angling site or taken home for further use. The release of such fish can have an important local impact, as discussed previously.

The fact that fish are being caught by and subjected to methods of treatment which inevitably damage them is an area of major controversy and this raises the whole question of pain in fish. Many anglers dispute that fish feel pain, but anyone who has ever kept fish in aquaria or handled live fish regularly in the field knows that they do. The attitude of the state in this matter is clear. Fish are vertebrates and as such are clearly included in the Animals (Scientific Procedures) Act 1986 (Home Office, 1990); the objective of this Act is to control scientific procedures applied to an animal "which may have the effect of causing that animal pain, suffering, distress or lasting harm". Thus no scientist can carry out any potentially painful experiment on a vertebrate without an appropriate licence and approval for the work involved which must usually be carried out within a designated establishment.

An experiment involving impaling live fish on hooks and then leaving them in the water for extended periods would definitely come within the Act and probably would only be allowed under anaesthetic. Yet in the wild, anglers consign thousands of fish every year to a painful and lingering death. This is something which brings angling into considerable disrepute nationally and, in the author's opinion, should be stopped forthwith — if not voluntarily then by legislation. Such a ban would have little effect on angling, since dead-baiting is a suitable alternative method.

In response to criticism from various quarters (Maitland and Turner, 1987), a number of codes of conduct have been produced by angling bodies in recent years, the most recent being a code for game anglers (Anon., 1990c). This covers some of the points discussed above but neglects others, notably the issue of live-baiting, "put and take" fisheries and the ethical issues of "catch and release".

It is clear that there are extensive and increasing pressures on our fresh waters — some of them from elsewhere in Europe (Bongers, 1990) — and that much discussion and many compromises are needed if this precious resource is to be used wisely. One recent example which highlights the kinds of conflicts involved relates to ducks and fish, where Giles et al. (1989) have suggested that because fish compete with some duck species for food, fish should be removed from some waters in order to increase the production of ducks for shooting. Anglers have yet to respond to this proposal.

CONCLUSIONS

The general conclusion arising from the above is that, although anglers are a powerful voice in the protection of water quality and some species of fish, they undoubtedly create a number of ecological and ethical problems. On the one hand, angling can mean the pursuit of wild fish in their natural habitats for sport and food, which seems to be an entirely legitimate activity. On the other, it may involve the capture of tame farmed fish, possibly to be held for hours in a keep net and released, if still alive, at the end of the day after further handling and weighing — perhaps to be caught again the next day and so on. The ethics of this part of the "sport" are debatable to many conservationists; to most the practice of live-baiting is entirely repugnant.

Thus anglers must give further consideration to such issues and strengthen their codes of practice where necessary. In addition, as with other groups in the population, in an overpopulated world with limited, even diminishing, resources, anglers may have to lower their expectations — especially where these may have included hopes for an expansion of fishing. A move towards reduced angling on natural sustainable fish populations, with a minimum amount of cruelty would be welcomed by conservationists. The general restriction of "put and take" fisheries to artificial waters only would also be a positive step, while the whole issue of "catch and release" requires further debate not only by anglers but by society as a whole.

ACKNOWLEDGEMENTS

I am grateful to Niall Campbell for useful comments on a draft of this paper.

36

The Ecological Basis for Catchment Management. A New Dutch Project: The Water System Explorations

J. P. A. LUITEN

Institute for Inland Water Management and Waste Water Treatment, The Netherlands

INTRODUCTION

Sustainable development is the target of almost every modern policy dealing with water or the environment. Sustainability is focused on human lives, but also on the ecological quality of our environment. Both aspects are essential for life on earth.

The ecological quality of aquatic systems can be expressed by biotic and abiotic parameters. Monitoring and research give information about these parameters and a comparison with the targets brings us to the necessity of supplementary policy-measures.

Human activities are the cause of the environmental problems. All kinds of social and economical activities influence water bodies. Reaching the goals is only possible if human influence on the environment is changed or reduced.

Within a catchment area relations can be made between human activities and ecological problems in the reception areas. Policy analysis looks for the most efficient way to solve the bottlenecks.

For an effective policy analysis in river catchment areas the following elements are essential:

(i)   quantitative data have to be available; monitoring programmes have to be adjusted to the needs of establishing the ecological basis;
(ii)   target values have to be determined in the laboratory or by expert opinion;
(iii)   the bottleneck and its causes should be known;
(iv)   an inventory of guiding measures for influencing the actual or future situation should be available;
(v)   sufficient knowledge about the relation between the guiding measures and ecological parameters are necessary — a research programme should be focused on this;

The Ecological Basis for River Management. Edited by D.M. Harper and A.J.D. Ferguson. © 1995 John Wiley & Sons Ltd

(vi)   a research programme should be focused on the relation between the steering points
       and the social and economical activities in society;
(vii) a policy analysis has to be carried out to determine options for future policies;
(viii) feedback is examined after restorative measures have been taken.

All these studies collect large quantities of data; presenting them in a clear way to give only
relevant information is one of the goals of the project "The Water System Explorations"
(Ten Brink and Woudstra, 1991).

BACKGROUND

Within a catchment area, several countries may be responsible for the water management. In
the Netherlands, which are part of the catchment areas Rhine, Meuse, Scheldt and Eems,
water management is controlled at three levels: the Ministry of Transport, Public Works and
Water Management, the 12 provinces, and the water boards. Each of them looks after the
management of waters in their area, while committing themselves to mutual agreements on
the scope of their responsibilities.

The Ministry looks in particular after the main water systems in the Netherlands. It also
controls the main issues of the national water management policy for all waters in the
country, including the regional surface waters and groundwater.

Thus many authorities, research institutes and policy-makers are involved in the control of
water management within the scope of a catchment area. In order to enable to make well-
founded decisions, the Ministry of Transport, Public Works and Water Management has
initiated the Water System Explorations project (Ten Brink and Woudstra, 1991). The aim is to
obtain an insight into the biological and chemical, physical and economic values of, for
instance, the Dutch water systems, including the related surroundings (channel bottom, shore,
banks, air, etc.). All relevant information will be gathered from mainly monitoring and
modelling activities. The approach is, in principle, suitable for each area, of each dimension.

The project offers a technical–scientific basis for policy preparation as part of an
integrated water management plan. Boards which are responsible for the implementation of
water management will derive their information from the Water System Explorations. It
should be feasible to use this information as a basis for policy preparation. Water
management plans, as defined by responsible regional departments as well as nationwide
policy notes on water management, will all be prepared on the basis of the same
technical–scientific information.

The Water System Explorations also represent a valuable support for the water
component, included in the coming national Environmental Policy Plan and the Nature
Policy Plan.

PROBLEMS IN WATER MANAGEMENT

During the work on the third Water Management Plan (Ministry of Transport, Public Works
and Water Management, 1989) from 1987 to 1990, a good overall picture of the Dutch water
management practice was achieved. We found out, however, that for an optimal policy on
water management within a catchment, relevant parts of basic elements were lacking. For
effective management, three basic elements are needed: verifiable objectives, knowledge

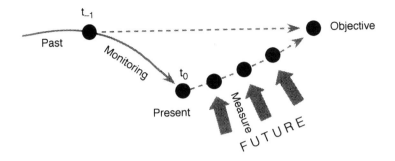

FIGURE 36.1. Basic elements of an effective management system

of the water systems and sufficient measures (Figure 36.1). If these elements are not optimal, the management is not optimal in effectiveness.

(i) *The absence of verifiable objectives*, especially for biological parameters. At present a coherent system of verifiable objectives for the Dutch policy on water management is lacking. In the actual policy note a qualitative description of the desirable situation is given, without any data concerning the ecological targets. The consequence is that the objectives are difficult to quantify and to verify. Which are the "desired" or "sustained" uses? What is "healthy water" and what are "ecological values" to be "preserved"? Are they, for instance, algae and bacteria, or seals and otters? Neither the species nor the desired numbers are stated explicitly. Choosing algae leads to a totally different policy than if otters or seals were chosen.

(ii) *Inadequate knowledge about the water systems*. This problem deals with monitoring of especially biological parameters, but also with prognosis models. Although monitoring provides us with a good deal of data, relevant information is rather scarce (Laane and Ten Brink, 1990). Most information about water systems is chemical information. Habitat and biological information is usually insufficient and there is a time-lag of years. Furthermore, the monitoring programmes, objectives and research programmes do not interact with each other, but handle different subjects.

The lack of reliable dose−effect relations is a common problem. Models of physical and chemical processes and population dynamics are scarce and the uncertainties on the prognoses are still high, so that it is hard to get any clear picture of the effectiveness of considered measures in advance. The development of models costs a huge amount of time and money in most cases.

Information about the kind and intensity of uses of the systems by society is only partly known.

(iii) *There are not enough measures available for reaching the desired goals*. There are several possibilities why a responsible authority is not capable of taking adequate measures to solve particular problems. First, some activities of society now and in the future are so essential for human life (e.g. primary human consumption, heating, recreation) in this area that leaving would be the only effective solution.

It is also possible that activities in the past may cause present and future negative impacts to water systems. For example, groundwater flow entering surface waters contains the effects of cattle manure deposition on land from several years ago. The only possibility for a solution to this problem is to wait, maybe for a very long time. A third problem could be that the water authority itself cannot take the measures, but some other authority has responsibility, e.g. for atmospheric deposition and other kinds of diffuse pollution.

Management by accident

From the above it can be seen that information for rational and efficient water management is inadequate. Co-operation between policy-makers and scientists is not effective (Ten Brink and Woudstra, 1991). A consequence of this is that calamities are taking over the role of science: "management by accident" instead of "management by knowledge and prediction" (see Figure 36.2).

Many examples can be given:

(i)   strict emission regulations for "drins", cadmium, mercury and PCBs only after cases of lethal poisoning (Japan and the Netherlands);
(ii) strict shipping regulations (MARPOL) only after disasters such as the Torrey Canyon and the Amoco Cadiz;
(iii) a 50% load reduction to the Rhine of several contaminants only as a result of the Sandoz affair (the Rhine Action Program);
(iv) a 50% load reduction to the North Sea of several contaminants and the acceptance of the "precautionary principle" at the Second North Sea Ministerial Conference in response to such alarming phenomena as toxic algae, anaerobic conditions in the German Bight in 1981–1983, fish diseases and large-scale fish mortality (the North Sea Action Program).

How can this deadlock be broken?

A new project to deal with this problem has been started in the Netherlands, which is described in the next section.

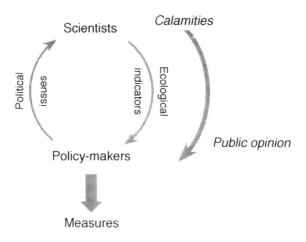

FIGURE 36.2. Management by accident

A SYSTEMATIC APPROACH: "THE WATERSYSTEM EXPLORATIONS"

There are several reasons for the inefficient co-operation between policy-makers and environmental scientists, as mentioned above. First, environmental policy is still relatively new in comparison with, for instance, economic sciences and socio-economic policy. Secondly, several decades passed before the "environmental problem" was generally accepted by society and by government institutions. This was mainly due to the economic costs of improving the environment. Third, both the political and the scientific worlds are characterized by improvization and *ad hoc* solutions to environmental problems.

Now that environmental problems are growing larger, knowledge is increasing rapidly and environmental issues stand high on the political agenda, the time has come for a more coherent approach. A continuous flow of adequate information must be created. Since the main goal for water management is the maintenance of a balance between the ecosystem and its economic use, in order to achieve a state of sustainable development, it is necessary to acquire information on both (Figure 36.3). The essence of the Water System Explorations is, therefore, to obtain quantitative information on the physical, chemical and biological components as well as on the uses of Dutch water systems, related to quantitative objectives and natural background values, and expressed in a small number of variables. This information system has to be short, accurate, accessible and understandable for policy-makers and the public.

For this the Water System Explorations project (Ten Brink and Woudstra, 1991) make use of target variables and a measuring gauge, both dealing with information from monitoring

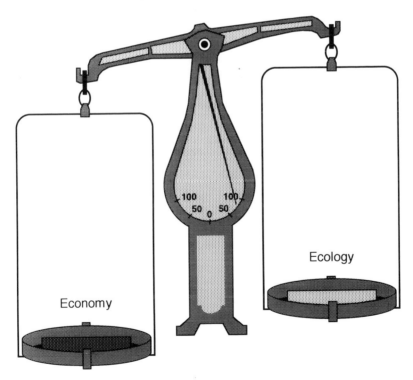

FIGURE 36.3. Searching for sustainable use of the water systems

and modelling studies, in relation to a definition of the reference system and management goals.

Target variables

A maximum of about 30 variables have been chosen for each of the following categories: physical, chemical, biological and use, which will indicate something about the sustainability of the water systems and their uses. These variables or parameters we have called "target variables" (total about 150). For example:

● Physical: radioactivity, turbidity, depth, intertidal area, etc.
● Chemical: P, N, Hg, PCB, etc.
● Biological: algae, sea-grass, sandwich tern, common seal, otter, etc.
● Uses: water supply, groundwater extraction, discharge of contaminants, fish catches, volume of sand extraction, recreation days, shipping, etc.

The following considerations contributed to the choice of target variables:

(i)   Past and present quantitative data must be available.
(ii)   The variables must be susceptible to human influence. It is pointless to select variables without knowing how to influence them. So each target variable must be linked to at least one steering variable: a measure.
(iii)   They must be accessible to easy, affordable and accurate measurement.
(iv)   They should have some indicator value for the condition of the water system or for the condition of its uses.
(v)   They should have some political value and social appeal.
(vi)   The target variables as a whole have to be a reasonable cross-section of the entire water system, its uses and its problems. For example, the biological variables should include species from the benthos, water column, water surface and shores; from high and low parts of the food web — carnivores and herbivores, plants and animals, etc. Chemical variables should include nutrients, heavy and organic micropollutants.
(vii)   They must correspond as much as possible to the parameters of the actual monitoring programme, objectives and models.
(viii) They must be useful for a 10−20 year period.
(ix)   The total number of target variables must be as small as possible.

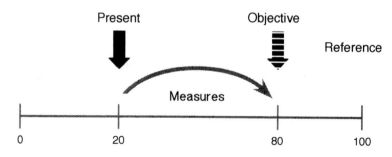

FIGURE 36.4. The measuring gauge in the water system explorations

Measuring gauge

Water System Explorations tries to use the fewest data to gain the best possible impression of the quality and use of water systems. Important characteristics (target variables) of the water systems and their use are defined by a measuring gauge (Figure 36.4).

The measuring gauge approach ensures:

(i)   a link between information and policy;
(ii) definition of the natural background levels and possible objectives;
(iii) the accumulation of information to enable adequate policy and management.

By means of a limited and fixed set of target variables (chemical, physical, biological and use), disorganized measurements are converted to useful packets of information to assist policy and management.

Reference system

Selecting the reference system is a crucial step in the formulation of ecological objectives. To obtain a reliable picture of the uninfluenced system, it makes sense to go as far back in time as possible. However, insufficient knowledge of the system at a particular time makes this impossible. For the North Sea and its neighbouring waters, for example, the condition as it was around the year 1930 has been selected. This is a pragmatic compromise between, on the one hand, the available knowledge and, on the other hand, a relatively low level of human interference. However, in several cases a correction was necessary for human influences.

Three sources are used to determine the reference system:

(i)   old inventories;
(ii) comparative research involving other systems;
(iii) ecological theory.

Management Goals

The next question to be addressed is, which ecosystem offers acceptable guarantees for sustainable development? In short, what is the ecological objective, expressed in terms of the maximum acceptable distance from the reference?

In the Netherlands sustainable development will be considered to have been achieved when the numbers of the organisms of the target variables approach those of the reference situation. It is not necessary to return to the reference situation completely, but when the distance to the reference situation has decreased enough, the situation will also be sustainable. That level has to be fixed on the basis of knowledge of the system and on the policy decision concerning the minimum environmental value.

The government must choose their objectives for the set of target variables based on policy analysis. The policy analysis investigates a number of strategies and looks, in essence, for a balance between the economic costs of measures and the loss of environmental values within the scope of sufficient guarantees for sustainable development in the long term.

Facts and Figures

The design and maintenance of an up-to-date information flow is an absolute pre-condition for adequate policy preparation with respect to water systems. One important objective of

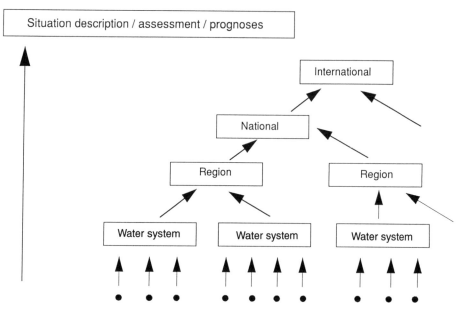

FIGURE 36.5.   The aggregation levels in the water system explorations

the project is to gather sufficient information as required by the water authorities, the province and the State to allow them to realize their tasks in water management. Different areas may need information at different levels; for example:

(i)   at a local level, i.e. a part of the River Rhine;
(ii)   at a water system level, i.e. the River Rhine;
(iii) at a regional level, i.e. the entire national Rhine catchment area;
(iv)   at a national level, i.e. all Dutch water systems, including the Rhine;
(v)   at an international level, i.e. the entire international Rhine catchment area;
(vi)   at a catchment level (national or international), i.e. the catchment area of the river Rhine.

The Water System Explorations aims for an information system in which a status can be presented for all aggregation levels. This is done by aggregating the monitoring data by standardized protocols for each aggregation level. Computers can be a great help. They provide the opportunity for "zooming in and out"; from a status of the River Rhine to a status of a specific part of it, and back to a status of the total Rhine catchment area, depending on the policy question and the level of authority (Figure 36.5).

MODELLING

Monitoring data show how water systems presently function. Models are essential for predicting the future. A central part of the knowledge deals with the relations between the elements within the water systems. Also the influences from outside should be clearly known. Figure 36.6 gives a rough overview of the relevant relations. This knowledge gives us the opportunity to choose the effective steering variables, with which the authorities could

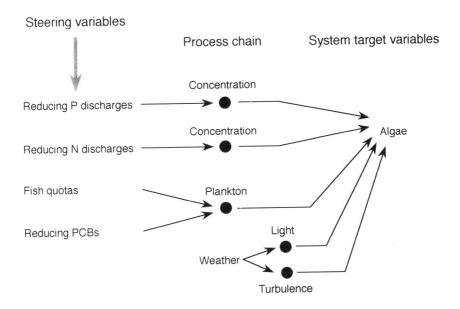

FIGURE 36.6. Relation between steering variables and system variables

influence the target variables. These steering variables also influence several aspects of uses (Figure 36.7).

Analyses provide an insight into the relationship between hydrological, ecological and user functions for water systems. They should form the basis of a definition of the problem

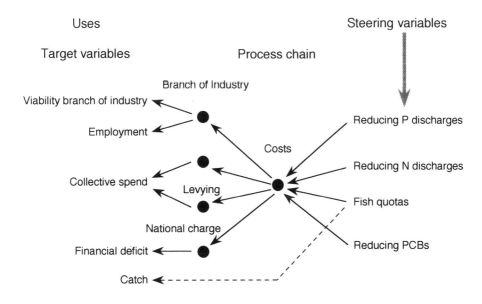

FIGURE 36.7. Relation between steering variables and uses variables

areas and formulation of measures for solutions. It has become clear, however, that policy-makers and managers do not have sufficient coherent qualitative data at their disposal about the fundamental status of water management. In addition, little is currently known about the effects and efficiency of measures. It is intended that the Water System Explorations project will carry out evaluations and perform research on a nationwide and regional level. The results will be entered into accessible databases that will allow prognoses to be made.

Computational framework

One of the aids in the realization of Water System Exploration is an adequate set of models (Luiten and Groot, 1992). These will be developed as part of the project, building on those models already available. A calculation and information system will be organized in support of the set-up and implementation of policy analyses. It is easily accessible, and indicates whether information systems, calculation models and persons or organizations are relevant to the analysis of policy issues.

These models will offer information about:

(i)   sources, paths and effects of natural pollution and pollution causes by man;
(ii) nature, the use and physical disruption of water systems;
(iii) the effect of interference with the environment and the effect on the ecosystem.

Because of the many aspects that had to be incorporated, and the often complex relations between the various components of the water management system, the policy analysis made

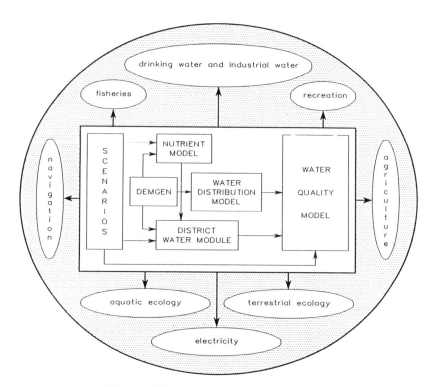

FIGURE 36.8. General modelling diagram

FIGURE 36.9. Schematization districts

extensive use of mathematical models. A review of the general modelling structure is given in Figure 36.8, showing the most important components and their interrelations. Each box in the diagram represents one or more models.

The central model system contains the following models:

(i) SCENARIOS: this model computes the emissions into the surface waters, starting out with several scenario assumptions. These emissions will be input for the district water module and also for the national water-quality model.

(ii) The water-quantity model DEMGEN (DEMand-GENerator), which computes the

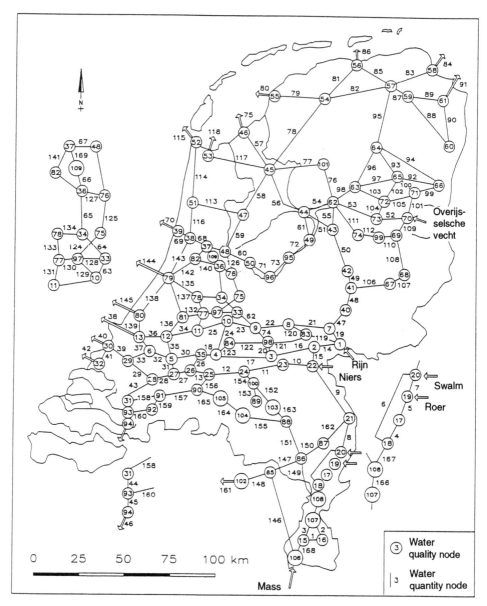

FIGURE 36.10. Schematization network

water balances of the 80 districts (Figure 36.9). The rural area is described by these 80 districts. DEMGEN determines the water demand of the various water users, supplies this information to the Water Distribution Model and then computes the consequences for the users when not all of the demand can be met. This schematization of the rural area has been connected with the distribution model using the same of discharge and withdrawal rules as in reality.

(iii) The Water Distribution Model, which computes the water distribution in the main

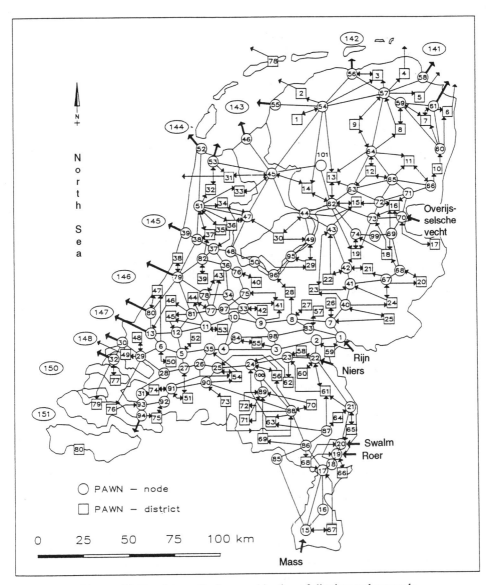

FIGURE 36.11.   Schematization combination of districts and network

water bodies in the country and over the water use categories (see Figure 36.10). The main rivers, canals, lakes and other main waters are represented as a network of links and nodes. Basic inputs to this model are the external supply, the various demands and the allocation rules. The results of the distribution model are: much information about discharges, water use, shortage of water, levels in lakes, and so on. In general: this model supplies all information for quantitative water management. Figure 36.11 gives an impression of the combination of the two schematizations.

(iv) The nutrient model for the simulation of the manure and other deposits in rural areas computes the loads of nutrients in surface waters.

(v)   A water quality model to compute the behaviour of substances in the 80 districts and the loads that enter the main water systems (DISTRICT WATER MODULE).

(vi)   The Water Quality Model, computing the behaviour of substances in the surface water. This model is an implementation of the connected models DELWAQ and BLOOM in the Dutch national water system (Figure 36.12).

Connected to this system were a series of models, of which the major ones are:

(vii) The Drinking water Simulation Model (DRISIM) which, given groundwater extraction quotas and groundwater charges (if any), determines the most economically efficient allocation from a national point of view of the available groundwater among industrial firms and public water supply companies.

(viii) The Electric Power Reallocation and Cost Model (EPRAC) which computes the least expensive way to use the inventory of power plants to satisfy the demand for electric power without discharging enough heat at any node to violate the thermal standards. It uses the excess temperature table computed by the DM, and computes the additional costs to the power companies that can be attributed to the thermal standards.

(ix)   The Shipping Economic Cost model deals with navigation: various models were developed to figure out the economic consequences for navigation due to changes in water depth and lock cycles (costs of delay and extra storage, and the required long-term fleet capacity).

(x)   The District Hydrological and Agricultural Model (DISTAG) which is a part of the earlier-mentioned model DEMGEN. It computes the surface water and salt balances of the 80 districts. It also computes the district groundwater levels and the sprinkling costs, physical damage to crops and loss of income due to lack of water and/or high salinity.

(xi)   The computation of the impacts on terrestrial ecology occurs using DEMNAT, a model that predicts such consequences on a nationwide scale.

(xii) For the aquatic ecology available knowledge will be processed into models.

All models described above were linked to the central models in one of the following ways:

(i)   on-line connection (such as DISTAG, the salt wedge model, the navigation cost models, and the pollutant transport models): the models were run as subroutines of the Water Distribution Model;

(ii)   pre-processing: results of a model or study were used as input in the distribution model (e.g. water demands by drinking water companies and industry);

(iii) post-processing: results coming from the distribution model were used in the relevant models (e.g. EPRAC).

With this set of models the impact of any infrastructural measures and/or particular water distribution policy could be determined.

REPRESENTATION: A TOOL FOR MANAGEMENT

"Amoeba"

When the relevant measuring gauges are placed in a "radardiagram", a clear and simple presentation will be the result (Figure 36.12). The essential elements and theories of this "amoeba-approach" have been published by Ten Brink, Hosper and Colijn (1991).

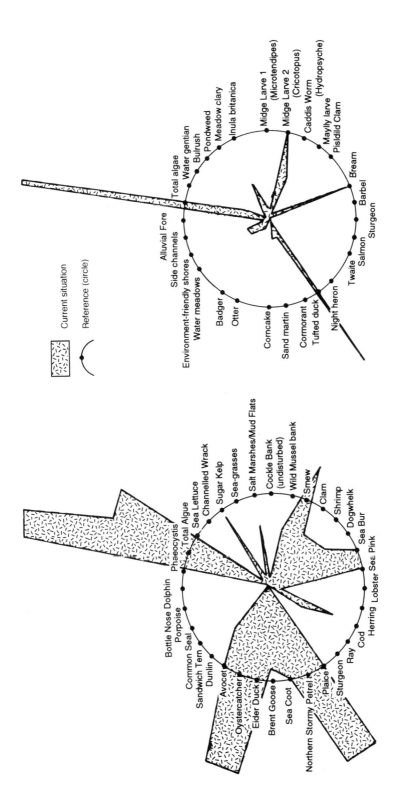

FIGURE 36.12.   The North Sea and river amoeba representation

468

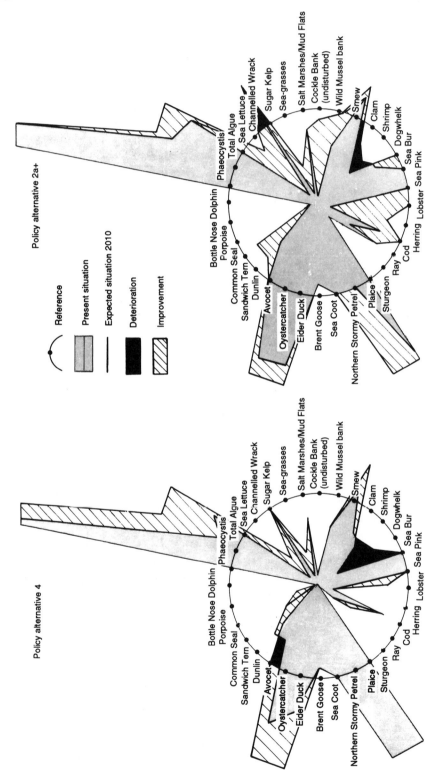

FIGURE 36.13. The impact amoebas for the North Sea

The target variables are arranged in systematic order in the form of a circle. The distance from the edge of the circle to the centre represents the numbers in the reference situation for each species. The actual numbers are superimposed on this circle. For visualization purposes, all points are connected by a line, which produces the two amoeba-like figures. These figures, the "amoebas" for a catchment area, for example, present a relatively simple picture of the actual situation of the ecosystems in that area. They show how mankind made use of the environment and, unintentionally, changed the ecological situation.

Impact amoebas

To achieve the objective, a strategy has been drawn up. Measures will be taken that will influence the levels of the target variables mostly in a positive fashion (but sometimes negatively). Conceptual and mathematical models were developed to assess changes of all species in the "amoeba" as a result of changing steering variables like water quality, fish catches, restoration of biotopes, etc. The impact of several policy alternatives (packages of measures) can be calculated. The impact of each alternative will be illustrated by "impact-amoebas". One notable 'impact-amoeba' for the sea is shown in Figure 36.13.

Status

A combination of several amoebas and other relevant information about the actual or calculated levels of the target variables for the system and the uses of the system can be placed in one simple overview: the status (Figure 36.14).

Environment Mondriaan for water

The Environment Mondriaan is a diagrammatic map of the catchment basins in the Netherlands which indicates the environmental conditions of the fresh, brackish and salt water areas (Figure 36.15). All the available information is brought together in this map, making it possible to see at a glance how a basin or a water system (stream, canal, river, estuary, etc.) is faring. A combined evaluation of the state of all target variables is represented by a corresponding colour on the chart, indicating the local situation. Each defined section on the Environment Mondriaan stands for biological, chemical, physical and user amoebas. For instance, if the section is red, this signifies serious pollution, unprofitable use, a disrupted ecosystem, or inadequate safety. The Environment Mondriaan is able to depict a maximum amount of relevant information using a minimum amount of figures.

Time development graph

While the status represents several variables for one system and the mondriaan gives information of one variable for several systems, there is a third representation method, which we have known for a long time: the two-dimensional time development graph. It gives information of one variable concerning one water-system (Figure 36.16).

Overview representation methods

With the use of the representation methods it is possible to represent quantitative information of ecological and economical aspects of water-systems, and to assess the extent to which the

470

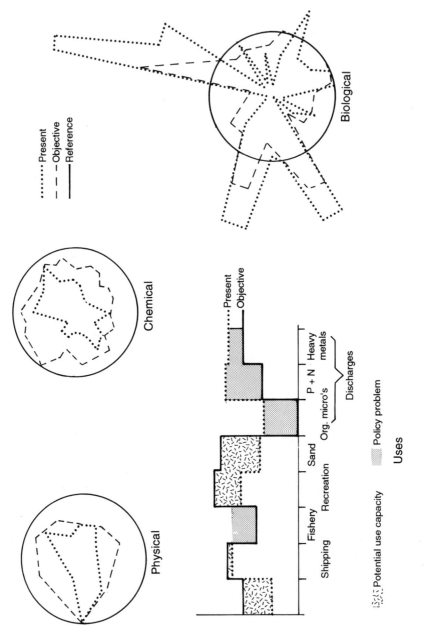

FIGURE 36.14. The status of a water system

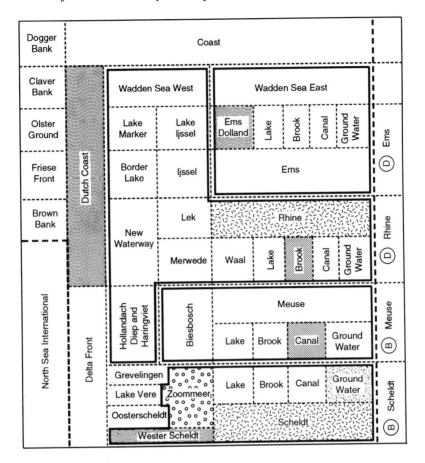

FIGURE 36.15. The Environment Mondriaan for water

objectives have been achieved. It thus makes more integrated and balanced decisions possible.

The representation is simple and easily visualized. It makes the problems of ecological decline more accessible for waters authorities and the public alike. Therefore it can function as a vehicle of communication between policy-makers and scientists. A better understanding of the ecological objectives, the problem areas and the measures that have to be taken will increase the collective will to act. At the same time it also visualizes mans's impact on the environment and makes it clear that the ecosystems may deviate even further from the ecological objectives.

The amoeba appears to be of reasonable universal application. In principle, it can be used on all scales and for every system or catchment area.

CONCLUSIONS

The Water System Explorations are a simple way of obtaining adequate policy on water management, resulting in a periodic, standardized report. The status can act as a

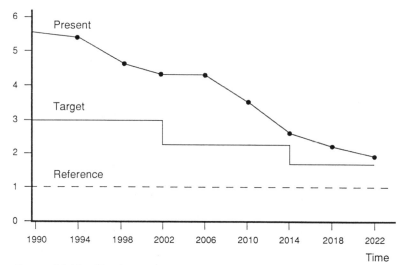

FIGURE 36.16.   The time—development graph of one variable for one system

thermometer for the condition and development of the entire water system and its uses, presented in an accessible form. Once target variables have been selected, policy makers can give direction to policy-orientated monitoring and research programmes.

Research and monitoring are focused on approximately 150 physical, chemical, biological, health and use variables. The Water System Explorations can, therefore, serve as a good vehicle for communication between policy-makers and scientists.

With the aid of this information system on water management, a balance can be found between the ecological aspects of the water system and the use of the systems by society in a rather systematic way.

These ecological and economic figures of the status are the two compasses by which water managers can plot a course towards sustainable development. The success and failure of each governmental period can be read from these compasses. Using these methods, the socio-economic policy of today is growing towards the socio-economic policy of tomorrow.

The biological component of the amoeba was already worked out in approach for the third Water Management Plan. In conclusion, it is of the highest importance that the water systems are preserved to reach a state of sustainable development. The model of the Water System Explorations can serve as a valuable instrument to allow water managers to achieve this.

A new policy strategy

Based on the described approach, the Water Management Plan (Ministry of Transport, Public Works and Water Management, 1989) concluded:

(i)   Dutch watersystems are incomplete and unbalanced in composition. With continuation of the present use ecological decline will proceed and there are no guarantees for sustainable development.

(ii) The present-day policy of a 50% reduction of the discharges of contaminants — Rhine and North Sea Action Programmes — do not produce any improvement; they merely prevent further decline.

(iii) The exclusive reduction of contaminants, even by 90%, has relatively limited effect. Nevertheless an ongoing reduction is a necessary condition for recovery.

(iv) Reduction of contaminants in combination with supplementary measures is, at this moment, the best policy option.

These conclusions lead to a new policy and management strategy for the North Sea: the multi-track approach.

37

The Use of Biological Techniques in Catchment Planning

ALUN S. GEE and FRANK H. JONES

National Rivers Authority, Cardiff, UK

INTRODUCTION

The demand for water in England and Wales is increasing: the present demand for potable water supply is 17×10^6 m^3 day^{-1}, whilst projections for 2021 average 21×10^6 m^3 day^{-1} (NRA, 1992a). Present and future requirements must take account of in-river needs for water however, as well as the legitimate demands of abstractors, in contrast to the historical situation. These in-river needs include the requirements of plants and animals living in, or dependent upon, the river and the people who use the river for recreation.

For the past 20 years the management of water in England and Wales has been organized on a catchment basis. When the National Rivers Authority (NRA) was formed in 1989, with the statutory responsibility for managing the whole of the water cycle, it inherited the regional structure of the former water authorities, thereby maintaining the ability to manage river catchments in an integrated way. Recently, the NRA has developed the concept of "Catchment Management Planning" as a vehicle for expressing its vision for these catchments. It was recognized that a means was needed of formally assessing present and future uses of water and associated land, so that interactions and potential conflicts could be identified and an action plan drawn up, which allocates responsibilities for achieving improvements.

As well as enabling the NRA to develop and express its vision for a catchment, Catchment Management Plans (CMPs) provide the focus for its numerous activities. The 1989 Water Act, as consolidated into the Water Resources Act 1991, allows for the classification of rivers and other waters and the setting of statutory Water Quality Objectives (WQOs). The NRA has proposed that these WQOs, and any more general classifications, should be incorporated into its CMPs.

A recent consultation exercise (NRA, 1992b) demonstrated that there is widespread support for the inclusion of biological information in national water-quality classification schemes. The advantages of using biological information in conjunction with the more traditional chemical assessment were widely recognized. This chapter outlines the way in which biological information can be used to great advantage not only in local investigations,

---

The Ecological Basis for River Management. Edited by D.M. Harper and A.J.D. Ferguson. © 1995 John Wiley & Sons Ltd

but also in assessing trends in quality and in supporting the planning of catchments for all water users.

CATCHMENT MANAGEMENT PLANNING IN THE NRA

What is Catchment Management Planning?

The water environment is increasingly becoming the focus for a variety of uses and activities. The NRA needs to examine the interaction between these uses, their effect on the environment, reconciling any conflicts that may arise, and generally protecting and improving the natural resource. Catchment Management Planning is the process of ensuring that all the problems and opportunities resulting from the uses within a catchment are presented within a well-defined, flexible framework capable of maximizing the overall well-being of the water environment.

For the purpose of a CMP, a catchment is defined as a discrete geographical unit within boundaries derived primarily from surface water considerations. One or more hydrometric sub-catchments may be included depending on practical convenience. Groundwater and other inter-catchment issues are treated as inputs to and exports from the CMP catchment. CMPs are also prepared for coastal zones, usually including the river catchments which drain into them.

The CMP process involves a considerable amount of consultation with interested parties. The final document produced is seen as an agreed strategy for realizing the environmental potential of the catchment within prevailing economic and political constraints.

How is a CMP produced?

CMPs involve the NRA and others with an interest in the water environment, in a series of activities, which include:

(i)   identifying current and future catchment uses and activities,
(ii)   setting environmental targets which are necessary to protect these uses and activities,
(iii)   comparing these targets with the current status of the water environment,
(iv)   identifying issues and problems resulting from the interaction of uses or failure to meet the targets,
(v)   identifying possible solutions,
(vi)   undertaking consultation on the uses, targets, issues and options,
(vii)   preparing an action plan to address the issues, setting out the means by which and the time-scale within which the agreed "vision" will be achieved,
(viii) implementing the action plan, and
(ix)   reviewing the CMP and update it as necessary.

In drawing up a CMP an attempt is made to accommodate the reasonable requirements of all the parties concerned, having due regard for the relative importance of the issues and uses. Inevitably, difficult decisions have to be made, with economic and political considerations sometimes overriding scientific judgements.

One of the key steps in the CMP procedure is the determination of the suitability of the environment to support the use or activity. This, in turn, is dependent upon an understanding of appropriate environmental standards. Most uses will be dependent upon the provision of a finite range of water-quality conditions, the availability of a particular flow regime and certain physical features of the environment. For example, to be suitable for use by migratory salmonid fish, a river must have good water quality, characterized by high oxygen content and low pollutant concentrations; a flow regime providing a stimulus to migrate upstream (and down); and a physical environment providing cover from predators, free access between the sea and their nursery grounds, and gravel beds suitable for the incubation of their eggs. Scientifically, one of the greatest challenges in catchment planning is the identification of these environmental requirements, without which the work necessary to allow achievement of the uses in the catchment cannot be determined. In the absence of process-based knowledge of the requirements, empirically-derived biological indicators can often be extremely useful.

STATUTORY WATER QUALITY OBJECTIVES AND CLASSIFICATION

Within the context referred to in the Introduction, the Government published a consultation document outlining a scheme of statutory Water Quality Objectives (WQOs) and proposals for a general classification of rivers (Department of the Environment and Welsh Office, 1992). This followed an extensive consultation process carried out by the NRA (NRA, 1991a), whose subsequent recommendations have been incorporated in the government proposals.

The WQO scheme centres around the concept of Use Classes (UCs) which would be applied to specific bodies of water. These would provide the principal driving force to maintain, or improve where necessary, water quality such that the discharger concerned, and the public, could see the specific advantages to be obtained from the improvement. Because these would vary from one body of water to another, it is also proposed to have a general classification scheme which would allow the overall quality of water to be assessed and compared on a common basis every five years.

The close similarity between the WQO proposals and the CMP philosophy is obvious. Both have as their foundation, the concept of managing the water environment in order to make it suitable for particular uses. However, the use-related classes which comprise WQOs are defined by water-quality standards alone, in contrast to CMP uses for which a more comprehensive set of environment requirements must be met. The proposals for WQOs are summarized in Table 37.1, taken from the Government's consultation document (Department of the Environment and Welsh Office, 1992). Full details of the standards for the River Ecosystem Use Class are given in the Government's Regulations (Department of Environment and Welsh Office, 1994). Table 37.2 demonstrates that these WQO uses are a subset of the wider range of uses and activities which are used in catchment planning.

In its original proposals (NRA, 1991a), the NRA suggested that biological indices should be used as standards for targeting environmental improvement. Consultees subsequently argued, however, that the proposed biological model, the River Invertebrate Prediction And Classification System (RIVPACS), because of its computer modelling origin, could not stand up to challenge in the courts. RIVPACS was developed by the Freshwater Biological Association (now the Institute of Freshwater Ecology, IFE) from the analysis of data from an extensive invertebrate survey of unpolluted British rivers (Wright et al., 1984). The system can predict from a set of environmental data (e.g. altitude, river width, water hardness, etc.), the invertebrate fauna that should be present under pristine water-quality conditions. A comparison of the actual fauna found at any site with that predicted by this system therefore provides a means of assessing the biological quality of a river which is independent of geographical variation.

TABLE 37.1. Proposed use classes for incorporation in a scheme of Statutory Water-Quality Objectives

Use classes (UCs)	
Fisheries ecosystem: (now River Ecosystem) Water quality intended to be suitable for:	Class 1: high class salmonid/coarse fishery Class 2: sustainable salmonid/high class coarse fishery Class 3: high class coarse fishery Class 4: sustainable coarse fishery Class 5: some fish may be present, but not a sustainable fishery Class 6: fish unlikely to be present
Abstraction for drinking water supply	Standards related to EC Directive
Agricultural abstractions	Sets of standards proposed for livestock watering; and for irrigation
Industrial abstraction	Suitable standards being considered
Special ecosystem	Site-specific standards selected from matrix of possibilities in Regulations
Watersports	Standards related to health risk

It was also argued that biological targets were not suitable for planning water-quality improvements because there could be no guarantee that a higher ecological quality would result from an improvement in water quality following investment in pollution control. Nevertheless, the use of a necessarily limited range of chemical standards is clearly inadequate to describe the overall quality of a water body.

In an attempt to derive a practical methodology for targeting environmental improvement, it should not be forgotten that the chemical quality of the water *per se* is of less relevance than its ability to sustain the required uses or activities. Indeed, some of the WQO use classes are clearly ecologically based (Table 37.1). In the case of the use of a river as a habitat for a diverse and self-sustaining ecosystem, a biological indicator may be a better gauge than a suite of sanitary chemical determinands.

Current proposals for the use of biological indicators in river classification recognize the above concerns. Whilst the WQO use classes would have chemical standards only, the generalized classification used for periodic assessment would include a biological component based on RIVPACS. There is also much scope for other biologically-based classification schemes which would help quantify the ecological state of catchments. These include fisheries and conservation indices which are the subject of intensive research and development at the present time.

THE ROLE OF BIOLOGY

Classification

There is a substantial amount of contemporary biological data for classified river reaches in the UK as a result of the 1990 River Quality Survey (RQS) and the level of monitoring that

TABLE 37.2. Typical (non-exhaustive) list of catchment uses for catchment management planning

Catchment uses related to SWQOs
River ecosystem
Abstraction for drinking water supply
Watersports
Industrial/agricultural abstraction
Special ecosystem

Other catchment uses
Basic amenity
Landscape
Angling
Boating
Recreation
Water transfer
Mine working
Solid waste disposal
Industrial effluent disposal
Sewage effluent disposal
Urban development
Road, rail and airport development
Agricultural activity
Wet fencing
Water power (including Mill Rights)
Archaeology and heritage
Flood defences
Flood water storage
Navigation
Commercial harvesting of marine fish/shellfish

has been subsequently sustained. A biological classification based on RIVPACS has been applied to these data in order to make general assessments of river quality. By this means, impacted areas within catchments can be identified which may assist in the targeting of pollution control efforts.

Further benefits can be obtained from biological surveys by comparison with the chemical data. Recent studies have demonstrated relationships between the biological and chemical data from the 1990 RQS (WRc, 1992; NRA, 1992c). For instance, it was demonstrated that the biological data could place 93% of sites within plus or minus one class of their true chemical grade (NRA, 1992c). Circumstances where biological quality was significantly poorer than may be expected from the chemical class may highlight episodic pollution or where non-sanitary chemical determinands may be operating. This may highlight the need for more comprehensive chemical analyses or continuous water-quality monitoring which may, in turn, assist in identifying a pollution source, leading to remedial action. Biological surveys can also provide cost-effective indications of water quality for those reaches which are not included in routine classification exercises.

Impact assessment

Biological techniques can play a major role in the assessments of the impact of discharges. Conventional chemical monitoring based on the collection of discrete samples may not

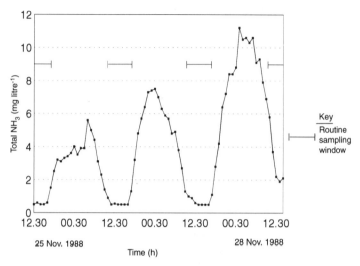

FIGURE 37.1. Continuous monitoring of sewage treatment works discharge showing fluctuations in quality outside the normal "sampling window" (from NRA, 1991e)

adequately characterize the impact of a discharge either because of fluctuations in quality (particularly outside the routine chemical sampling window) or because of the presence of biologically active substances that are not routinely analysed (Figure 37.1). Such regimes may be detected by examination of the invertebrate fauna since they provide a time-integrated monitor of water quality. Studies of simulated farm pollution episodes have shown that the invertebrate fauna takes about two months to recover to its pre-episode quality (Figure 37.2)

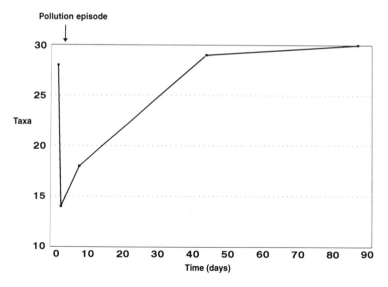

FIGURE 37.2. The rate of recovery of invertebrate fauna from a simulated farm pollution episode (from Turner, 1989)

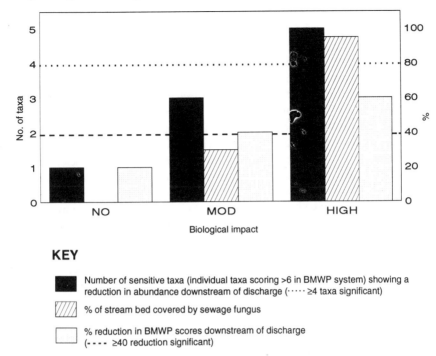

KEY

▪ Number of sensitive taxa (individual taxa scoring >6 in BMWP system) showing a reduction in abundance downstream of discharge (····· ≥4 taxa significant)

▨ % of stream bed covered by sewage fungus

☐ % reduction in BMWP scores downstream of discharge
(- - - - ≥40 reduction significant)

FIGURE 37.3. Example of a range of impacts of sewage treatment works effluents on the macroinvertebrate fauna and sewage fungus communities in receiving watercourses (NRA, 1992d, reproduced by permission)

(Turner, 1989). Sewage treatment works (STW) effluents can be subject to these influences due to the periodicity in the quantity and concentration of influent raw sewage which may also contain toxic substances from industry. Systems for assessing the biological impact of STWs which are based on the effects on invertebrate and microbial communities, have been developed in some NRA Regions and River Purification Boards (Extence and Ferguson, 1989; NRA, 1992d; ADRIS, 1993). Samples upstream and downstream of the discharge are collected and the degree of impact is determined from a number of criteria including Biological Monitoring Working Party (BMWP) score differences, reductions in the abundance of pollution-sensitive invertebrate taxa and the presence of sewage fungus in and below the effluent mixing zone (Figure 37.3). From these, an overall biological impact rating is derived which, together with other criteria (including aesthetic impact and the effect on downstream water-quality objectives), is used to determine an overall environmental impact. In the Welsh Region this information is used to determine the need for revisions to discharge consents and to present to the local water company the NRA's view of priorities for improvements in discharge quality, thereby influencing the investment programme.

Similar approaches to that adopted for STWs could be applied to other types of discrete discharges. The measure of impact on the invertebrate fauna could be improved by basing the assessment on the relative difference between the fauna actually observed and that predicted by RIVPACS upstream and downstream of the discharge. However, for this technique to be applied extensively there would have to be further development of the RIVPACS model to include lower order streams; at present it can only be applied reliably to streams more than 5 km from their source.

Rapid indicators

The biological techniques described above rely on the collection of samples and their subsequent transport to the laboratory for processing, which typically requires 2−3 h per sample. The time taken, therefore, places restrictions on the size of sampling programmes and inevitable delays occur in the reporting of results. There is clearly a need to develop rapid methods of assessment of stream quality whereby large numbers of sites can be surveyed and the results reported quickly. This approach is particularly valuable for extensive catchment surveys to detect pollution sources which would be difficult to pinpoint by conventional means. To date, systems have been developed to detect watercourses affected by farm pollution and acidic precipitation (Rutt, Weatherley and Ormerod, 1989; WRc, 1991). These systems have been developed using similar principles. Initially, an extensive invertebrate survey is undertaken covering an area of similar topography which as well as including impacted sites also encompasses streams of pristine quality. TWINSPAN community analysis (Hill, 1979) is then applied to the results to identify key indicator taxa. The presence or absence of these can then be used to assess the impact of agricultural pollution or to assess the degree of acidification (Figures 37.4 and 37.5).

The use of the rapid indicator method has a key role in the strategy employed by the NRA in tackling the problem of farm pollution in areas where there are a large number of farms. Biological surveys are used to target efforts of pollution control at those farms actually having an adverse impact on stream ecology. A relatively large number of sites can be evaluated by applying the indicator key (Figure 37.4) to samples examined on the bankside and the results simply presented on catchment maps to highlight the location of problem farms. A major feature of this approach is that stream quality is re-evaluated following the

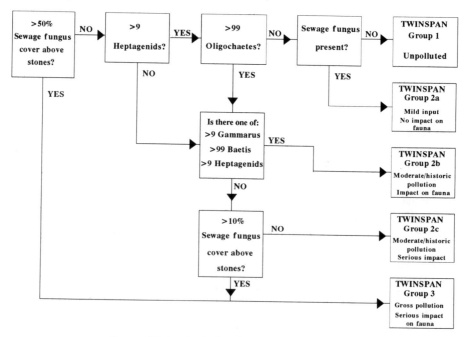

FIGURE 37.4.   Agricultural Pollution Indicator Key

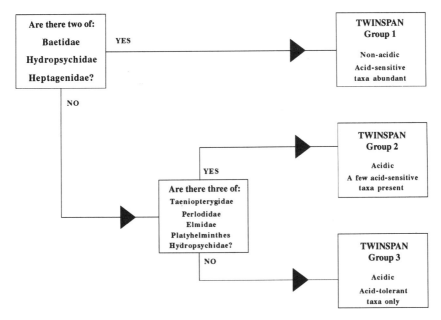

FIGURE 37.5. Acidification Indicator Key (modified from Rutt, Weatherley and Ormerod, 1990)

implementation of remedial pollution prevention measures so that their benefit can be directly assessed.

The acidification key is similarly applied in the field (Figure 37.5) and is a cost-effective method of determining affected catchment areas. These assessments can be important in determining land-use policies such as responses to afforestation proposals and in determining the need for catchment liming. It also has applications in providing evidence of fish mortalities arising from acidification.

Risk assessment

Biological information in the form of toxicity data has traditionally been invaluable in dealing with pollution investments. Such data could be incorporated formally into an assessment of the risk of a chemical spillage having an unacceptable effect on a receiving water. The NRA is currently exploring the ways in which such techniques could be used to support the promotion of water protection zones under Section 93 of the Water Resources Act 1991. Information would need to be collected on the quantities of chemicals stored, spill-containment facilities, the proximity of plant to surface waters, the prediction of pollutant concentrations at defined points downstream of the industrial site and the acceptability of these based on toxicity data. Pollution prevention measures could then be required at industrial premises to reduce the risk of incidents to an acceptable level.

CASE STUDIES

The application of some of the biological techniques outlined can be illustrated by reference to three Catchment Management Plans currently being prepared in the Welsh Region, namely those for the Ogmore, Cleddau and Wye catchments (Figure 37.6).

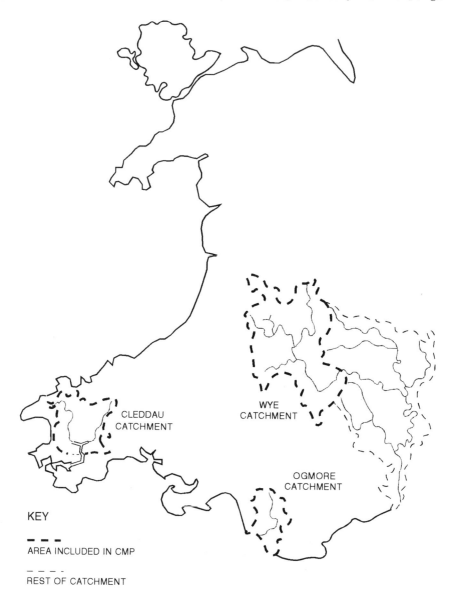

FIGURE 37.6.   Location of Ogmore, Cleddau of Wye catchments

The Ogmore catchment

This is a heavily urbanized and industrialized catchment within the South Wales coalfield. It has been subject to a long history of pollution but within recent years there has been a substantial improvement in quality due to a reduction in heavy industry and improvements in the quality of sewage discharges and the remaining industrial effluents.

The social and economic changes which have occurred in the catchment have combined to require and allow for a wide range of water uses. In particular, improvements in water

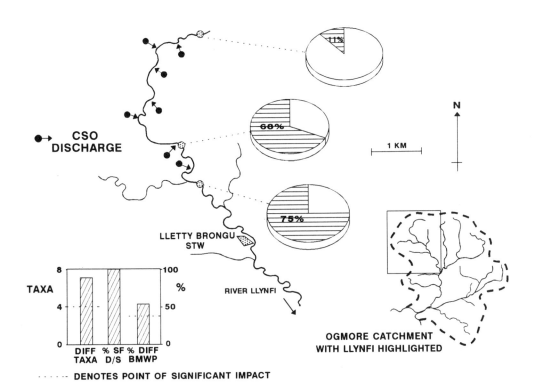

FIGURE 37.7.   River Ogmore Catchment: differences between biological quality of the River Llynfi in spring and autumn due to Combined Sewer Overflow (CSO) impact. The effect of Lletty Brongu STW on macroinvertebrates and sewage fungus is also illustrated

quality have resulted in improved runs of migratory fish and an increasing demand for fishing and other water-related activities. Nevertheless, some significant problems remain and are addressed in the CMP. Conventional biological monitoring within the catchment has demonstrated substantial seasonal fluctuations in the quality of the water in some classified reaches. Typically, biological quality in the spring has been found to be markedly poorer than in the summer or autumn (Figure 37.7). Studies in other catchments in South Wales had attributed this feature to the adverse effects of the operation of combined sewer overflows (CSO) during the winter period (NRA, 1991b). This diagnosis was subsequently confirmed in the Ogmore catchment by the deployment of instrumentation to determine the frequency of operation of CSOs. This demonstrated that the CSOs responded very rapidly to rainfall before there was adequate dilution in the receiving stream, reflecting the inadequate capacity of the sewerage system (NRA, 1992e). This information is being used to determine investment priorities for improving the operation of CSOs.

Comparisons between the biological and chemical quality of the classified reaches in recent years identified two reaches where biological quality was markedly poorer than expected from the chemical class. In one case, this was ascribed to the periodic drying of stream as a result of seepage through the stream bed. In the other, it was related to the influence of a paper mill discharge. Recently, there have been major improvements in the quality of the effluent discharged from this industry which have resulted in corresponding improvements in both the chemical and biological quality of the river. Consequently there are no longer any major discrepancies between the chemical and biological assessments.

Biological surveys have repeatedly shown that a large sewage works in the Ogmore catchment is having a major environmental impact. In conjunction with the investigations of the operation of CSOs, this has demonstrated the need for a revision of discharge consents and the need for substantial improvements in the operation of the sewage works and the supporting sewerage system.

Although the Ogmore catchment is not one that would be generally considered to be at risk from acidification because of the nature of its dominant geology, application of the acid indicator key has demonstrated that one of its tributaries is impacted. This stream is poor in acid-neutralizing capacity and receives enhanced acidic deposition due to scavenging by conifers (NRA, 1992f). Remedial measures such as liming are being considered for this area if funding can be obtained and conservation interests protected, pending an adequate reduction in atmospheric pollution.

Historically, there has been a high frequency of pollution incidents in the Ogmore catchment and some of the more serious incidents have occurred as a result of the release of toxic chemicals from industrial sites. The most serious incident occurred in 1987 when about 50 000 salmonids were killed as a result of a leakage of a chemical called Kymene, a polymer used in the manufacture of tissue paper. It is intended that this catchment will have a high priority for the application of a risk assessment analysis and industrial site visit programme.

The Cleddau catchment

This is a predominantly rural catchment situated in western Wales where the principal land use is dairying and cattle rearing. Cattle densities in the catchment are amongst the highest in the UK (about 200 km^{-2}), which, coupled with a high annual rainfall ($c.$ 1270 mm ann^{-1}), leads to problems in the safe disposal of cattle slurry and silage liquor. These problems need to be overcome if the full potential of the catchment to sustain the wide range of water uses is to be realized. In particular, important water abstractions and local fisheries are at risk from episodic farm pollution.

Despite the high organic loading within the catchment, the biological and chemical quality of classified river reaches as determined by routine monitoring is high because of the adequate dilution afforded. However, application of the agricultural indicator key has demonstrated that many streams within sub-catchments are adversely affected by organic pollution (Figure 37.8). Of 75 sites surveyed within 17 sub-catchments between 1989 and 1991, 28% were found to be grossly impacted and 27% moderately impacted. These sub-catchment areas are important nursery areas for salmonid fish and low densities of juvenile salmonids have been found to be associated with organically polluted sites. There has also been a major decline in the catches of adult migratory salmonids in the catchment in recent years. The biological techniques are playing an important role in identifying the problem farms and a major programme of farm visits is now underway.

The Upper Wye catchment

The CMP covers the part of the catchment upstream of Hay-on-Wye, equivalent to 40% of the catchment area. This is a rural area with afforestation and sheep rearing the predominant

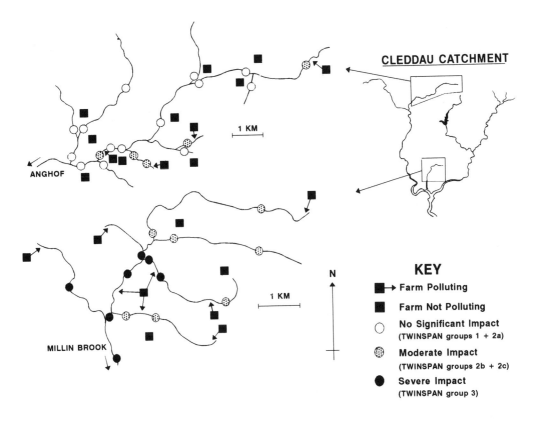

FIGURE 37.8. Application of the Agricultural Pollution Indicator Key within the Cleddau catchment

land use in the uplands and some cattle rearing in lowland areas. The nature of the underlying geology and the thin soils in the uppermost part of the catchment renders it vulnerable to acid precipitation. This is a significant factor in limiting the spawning success of salmon in some of the upper tributaries and a contributory factor in the decline of salmon in the Wye. Extensive surveys using the acid indicator key have established that 51 km of the streams surveyed (approximating to 89 km^2 of the catchment) are impacted (i.e. within TWINSPAN groups 2 and 3) (see Figure 37.9). Previous investigations have established that fish are either absent or in very low densities at sites within these TWINSPAN groups (Weatherley and Ormerod, 1990). A combination of emission controls, effective land use planning and ameliorative management is needed to overcome these problems.

FUTURE CHALLENGES

It is becoming increasingly apparent that the effective management of the water environment depends both on an appreciation of the interactions between its various components and on the effects of external factors, particularly the influence of man. There has, historically, been a tendency to plan for the use of water in a uni-functional, blinkered way. Those planning the abstraction of water from a river for domestic or industrial purposes tended not

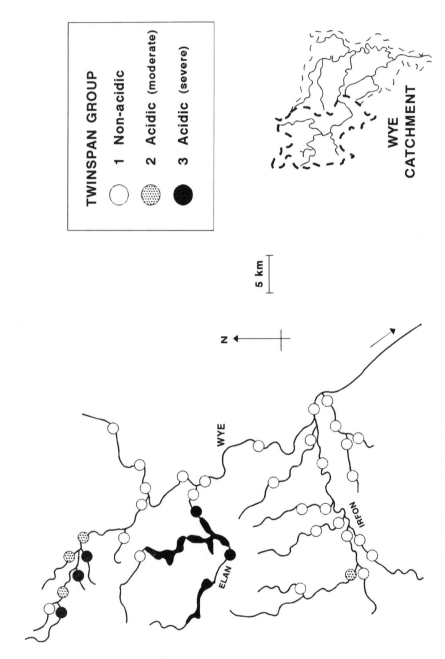

FIGURE 37.9.   Application of the Acidification Indicator Key within the Wye catchment

to consider the needs of the plants and animals living there. Water-quality assessments had little regard for the river's wider, biological condition. Pollution control activity tended to concentrate on discrete discharges and had less regard for the effects of land use. The problem of acidic precipitation has illustrated how external influences, sometimes spanning countries and continents, can materially affect local conditions. Today, catchment planning facilitates the consideration of all the relevant factors which must be understood and managed if a river is to sustain the demands which are placed upon it.

In England and Wales, the NRA is well placed to manage the water environment in an integrated way. In future, under the regime of an environment agency for land, air and water, the challenges are even greater. Our understanding of the interactions between these various media must be increased and practical tools developed to assist the process. Biological information has yet to be routinely used in river quality classification due to an incomplete understanding of the relationships between chemical and biological information. Greater challenges await those who attempt to take into account the effects of water quantity and habitat features.

This chapter has concentrated on the way in which biological tools, such as simple invertebrate indicators, can help pinpoint problems which would otherwise require time-consuming and expensive chemical analyses. Clearly, catchment planning requires more of these tools if environmental problems constraining the many catchment uses are to be identified and overcome.

ACKNOWLEDGEMENTS

This chapter is published with the permission of the National Rivers Authority but the views expressed within it are those of the authors and do not necessarily represent those of the National Rivers Authority.

38

Managing Rivers and their Water Resources as Sustainable Ecosystems: The South African Experience

JAY O'KEEFFE

Rhodes University, Grahamstown, South Africa

INTRODUCTION

In South Africa, adequate supplies of good quality freshwater are often the limiting resource for economic and social development. Much of the country has a mean annual rainfall of less than 500 mm, and on average, only 8.6% of precipitation is converted to runoff in rivers. Typically for arid areas, rainfall is erratic and unpredictable, so that much of the runoff appears in large floods which cannot be stored for human use. There are practically no freshwater lakes in South Africa, and limited groundwater supplies, so that the country depends for its water supplies on rivers. The mean annual runoff of the country is 51.1×10^9 m^3 ann^{-1}, but only 33×10^9 m^3 ann^{-1} of this can practically be used by people. Complicating the problems of managing this inadequate and erratic supply, the distribution of high rainfall in the country is not well matched to the areas of major demand, such as the Pretoria/Johannesburg area, which was sited, for mineral and industrial reasons, on a watershed between the Limpopo and Orange/Vaal River systems. An ill-defined wet/dry cycle of 18–20 years (Preston-Whyte and Tyson, 1988) further complicates the management of water resources. In a developing, but still largely Third World country such as South Africa, there is an urgent need to upgrade domestic water supplies and to encourage economic growth, but rapid population growth (averaging 2.85% per annum) shows no sign of slowing down, and threatens to outrun all attempts to raise standards of living. There has also been a desire to decentralize development from the overpopulated and over-exploited mining-based centres.

Since the early 1950s, there has been growing concern about the limited quantity and deteriorating quality of the country's water supply (e.g. Stander, 1952). However, this concern has been manifested in a determination to exploit all available river systems to the limit, and river channels have been viewed simply as conduits for leading water from "A" to "B", and for the disposal of effluents. The prevalent view was that any water which was allowed to flow into the sea was a waste, and that plans should be made to intercept it.

The Ecological Basis for River Management. Edited by D.M. Harper and A.J.D. Ferguson. © 1995 John Wiley & Sons Ltd

Biological research on rivers during the 1950s and 1960s (e.g. Harrison and Elsworth, 1958; Oliff, 1960; Allanson, 1961; Chutter, 1971) still provides the basis of our understanding of the ecology of South African rivers, but was aimed largely at providing indices of pollution such as Chutter's Biotic Index (Chutter, 1972), rather than defining sustainable limits to exploitation.

Since the early 1980s, a realization has developed for the need to allocate "water for nature" in the country's rivers (Roberts, 1983), and to set special effluent standards in heavily exploited catchments to maintain water quality (Department of Water Affairs, 1986). In the case of water-quality standards, management objectives have always been linked closely to user requirements, and have automatically incorporated a concern for the riverine biota, which have been seen as first-line indicators when conditions are unsatisfactory. (Fish kills, for example, have always been seen as extreme examples of unacceptable water quality.) The incorporation of environmental standards for water quality has therefore not been a problematic concept, since the needs of human users and the natural biota have always been seen as compatible. The recent adoption by the Department of Water Affairs of the policy of Receiving Water Quality Objectives (RWQO) (van der Merwe and Grobler, 1990) now routinely includes an assessment of the requirements of the riverine biota. The RWQO method, developed in the USA, aims to set specific water-quality standards for each catchment, based on the requirements of users and the assimilative capacity of rivers for particular water-quality variables.

The policy towards providing sufficient quantity of water to maintain the riverine environment is rather more confused. The Department of Water Affairs have accepted, in principle, the need for "the release of water to maintain the riverine or estuarine ecology" (Department of Water Affairs, 1986), but this need is seen as separate from, and competing with, all the requirements of other users. The natural biota is therefore seen as one amongst a number of potential users of a river's resources, and ecologists are charged with assessing the quantity and seasonal distribution of water required to maintain the biota. The provision of this water then has to be motivated and negotiated in competition with the requirements of agriculture, municipalities and industries. The implication is that, if the motivation is not strong enough, "the ecology" of the river will have to make do with a reduced allocation, or possibly no allocation at all. This view of the biota/ecology/nature/environment (all these words are used interchangeably by non-ecologists) as separate from the resource, and competing with other users, prevents the development of an integrated management plan for rivers, and will usually result in the environment losing out in the competition for water, on short-term economic grounds.

The purpose of this chapter is to examine the user requirements in two contrasting river catchments in South Africa, and to describe recent assessments of water quantity and quality conditions, comparing historical and present conditions with the needs expressed by present users. The two rivers, the Sabie in the eastern Transvaal and the Buffalo in the eastern Cape Province, have both been the subject of intensive research, the results of which have been supplied to the South African Department of Water Affairs to assist them in drawing up management plans for the two catchments.

A BRIEF DESCRIPTION OF THE SABIE AND BUFFALO RIVERS

The Sabie River (Figure 38.1) rises on the eastern Transvaal escarpment, in extensive areas of commercial forestry, and flows through the small town of Sabie, into an area of largely

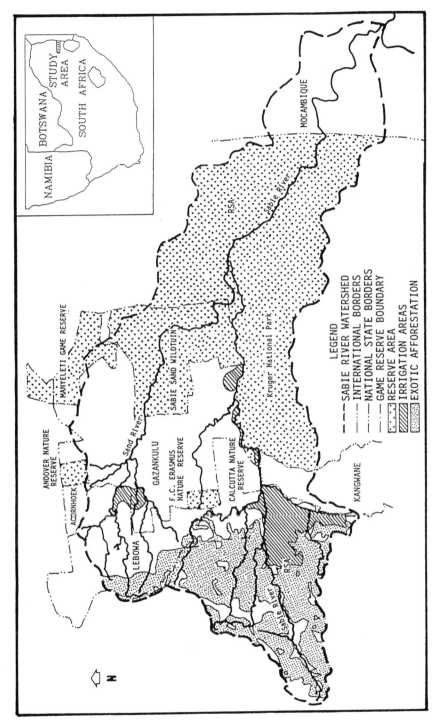

FIGURE 38.1. A map of the Sabie River system, eastern Transvaal, showing the main types of land use, and the boundaries of the self-governing areas

TABLE 38.1. Basic characteristics of the Sabie and Buffalo catchments, including their tributaries

	Sabie	Buffalo
Catchment area (km²)	6252	1230
Mean annual rainfall (mm)	833	736
Present mean annual runoff (m³ × 10⁶)	577	85
Length of mainstream (km)	175	140
Number of mainstream dams	0	4
Catchment population (× 10³)	420	600
Present water use (m³ × 10⁶ ann⁻¹)	250	67
Projected water use in 2010	520	124

irrigated agriculture (mainly bananas, citrus and tobacco). The middle catchment is mainly situated in the self-governing areas of Lebowa, Gazankulu and Kangwane. The lower catchment within South Africa is devoted to wildlife reserves, the largest of which is the Kruger National Park. The river flows across the Mocambique border into the Corumana dam, and then for 60 km to its confluence with the Incomati River. This chapter concentrates on the catchment within South Africa, since little information is available for conditions in Mocambique. Table 38.1 lists some basic information about the Sabie, in comparison to the Buffalo River.

The mainstream of the Sabie River is not at present impounded within South Africa, although several potential dam sites have been identified, and may be built in the near future (O'Keeffe and Davies, 1991). Detailed information on the catchment, land use, and water resources is available from Chunnett, Fourie and Partners (1987). The river is currently the subject of intensive ecological investigation as part of the Kruger National Park Rivers Research Programme. The fish communities have been described by Pienaar (1978) and Russell and Rogers (1989), and invertebrates by Moore and Chutter (1988). Preliminary attempts to assess the environmental water requirements for the river within the Kruger Park have been made by Davies et al. (1991), Gore, Layzer and Russell (1992), and O'Keeffe and Davies (1991).

The Buffalo River (Figure 38.2) rises in the Amatole Mountains in Ciskei, and flows through indigenous montane forest into two small impoundments, below which the upper middle catchment is mainly used for irrigated market gardening, and dry-land farming. The river then flows through the urban/industrial complex of King William's Town/Zwelitsha, and into Laing Dam, the main supply reservoir for the two towns. Downstream of Laing Dam the catchment is rural, with small villages, grazing and dry-land farming. Bridle Drift Dam, the largest in the river, is situated in the lower catchment dominated by the town of Mdantsane, from which it receives urban runoff. The catchment below Bridle Drift is steep and largely inaccessible coastal forest, and the estuary forms East London's harbour. The effects of the impoundments on the Buffalo River have been described by Palmer and O'Keeffe (1989, 1990a,b) and O'Keeffe et al. (1990), and invertebrate communities have been described by Palmer et al. (1991) and Palmer, O'Keeffe and Palmer (1992). Catchment zonation and land use have been described by O'Keeffe (1989). Table 38.1 summarizes basic catchment and river data in comparison to the Sabie River.

The Sabie catchment, including that of its main tributary the Sand River, is considerably larger than that of the Buffalo River, and produces 7.5 times as much runoff. Catchment

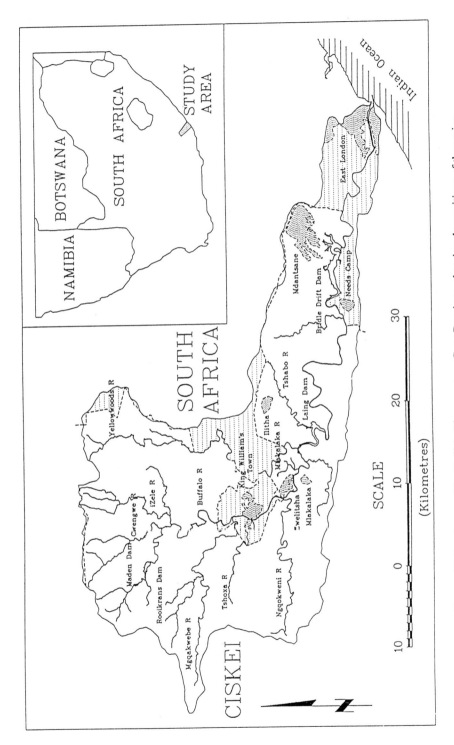

FIGURE 38.2. A map of the Buffalo River, eastern Cape Province, showing the positions of the major towns and dams, and the boundaries between South Africa (shaded areas) and Ciskei

TABLE 38.2. Present and projected water use in the Sabie catchment (modified from Chunnett, Fourie and Partners, 1987). Figures in $m^3 \times 10^6$ ann^{-1}

	1987	2010	% increase
Urban	6.9	61.6	793
Livestock	1.8	1.8	0
Irrigation	107.7	327.9	204
Afforestation	124.9	128.8	3
Total	241.3	520.0	115

populations are of similar size, but the present water storage capacity in the Buffalo catchment is considerably greater (103×10^6 m^3) than in the Sabie, which at present has only small dams on tributaries, with capacities totalling 29×10^6 m^3.

WATER USE IN THE SABIE CATCHMENT

At present, commercial forestry using pine and eucalypt is the major water user in the Sabie catchment. However, planned increases in irrigation, and urban water consumption (Table 38.2), could cause an increase of 115% in water consumption from the catchment by the year 2010. All of the major users have a fairly steady requirement throughout the year, so that critical low-flow periods are not linked with corresponding low demand. Present water use of 250×10^6 m^3 ann^{-1} is equivalent to an average flow of 7.93 m^3 s^{-1}. According to hydrological simulations using Pitman's (1973) model, this is higher than the mean flow of the river under natural conditions during July to October (the driest months) (see Figure 38.3). During the driest months in a 60 year simulation (Department of Water Affairs, 1990), flow during all months was below 7 m^3 s^{-1}. The present water uses in the upper and middle catchment, if met in full, are therefore sufficient to intercept all surface flow upstream of the Kruger National Park during low-flow months. With the envisaged 5% increase per annum in water use, it is inevitable that the river will be reduced to seasonal flow unless some provision for additional storage can be made.

It is within this framework that the South African Department of Water Affairs initiated research to assess the environmental water requirements of the Sabie River. These requirements are seen as catering primarily for the maintenance of the natural perennial river fauna and flora of the Sabie River in the Kruger National Park. However, it is equally important for the river to remain perennial upstream of the nature conservation areas. The river is extensively used by the people and livestock in the catchment for direct water use — abstraction, laundry and washing. Not only would the reduction of the river to a series of pools cause water shortages, but accumulation of pollutants would lead to health risks, and the availability of standing water as habitat for malarial mosquitoes and bilharzia snails would increase considerably. Fortunately, if the requirements of the conservation areas and Mocambique are met, by definition the upstream reaches of the river will remain perennial.

To date, three independent methods have been used to assess the environmental water requirements of the Sabie River. First Davies et al. (1991) used a water budget method in which the consumptive and non-consumptive water uses were estimated, to provide a seasonally distributed flow requirement. Major consumptive uses were evaporation from the

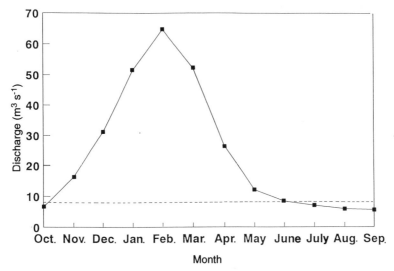

FIGURE 38.3. Mean flow rates in the Sabie River at the Mozambique border under natural conditions (■ ___ ■), compared with present water demand, averaged over the year (----).

river surface, and evapotranspiration by riparian vegetation, amounting to 30×10^6 m³ ann^{-1} for the Sabie in the KNP. Additional uses for animal drinking and staff/tourist consumption within the Park amounted to 0.5×10^6 m³ ann^{-1}. To this consumption was added the limiting non-consumptive requirement, which was assessed as a 0.1 m depth of flow over riffles to ensure the maintenance of riffle habitat and passage for fish.

Total baseflow requirements using this method amounted to 136.5×10^6 m³ ann^{-1}, distributed as monthly average flows of $0.6-5$ m³ s^{-1}. Additional biannual and longer term flood requirements were also specified.

Secondly, Gore, Layzer and Russel (1992) used the PHABSIM model to provide preliminary estimates of hydraulic habitat requirements for four key species in the Sabie River. They defined *Barbus viviparus*, a small cyprinid fish; *Serranochromis meridianus*, an endemic predatory cichlid fish; *Chiloglanis swierstrai*, a small rheophilic catfish; and the hippo (*Hippopotomas amphibius*), as key species covering the main types of hydraulic habitat requirements. Gore, Layzer and Russell (1992) show habitat preference curves which imply significantly increased habitat loss rates for each species at the discharges indicated in Table 38.3. From these measurements, Gore, Layzer and Russell (1992) conclude that the minimum discharge necessary to maintain a "minimum diversity of fish fauna" would be 2 m³ s^{-1}. For three of the four key species, a flow of around 3.5 m³ s^{-1} would provide acceptable habitat availability (Table 38.3). Gore, Layzer and Russell (1992) found that the relationship between increased species diversity and hydraulic diversity peaked at 6 m³ s^{-1}. Translated into annual water volumes to maintain survival, acceptable and ideal conditions, these flows would require 63, 110 and 189 m³ $\times 10^6$ respectively.

The third assessment of environmental flow needs used hydrological simulations of natural, present and various future impounded conditions of the Sabie River to predict possible ecological conditions (O'Keeffe and Davies, 1991). Ecological consequences of successive reductions in discharge were then assessed using the River Conservation System (O'Keeffe, Danilewitz and Bradshaw, 1987), an expert-system based model, to determine

TABLE 38.3. Discharges in the Sabie River below which habitat loss rates for selected species increase rapidly. (Discharges inferred from habitat preference curves in Gore, Layzer and Russell, 1992)

Species	Critical discharge ($m^3 s^{-1}$)
Barbus viviparus	3.3
Serranochromis meridianus	3.6
Chiloglanis swierstrai	6.8
Hippotamus amphibius	3.5

TABLE 38.4. Recommendations by O'Keeffe and Davies (1991) (in association with Chunnett Fourie and Partners), for annual flow volumes in the Sabie River at different probabilities of exceedence (e.g. it is recommended that annual flow should exceed $127 \times 10^6 m^3$ for at least 19 years in 20)

% Exceedence	Annual volume ($m^3 \times 10^6$)
95	127
75	211
50	295

the conservation status of the river under different management conditions. Recommended flows were presented in the form of a 60 year month-by-month simulation, relating recommended flows as a percentage of natural flows (O'Keeffe and Davies, 1991). These recommendations, if implemented, would ensure that worst historical drought conditions during the 60 year period of record would not be exceeded, and there would be a very low probability of their being repeated in consecutive years, or at short intervals. Table 38.4 shows the annual exceedance volumes at various levels of probability.

The worst drought in the 60 year record (1982/83) produced a runoff of $74 \times 10^6 m^3$, and this was taken as the minimum, or survival, recommendation by O'Keeffe and Davies (1991). The 95% assured flow, $127 \times 10^6 m^3$, should ensure acceptable environmental maintenance, while the 75% assured flow, $211 \times 10^6 m^3$, should ensure desirable conditions.

Having converted these three assessments to a common currency, that of annual volumes of water at survival, acceptable and desirable levels for environmental maintenance, it is interesting to compare the different recommendations, as in Table 38.5. Bearing in mind that these assessments each used different methods, based on different data, analysed by different teams, the recommendations are very similar, and give some basis for confidence that they are appropriate base flows for the maintenance of the riverine ecosystem. Seen as a percentage of the present MAR of the Sabie River (Table 38.1), they are not unreasonable requirements, and should be accepted as targets for management of the river's water resources.

WATER USE IN THE BUFFALO CATCHMENT

The Buffalo River, with an MAR of $85 \times 10^6 m^3$, has four main storage dams with a total capacity of $103 \times 10^6 m^3$, but a firm annual yield of only $57.7 \times 10^6 m^3$ (Department of

TABLE 38.5. Base flow recommendations for the Sabie River, from three different methods, aimed at maintaining environmental conditions at a survival, acceptable or desirable level. Recommended volumes are for the Sabie River immediately upstream of the Sabie/Sand confluence (flow volumes in $m^3 \times 10^6$ ann^{-1})

	Survival	Acceptable	Desirable
Consumptive/non-consumptive (Davies et al., 1991)	—	136.5	—
PHABSIM (Gore et al., 1992)	63	110	189
Hydrological simulations (O'Keeffe and Davies, 1991)	74	121	211
% of mean annual runoff	12	21	35

Water Affairs, 1986). Of this, 4.4×10^6 m^3 is released for irrigation, leaving 53.3×10^6 m^3 ann^{-1} for urban and industrial use. This is at present rather less than water demand (see Table 38.1), but has not yet led to serious undersupply. This is probably in part because Laing Dam is immediately downstream of King Williams Town and Zwelitsha, and therefore receives effluent from industry and sewage treatment works — an informal recirculation system. The construction of the first phase of the Amatole water transfer scheme is now complete, and this could eventually provide an additional firm yield of 36×10^6 m^3 ann^{-1} from the neighbouring Kei River catchment, into Laing Dam via the Yellowwoods tributary. The immediate problem in the Buffalo catchment is not therefore water quantity, but water quality.

Considerable urban and industrial development in the middle and lower reaches of the river have resulted in increasing effluent loads into the river. The effects of these effluents have been exacerbated by the positioning of the largest two dams, Laing and Bridle Drift. Laing is situated immediately downstream of the major towns of King Williams Town and Zwelitsha, and receives treated sewage effluent from both sewage treatment works and largely untreated industrial effluent from a number of large companies. Mdantsane (population 167 000) is situated in the immediate northern catchment of Bridle Drift dam, and a number of small tributaries rise in the town and flow directly into the dam. Inadequate maintenance of sewer pipes has led to breakages, and direct inflow of sewage into Bridle Drift dam during 1991. Faecal coliform counts in parts of the dam have exceeded 400 000 cells per 100 ml, and exceed the general recreational criterion of 2000 cells per 100 ml for more than 50% of the time (O'Keeffe et al., in press).

A recent water-quality situation analysis (O'Keeffe et al., in press) has attempted to define user requirements for water quality, and to relate these to present conditions in the river. Municipal and commercial water users were canvassed for their water quality requirements, and Tables 38.6 and 38.7 summarizes some of their responses, and the conditions in the river over the past 15 years.

For the people and organisations using the river, the following conclusions can be drawn:

- *Salinity*: median values fall within the acceptable/tolerable range, but there are occasions when conditions are unacceptable. A closer examination of the historical record (O'Keeffe et al., in press) revealed that the salinity of water flowing into Laing Dam has been unacceptable (in terms of Table 38.7) in 42% of samples.

TABLE 38.6.   Most stringent user requirements for some water-quality criteria in the middle Buffalo
River (modified from O'Keeffe et al., in press)

Variable	Ideal	Acceptable	Tolerable	Unacceptable
Salinity (EC as mS m^{-1})	20	100	—	200
pH	7	7.5	8.5	<7>8.5
Total phosphate (mg litre^{-1})	1	20	45	55
Alkalinity (mg litre^{-1})	30	50	75	100
Chloride (mg litre^{-1})	50	100	—	200
Sodium (mg litre^{-1})	41	200	300	400

TABLE 38.7.   A summary of water-quality conditions in the Buffalo River at the inflow to Laing Dam,
from samples taken between 1975 and 1990 (modified from O'Keeffe et al., in press)

Variable	Minimum	Median	Maximum
Salinity (EC as mS m^{-1})	16	118	770
pH	5.8	8.0	10.2
Total phosphate (mg litre^{-1})	0.1	1.5	7.1
Soluble reactive phosphate (mg litre^{-1})	0	1.4	6.5
Alkalinity (mg litre^{-1})	0	241	1440
Chloride (mg litre^{-1})	31	204	2470
Sodium (mg litre^{-1})	8	214	1424

- *pH*: the river has on occasion been both too acid and too alkaline for some users, but median pH is within tolerable limits.
- *Total phosphate*: concentrations have always been within acceptable limits for users.
- *Alkalinity*: median concentrations are nearly 2.5 times the unacceptable limit.
- *Chloride*: median values are at the unacceptable level.
- *Sodium*: concentrations are acceptable to tolerable for most of the time, but extreme values are well above unacceptable levels.

Such an analysis may appear to provide clear management guidelines for water quality, even if these may be difficult to meet in practice. However, despite the fact that total phosphate concentrations are well within acceptable user requirements, Laing Dam suffers from intermittent cyanobacterial blooms which cause water purification problems, and phosphate appears to be the nutrient responsible (Selkirk and Hart, 1984). To try to define an ecological water-quality requirement, members of the Institute for Water Research have examined the riverine biota for signs of stress. Riffle-dwelling invertebrates were collected over two years from 17 sites down the river, and associated water-quality samples were collected (O'Keeffe et al., in press). The distribution of common taxa were examined in relation to the water-quality status at each site. Two sites had higher salinity and phosphate concentrations than all others. These were at the inflow to Laing Dam, and 5 km downstream of Bridle Drift Dam. In both cases, elevated concentrations resulted from industrial and/or sewage effluents.

TABLE 38.8. Invertebrate taxa from riffles in the Buffalo River, which were common and widespread except at the most polluted sites

Baetis type B	(Ephemeroptera)
Centroptiloides bifasciatum	(Ephemeroptera)
Pseudopannota maculosum	(Ephemeroptera)
Choroterpes elegans	(Ephemeroptera)
Neurocaenis reticulata	(Ephemeroptera)
Macrostemum capense	(Trichoptera)
Simulium damnosum	(Diptera)
Elmidae	(Coleoptera)

TABLE 38.9. Tolerance limits for common riffle-dwelling invertebrates in the Buffalo River

Salinity (EC as mS m^{-1})	77
Soluble reactive phosphate (mg litre^{-1})	0.4
pH	8.7

Twenty-two common and widely distributed taxa were analysed to ascertain whether they were missing from, or present in reduced numbers at more polluted sites. Eight of the 22 (36%) were absent or severely reduced at the two most polluted sites, but were common at both up- and downstream sites (Table 38.8). This represented a major reduction in diversity at these sites, and this reduction was taken as an indication that the general ecological functioning of the river at these sites was significantly modified. These species were all distributed in the river throughout the year, and it has therefore been assumed that they are able to live in conditions which applied in reaches of the river other than the two most polluted sites, for most of the time. O'Keeffe et al. (in press) therefore defined the highest values of different water-quality parameters which occurred for 90% of samples, at all sites other than the two most polluted, as being the tolerance limits for the natural biota of the river. Table 38.9 lists these tolerance limits for a number of water quality criteria. Unfortunately, soluble reactive phosphate (SRP) rather than total phosphate was measured, so that results for phosphate are not strictly comparable with the user requirements listed above, which were expressed in terms of total phosphate.

This empirical definition of environmental water-quality requirements provides an interesting comparison with the stated requirements of the users (Table 38.6). Salinity tolerances of the natural biota are within the acceptable range for users. pH tolerance is greater than the unacceptable level defined by users. Phosphate tolerances are not comparable, but are undoubtedly much more stringent than user requirements. The Buffalo River at present is subject to a special SRP standard of 1 mg litre^{-1} for all effluents (Department of Water Affairs, 1986). This has not in practice been met for much of the river, where yearly median concentrations in excess of 10 mg litre^{-1} have been recorded (Palmer and O'Keeffe, 1990). The tolerance limit of 0.4 mg litre^{-1} for the natural biota may be too low to be practically manageable, but is likely to be the requirement if secondary (ecological) water-quality problems such as algal blooms are to be avoided.

CONCLUSIONS AND DISCUSSION

The Sabie and Buffalo catchments, although not fundamentally different in their ecology, are subject to quite different levels of development and user priorities, and it is therefore difficult to envisage a common management plan that would be appropriate to both. Major priorities in the Sabie catchment are to provide sufficient water for domestic and agricultural development, while protecting the invaluable heritage and tourism potential of the wildlife reserves further downstream. In the Buffalo catchment, the main priority is to improve the water quality in the middle and lower reaches, where it has deteriorated for much of the time, to conditions unacceptable for the users. Both rivers may, in time, experience the same problems as the other, as water demand increases in the Buffalo catchment, and as effluent disposal increases in the Sabie River. In both rivers, however, it is of primary importance to make sure that the basic functions which make them valuable natural resources are maintained. If the Sabie is reduced to a series of stagnant pools, or the Buffalo to an open sewer, the incremental use of the rivers gained will not offset the costs in terms of disease, additional water purification costs, emergency water supply additions during droughts, and loss of amenity values and tourist revenues. The introduction to this chapter pointed out the difficulties inherent in treating the environment as a user of a river, in competition with other users. There is a real need for the environment to be recognized as inclusive of the river, its catchment, and its associated biota, including people. Rivers are natural resources, which need to be maintained in proper working order, if they are to provide continuously for the needs of people in their catchments. It is the job of ecologists to quantify how far riverine resources can be exploited yet still remain functionally acceptable. It is the responsibility of managers to ensure that exploitation is not taken beyond those limits. Environmental requirements for a river should therefore be met as the first priority. Water volumes, or assimilative capacity, in excess of environmental requirements can then be allocated to users without fear of permanent damage to the resource.

Dams in rivers have traditionally been seen by river ecologists as a disruption of the ecosystem (e.g. Ward and Stanford, 1983). This may be true in pristine rivers, but pristine rivers have become a rare luxury in South Africa. The Sabie River is one of the largest in South Africa to remain unimpounded, and in many ways it will lose some of its value if it is impounded. However, it is not a pristine river, and loses much of its water at low flows to forestry and direct abstraction for irrigation and domestic use. Without an impoundment, there are few management options through which to augment low flows. The impoundment most likely to be built on the Sabie River would be the Injaka Dam on the Marite tributary. Such off-mainstream storage would limit the barrier effect on the Sabie, and would have minimal effects on the physico-chemistry, geomorphology, and flood attenuation downstream. O'Keeffe and Davies (1991), suggested a management regime for the Injaka Dam which would ensure adequate compensation flows, and therefore cause minimum disruption, but provide maximum management potential, downstream.

In the Buffalo River, Laing Dam receives water high in nutrients and salts, but releases water relatively low in both (O'Keeffe et al., 1990). This is due to sedimentation and algal uptake of nutrients, and to dilution of salts mixing with low salinity flood water previously stored in the dam. Laing therefore acts "as a settling pond, improving downstream water quality" (O'Keeffe et al., 1990). It is important for ecologists to recognize the benefits, as well as the costs, of water development schemes, and to work co-operatively with managers and engineers to ensure sustainable development.

ACKNOWLEDGEMENTS

I would like to express my sincere appreciation to the South African Water Research Commission, who have funded much of the research described in this paper, and have been instrumental in encouraging and initiating the search for ways of assessing environmental water requirements in South Africa. In particular I would like to thank Peter Reid of the WRC for help and encouragement, and my colleagues Rob Palmer, Carolyn Palmer, Bryan Davies, Des Weeks and Carin van Ginkel, who carried out most of the research referred to in this paper.

39

Ecological Master Plan for the Rhine Catchment

ANNE SCHULTE-WÜLWER-LEIDIG

*International Commission for the Protection of the Rhine against Pollution, Koblenz,
Germany*

INTRODUCTION

On 1 November 1986 a serious industrial accidental occurred at Basel in Switzerland. Nearly 30 tonnes of toxic chemicals (insecticides, fungicides and herbicides) spilled into the Rhine. Pictures of thousands of dead eels floating between Basel (km 159) and the Loreley (km 560) were seen around the world. Besides the death of different fish species and macroinvertebrates, mercury contained in the fungicides polluted the sediments many kilometers downstream of the city of Basel.

Evaluating the effects of this spill, and seeing the vulnerability of the Rhine ecosystem, the Ministers of the countries bordering the Rhine decided to accelerate efforts to upgrade water quality and improve the state of this ecosystem. Therefore, in 1987, they started the "Rhine Action Programme" (International Commission for the Protection of the Rhine against Pollution, 1987, 1989). The goals to be achieved by the turn of the century are:

(i)   the ecosystem of the Rhine must become a suitable habitat to allow the return to this great European river of higher species which were once present here and have since disappeared, such as salmon (Figure 39.1);
(ii) the use of Rhine water for drinking water production must be guaranteed;
(iii) a substantial decrease of pollution by toxic agents must be achieved in particular with respect to the sediments;
(iv) the North Sea must be protected against pollution.

This "Rhine Action Programme" aims at a clear improvement of the water quality as well as of the ecosystem and comprises the following actions:

(i)   to reduce pollution originating from direct inputs (industry, municipalities) and non-point sources (atmosphere, agriculture);
(ii) to reduce accidental spills by increasing the security of industrial plants;
(iii) to improve ecological conditions for flora and fauna.

The Ecological Basis for River Management. Edited by D.M. Harper and A.J.D. Ferguson. © 1995 John Wiley & Sons Ltd

506

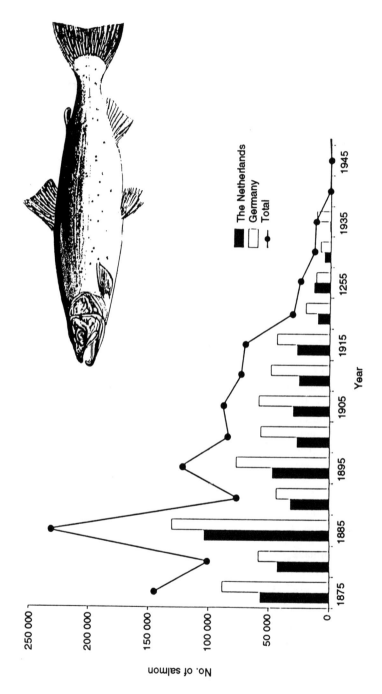

FIGURE 39.1. Number of salmon caught in Germany and the Netherlands between 1875 and 1950

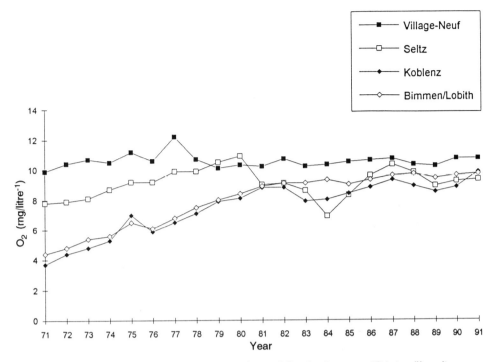

FIGURE 39.2. Average annual concentrations of dissolved oxygen (O_2) (mg/litre^{-1})

It has been recognized that measures aiming only at improving the water quality are not sufficient and will not lead to a lasting rehabilitation of the Rhine. Figures 39.2–39.4 represent some examples of the improvement of the Rhine water quality during the last two decades. These improvement measures, amounting to some £40–50 million, will not alone cause the salmon to return. They cannot turn the Rhine into an organic habitat with a great variety of species and considerable regenerative power as long as the river is mainly a navigable waterway from which water is drawn off and into which effluents are discharged as required.

In the view of this objective of environment policy, the ICPR approved the "Ecological Master Plan for the Rhine (Salmon 2000)" in 1991. This Master Plan focuses on two points:

(i) restoration of the main stream as the backbone of the ecosystem especially for the long-distance migratory fish like salmon, and
(ii) protection, preservation and improvement of ecologically important reaches.

This programme explains in detail the different measures necessary to achieve a long-lasting improvement of the Rhine's ecosystem. In the programme the salmon is to be regarded as the symbol for all endangered species of migrating fish whose habitat at the moment is far from being intact.

THE PLAN "SALMON 2000"

Migratory fish can only return to the Rhine when certain conditions are given. Expensive improvement measures including the re-establishing and re-stocking of previous habitats and

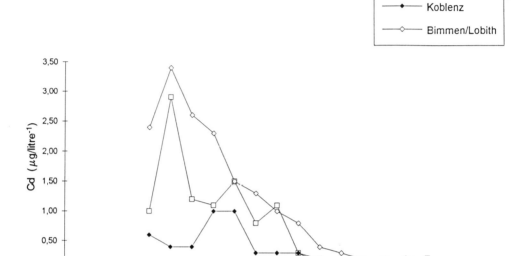

FIGURE 39.3.   Average annual concentrations of total cadmium (μg/litre^{-1})

the accessibility for migratory fish must be realized. For the support and realization of the project "Salmon 2000", two applications for support within the context of the European Union's "LIFE" programme were approved. The first project concerns various habitat improvements, re-establishing, re-stocking and success control measures in several Rhine tributaries. The total costs for this point project of all Rhine-bordering countries are estimated to amount to ECU 4 900 000 of which 50% will be met by EC subsidies. The second project, for which the EC has approved the sum of ECU 600 000 (5%), concerns the methods of improvements at barrages which are currently considerable obstacles to migration. At the barrages of Iffezheim and Gambsheim/Upper Rhine, the building of new fish ladders should commence with absolute priority. The investment costs amount to ECU 10 000 000. On the Lower Lahn, the adaptation of the barrages must start as soon as possible. The improvements necessary have been estimated to cost ECU 2 000 000.

The Ecological Master Plan demands an extension of the alluvial areas along the Rhine. Complete ecological interaction between these areas and the river must be allowed. Farming must cease or be significantly extensified. Alluvial areas must be extended. The ecological working group of the ICPR is working on a specific ecological network with the most important alluvial areas along the Rhine. These areas shall serve as "stepping stones" for the whole Rhine ecosystem.

ECOLOGICAL OBSERVATION SYSTEM "RHINE"

In view of the objective "Salmon 2000" of environment policy, the ICPR has introduced an

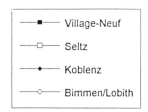

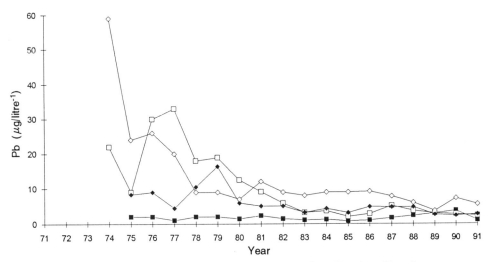

FIGURE 39.4. Average annual concentrations of total lead (μg/litre^{-1})

ecological observation system "Rhine" which complements the monitoring of the water quality practised so far (Internationale Kommission zum Schutze des Rheins, 1993). Apart from an inventory of fish fauna, macroinvertebrates and plankton, it includes analysis of toxic substances accumulated in aquatic organisms, particularly in fish, because only a systematic and continuing examination of the latest biological conditions of the river over its whole length will allow us to recognize changes in the ecosystem and links within it. Furthermore, it enables us to follow the effects of pollution, to judge their ecological consequences and to plan and implement appropriate environment protection measures.

FISH IN THE RHINE

According to analyses made by the Senckenberg Institut in Frankfurt, 40 of the 47 former indigenous fish species and cyclostomata have returned to the Rhine (Lelek and Buhse, 1992). However, some of the formerly frequently occurring migratory fish, such as salmon and sturgeon are still missing. Furthermore, up until now only single specimens of other migratory fish such as alice shad, sea trout and sea lamprey have been caught. These results show that, on the basis of water quality, all former species which have disappeared could live again in the Rhine. None the less, a few relatively unspecialized species such as European roach, bleak and bream predominate and represent 75% of animals present. The analysis also reveals that, compared to the middle of the river (i.e. the shipping channel), habitats near to their natural state (tributaries and oxbow lakes of the Rhine, lateral water bodies, etc.) have a significantly larger species diversity. There is no denying that these habitats are most important for the fish fauna.

Concentration of pollutants in fish caught in the Rhine

Since fish can be used as biological indicators for the pollution of water bodies and in addition also serve as food, the concentration of pollutants in fish caught in the Rhine has been analysed over a long time. Particular attention was and still has to be paid to the clearly measurable contamination of fish caught in the Rhine with HCB (hexachlorobenzene) and PCB (polychlorinated biphenyls) (Internationale Kommission zum Schutze des Rheins, 1993). The concentration of these substances in most of the analysed fish, especially eels, even exceeded the maximum amount for human consumption permitted by German law. For years, lead and cadmium concentrations proved to be very low. Though average mercury concentrations have been below the permitted maximum amount, in some individual cases this limit has been exceeded. These data lead to the conclusion that the concentration of pollutants in the Rhine must be reduced further and all efforts in this direction must be continued.

MACROFAUNA OF THE RHINE

The first analysis of the macrofauna available from the literature (Tittizer, Schöll and Dommermuth, 1993) was carried out by Lauterborn, who, between 1900 and 1920, managed with the techniques and knowledge then available to detect 80 species, of which about 40 were insect larvae. At the time of the highest pollution of the Rhine by wastewater (in the late 1960s and the early 1970s) only 27 species could be found. There has been a considerable increase in the number of species of the macrofauna population since then (Figure 39.5). About 150 species have now been detected in the last few years but there are a number of immigrants and sensitive species formerly resident in the Rhine (e.g. many insects, in particular stonefly larvae) that have disappeared — the community structure has changed.

PRESENTATION OF THE ECOLOGICAL MASTER PLAN AND THE INITIAL IMPLEMENTATION

In the view of the first goal of the RAP (Improvement of ecological conditions), the ICPR developed an "Ecological Master Plan" for the Rhine (Anon., 1989, 1991).

The restoration of the Rhine ecosystem pre-supposes efforts to improve water quality will be successful, as well as the protection and restoration of habitats for animal and plant species which have either disappeared or are endangered. Therefore, the "Ecological Master Plan" focuses on two points:

(i) restoration of the main channel as the backbone of the ecosystem, especially for the long-distance migratory fish (e.g. salmon, sea trout, alice shad), and
(ii) protection, preservation and improvement of ecologically important reaches of the Rhine and the Rhine valley floodplains with a view to increasing the diversity of the indigenous animals and plants.

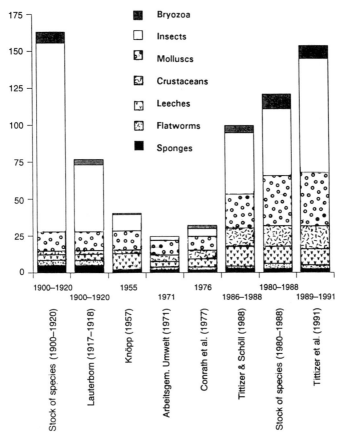

FIGURE 39.5. Changes in the stock of invertebrates in the Rhine (152−870 km)

Restoration of the main stream as the backbone of the ecosystem

The objective is that migratory fish which spawn in the upper reaches of a river system, can return to the Rhine. They require the habitats of the entire river to sustain their life cycle and are therefore particularly suitable indicator organisms. Since the salmon is the best-known example of migratory fish, it is being used as a symbol. It is well known that the salmon formerly passed through the Rhine catchment area and reached the Rhine Falls at Schaffhausen as well as the tributaries of the Aare (Switzerland) and that many fishermen depended essentially on salmon fishing for their livelihood.

As long as the main channel does not allow unhindered fish migration upstream towards the spawning grounds and downstream towards the sea it cannot offer an efficient habitat for migratory fish. The Rhine at present, with obstacles such as barrage weirs and power stations, has become a waterway solely for large ships.

During the last century, man began transforming the river into a navigable waterway. Even though fish could still pass up- and downstream, migration possibilities deteriorated on account of the uniform nature of the channel (lacking diversity) and the absence of resting places. Spawning grounds in the river were destroyed and the rapid flood drainage affected the survival of fry. The construction of dikes and bank stabilization systems cut off many meanders, fast-flowing tributaries and river meadows so that along today's Rhine between Basel and Karlsruhe, only remnants of the former alluvial areas have survived. This has resulted in a tremendous loss of spawning grounds (gravel banks) and nursery grounds.

At the very least, salmon should be reintroduced to all those reaches where intact habitats still exist. Initial investigations gave the number and size of the spawning and nursery grounds still remaining for repopulation. These show that:

(i)   in the lower Sieg and its tributaries, 14 ha of spawning grounds and 46 ha of nursery grounds still exist;
(ii) in the Sauer and the Our in Luxemburg, 5.3 ha of spawning grounds and 77 ha of nursery grounds could immediately be used once the barrage weirs along the Moselle have been made surmountable;
(iii) 4.1 ha of spawning grounds and 65 ha of nursery grounds have remained intact in the old river bed of the Rhine (Restrhein) between Basel and Breisach; and
(iv) on the French side, the Ill-Bruche area still presents 4 ha of spawning grounds and 57 ha of nursery grounds.

These areas still intact are considered as sufficient for an initial re-stocking. But the future range for migratory fish in the "Salmon 2000" target is the Rhine from its estuary to Basel including the tributaries Sieg, Saynbach, Lahn, Lauter, Bruche, Ill, lower Moder, Kinzig and Murg. In the longer term, further tributaries are to be integrated into the programme. For these, expensive improvement measures including the re-establishing and re-stocking of previous habitats and the accessibility for migratory fish have first to be achieved. Costs for the first focus — restoration of the main stream as the backbone of the ecosystem — (until the year 2000) are estimated to exceed £50 million.

"Salmon 2000" is supported by two projects within the context of the LIFE programme by the European Union. The first concerns various measures such as improvement and re-establishing habitats, re-stocking and success control in several Rhine tributaries. The total costs for this project to all Rhine-border countries are estimated as £4 million, of which 50% will be met by EC support.

Free upstream and downstream fish migration as well as free access to tributaries is required. That is why several existing obstacles must now be modified and efficient fish passages or sluices must be installed. Plate 3 gives a general survey of the large number of barrage weirs in the Rhine catchment area which are considered as possible obstacles to migration.

The control of the sluices on the Haringvliet and IJsselmeer (the estuary of the Rhine in the Netherlands) is to be adapted to the requirements of migratory fish in order to make it easier for them to surmount these obstacles. One of the basic requirements for the return of migratory fish is free passage through the estuary. The possibilities of surmounting three more barrage weirs along the Nederrijn/Lek must be improved. An undammed arm of the river does exist, but it is intensively used as a main navigation connection with Rotterdam. As far as Iffezheim, no further obstacles disrupt migration through the main stream, but obstacles exist in almost all tributaries. The Sieg, a tributary in the Middle Rhine area, is the major exception: four barrage weirs obstructing migration in its middle and lower reaches have been equipped with ramps.

A technical solution for surmounting the Iffezheim and the Gambsheim weir, the first and second sizeable barrage weir on the Rhine up from the estuary, has been found. It is known that the fish ladders existing at the Iffezheim and Gambsheim power-station are not

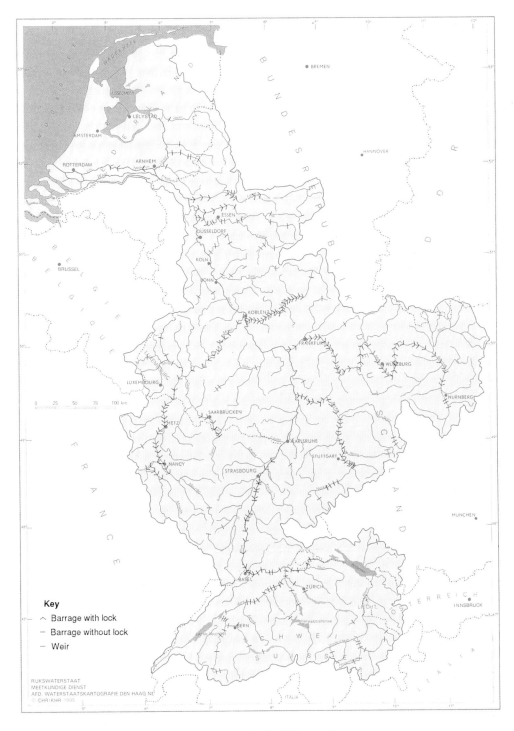

PLATE 3   Barrages in the Rhine catchment

effective. The investment costs for adapting them to the requirements of migratory fish amount to £8 million. At both barrages, new basin passes (vertical slot passes) will be built almost midstream, each consisting of 42 individual basins with an approximate capacity of 15 m^2 each. The EC has approved a sum of some £400 000 for this. If both weirs are equipped with effective fish passages, migratory fish will again be able to reach the spawning grounds in the Ill and Bruche (France) and Kinzig (Germany).

Protection, preservation and improvement of ecologically important reaches

Not only the river itself as the backbone of the ecosystem, but all connected habitats such as the river bed, the banks and alluvial areas must be restored, thus allowing self-regulating biocenoses (intact food chains) to develop. A greater diversity of habitats and living conditions also leads to a greater variety of species (see Chapter 19). The most desirable state of the ecosystem is a great variety of species in different habitats, as close as possible to the situation as it once was. The loss of alluvial areas is particularly threatening in the regions of the Upper and Lower Rhine.

Due to the construction of dikes and the straightening of the course of the river, 90% of the alluvial areas between Basel and Karlsruhe, for example, have vanished (Figure 39.6). This means that due to the loss of habitat, many animals and plants formerly living in the Rhine basin are reduced in number or have disappeared altogether. The remaining alluvial areas and other zones of ecological importance must be preserved, protected and, wherever possible, extended in order to restore living conditions for a greater variety of species. Fauna and flora must again have access to a large number of habitats.

The following action is necessary to improve the conditions of habitats:

(i)   Reaches of ecological importance along the Rhine and its valley which have already been placed under protection must be preserved at all costs.
(ii) Further zones of ecological importance should be placed under protection.
(iii) Plans which exist for the rehabilitation of river banks, alluvial areas or foreshores along certain stretches of the Upper and Lower Rhine should be promoted.
(iv) Various areas along the Rhine should be protected in accordance with the terms of the Ramsar Convention by declaration as wetlands of international importance and thus as habitats for bird species depending on the living conditions offered in such areas.
(v)   All applications for extended land use and new projects for land use along the Rhine and the alluvial areas of its valleys must be subjected to an environmental audit.

The ecological working group of the ICPR is working on a specific ecological network with the most important alluvial areas along the Rhine. Complete ecological interaction between these areas and the river must be allowed. The most important areas will thus serve as "stepping stones" for the whole Rhine ecosystem.

DISCUSSION AND CONCLUSION

The ecological Master Plan will allow in particular an improvement of the essential biotopes, not only the flowing waters in the narrow sense of the word (aquatic environment including the river bed), but equally the littoral and alluvial areas (amphibious and terrestrial zone).

Due to European Commission support, the ICPR member states will be able to carry out some of the necessary measures within the next few years. Further action mentioned in the

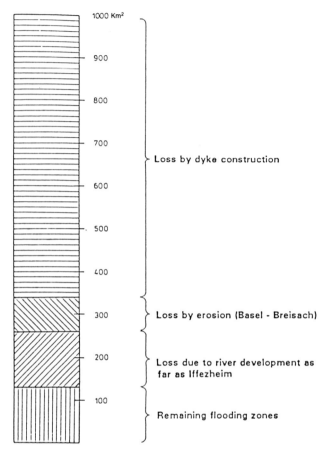

FIGURE 39.6.   Disappearance of the flooding zones between Märkt (Kembs) near Basel and Maxau
near Karlsruhe; natural flooding zone is 1000 km²

Master Plan will soon be initiated by the Rhine-bordering countries with a view to reaching
the objectives set in the RAP.

The comprehensive ecological objectives of the "Rhine Action Programme" are
innovative in the field of international policy towards the prevention of water pollution.
"Salmon 2000" has a pilot function in this respect.

40

The Interface Between Ecology and Economics in Catchment Management

JOHN BOWERS

University of Leeds, Leeds, UK

INTRODUCTION

The current wisdom in environmental economics is that environmental problems arise because the environment is unpriced and is in consequence ignored when decisions are taken. If correct pricing of the environment were introduced the result would not be no further environmental damage but rather the optimum mix of environmental and other goods would be achieved. Progress towards this nirvana according to the influential Pearce Report (Pearce, Markandya and Barbier, 1989) requires action on three fronts:

(i)   the introduction of Pigovian taxes or some alternative equivalent economic instrument on the environmental detriments arising from the production or consumption of goods and services that have negative effects on the environment;
(ii) valuation of the environmental effects of public investment projects in cost−benefit analysis;
(iii) the incorporation of the environment into national social accounts so that the depletion of "natural capital" in the course of income generation is revealed.

Within the water industry there are a number of pathways to the improvement or deterioration of the ecology of catchments. In addition, of course, catchment ecology can be affected by a range of economic changes which originate from outside of the water industry. These operate largely through the land-use planning system.

The water industry pathways are as follows:

(i)   discharge control from point pollution sources
(ii)   abstraction control
(iii) water containment and storage: reservoirs, etc.
(iv)   arterial land drainage and flood protection
(v)   controls on non-point pollution sources: agricultural runoff, etc
(vi)   engineering works relating to configuration of water courses and their containment
(vii) Other controls on use: boats, fishing, water sports, etc.

The Ecological Basis for River Management. Edited by D.M. Harper and A.J.D. Ferguson. © 1995 John Wiley & Sons Ltd

Pathways originating from outside the water industry include construction of housing, industry and roads; yielding alterations in drainage characteristics and in the volume and chemical composition of flows into the system.

The aim of this chapter is to discuss the application of the three Pearce propositions to the issue of protection of catchment ecology in the UK.

ENVIRONMENTAL TAXES

This first Pearce proposition is relevant to three of the pathways listed above, namely point and non-point pollution and water abstraction.

Point pollution

The standard textbook treatment of the economics of pollution starts with the definition of an optimum degree of pollution defined as that level of pollution at which marginal social cost of pollution control equals the marginal social benefit from controlling the pollution. In the Pigovian model, marginal social benefit is identified as the collective willingness to pay at the margin by the sufferers from that pollution for derogation, assuming that some "incentive compatible" mechanism exists to prevent tendencies to free ride. In a wide range of cases such an optimum will not exist since the implicit assumptions of the Pigovian model will not be satisfied (Bowers, 1991). I do not believe that it is possible to specify an optimum level of pollution in this economic sense to discharges to water catchments. However, the first Pearce proposition is meant to apply to any target pollution discharge level, however it is determined. I have more to say about the issue of valuing catchment environments below.

It is well established that any pollution target may be achieved by the use of non-market instruments such as physical controls backed by the force of law. Neo-classical economists favour economic instruments over non-market instruments because they believe that they achieve the objective at least resource cost. While this is demonstratively true in a simple model in which costs of monitoring and compliance are ignored or, what amounts to the same thing, treated as invariant between alternative instruments, the conclusion does not hold in a complex reality. Thus a recent authoritative study of alternative instruments concluded:

> "no general statements can be made about the relative desirability of alternative policy instruments once we consider such practical complications as that location matters, that monitoring is costly, and that exogenous change occurs in technology, regional economies and natural environment systems."
> (Bohm and Russell, 1985)

The existing system in the UK — control of point pollution through discharge consents, with the consents set to conform with statutory river water-quality targets — is, if properly implemented, arguably a socially efficient method of meeting the objective. Objections to it mainly centre around concern over the commitment of the authorities to the objective and the obstacles and difficulties of ensuring conformity of key polluters: water companies in their role as sewage undertakers and some sections of manufacturing; within an acceptable time-scale. A minority view, espoused notably by "Greenpeace", would impose tighter river water-quality standards and seek zero emissions of some industrial chemicals.

On the issue of compliance, some of the difficulties arise from the conditions and commitments undertaken to achieve privatization of the water industry (Bowers and O'Donnell, 1989). In any event economic instruments, including the current favourite of marketable permits, are not self-policing and improvements in the timetable of compliance can only be achieved at a cost. If extra resources are to be committed to improve compliance, it has yet to be demonstrated that they would not be best used for strengthening the existing system rather than by switching to an efficiently monitored market instrument.

Non-point pollution

With non-point pollution such as runoff of nitrates and pesticides from agriculture, the case for market instruments is stronger since the consent system is unworkable. Restriction on agricultural activities in nitrate-sensitive groundwater areas and controls on design and construction of silage and manure storage and disposal facilities are the chosen non-market control instruments. In my view the case for input taxes on fertilizer and pesticides as a means of reducing their use and hence their negative environmental impact is strong. The objection to them is that the price elasticities are low and the required tax rates are therefore at a level which would impose serious income losses on farmers with consequential social costs (Kumm, 1991; Hanley, 1991). The estimated elasticities are essentially short term: in the longer term elasticities may be expected to be much higher. Furthermore the tolerance of nitrate and pesticide pollution is part of the hidden social costs of agricultural support. These costs should be manifest and charged to the right budget. If input taxes are the appropriate means of controlling agricultural runoff then they should be imposed. I can see no argument for restricting the "Polluter Pays" principle to non-agricultural activities.

Abstractions

In meeting river water quality objectives, abstraction controls are needed to complement discharge controls for two reasons:

(i) The impact of a given rate of discharge of any pollutant on the water environment is dependent on the volume of the receiving water and its rate of flow. In setting discharge consents, NRA has to take a view of these parameters if it is to meet the quality objectives;

(ii) Aquatic flora and fauna are dependent on the maintenance of minimum rates of flow. Catchment ecology is modified and typically impoverished when the catchment is depleted.

One might conceive an ideal non-market system where the competent authority, in the present instance the NRA, controlled all discharge consents and abstraction licences so as to maintain the ecology and amenity of catchment waters according to predetermined and exogenous water-quality objectives, these objectives varying between catchments according to social and economic needs. The system currently achieves standards lower in aggregate than is desirable by the majority of opinion and a mechanism for raising these target standards over time and lowering abstraction and discharge consents in consequence is needed.

This is perhaps the vision underlying the recent Water Acts. My task as an economist would then be to ask whether the efficiency gains stemming from an alternative set of market instruments designed to achieve the same end would outweigh the additional costs of monitoring and enforcement that was entailed. But, of course, this system is being introduced against the background of an existing set of discharges and abstractions which were not constructed with this objective in mind. On the abstractions side, the constraint takes the form of an existing set of entrenched abstraction rights which pre-date the 1972 Water Act and have continued effectively unscathed through the privatization of the water industry. The alternative therefore is a set of abstraction charges. These should be based on the following principles:

(i) Charges could be set so as to equate demand for abstractions with the ecologically sustainable supply. On the previous argument economists would see such charges as a preferable alternative to abstraction quotas on efficiency grounds since it would allow reductions in abstractions to be met at least cost. However, in a typical situation where the vast bulk of abstractions are carried out by one or two operators and principally by a water supply company the efficiency gains over quotas would probably be small. The water supply company being a monopoly with relatively inelastic demand would simply pass these charges on to final consumers. The studies undertaken suggest that demand is more elastic in the presence of metering since the consumer pays for water consumption at the margin. In the long run the demand would be more elastic than in the short run because of the scope for water economizing investment. The opportunities for the water company to switch to alternative supplies is limited by absence of a national grid, and in south and east England, where the majority of cases of over-abstraction occur, alternative sources are limited. This solution is then probably ruled out because it imposes unacceptably high costs on consumers. Additionally, the NRA does not have the information required to set abstraction charges to meet this objective. The information on the water supply industry is available to the Office of Water (OFWAT) but the information is unlikely to be available to the NRA. The issue of whether investment in a grid for improved inter-regional transfer of water where a principal benefit is the quality of the water environment in the form of losses from over-abstraction in the South and East and possible catchment damage from the construction of additional water storage facilities (several major reservoir construction schemes are under consideration) is justified is considered below. It is, of course, possible for the NRA to set abstraction charges to users other than water undertakers so as to keep their abstractions within ecologically safe limits. This is particularly a problem for groundwater abstractions. There is no serious problem of equity in doing this.

(ii) If taxing to keep abstractions in line with ecological needs is deemed infeasible the alternative is for the NRA to undertake investment in compensating for losses of flows and to seek to recover this expenditure from the abstracters who cause the work to be undertaken. The NRA has in fact launched a programme for tackling the problem of low flows in England and Wales and is proposing a series of possible actions including:

> "revoking or varying existing licenses, constructing river flow augmentation schemes, constructing new river channels, lining river beds, and negotiating revised operating policies for existing abstractions."   (NRA, 1991)

I am not in a position to judge how feasible and effective these engineering works will be; they are presumably inferior in ecological terms to bringing abstractions down to environmentally sustainable levels in the existing system. Except where licence revocation or modifications of operation are involved, the cost of these engineering solutions will fall on the cost of water abstraction licences which under the new arrangements NRA has to set to cover for each region the cost of water resource management. This is clearly economically inefficient since the costs of maintaining river ecology are spread across all abstractors in the region rather than falling on those who cause the problem. If some abstractors are forced to cease to abstract or to reduce their abstractions, while others have the costs of the works to counter their activities spread across all licence holders in the region, then there is obviously a problem of equity as well as one of economic efficiency.

COST – BENEFIT ANALYSIS

Cost – benefit analysis is used as aid to decision taking in public sector investment appraisal. Within the water industry, cost – benefit analysis (CBA) has been routinely used in decisions on arterial land drainage and flood protection schemes, in reservoir construction, major abstraction schemes, and major schemes for improving discharges (e.g. sewage works); that is, in major programmes of capital works where demonstrating that benefits exceed costs has typically been a condition for grant aid. CBA has properly not been used as a decision tool in improvements consequent on the implementation of established policy. Thus it is not required for decisions on individual discharge consents and abstraction licences which are the consequence of statutory powers and procedures. CBA would be appropriate for assessing the policies under which these decisions are taken but not the implementation of those policies in specific instances. Indeed it would not be possible to carry out CBA in these cases since there is in effect a subsidiarity problem. This distinction between policies and their implementation is of heuristic value but is not always clear. Thus in the case of groundwater protection zones it is not clear whether the decision to schedule a particular zone should be subject to CBA. The degree of discretion implied in the policy suggests to me that it should.

CBA is not used typically in decisions concerning the maintenance or renewal of existing land drainage and flood protection operations and structures. In fact, the Treasury rules under which the arterial drainage has been funded since the 1972 Water Act have specifically been confined to investment leading to drainage or flood protection improvement, where latter term must be interpreted as referring to reductions in flood risk or lowering of water tables! One consequence of this has been the incorporation of an element of improvement into major renewal schemes in order to attract grant aid. Thus the five schemes constituting the drainage of Halvergate Marshes in the Norfolk Broads arose because the steam-driven pumps which maintained the levels were in need of replacement and the Internal Drainage Board responsible could not afford to replace them without assistance from the public purse. While this particular form of economic idiocy and ecological vandalism is of historical interest only, maintenance and renewal of drainage structures constructed in the last 30 or so years remains an impediment to ecological improvements in many catchments and still constitutes a cause of further deterioration. In these cases it can be argued that CBA should be a requirement for maintenance and renewal schemes since new public resources are being committed to further decisions whose rationale has disappeared. In fact, in many cases there was no proper rationale to start with. Thus my study of some major arterial land drainage schemes in England and Wales, where the principal benefit was increases in agricultural production and/or productivity, found that none would have been viable on a properly conducted CBA even if the ecological consequences were ignored, as typically they were (Bowers, 1988). A substantial number of expedients for raising the measured cost – benefit ratios were identified, including use of inappropriate subsidized farm-gate prices, arbitrary and over-optimistic assumptions about take-up of benefits by farmers, omission of farm fixed costs, choice of cropping patterns, use of theoretical crop yields and stocking rates which maximized measured benefits but which had no exogenous justification and would not have been predicted from observation of farming practices in the areas affected; and failure to take account of expected maintenance costs of structures. It was also shown, notably in the case of the Yare Barrier but also for schemes in the Somerset Levels, that when an initial appraisal failed to produce sufficient benefits to justify the scheme, a search was instigated to "find" additional benefits. Finally, and again the Yare Barrier provided the classic example, it was shown that pre-scheme flood risk was such as to make profitable the agricultural improvements that schemes were supposed to make possible. In short the question set was "how do we justify this scheme on a CBA (and thus attract the necessary grant aid)?" and not "what is the socially efficient method of bringing about these desired agricultural improvements?" The latter question of course was outside of the competence of the drainage authorities, but they implicitly assumed that the answer to it was to improve drainage or flood protection. That more drainage and flood protection for agricultural land was desirable and constituted an efficient and proper use of tax-payers' money was, I suspect, never questioned by those who were designing and implementing the schemes. The results in terms of losses of ecologically valuable grazing marsh and wet neutral grassland have been documented. These continue; I have currently on my desk two agricultural schemes promoted by Internal Drainage Boards in Yorkshire.

This misuse of CBA has not been confined to the land drainage section of the water industry. In the case of reservoir construction, Herrington has argued that the option of demand management (e.g. by the introduction of metering), and alternative strategies for increasing delivered supply from existing sources (e.g. by reducing distribution losses) were ruled out of the appraisal process so that the issue became simply one of reservoir location and minor details of construction and operating processes. Thus, as in land drainage, the essentials of the solution were decided before the appraisal and the objective of the scheme were specified in a narrow technical manner.

Of course, this practice is not confined to the water industry. Similar processes are observable in the road construction programme in the UK where the issue is specified as accommodating exogenous forecasts of road traffic demand and not meeting determined mobility or other social objectives. In this case too the environmental costs of the investment are often severe: over 160 scheduled Sites of Special Scientific Interest (SSSIs) are threatened, by current road proposals (Bowers, Hopkinson and Palmer, 1992). Except where SSSIs are threatened catchment ecology is affected probably only in minor ways by road construction.

The conclusion to be drawn from this discussion is that Pearce, Markandya and Barbier, 1989, are wrong in believing that environmental damage would be avoided by the measurement of environmental consequences of major public works and the incorporation of these values into CBA. Given the ways in which CBA has been used, any values for ecological damage produced would simply be neutralized by adjustment of other measured benefits. The objective of the appraisal in the minds of those conducting it is not to determine whether the scheme should be carried out, but to justify it and to attract public money to it. Unless CBA is carried out by a wholly uncommitted and independent body immune from institutional capture, seeking guidance on the social interest (which is what the theory assumes is done), specifying that environmental effects should be monetized will have no effect. As an example of how things can get distorted, a later appraisal in the saga of the Yare Barrage cited the protection of ecologically valuable grazing marshes from periodic flooding as a benefit of the scheme! CBA should not be left in the hands of specialist agencies with a clear but narrow sectoral remit.

While eliminating the institutional capture of the CBA process will remove many threats to catchment ecology from uneconomic development, there will still be cases of developments from within the water industry and still more of cases of non-water industry policies (housing, roads, etc.) which except for the impact on catchment ecology would be deemed acceptable on a properly conducted CBA so that the economic benefits have to be compared with the ecological costs. How should these be dealt with?

ALTERNATIVES TO CBA

Economists have developed a number of approaches to the valuation of non-market effects, or "intangibles" as they are termed, of investment projects. They may be classified as follows:

(i) *Surrogate market approaches*. Here a market is found where the intangible is traded implicitly as a complement to another good. Observation of variations in the prices obtained by the good with varying quantities or qualities of the intangible attached allows inferences to be drawn about the value attached to the intangible. The usual example given is the determination of the negative value placed on disturbance form aircraft noise by observing the prices of houses in various noise contours around airports.

(ii) *Travel cost approaches*. The Clawson technique is based on the assumption that travel to a recreational or similar site is a perceived by consumers as a cost yielding no direct utility and that the value derived from a site visit must for the marginal visitor be equal to the costs incurred. Observations of visitor density by travel cost bands then permits inference of a demand curve for the site. This is a frequently used technique for valuing recreational facilities. It is simply modified where the site carries an entrance or other user charge.

(iii) *Questionnaire determinations of consumer willingness to pay*. The commonest technique is contingent valuation where the consumer is asked to indicate her/his willingness to pay to retain a specified asset contingent on a set of specified conditions including typically the payment vehicle. The other less common technique in this class is stated preference analysis where respondents are asked to indicate rankings between various combinations of the intangible and market goods. This is an experimental version of (i) above.

(iv) *Cost of replacement or re-creation of the intangible*. This is obviously confined to cases where the means of replacement or re-creation are known.

(v) *Expert judgement*. Since the experts are never experts in the valuation of environmental or other assets, this technique is now generally discredited.

(vi) *Contingent costs*. Where the loss or damage to the intangible gives rise to known or estimatable costs it can be valued at these consequential costs. Thus if additional air pollution arising from a project has predictable consequences for human health then it can be valued in terms of the additional medical expenditure plus the lost production of those affected, etc. This is usually at best a partial valuation since pollution can be perceived as a nuisance independently of any effects on health, damage to building fabric or whatever.

With the exception of questionaire techniques and especially, contingent valuation analysis, these approaches have limited application. The Clawson technique is useful for assessing the recreational and amenity consequences of investment decisions in river catchments and contingent cost calculations are likely also to have a function. But only contingent valuation has the potential for general application to the valuation of ecological effects of investments and indeed it is used widely as a technique for attaching monetary values to wildlife resources. Apart from its flexibility — you can ask consumers how much they are willing to pay for almost anything — its proponents argue that it is the only technique that allows measurement of all components of consumer valuation. Apart from direct user benefit which is all that the other techniques purport to measure, contingent valuation permits the measurement of option value (the willingness of the consumer to pay to retain the option of using the asset) the existence value (the willingness to pay for the knowledge that the asset exists even though there is no expectation of using/experiencing it); and the bequest value (the value attached to passing the asset on to succeeding generations). Appropriate sample selection strategies and questionnaire design allows these components of total consumer valuation to be distinguished.

Despite these apparent advantages, I believe that contingent valuation when applied to obtain valuations of effects of investments on complex catchment habitats and ecosystems is a snare and delusion. It is possible by using the technique to obtain monetary valuations, but the values so obtained have no objective status or meaning (Bowers, 1993).

Where respondents have direct experience or understanding of the environmental effect to be valued, such as with noise pollution or offensive smells, contingent valuation will yield meaningful valuations, but it is in these situations that other techniques can also be used. These are the true Pigovian externalities and if markets do not exist it is possible to create them. Where this direct experience is lacking, then responses are dependent on the information supplied, on the understanding of that information, and on perceptions of the motives of the supplier. There is no correct way of describing an ecosystem or a habitat: descriptions can vary from the popular ("it contains otters and kingfishers — here are pictures") to the scientific. Nor of course is it possible to provide a complete description. Different descriptions are suitable for different purposes and may be understood by different individuals depending on their training and motivation: what will excite a freshwater biologist may not interest a yachtsman. Different information packages will elicit different responses including expressions of willingness to pay. There is no description that is uniquely appropriate for contingent valuation and hence no proper or true valuation. Proponents of the technique recognize that responses vary with the presentation of the problem and talk of sources of "bias". This language predicates the existence, at least theoretically, of a true response. It does not exist. The analogy with the perceived nuisances, the Pigovian externalities, is false. The source of market failure is different and indeed it is inappropriate to talk of a market at all.

If we are unable to place meaningful monetary valuations on the ecology of catchments or on modifications to the ecology consequent on policy proposals then how are we to proceed? The answer is that we operate with environmental standards. These require the protection of specified valuable habitats such as SSSIs, and that certain requirements for environmental quality act as constraints on investment proposals: the environmental standards have to be met if the investment is to proceed. Treating the environment as a constraint solves the valuation problem since the value that properly enters into the CBA is the cost of meeting the standard. Technique (iv) above is the proper basis of valuation in the presence of a sustained constraint. In other circumstances it is inappropriate.

The case for environmental standards to protect catchments has been strengthened by the recent commitment to sustainability. Sustainability entails the imposition of constraints on current economic choices in the interests of future generations. Constraints aimed at conserving genetic diversity and on protecting valuable landscapes including riverine landscapes as well as on protecting the water quality of catchments are examples of such constraints (Barker et al., 1993). Sustainability entails modifications to CBA but not in the direction that Pearce suggests, towards placing spurious monetary values on things which cannot from their nature carry them. The market is a powerful human institution but it has its limitations.

ENVIRONMENTAL ACCOUNTS

The function of environmental accounts would be as a supplement to conventional social accounts to give a broad macro-view of the state of the environment. If they could be constructed and were reasonably comprehensive then they could be used to indicate where resources for environmental protection should be allocated and to give an insight into the interaction between environmental quality and economic activity. It is clear that to have any value these accounts would have to be in physical and not monetary units so that no environmental income akin to national income could be calculated. There have been some attempts in France to calculate environmental accounts for the water system. They are extremely complex and it is unclear how they should be used. While progress on the issue is to be encouraged, they have no present relevance to the issue of protection of catchment ecology.

CONCLUSIONS

(i)   There is no case for the introduction of market instruments for the control of point pollution of catchments.
(ii) The case for the use of tax instruments for the control of non-point pollution from agricultural chemicals should be re-assessed.
(iii) A market instrument is used for abstraction control but it is technically inefficient in that the costs of investment to safeguard catchment flows are spread across abstractors rather than bourne by those who cause the problem. The economics of investment in water transfer are distorted by the pricing system used.
(iv) Catchment ecologies would not be protected by a requirement to place monetary valuations on their modification in the context of cost−benefit analysis.
(v)   In any event, meaningful monetary valuations could not be made.
(vi) The appropriate procedure would be the implementation of environmental standards as a constraint on investments. Standards lead directly to a valuation of effects on catchment ecology in terms of the costs of meeting the standards.
(vii) An environmental standards approach is consistent with a commitment to sustainable development.

References

Adams, C.E. (1993). Environmentally-sensitive predictors of boat traffic loading on inland waterways, *Leisure Studies*, **12**, 71–79.

Adams, C.E., Tippett R., Nunn, S. and Archibald, G. (1992). The utilization of a large inland waterway (Loch Lomond, Scotland) by recreational craft, *Scottish Geographical Magazine*, **108**, 113–118.

Adams, J., Gee, J., Greenwood, P., McLelvey, S. and Perry, R. (1987). Factors affecting the microdistribution of *Gammarus pulex* (Amphipoda): an experimental study, *Freshwater Biology*, **17**, 307–316.

Admiraal, W., van der Velde, G. and Cazemiek, W.G. (1992). The rivers Rhine and Meuse in the Netherlands: present state and signs of ecological recovery, *Hydrobiologia*, 1–32.

Adrian, M.I., Wilden, W. and Delibes, M. (1985). Otter distribution and agriculture in southwestern Spain, in *Proceedings of the 17th Congress of the International Union of Game Biologists*, Brussels, 17–21.

ADRIS (1993). Biological assessment of discharge impact, Association of Directors and River Inspectors Scotland, Biologists Group, unpublished Report.

Ahern, J. (1991). Greenways and ecology, in *Proceedings from selected Educational Sessions of the 1991 ASLA Meeting*, Kansas City, Missouri, pp. 75–89.

Alabaster, J.S., Garland, J.H.N., Hart, I.C. and Solbe, J.F. (1972). An approach to the problem of pollution and fisheries, *Symposium of the Zoological Society of London*, **29**, 87–114.

Alderson, R. (1969). Studies on the larval biology of caddis flies of the family Psychomyiidae. PhD Thesis, University of Wales.

Aldridge, D.W., Payne, B.S. and Miller, A.C. (1987). The effects of intermittent exposure to suspended solids and turbulence on three species of freshwater mussels, *Environmental Pollution*, **45**, 17–28.

Alexander, G.R. (1979). Predators of fish in coldwater streams, in *Predator-Prey Systems in Fisheries Management* (Ed. H. Clepper), pp. 153–170, Sport Fishing Institute, Washington, DC.

Alimoso, S. (1991). Catch effort data and their use in the management of fisheries in Malawi, in *Catch-effort Sampling Strategies* (Ed I.G. Cowx), pp. 393–403, Fishing News Books, Blackwell Scientific Publications, Oxford.

Allanson, B.R. (1961). Investigations into the ecology of polluted inland waters in the Transvaal, Part I, The physical, chemical and biological conditions in the Jukskei–Crocodile River System, *Hydrobiologia*, **18**, 1–76.

Allard, M. and Moreau, G. (1985). Short-term effect on the metabolism of lotic benthic communities following experimental acidification, *Canadian Journal of Fisheries and Aquatic Sciences*, **42**, 1676–1680.

Allen, G.R. (1989). *Freshwater Fishes of Australia*, TFH Publications, New Jersey.

Allen, K.O. and Hardy, J.W. (1980). *Impacts of navigational dredging on fish and wildlife: a literature review*, Report No. FWS/OBS-80/07, US Fish and Wildlife Service, Biological Services Program, National Coastal Ecosystems Team, Slidell, La., USA.

Allen, K.R. (1985). Comparison of the growth rate of brown trout (Salmo trutta) in a New Zealand stream with experimental fish in Britain, *Journal of Animal Ecology*, **54**, 487–495.

Allendorf, F.W. and Phelps, S.R. (1981). Isozymes and the preservation of genetic variation in salmonid fishes, in *Fish Gene Pools* (Ed. N. Ryman), *Ecological Bulletin* (Stockholm), **34**, 37–45.

Allendorf, F.W., Ryman, N. and Utter, F.M. (1987). Genetics and fishery management: past, present and future, in *Population Genetics and Fishery Management* (Ed. N. Ryman and F.M. Utter), pp. 1–19, University of Washington Press, Seattle.

Altaba, C.R. (1992). Les Naïades (Mollusca: Bivalva: Unionida) dels Països Catalans, *Butlletí Institució Catalana Història Natural*, **60**, 23–44.

Anderson, F.E. (1975). *The short-term variation in suspended sediment concentration caused by the passage of a boat wave over a tidal flat environment*, Technical Report No. 2, Dept. of Earth Sciences and Jackson Estuarine Laboratory, University of New Hampshire, Durham, USA.

Anderson, J. (1969). Habitat choice and life history of Bembidiini (Coleoptera, Carabidae) on river banks in central and northern Norway, *Norsk ent. Tidsskr.*, **17**, 17–65.

Anderson, J. (1983). The habitat distribution of species of the tribe Bembidiini (Coleoptera, Carabidae) on banks and shores in northern Norway, *Notulae Entomologicae*, **63**, 131–142.

Anderson, J.R. (1991). *The Implications of Salinity, and Salinity Management Initiatives, on Fish and Fish Habitat in the Kerang Lakes Management Area*, Arthur Rylah Institute for Environmental Research, Technical Report Series No. 103, Department of Conservation and Environment, Victoria.

Anderson, J.R. and Morison, A.K. (1989). *Environmental Flow Studies for the Wimmera River, Victoria — Part C, Water Quality and the Effects of an Experimental Release of Water*, Arthur Rylah Institute for Environmental Research, Technical Report Series No. 75.

Anderson, N.H. and Grafius, E. (1975). Utilization and processing of allochthonous material by stream Trichoptera, *Verhandlungen Internationale Vereinigung für Theoretische und Angewandte Limnologie*, **19**, 3022–3028.

Anderson, N.H. and Lehmkuhl, D.M. (1968). Catastrophic drift of insects In a woodland stream, *Ecology*, **49**, 198–206.

Anderson, N.H. and Sedell, J.R. (1979). Detritus processing by macroinvertebrates in stream ecosystems, *Annual Review of Entomology*, **24**, 351–377.

Anderson, N.H., Steedman, R.J. and Dudley, T. (1984). Patterns of exploitation by stream invertebrates of wood debris (xylophagy), *Verhandlungen Internationale Vereinigung für Theoretische und Angewandte Limnologie*, **22**, 1847–1852.

Anderson, R.O. (1959). A modified flotation technique for sorting bottom fauna samples, *Limnology and Oceanography*, **4**, 223–5.

Anderson, R.V. and Day, D.M. (1986). Predictive quality of macroinvertebrate–habitat associations in lower navigation pools of the Mississippi River, *Hydrobiologia*, **136**, 101–112.

Andersen, U.V. (1994). Sheep grazing as a method for controlling *Heracleum mantegazzianum*, in *Ecology and Management of Invasive Riverside Plants* (eds L.C. de Waal, L.E. Child, P.M. Wade, and J.H. Brock), John Wiley, Chichester.

Angermeier, P.L. and Karr, J.R. (1983). Fish communities along environmental gradients in a system of tropical streams, *Environmental Biology of Fishes*, **9**, 117–135.

Angermeier, P.L., Neves, R.J. and Nielson, L.A. (1991). Assessing stream values: perspectives of aquatic resource professionals, *North American Journal of Fisheries Management*, **11**, 1–10.

Anon. (1982). *The Logical Framework Approach (LFA)*, Norwegian Agency for Development Corporation.

Anon. (1985). Report of meeting of working group on North Atlantic Salmon, ICES CM 1985/ASSESS, 11, 1–67.

Anon. (1990a). Annual Report No. XXXV Salmon Research Agency of Ireland Inc., 1–66.

Anon. (1990b). *Integrated Fishery Management Plan for the Tweed Fishery District*, Tweed Foundation Special Publication, 1–18.

Anon. (1990c). *Game Angling Code*, Salmon and Trout Association, London.

Anon. (1991). *Angling: An independent review of the sport of angling*, commissioned by The Sports Council and the National Anglers' Council, The Sports Council, London.

ANZECC, (1992). *Australian Water Quality Guidelines for Fresh and Marine Waters*, Australian and New Zealand Environment and Conservation Council, Melbourne.

Aragon, J. (1943). *Informe Sobre La Salinidad Del Agua Del Ebro*, Instituto Nacional de Colonización, Spain.

Arbeitsgemeinschaft Umwelt Mainz (1972). Bestandsrückgang der Schneckenfauna des Rheins zwischen Straßburg und Koblenz, *Natur und Museum*, **102** (6), 197–206.

Armitage, P.D. (1976). A quantitative study of the invertebrate fauna of the River Tees below Cow Green reservoir, *Freshwater Biology*, **6**, 229–240.

Armitage, P.D. (1977·). Invertebrate drift in the regulated river Tees, and an unregulated tributary Maize Beck, below Cow Green dam, *Freshwater Biology*, **7**, 167–183.

Armitage, P.D. (1978). Downstream changes in the composition, numbers and biomass of bottom fauna in the Tees below Cow Green reservoir and in an unregulated tributary Maize Beck, in the first five years after impoundment, *Hydrobiologia*, **58**, 145–156.

Armitage, P.D. (1980). Stream regulation in Great Britain, in *The Ecology of Regulated Streams*, (Eds J.V. Ward and J.A. Stanford), Plenum Press, New York.

Armitage, P.D. (1994). Prediction of biological responses, in *The Rivers Handbook*, Vol. 2 (Eds P. Calow and G.E. Petts), pp. 254–275, Blackwell Scientific, Oxford.

Armitage, P.D. and Capper, M.H. (1976). The numbers, biomass and transport downstream of micro-crustaceans and Hydra from Cow Green Reservoir (Upper Teesdale), *Freshwater Biology*, **6**, 425–432.

Armitage, P.D. and Ladle, M. (1991). Habitat preferences of target species for application in PHABSIM testing, in *Instream Flow Requirements Of Aquatic Ecology in Two British Rivers*, (Eds A. Bullock, A. Gustard and E.S. Grainger), Institute of Hydrology, Report 115.

Armitage, P.D. and Petts, G.E. (1992). Biotic score and prediction to assess the effects of water abstractions on river macroinvertebrates for conservation purposes, *Aquatic Conservation: Marine and Freshwater Ecosystems*, **2**, 1–17.

Armitage, P.D., Moss, D., Wright, J.F. and Furse, M.T. (1983). The performance of a new biological water quality score system based on macro-invertebrates on wide range of unpolluted running water sites, *Water Research*, **17**, 333–347.

Armour, C.L. and Taylor, J.G. (1991). Evaluation of the instream flow incremental methodology by US Fish and Wildlife Service field users, *Fisheries*, **16**, 36–43.

Arner, D.H., Robinette, H.R., Frasier, J.E. and Gray, M.H. (1976). *Effects of Channelization of the Luxapalila River on Fish, Aquatic Invertebrates, Water Quality and Fur Bearers*. Report no FWS/OBS-76/08, Office of Biological Services, Fish and Wildlife Service, US Department of the Interior, Washington DC.

Arthington, A.H. and Mitchell, D.S. (1986). Aquatic invading species, in *Ecology of Biological Invasions: An Australian Perspective* (Eds R.H. Groves and J.J. Burdon), Australian Academy of Science, Canberra.

Arthington, A.H., Conrick, D.L. and Bycroft, B.M. (1992). *Environmental Study Barker–Barambah Creek. Vol. 2*: Scientific Report, Centre for Catchment and Instream Research, Griffith University, and Water Resources Commission, Department of Primary Industries, Brisbane.

Arthington, A.H., King, J.M., O'Keeffe, J.H., Bunn, S.E., Day, J.A., Pusey, B., Bluhdorn, D.R. and Tharme, R. (1992). Development of an holistic approach for assessing environmental flow requirements of riverine ecosystems, in *Water Allocation for the Environment, Proceedings of an International Seminar and Workshop* (Eds J.J. Pigram and B.P. Hooper), pp. 69–76, Centre for Water Policy Research, University of New England, Armidale, Australia.

Arunachalam, M., Madhusoodanan Nair, K.C., Vijverberg, J., Kortmulder, K. and Suriyanarayanan, H. (1991). Substrate selection and seasonal variation in densities of invertebrates in stream pools of a tropical river, *Hydrobiologia*, **213**, 141–148.

Association of River Authorities (1974). *Coarse Fisheries*, A Report of the Working Party, The Association of River Authorities, London.

Aukes, P. (1992). Criteria for ecological rehabilitation, in *Contributions to the European Workshop Ecological Rehabilitation of Floodplains*, Arnhem, Netherlands, pp. 75–78.

Avery, L. and Hunt, R.L. (1981). *Population dynamics of wild brown trout and associated sport fisheries in four central Wisconsin streams*. Wisconsin Department of Natural Resources Technical Bulletin, 121.

Bachman, R.A. (1984). Foraging behavior of free-ranging wild and hatchery brown trout in a stream, *Transactions of the American Fisheries Society*, **113**, 1–32.

Baglinière, J.L. (1991). La truite commune (*Salmo trutta* L.), son origine, son aire de répartition, ses intérats économique et scientifique, in *La Truite, Biologie et Écologie* (Eds J.L. Baglinière and G. Maise), pp. 11–22, INRA, Paris.

Baglinière, J.L. and Arribe-Moutounet, D. (1985) Microrépartition des populations de truite commune (*Salmo trutta* L.) de juvenile de saumon atlantique (*Salmo salar* L.) et des autres espèces présentes dans la partie haute du Scorff (Bretagne), *Hydrobiologia*, **120**, 229–239.

Baglinière, J.L. and Maisse, G. (1990). La croissance de la truite commune (*Salmo trutta* L.) sur le bassin du Scorff, *Bulletin Française de la Pêche et Pisciculture*, **318**, 89–101.

Baglinière, J.L., Maisse, G., Lebail, P.Y. and Nihouarn, A. (1989). Population dynamics of brown trout, *Salmo trutta* L., in a tributary in Brittany (France): spawning and juveniles, *Journal of Fish Biology*, **34**, 97–110.

Bailey, J.P. (1989). *Cytology and breeding behaviour of giant alien* Polygonum *species in Britain*, PhD Thesis, Leicester University.

Bain, M.B., Finn, J.F. and Booke, H.E. (1988). Streamflow regulation and fish community structure, *Ecology*, **69**, 382–392.

Baker, R.M. (1988). Mechanical control of Japanese Knotweed in an SSSI, *Aspects of Applied Biology*, **16**, 189–192.

Baldi, E. and Moretti, G. (1938). La vita nell'Olona e nel Lambro, storia della deformazione di un carico biologico, *Atti Soc. It. Sci. Nat.*, **77**, 79–124.

Balfour-Browne, F. (1950) *British Water Beetles*, Vol. 2, Ray Society, London.

Balon, E.K. (1956). Breeding and postembryonic development of the roach (*Rutilus rutilus* ssp), *Biologicke prace*, **3**, 1–60.

Balon, E.K. (1975). Reproductive guilds of fishes: a proposal and definition, *Journal of the Fisheries Research Board of Canada*, **32**, 821–864.

Balon, E.K. (1981). Additions and amendments to the classification of reproductive styles in fishes, *Environmental Biology of Fishes*, **6**, 377–389.

Banks, J.W. (1990). Fisheries management in the Thames Basin, England, with special reference to the restoration of a salmon population, in *Management of Freshwater Fisheries* (Ed. W.L.T. Densen, B. Steinmetz and R.H. Hughes), pp. 511–519. Pudoc, Wageningen.

Barak, N.A.-E. and Mason, C.F. (1990). A survey of heavy metal levels in eels (*Anguilla anguilla*) in some rivers in East Anglia, England: the use of eels as pollution indicators, *Internationale Revue für die gesamte Hydrobiologie*, **75**, 827–833.

Barclay, J.S. (1980). *Impact of stream alterations on riparian communities in south-central Oklahoma*, Report no FWS/OBS- 80/17, Office of Biological Services, Fish and Wildlife Service, US Department of the Interior, Washington DC.

Bardonnet, A., Gaudin, P. and Persat, H. (1991) Microhabitats and diel downstream migration of young grayling (*Thymallus thymallus* L.), *Freshwater Biology*, **26**, 365–376.

Barker, J., Bowers, J. Hopkinson, P. and Lyall, K. (1993). *Environmental Standards, Issues for the Commission*, Environmental Policy Unit, School of Business and Economic Studies, University of Leeds.

Barko, J.W. and Smart, R.M. (1981). Sediment based nutrition of submersed macrophytes, *Aquatic Botany*, **10**, 339–352.

Bärlocher, F. and Kendrick, B. (1973) Fungi and food preferences of *Gammarus pseudolimnaeus*, *Archiv für Hydrobiologie*, **72**, 501–516.

Barmuta, L.A. (1989) Habitat patchiness and macrobenthic community structure in an upland stream in temperate Victoria, Australia, *Freshwater Biology*, **21**, 223–236.

Barmuta, L.A., Marchant, R. and Lake, P.S. (1992). Degradation of Australian streams and progress towards conservation and management in Victoria, in *River Conservation and Management* (Eds P.J. Boon, P. Calow and G.E. Petts), pp. 65–79, Wiley, Chichester.

Barnes J.R. and Minshall, G.W. (Eds) (1983). *Stream Ecology: Application and Testing of General Ecological Theory*, Plenum Press, New York.

Barrett, J.C., Grossman, G.D. and Rosenfeld, J. (1992). Turbidity-induced changes in reactive distance of rainbow trout. *Transactions of the American Fisheries Society*, **121**, 437–443.

Bartholomew, G.A. (1958). The role of physiology in the distribution of terrestrial vertebrates, in *Zoogeography* (Ed. C.L. Hubbs), pp. 81–95, American Association for the Advancement of Science, Washington DC.

Bas, N., Jenkins, D. and Rothery, P. (1984). Ecology of otters in northern Scotland, V. The distribution of otter *Lutra lutra* faeces in relation to bankside vegetation on the River Dee in summer 1981, *Journal of Applied Ecology*, **21**, 507–513.

Basingstoke Canal Authority (1993). *Basingstoke Canal SSSI Management Plan, Third Draft*, BCA,

Basingstoke.

Bass, D. (1986). Habitat ecology of chironomid larvae of the Big Thicket streams, *Hydrobiologia*, **134**, 29–41.

Battarbee, R.W. (1984). Diatom analysis and the acidification of lakes, *Philosophical Transactions of the Royal Society, B*, **305**, 451–477.

Battarbee, R.W. (1992). *Critical Loads and Acid Deposition for UK Freshwaters*; an interim report to the DoE from the Critical Loads Advisory Group, Environmental Change Research Centre, University College London, Research Papers No. 5.

Battegazzore, M., Petersen, R.C., Moretti, G. and Rossaro, B. (1992). An evaluation of the environmental quality of the River Po using benthic macroinvertebrates, *Archiv für Hydrobiologie*, **125**, 175–206.

Batten L.A. (1977). Sailing on reservoirs and its effects on waterbirds, *Biological Conservation*, **11**, 49–58.

Battiker, B. (1989). Analyse de la peche de la truite (*Salmo trutta*) dans les rivières du canton de Vaud (Suisse). *Bull. Soc. Vaudoise Sci. Nat.*, **79**, 161–170.

Baxter, G. (1961). River utilization and the preservation of migratory fish life, *Minutes of the Proceedings of the Institution of Civil Engineers*, **18**, 225–244.

Bayley, P.B. (1983). *Central Amazon fish populations: biomass, production and some dynamic characteristics*, PhD Thesis, Dalhousie University, Halifax, Nova Scotia.

Bayley, P.B. (1985). Sampling problems in freshwater fisheries, in *Proceedings of the 4th British Freshwater Fisheries Conference*, pp. 3–11, University of Liverpool.

Bayley, P.B. (1993). Quasi-likelihood estimation of marked fish recapture. *Canadian Journal of Fisheries and Aquatic Science* (in press).

Bayley, P.B. and Austen, D.J. (1988). Comparison of detonating cord and rotenone for sampling fish in warmwater impoundments, *North American Journal of Fisheries Management*, **8**, 310–316.

Bayley, P.B. and Austen, D.J. (1990). Modeling the sampling efficiency of rotenone in impoundments and ponds, *North American Journal of Fisheries Management*, **10**, 202–208.

Beard, T.D. and Carline, R.F. (1991). Influence of spawning and other stream habitat features on spatial variability of wild brown trout, *Transactions of the American Fisheries Society*, **120**, 711–722.

Beaudou, D. and Cuinat, R. (1990). Relationship between growth of brown trout, *Salmo trutta fario* L., and environment, in Massif Central rivers, *Bulletin Française de la Pêche et Pisciculture*, **318**, 82–88.

Beerling, D.J. (1990). *The ecology and control of Japanese knotweed and Himalayan balsam on River banks in South Wales*, PhD Thesis, University of Wales, Cardiff.

Beerling. D.J. and Perrins, J.M. (1993). Biological flora of the British Isles, *Impatiens glandulifera* Royle, *Journal of Ecology*, **81** 367–382.

Behmer, D.J. and Hawkins, C.P. (1986). Effects of overhead canopy on macroinvertebrate production in a Utah stream, *Freshwater Biology*, **16**, 287–300.

Behrendt, A. (1977). *The Management of Angling Waters*, Andre Deutch, London.

Bell, D.V., Odin, N. and Torres, E. (1985). Accumulation of angling litter at game and coarse fisheries in South Wales, UK, *Biological Conservation*, **34**, 369–379.

Bellinger, E.G. and Belham, B.R. (1978). The levels of metals in dockyard sediments with particular reference to the contribution from ship-bottom paints. *Environmental Pollution*, **15**, 71–81.

Bencala, K.E. and Walters, R.A. (1983). Simulation of solute transport in a mountain pool and riffle stream: a transient storage model, *Water Resources Research*, **19**, 718–724.

Bendell, B.E. and McNicol, D.K. (1987). Fish predation, lake acidity and the composition of aquatic insect assemblages, *Hydrobiologia*, **150**, 193–202.

Benke, A.C., Van Arsdall, T.C., Gillespie, D.M. and Parrish, F.K. (1984). Invertebrate productivity in a subtropical blackwater river: the importance of habitat and life history, *Ecological Monographs*, **54** (1), 25–63.

Berg, C.O. (1950). *Hydrellia* (Ephydridae) and some other related acalypterate Diptera reared from *Potamogeton*, *Annals of the Entomological Society of America*, **43**, 374–398.

Bergey, E.A., Balling, S.F., Collins, J.N., Lamberti, G.A. and Resh, V.H. (1992) Bionomics of invertebrates within an extensive *Potamogeton pectinatus* bed of a California marsh, *Hydrobiologia*, **234** (1), 15–24.

Bergman, A. and Olsson, M. (1985). Pathology of Baltic grey seal and ringed seal females with special reference to adrenocortical hyperplasia: is environmental pollution the cause of a widely distributed disease syndrome?, *Finnish Game Research*, **44**, 47–62.

Beschta, R.L. and Platts, W.S. (1986). Morphological features of small streams: significance and function, *Water Resources Bulletin*, **22**, 369–379.

Bhowmik, N.G. and Adams, J.R. (1989). Successional changes in habitat caused by sedimentation in navigation pools, *Hydrobiology*, **176/177**, 17–27.

Bickerton, M. (1992). Changes of the invertebrate fauna of the river Glen, in *The River Glen* (Ed G.E. Petts), National Rivers Authority, Anglian Region, Peterborough. Annex C.

Biggs, J.F. and Close, M.E. (1989). Periphyton biomass dynamics in gravel bed rivers: the relative effects of flows and nutrients, *Freshwater Biology*, **22**, 209–231.

Bilby, R.E. (1981). Role of organic debris dams in regulating the export of dissolved and particulate matter from a forested watershed, *Ecology*, **62**, 1234–1243.

Bilby, R.E. and Likens, G.E. (1980). Importance of organic debris dams in the structure and function of stream ecosystems, *Ecology*, **61**, 1107–1113.

Billingham, N. (1993). Less wash, at last, the launch of the Ostec hull, *Canal and Riverboat*, **16**, 52–54.

Billington, N. and Hebert, P.D.N. (1991). Mitochondrial DNA diversity in fishes and its implications for introductions, *Canadian Journal of Fisheries and Aquatic Sciences*, **48** (Suppl. 1), 80–94.

Binns, N.A. and Eiserman, F.M. (1979). Quantification of fluvial trout habitat in Wyoming, *Transactions of the American Fisheries Society*, **108**, 215–228.

Bird, S.C., Walsh, R.P.D. and Littlewood, I.G. (1990). Catchment characteristics and basin hydrology: their effects on stream acidity, in *Acid Waters in Wales* (Eds R.W. Edwards, A.S. Gee and J.H. Stoner), pp. 203–221, Kluwer Academic, Holland.

Blaxter, J.H.S. (1970). Light. Fishes, in *Marine Ecology, Vol. 1. Environmental Factors* (Ed O. Kinne), pp. 213–320.

Bledisloe, (1961). Report of the Committee on Salmon and Freshwater Fisheries, Command Paper 1350, Her Majesty's Stationery Office, London.

Blindow, I. (1987). Composition and density of epiphyton on several species of submerged macrophytes — the neutral substrate hypothesis tested, *Aquatic Botany*, **29**, 157–168.

Blondel, J., Ferry, C. and Frochot, B. (1981). Point counts with unlimited distance, *Studies in Avian Biology*, **6**, 414–420.

Blühdorn, D.R. and Arthington, A.H. (1994). *The Effects of Flow Regulation in the Barker-Barambah Catchment*, Centre for Catchment and In-stream Research, Griffith University, Brisbane, Australia.

Blyth, J. D. (1983). Rapid stream survey to assess conservation value and habitats available for invertebrates, in *Survey Methods for Nature Conservation*, Proceedings of a workshop held at Adelaide University, 31 Aug–2 Sept 1983 (Eds K. Myers, C. R. Margules and I. Musto), Vol. I, pp. 343–375, CSIRO Division of Water and Land Resources, Canberra.

Bohlin, T. and Cowx, I.G. (1990). Implications of unequal probability of capture by electric fishing on the estimation of population size, in *Fishing with Electricity* (Ed I.G. Cowx and P. Lamarque), pp. 145–155, Fishing News Books, Blackwell Scientific Publications, Oxford.

Bohlin T., Heggberget, T.G. and Strange, C. (1990). Electric fishing for sampling and stock assessment, in *Fishing with Electricity* (Ed I.G. Cowx and P. Lamarque) pp. 112–139, Fishing News Books, Blackwell Scientific Publications, Oxford .

Bohlin T., Hamrin, S., Heggberget, T.G., Rasmussen, G. and Saltveit, S.J. (1989). Electrofishing — Theory and practice with special emphasis on salmonids, *Hydrobiologia*, **173**, 9–43.

Bohm, P. and Russell, C. S. (1985). Comparative analysis of alternative policy instruments, in *Handbook of Natural Resource and Energy Economics* (Eds A.V. Kneese and J.L. Sweeny), North Holland Press, Amsterdam.

Bolas, P.M. and Lund, J.W.G. (1974). Some factors affecting the growth of *Cladophora glomerata* in the Kentish Stour, *Water Treatment and Examination*, **23**, 25–51.

Boles, G.L. (1981) Macroinvertebrate colonization of replacement substrate below a hypolimnial release reservoir, *Hydrobiologia*, **78**, 133–146.

Bongers, J.J.A. (1990). A unified market for European angling, in *Proceedings of the Institute of Fisheries Management 21st Anniversary Conference*, pp. 105–108.

Bonham, A.J. (1980). *Bank Protection Using Emergent Plants Against Boat Wash In Rivers And Canals*, Report No. IT 206, Hydraulics Research Station, Wallingford, UK.

Boon, P.J. (1991). The role of Sites of Special Scientific Interest (SSSIs) in the conservation of British rivers, *Freshwater Forum*, **1**, 95–108.

Boon, P.J. (1992a). Essential elements in the case for river conservation, in *River Conservation and Management* (Eds P.J. Boon, P. Calow and G.E. Petts), pp. 11–33, John Wiley, Chichester.

Boon, P.J. (1992b). Channelling scientific information for the conservation and management of rivers, *Aquatic Conservation: Marine and Freshwater Ecosystems*, **2**, 115–123.

Boon, P.J. (1994). The conservation of freshwater habitats and species, in *The Fresh Waters of Scotland: A National Resource of International Significance* (Eds P.S. Maitland, P.J. Boon and D.S. McLusky), John Wiley, Chichester.

Boon P.J., Calow, P. and Petts, G.E. (Eds) (1992). *River Conservation and Management*, John Wiley, Chichester.

Boon, P.J., Holmes, N.T.H., Maitland, P.S. and Rowell, T.A. (1994). A system for evaluating rivers for conservation ("SERCON"): An outline of the underlying principles, *Verhandlungen der Internationalen Vereinigung für theoretische und angewandte Limnologie*, **25**, 1510–1514.

Bormann, F.H., Likens, G.E., Siccama, T.G., Pierce, R.S. and Eaton, J.S. (1974). The export of nutrients and recovery of stable conditions following deforestation at Hubbard Brook, *Ecological Monographs*, **44**, 255–77.

Bosch, J.M. and Hewlett, J.D. (1982). A review of catchment experiments to determine the effect of vegetation changes on water yield and evapotranspiration, *Journal of Hydrology*, **55**, 3–23.

Bouchardy, C. (1986). La loutre, Sang de la terre, Paris.

Bournaud, M. and Cogerino, L. (1986) Les microhabitats aquatiques des rives d'un grand cours d'eau: approche faunistique, *Annlales de Limnologie*, **22**, 285–294.

Boussard, A. (1981). The reactions of roach (*Rutilus rutilus*) and rudd (*Scardinius erythrophthalmus*) to noises produced by high speed boating, in *Proceedings of the 2nd British Freshwater Fisheries Conference*, pp. 188–196, University of Liverpool, UK.

Boussard, A. (1984). Convention de recherche relative à l'influence de la navigation fluviale sur le frai dans les cours d'eau publics: ski-nautique et navigation de plaisance, Synthagese des conclusions finales (1977–1983), Publication du Ministagere des Travaux Publics, Bruxelles, Belgium.

Boussard, A. and Falter, U. (1982). Influence de la turbidité de l'eau sur le taux d'éclosion du gardon (*Rutilus rutilus* L.). *Annls. Soc. Roy. Zool. Belg.*, **T112**, Fasc. 2, 237–250.

Bovee, K.D. (1978). The incremental method of assessing habitat potential for coolwater species, with management implications, *American Fisheries Society Special Publication*, **11**, 340–346.

Bovee, K.D. (1982). A Guide to Stream Habitat Analysis using the Instream Flow Incremental Methodology, Instream Flow Information Paper 12, FWS/OBS-82/26 Office of Biological Services, US Fish and Wildlife Service.

Bovee, K.D. (1986). Development and evaluation of habitat suitability criteria for use in the instream flow incremental methodology. Instream flow information paper 21, *U.S. Fish Wildl. Serv. Biol. Rep.* **86**, US Fish and Wildlife Service.

Bowers, J. (1988) Cost–benefit analysis in theory and practice: the case of agricultural land drainage, in *Sustainable Environmental Management* (Ed. R. Kerry Turner), Belhaven Press for the ESRC.

Bowers, J. (1991). Environmental Problems and the Limits of the Market, School of Business and Economic Studies Discussion Paper 1/91, University of Leeds.

Bowers, J. (1993) Pricing the environment: a critique, *International Review of Applied Economics*, **7**, 91–107.

Bowers, J. and O'Donnell, K. (1989). *Liquid Costs*, Media Natura.

Bowers, J., Hopkinson P. and Palmer, A.P. (1992). *The Implementation of Sustainability in the Roads Programme*, English Nature, Peterborough, UK.

Bowker, D.W., Wareham, M.T. and Learner, M.A. (1985). A choice chamber experiment on the selection of algae as food and substrata by *Nais elinguis* (Oligochaeta: Naididae), *Freshwater Biology*, **15**, 547–557.

Bowlby, J.N. and Roff, J.C. (1986). Trout biomass and habitat relationships in southern Ontario streams, *Transactions of the American Fisheries Society*, **115**, 503–514.

Bowles, F.J., Frake, A.A. and Mann, R.H.K. (1990). The use of boom-mounted multi-anode electric fishing equipment for a survey of the fish stocks of the Hampshire Avon, in *Fishing with Electricity*

(Ed. I.G. Cowx), Fishing News Books, Blackwells Scientific Publications, Oxford.

Braña, F., Nicieza, A.G. and Toledo, M.M. (1992). Effects of angling on population structure of brown trout, *Salmo trutta* L., in mountain streams of Northern Spain, *Hydrobiologia*, **237**, 61–66.

Braum, E. (1978). Ecological aspects of the survival of fish eggs, embryos and larvae, in *Ecology of Freshwater Fish Production* (Ed. S.D. Gerking), pp. 102–131, Blackwell Scientific Publications, London.

Brenniman, G.R., Anver, M.R., Hartung, R. and Rosenburg, S.H. (1979). Effects of outboard motor exhaust emissions on goldfish (*Carassius auratus*), *Journal of Environmental Pathology and Toxicology*, **2**, 1267–1281.

Brett, J.R. (1979). Environmental factors and growth, in *Fish Physiology*, Vol. 8 (Eds W.S. Hoar, D.J. Randall and J.R. Brett), pp. 599–675, Academic Press, New York.

Brett, J.R. and Groves, T.D.D. (1979). Physiological energetics, in *Fish physiology* Vol. 8 (Eds W.S. Hoar, D.J. Randall and J.R. Brett), pp. 279–352, Academic Press, New York.

Brinkhurst, R.O. (1982). *British and other Marine and Estuarine Oligochaetes Keys and Notes for the Identification of the Species*, The Linnean Society of London and The Estuarine and Brackish-water Sciences Association, Cambridge University Press, Cambridge, England.

Brinson, M.M. (1977) Decomposition and nutrient exchange of litter in an alluvial swamp forest, *Ecology*, **58**, 601–609.

British Marine Industries Federation (1992). *A Guide to Boating and the Environment*, BMIF, Surrey, UK.

British Transport Docks Board (1972). *Creation of Wash by Pleasure Craft*, British Transport Docks Board Research Station Report No. R236, London.

British Waterways Board (1993a). *Bridgwater and Taunton Canal: Strategy for Leisure and Tourism Development*, British Waterways Board, Watford, England.

British Waterways Board (1993b). *Leisure and Tourism Strategy*, British Waterways Board, Watford, England.

Brittian, J.E. and Eikeland, T.J. (1988). Invertebrate drift — a review, *Hydrobiology*, **166**, 77–83.

Broads Authority (1982). *Strategic Management Plan for Broadland*, Boards Authority, Norwich, England.

Broads Authority (1987). *Broads Plan*, Broads Authority, Norwich, England.

Broads Authority (1993). *No Easy Answers*, Draft Broads Plan 1993, Broads Authority, Norwich, England.

Brock, J.H. and Wade, P.M. (1992). Regeneration of *Fallopia japonica*, Japanese knotweed from rhizome and stems: Observations from greenhouse trials. *IXe Colloque International sur la Biologie des Mauvaises Herbes*, Dijon, (France), pp. 85–94.

Brock, J.H., Child, L.E., Waal, L.C. de and Wade, P.M. (in press). The invasive nature of *Fallopia japonica* is enhanced by vegetative regeneration from stem tissues, in *Plant Invasions — Theory and Applications* (Eds P. Pysek, K. Prach,. M. Rejmánek and P.M. Wade), SPB — Academic Publishing, The Hague.

Broekhuizen, S. (1989). Belasting van otters met zware metalen en PCBs, *De Levende Natuur*, **90**, 43–47.

Brooker, M. P. (1981). The impact of impoundments on the downstream fisheries and general ecology of rivers, in *Advances in Applied Ecology*, (Ed. T.H. Coaker), Vol. 6, pp. 91–152, Academic Press, New York.

Brooker, M.P. and Edwards, R.W. (1975). Review paper: aquatic herbicides and the control of water weeds. *Water Research*, **9**, 1–15.

Brooker, M.P. and Hemsworth, R.J. (1978). The effect of a release of an artificial discharge of water on invertebrate drift in the R. Wye, Wales, *Hydrobiologia*, **59**, 155–163.

Brooker, M.P., Morris, D.L. and Hemsworth R.J. (1977). Mass mortalities of adult salmon (Salmo salar) in the R. Wye, 1976, *Journal of Applied Ecology*, **14**, 409–417.

Brooker, M.P., Morris, D.L. and Wilson, C.J. (1978). Plant–flow relationships in the River Wye catchment, in *Proceedings of the European Weed Research Society 5th Symposium on Aquatic Weeds*, pp. 63–69, European Weed Research Society, Amsterdam.

Brookes, A. (1983). *River channelization in England and Wales: downstream consequences for the channel morphology and aquatic vegetation*, PhD thesis, University of Southampton, UK.

Brookes, A. (1986). Response of aquatic vegetation to sedimentation downstream from river channelisation works in England and Wales, *Biological Conservation*, **38**, 351−367.

Brookes, A. (1987). River channel adjustments downstream from channelization works in England and Wales, *Earth Surface Processes and Landforms*, **12**, 337−351.

Brookes, A. (1988). *Channelized Rivers: Perspectives for Environmental Management*, John Wiley, Chichester.

Brookes, A. (1991). *Design practices for channels receiving urban runoff: examples from the River Thames catchment, UK*, Paper given at Engineering Foundation Conference, Mt Crested Butte, Colorado, August 1991. To be published as an ASCE proceedings.

Brookes, A. (1992) Recovery and restoration of some engineered British river channels, in *River Conservation and Management* (Eds P.J. Boon, P. Calow and G.E. Petts), 337−352, Wiley, Chichester.

Brookes, A. (1994). River channel change, in *The Rivers Handbook*, Vol. 2 (Eds P.Calow and G.E. Petts), pp. 55−75,Blackwell Scientific Publications, Oxford.

Brookes, A. and Hanbury, R.G. (1990). Environmental impacts on navigable river and canal systems: a review of the British experience, *PIANC, S.I.*, **4**, 91−103.

Brookes, A. and Long, H.J. (in press). A method for the geomorphological assessment of river channels at a catchment scale.

Brown, A.V. and Brussock, P.P. (1991). Comparisons of benthic invertebrates between riffles and pools, *Hydrobiologia*, **220**, 99−108.

Brown, D.J.A. and Sadler, K. (1989). Fish survival in acid waters, in *Acid Toxicity and Aquatic Animals* (Eds R. Morris, E.W. Taylor, D.J.A. Brown and J.A. Brown), pp. 31−44, Cambridge University Press, Cambridge.

Brown, G.W. and Krygier, J.T. (1970). Effects of clear-cutting on stream temperature. *Water Resources Research*, **6**, 1131−1139.

Brown, H.P. (1987) Biology of riffle beetles, *Annual Review of Entomology*, **32**, 253−273.

Brown, S.M. (Ed.) (1988). *Kennett and Avon Canal Restoration Project: Ecology Report for Berkshire, Report of the Kennett and Avon Canal Survey April 1983−August 1987*, British Waterways Board: Kennett and Avon Canal Project, BWB, Gloucester.

Browne, J. (1990). Salmon research in the Corrib catchment, western Ireland, *Proceedings of the IFM 20th Annual Study Course*, Galway, 1989, pp. 166−178.

Brusven, M.A., Meeham, W.R. and Biggam, R.C. (1990). The role of aquatic moss on community composition and drift of fish-food organisms, *Hydrobiologia*, **196**, 39−50.

Bryan, G.W. (1976). Some aspects of heavy metal tolerance in aquatic organisms, in *Effects of Pollutants on Aquatic Organisms* (Ed. A.P.M. Lockwood), pp. 7−34, Cambridge University Press, Cambridge.

Buck, R.J. and Hay, D.W. (1984). The relation between stock size and progeny of Atlantic salmon, *Salmo salar* L., in a Scottish stream, *Journal of Fish Biology*, **23**, 1−11.

Buckley, D.R. and Stott, B. (1977). Grass carp in a sport fishery, *Fisheries Management*, **15**, 9−14.

Bullock, A. and Gustard, A. (1992). Application of the instream flow incremental methodology to assess ecological flow requirement in a British lowland river, in *Lowland Floodplain Rivers: Geomorphological Perspectives* (Eds P.A. Carling and G.E. Petts), Wiley, Chichester.

Bunn, S.E. (1993). Riparian-stream linkages: research needs for the protection of instream values, in *An Endangered National Resource* (Ed. A.H. Arthington), *Australian Biologist*, **6**, 46−51.

Bunt, D.A. (1991). Use of rod catch effort data to monitor migratory salmonids in Wales, in *Catch-Effort Sampling Strategies* (Ed. I.G. Cowx), pp. 15−32, Fishing News Books, Blackwell Scientific Publications, Oxford.

Burgess, S.A. and Bider, J.R. (1980). Effects of stream habitat improvements on invertebrates, trout populations, and mink activity, *Journal of Wildlife Management*, **44**, 871−880.

Burrough, P.A. (1986). *Principles of Geographical Information Systems for Land Resources Assessment*, Monographs on Soil and Resources Survey No. 12, Oxford University Press, Oxford.

Burrows, I.G. and Whitton, B.A. (1983). Heavy metals in water, sediments and invertebrates from a metal-contaminated river free of organic pollution, *Hydrobiologia*, **106**, 263−73.

Bussell, R.B. (1979). *Changes of river regime resulting from regulation which may affect ecology: a preliminary approach to the problem*. Central Water Planning Unit, UK.

Büttiker, B. (1989). Analyse de la pêche de la truite (*Salmo trutta*) dans les rivières du canton de Vaud (Suisse), *Bull. Soc. Vaudoise Sci. Nat.*, **79**, 161−170.

Bye, J.V. (1984). The role of environmental factors in the timing of reproductive cycles, in *Fish Reproduction: Strategies and Tactics* (Eds. G.W. Potts and R.J. Wootton), pp. 187–205, Academic Press, London.

Bylander, C.B. (1990). Walleye tournament study sheds new light on catch and release, *Minnesota DNR Region III News*, **6**, 2–14.

Byrd, J.E. and Perona, M.J. (1980). The temporal variations of lead concentrations in a freshwater lake, *Water, Air and Soil Pollution*, **13**, 207–220.

Cadwallader, P.L. (1986). Flow regulation on the River Murray system and its effects on the native fish fauna in *Stream Protection: The Management of Streams for Instream Uses* (Ed. I.L. Campbell), pp. 115–133, Chisholm Institute of Technology, Victoria.

Cairns, J. (1982), *Artificial Substrates*, Ann Arbor Science, Ann Arbor, MI.

Cairns, J. Jr, Dickson, K.L. and Herricks, E.E. (1977). *Recovery and Restoration of Damaged Ecosystems*, University of Virginia Press, Charlottesville, Virginia, USA.

Calder, I.R. and Newson, M.D. (1979) Land-use and upland water resources in Britain — a strategic look, *Water Resources Bulletin*, **15**.

Calow, P. and Petts, G.E. (Eds) (1992). *The Rivers Handbook*, Vol. I, Blackwell Scientific Publications, Oxford.

Campbell, R.N. (1967a). A method of regulating Brown Trout (*Salmo trutta* L.) populations in small lakes, *Salmon and Trout Magazine*, May, 135–142.

Campbell, R.N. (1967b). Improving highland trout lochs, *The Flyfishers' Journal*, **56**, 61–69.

Candeias, D. (1981). *As colónias de garalcas em Portugal*, C.E.M.P.A., Secretaria de Estado do Ambiente, Lisboa.

Capelli, G.M. and Magnuson, J.J. (1983). Morpho-edaphic and biogeographic analysis of crayfish distribution in northern Wisconsin, *Journal of Crustacean Biology*, **3**, 548–564.

Carle, F.L. and Strub, M.R. (1978) A new method for estimating population size from removal data, *Fish Management*, **14**, 67–82.

Carling, P.A. (1984). Deposition of fine and coarse sand in an open-work gravel bed, *Canadian Journal of Fisheries and Aquatic Sciences*, **41**, 263–270.

Carling, P.A. (1987). Bed stability in gravel streams, with reference to stream regulation and ecology, in *River Channels: Environment and Process* (Ed. K. Richards), pp. 321–347, Blackwell, Oxford.

Carling, P.A. (1988). The concept of dominant discharge applied to two gravelbed streams in relation to channel instability thresholds, *Earth Surface Processes and Landforms*, **13**, 355–367.

Carling, P.A. (1988). Channel change and sediment transport in regulated U.K. rivers, *Regulated Rivers: Research and Management*, **2**, 369–387.

Carling, P.A. (1991). An appraisal of the velocity-reversal hypothesis for stable pool–riffle sequences in the River Severn, England, *Earth Surface Processes and Landforms*, **16**, 19–31.

Carling, P.A. (1992). The nature of the fluid boundary layer and the selection of parameters for benthic ecology, *Freshwater Biology*, **28**, 273–284.

Carling, P.A. (1992). In-stream hydraulics and sediment transport, in *The Rivers Handbook*, Vol. 1 (Eds P. Calow and G.E. Petts), pp. 101–125, Blackwell Scientific, Oxford.

Carling, P.A. and McCahon, C.P. (1987). Natural siltation of brown trout (*Salmo trutta* L.) spawning gravels during low-flow conditions, in *Regulated Streams: Advances In Ecology* (Eds J.F. Craig and J.B. Kemper), pp. 229–244, Plenum, New York.

Carling, P.A., Orr, H.G. and Glaister, M.S. (1994). Preliminary observations and significance of dead zone flow structure for solute and fine particle dynamics, in *Mixing and Transport in the Environment* (Eds K.J. Beven, P.C. Chatwin and J.H. Millbank), pp. 139–157, Wiley, Chichester.

Carlton, F.E. (1975). Optimum sustainable yield as a management concept in recreational fisheries, in *Optimum Sustainable Yield as a Concept in Fisheries Management* (Eds P.M. Roedel), American Fisheries Society Special Publication, **9**, 45–49.

Carson, M.A. (1984). The meandering-braided threshold: a reappraisal, *Journal of Hydrology*, **73**, 315–334.

Carss, D.N. and Marquiss, M. (1991). Avian predation at farmed and natural fisheries, *Proceedings of the Institute of Fisheries Management Annual Study Course*, **22**, 179–196.

Cattaneo, A. (1983). Grazing on epiphytes, *Limnology and Oceanography*, **28**, 124–132.

Cattaneo, A. and Kalff, J. (1980). The relative contribution of aquatic macrophytes and their epiphytes to the production of macrophyte beds, *Limnology and Oceanography*, **25** (2), 280–289.

Cerny, K. (1975). Mortality of the early developmental stages of roach — *Rutilus rutilus* (Linnaeus, 1758), *Vestnik Ceskoslovenske Spolecnosti Zoologicke*, **39**, 81–93.

Chadwick, E.M.P. (1982). Stock–recruitment relationships for Atlantic salmon (*Salmo salar*) in Newfoundland rivers, *Canadian Journal of Fisheries and Aquatic Sciences*, **39**, 1496–1501.

Chadwick, E.M.P. (1985a). The influence of spawning stock on production and yield of Atlantic salmon (*Salmo salar* L.) in Canadian rivers, *Aquaculture and Fisheries Management*, **16**, 111–119.

Chadwick, E.M.P. (1985b). Fundamental research problems in the management of Atlantic salmon. *Salmo salar* L., in Atlantic Canada, *Journal of Fish Biology*, **27**, (Suppl. A), 9–25.

Chadwick, E.M.P. (1988). Relationship between Atlantic salmon smolts and adults in Canadian rivers, in *Atlantic Salmon: Planning for the Future* (Eds D. Mills and D. Piggins) pp. 301–324, Proceedings of the 3rd International Atlantic Salmon Symposium, Biarritz, 1986. Croom Helm.

Chadwick, E.M.P. (1991). *Stock–recruitment of Atlantic salmon*, Atlantic Salmon Trust/Royal Irish Academy Workshop, Dublin, 1–4.

Chadwick, E.M.P. and Green, J.M. (1985). Atlantic salmon (*Salmo salar* L.) production in a largely lacustrine Newfoundland watershed, *Verhandlungen der Internationalen Vereinigung für Theoretische und Angewandte Limnologie*, **22**, 2509–2515.

Chambers, P.A., Prepas, E.E., Bothwell, M.L. and Hamilton, H.R. (1989) Roots versus shoots in nutrient uptake by aquatic macrophytes in flowing waters, *Canadian Journal of Fisheries and Aquatic Sciences*, **46**, 435–439.

Chambers, P.A., Prepas, E.E., Hamilton, H.R. and Bothwell, M.L. (1991) Current velocity and its effect on aquatic, macrophytes in flowing waters, *Ecological Application*, **1**, 249–257.

Chamier, A.-C. (1987). Effect of pH on microbial degradation of leaf litter in seven streams of the English Lake District, *Oecologia*, **71**, 491–500.

Champigneulle, A., Büttiker, B., Durand, P. and Melhaoui, M. (1991). Principales caractéristiques de la biologie de la truite (*Salmo trutta* L.) dans le Léman et quelques affluents, in *La Truite, Biologie et Écologie* (Eds J.L. Baglinière and G. Maise), pp. 153–182, INRA, Paris.

Champion, A.S. (1992). The recovery of salmon stocks in the River Tyne, *Salmon Net*, **24**, 25–30.

Chandler, J.R. (1990). Integrated catchment management in England and Wales, in *Management of Freshwater Fisheries* (Ed. W.L.T. Densen, B. Steinmetz and R.H. Hughes), pp. 520–525, Pudoc, Wageningen.

Chanin, P.R.F. and Jefferies, D.J. (1978). The decline of the otter *Lutra lutra* L. in Britain: an analysis of hunting records and discussion of causes, *Biological Journal of the Linnean Society*, **10**, 305–328.

Chapman, V.J. (1964). *The Algae*. Macmillan, London.

Charlton, S.E.D. and Bayne, D. (1986). *Phosphorus removal: the impact upon water quality in the Bow River downstream of Calgary, Alberta, Bow River data base 1980–1985*, Pollution Control Division, Alberta Environment, Edmonton.

Chatwin, P.D. and Allen, C.M. (1985). Mathematical models of dispersion in rivers and estuaries, *Annual Review in Fluid Mechanics*, **17**, 119–149.

Chaveroche, P. and Sabaton, C. (1989). An analysis of brown trout (*Salmo trutta*) habitat: the role of qualitative data from expert advice in formulating probability-of-use curves, *Regulated Rivers: Research and Management*, **3**, 305–321.

Child, L.E., de Waal, L.C., Wade, P.M. and Palmer, J.P., (1992). Control and management of *Reynoutria* species (knotweed), *Aspects of Applied Biology*, **29**, 295–307.

Chin, A. (1989). Step pools in stream channels, *Progress In Physical Geography*, **13**, 391–407.

Chunnet, Fourie and Partners (1987). *Sabie River Water Resources Development Study*, Briefing paper to Department of Water Affairs.

Church, M. (1992). Channel morphology and typology, in *The Rivers Handbook* Vol. 1 (Eds P. Calow and G.E. Petts), pp. 126–143, Blackwell Scientific, Oxford.

Churchfield, S. (1990). *The Natural History of Shrews*. Helm, London.

Churchward, A.S. and Hickley, P. (1991). The Atlantic salmon fishery of the River Severn (UK), in *Catch-Effort Sampling Strategies* (Ed. I.G. Cowx), pp. 1–14, Fishing News Books, Oxford.

Chutter, F.M. (1971). Hydrobiological studies in the catchment of Vaal Dam, South Africa: Part 2: The effects of stream contamination on the fauna of stones-in-current and marginal vegetation biotopes, *Internationale Revue der gesamten Hydrobiologie*, **56**, 227–240.

Chutter, F.M. (1972). An empirical biotic index of the quality of water in South African streams and rivers, *Water Research*, **6**, 19–30.

Ciborowski, J.J.H., Pointing, P.J. and Corkum, L.D. (1977). The effect of current velocity and sediment on the drift of the mayfly *Ephemerella subvaria* McDunnough, *Freshwater Biology*, **7**, 567–572.

Claassen, T.H.L. and de Jongh, J. (1988). De otter als normsteller voor kwaliteit van het oppervlaktewater (Otter as a quality indication for surface water), H_2O, **21**, 16, 432–436.

Clapham, A.R., Tutin, T.G. and Warburg, E.F. (1954). *Excursion Flora Of The British Isles*, Cambridge University Press, Cambridge.

Clark, W.R. (1981). *Assessment of navigation effects on muskrats in Pool G of the Upper Mississippi River*, Final Report for the Environment Works Team, Upper Mississippi River Basin Committee Master Plan, Minnesota, USA.

Clarke, J.M. (ed.) (1978). *Hampshire/Surrey border bird report for 1978*, private publication.

Clark, M.J., Gurnell, A.M., Davenport, J. and Azizi, A. (1993). Integrated river channel management through GIS, in *Water and the Environment* (Eds J.C. Currie and A.T. Pepper), Ellis Horwood, Chichester.

Clarke, R.T. and Newson, M.D. (1978). Some detailed water balance studies of research catchments, *Proceedings of the Royal Society of London, Series A*, **363**, 21–42.

Clegg, L. and Grace, J. (1974). The distribution of *Heracleum mantegazzianum* (Somm. and Levier) near Edinburgh, *Transactions of the Botanical Society of Edinburgh*, **42**, 223–229.

Clements, W.H., Cherry, D.S. and Cairns, J. (1988). Impact of heavy metals on insect communities in streams: a comparison of observational and experimental results, *Canadian Journal of Fisheries and Aquatic Sciences*, **45**, 2017–2025.

Clifford, N.J., Hardisty, J., French, J.R. and Hart, S. (1994). Downstream variation in bed material characteristics: a turbulence-controlled form-process feedback mechanism, in *Braided Rivers: Form, Process and Economic Applications*.

Cline, L.D., Short, R.A. and Ward, J.W. (1982). The influence of highway construction on the macroinvertebrates and epilithic algae of a high mountain stream, *Hydrobiologia*, **96**, 149–159.

Close, A. (1990). River Salinity, in *The Murray* (Eds N. Mackay and D. Eastburn), Murray Darling Basin Commission, Canberra, Australia.

Cobb, D.G., Galloway, T.D. and Flannagan, J.F. (1992). Effect of discharge and substrate stability on density and composition of stream insects, *Canadian Journal of Fisheries and Aquatic Sciences*, **49**, 1788–1795.

Coffman, W.P., Cummins, K.W. and Wuycheck, J.C. (1971). Energy flow in a woodland stream ecosystem. I. Tissue support trophic structure of the autumnal community, *Archiv für Hydrobiologie*, **68**, 232–276.

Coles, T.F., Wortley, J.S. and Noble, P. (1985). Survey methodology for fish population assessment within Anglian Water, *Journal of Fish Biology*, **27** (Suppl. A), 175–186.

Collier, K.J. and McColl, R.H.S. (1992). Assessing the natural value of New Zealand rivers, in *River Conservation and Management* (Eds P.J. Boon, P. Calow, and G.E. Petts), pp. 195–211, John Wiley, Chichester.

Collier, K.J. and Winterbourn, M.J. (1990). Structure of epilithon in some acidic and circumneutral streams in South Westland, New Zealand, *New Zealand Natural Sciences*, **17**, 1–11.

Conolly, A.P. (1977) The distribution and history in the British Isles of some alien species of *Polygonum* and *Reynoutria*, *Watsonia*, **11**, 291–311.

Conrath, W., Falkenhage, B. and Kinzelbach, R. (1977). Übersicht über das Makrozoobenthon des Rheins im Jahre 1976. *Gewässer und Abwässer*, **62/63**, 63–84.

Cooke, A.S. (1987). Disturbance by anglers of birds at Grafham Water, *Institute of Terrestrial Ecology Symposium*, **19**, 15–22.

Cooper, M.J. and Wheatley, G.A. (1981). An examination of the fish population in the River Trent, Nottinghamshire, using anglers' catches, *Journal of Fish Biology*, **19**, 539–556.

Copp, G.H. (1989). The habitat diversity and fish reproductive function of floodplain ecosystems, *Environmental Biology of Fishes*, **26**, 1–27.

Copp, G.H. (1990a). Effects of regulation on 0+ fish recruitment in the Great Ouse, a lowland river, *Regulated Rivers: Research and Management*, **5**, 251–263.

Copp, G.H. (1990b). Shifts in the microhabitat of larval and juvenile roach *Rutilus rutilus* (L.) in a floodplain channel, *Journal of Fish Biology*, **36**, 683–692.

Copp, G.H. (1992). Comparative microhabitat use of cyprinid larvae and juveniles in a lotic floodplain channel, *Environmental Biology of Fishes*, **33**, 181–193.

Corbet, G.B. and Harris, S. (1991). *The Handbook of British Mammals*, 3rd edn, Blackwell, Oxford.

Corbet, G.B. and Ovenden, D. (1980). *The Mammals of Britain and Europe*, Collins, London.

Corbet, P.S. (1980) Biology of Odonata, *Annual Review of Entomology*, **25**, 189–217.

Cormack, R.M. (1968). The statistics of capture–recapture methods, *Oceanography and Marine Biology Annual Review*, **6**, 455–506.

Cosgrove, D.E. and Petts, G.E. (Eds) (1990). *Water, Engineering and Landscape*, Belhaven, London.

Cowx, I.G (1983). Review of the methods for estimating fish population size from survey removal data, *Fisheries Management*, **14**, 67–82.

Cowx, I.G. (1990). Application of creel census data for the management of fish stocks in large rivers in the United Kingdom, in *Management of Freshwater Fishers* (Eds W.L.T. van Densen, B. Steinmetz and R.H. Hughes), pp. 526–534, Pudoc, Wageningen.

Cowx, I.G. (Ed.) (1991). *Catch-Effort Sampling Strategies*, Fishing News Books, Blackwell Scientific Publications, Oxford.

Cowx, I.G. and Broughton, N.M. (1986). Changes in the species composition of angling catches in the River Trent (England) between 1969 and 1984, *Journal of Fish Biology*, **28**, 625–636.

Cowx, I.G. and Harvey, J. (1993). *Electric fishing in large deep rivers*, National Rivers Authority R&D Report No 0334/4/ST.

Cowx, I.G., Wheatley, G.A. and Hickley, P. (1988) Developments of boom electric fishing equipment for use in large rivers and canals in the United Kingdom, *Aquaculture and Fisheries Management*, **19**, 205–212.

Cowx I.G., Young, W.O. and Hellawell, J.M. (1984). The influence of drought on the fish and inveretebrate populations of an upland stream in Wales, *Freshwater Biology*, **14**, 165–167.

Cowx, I.G., Wheatley, G.A., Hickley, P. and Starkie, A.S. (1990). Evaluation of electric fishing equipment for stock assessment in large rivers and canals in the United Kingdom, in *Fishing with Electricity* (Ed. I.G. Cowx) Fishing News Books, Blackwell Scientific Publications, Oxford.

Cragg, B.A., Fry, J.C., Bachus, Z. and Thyrley, S.S. (1980). The aquatic vegetation of Llangorse Lake, Wales, *Aquatic Botany*, **8**, 187–196.

Craig, J.F. and Kemper, J.B. (Eds) (1987). *Regulated Streams: Advances in Ecology*, Plenum Press, New York.

Crane, M. and Maltby, L. (1991). The lethal and sublethal responses of *Gammarus pulex* to stress:sensitivity and sources of variation in an *in situ* bioassay, *Environmental Toxicology and Chemistry*, **10**, 1331–339.

Cranston, P.S. (1982) A key to the larvae of the British Orthocladiinae (Chironomidae), *Freshwater Biological Association Scientific Publications*, **45**, 1–152.

Cranston, P.S. (1984) The taxonomy and ecology of *Orthocladius* (*Eudactylocladius*) *fuscimanus* (Kieffer), a hygropetric chironomid (Diptera), *Journal of Natural History*, **18** (6), 873–895.

Cranston, P.S., Oliver, D.R. and Sæther, O.A. (1983) The larvae of Orthocladiinae (Diptera: Chironomidae) of the Holarctic region — Keys and diagnoses, *Entomologica Scandinavica Supplement*, **19**, 149–291.

Crean, K. (1993). Planning and development of inland fisheries, in *Rehabilitation of Freshwater Fisheries* (Ed. I.G. Cowx), pp. 21–33, Fishing News Books, Blackwell Scientific Publications, Oxford.

Crisp, D.T. (1966). Input and output of materials for an area of Pennine moorland: the importance of precipitation, drainage, peat erosion and animals, *Journal of Applied Ecology*, **3**, 327–340.

Crisp, D.T. (1977). Some physical and chemical effects of the Cow Green (Upper Teesdale) impoundment, *Freshwater Biology*, **7**, 109–120.

Crisp, D.T. (1981). A desk study of the relationshilp between temperature and hatching time for eggs of five species of salmonid fishes, *Freshwater Biology*, **11**, 361–368.

Crisp, D.T. (1988) Prediction from temperature of eyeing, hatching and swim-up times for salmonid embryos, *Freshwater Biology*, **19**, 41–48.

Crisp, D.T. (1989a). Some impacts of human activities on trout, *Salmo trutta*, populations, *Freshwater Biology*, **21**, 21–33.

Crisp, D.T. (1989b). Use of artificial eggs in studies of washout depth and drift distance for salmonid eggs, *Hydrobiologia*, **178**, 155–163.

Crisp, D.T. (1991). Stream channel experiments on downstream movement of newly emerged trout, *Salmo trutta* L., and salmon, *S. salar* L. — III Effects of developmental stage and day and night upon dispersal, *Journal of Fish Biology*, **39**, 371–381.

Crisp, D.T. (1992). Measurement of stream water temperature and biological applications to salmonid fishes, grayling and dace. *Freshwater Biological Association Occasional Publication*, **29**, 1–72.

Crisp, D.T. (1993). The environmental requirements of salmon and trout in freshwater, *Freshwater Forum*, **3**, 176–202.

Crisp, D.T. and Carling, P.A. (1989). Observations on siting, dimensions and structure of salmonid redds, *Journal of Fish Biology*, **34**, 119–134.

Crisp, D.T. and Hurley, M.A. (1991a). Stream channel experiments on downstream movement of newly emerged trout, *Salmo trutta* L. and salmon, *S. salar* L. — I. Effect of four different water velocity treatments upon dispersal rate, *Journal of Fish Biology*, **39**, 347–361.

Crisp, D.T. and Hurley, M.A. (1991b). Stream channel experiments on downstream movement of newly emerged trout, *Salmo trutta* L. and salmon, *S. salar* L. II. — Effects of constant and changing velocities and of day and night upon dispersal rate, *Journal of Fish Biology*, **39**, 363–370.

Crisp, D.T. and Robson, S. (1979). Some effects of discharge upon the transport of animals and peat in a north Pennine headstream, *Journal of Applied Ecology*, **16**, 721–736.

Crisp, D.T., Mann, R.H.K. and Cubby, P.R. (1983). Effects of regulation of the river Tees upon fish populations below Cow Green Reservoir, *Journal of Applied Ecology*, **20**, 371–386.

Crisp D.T., Mann, R.H.K. and McCormack, J.C. (1978). The effects of impoundment and regulation upon the stomach contents of fish at Cow Green, upper Teesdale, *Journal of Fish Biology*, **12**, 287–301.

Cross, D.G. and Stott (1975). The effect of electric fishing on the subsequent capture of fish, *Journal of Fish Biology*, **7**, 349–357.

Cross, T.F. (1989). *Genetics and the Management of the Atlantic Salmon*. Atlantic Salmon Trust, Pitlochry, pp. 1–74.

Cross, T.F., Mills, C.P.R. and De Courcy Williams, M. (1992). An intensive study of allozyme variation in freshwater resident and anadromous trout, *Salmo trutta* L. in western Ireland, *Journal of Fish Biology*, **40**, 25–32

Crowder, L.B. and Cooper, W.E. (1982). Habitat structural complexity and the interaction between bluegills and their prey, *Ecology*, **63**, 1802–1813.

Crozier, W.W. and Kennedy, G.J.A. (1991). Salmon research on the River Bush, in *Irish Rivers: Biology and Management* (Ed. M.W. Steer), pp. 29–46, Royal Irish Academy, Dublin.

Crozier, W.W. and Kennedy, G.J.A. (1993). Marine survival of wild and hatchery reared Atlantic salmon (*Salmo salar* L.) from the R Bush, Northern Ireland, in *Salmon in the Sea and New Enhancement Strategies* (Ed. D.H. Mills), pp. 139–162, Fishing News Books, Blackwell Scientific Publications, Oxford.

Crozier, W.W., and Kennedy, G.J.A. (1994). Marine exploitation of Atlantic salmon (*Salmo salar* L.) from the R Bush, Northern Ireland, *Fishery Research*, **19**, 141 –155.

Cryer, M. and Edwards, R.W. (1987). The impact of angler ground bait on benthic invertebrates and sediment respiration in a shallow eutrophic reservoir, *Environmental Pollution*, **46**, 137–150.

Cryer, M., Corbett, J.J. and Winterbotham, M.D. (1987). The deposition of hazardous litter by anglers at coastal and inland fisheries in South Wales, *Journal of Environmental Management*, **25**, 125–135.

Cryer, M., Linley, N.W., Ward, R.M., Stratford, J.O. and Randerson, P.F. (1987). Disturbance of overwintering wildfowl by anglers at two reservoir sites in South Wales, *Bird Study*, **34**, 191–199.

CSIRO (1992). *Towards healthy rivers*: a report from CSIRO to the Honourable Ros Kelly, Minister for Arts, Sport, the Environment and Territories. CSIRO Division of Water Resources Consultancy Report No. 92/44, CSIRO, Australia.

Culp, T.M., Wrona, F.J. and Davies, R.W. (1986). Response of stream benthos and drift to fine sediment deposition versus transport. *Canadian Journal of Zoology*, **64**, 1345–1351.

Cummins, K.W. (1973). Trophic relations in aquatic insects, *Annual Review of Entomology*, **18**, 183–206.

Cummins, K.W. (1974). Structure and function of stream ecosystems, *BioScience*, **24**, 631–641.

Cummins, K.W. (1975) Macroinvertebrates. In *River Ecology, Studies in Ecology* Vol. 2, (Ed. B.A. Whitton), pp. 181–186, University of California Press, Berkeley.

Cummins, K.W. (1988) The study of stream ecosystems: a functional view, in *Ecosystem Processes* (Eds L.R. Pomeroy and J.J. Alberts), pp. 240–245, Springer-Verlag, New York.

Cummins, K.W. (1992). Invertebrates, in *The Rivers Handbook*, Vol. 1 (Eds P. Calow and G.E. Petts), pp. 234–50, Blackwell Scientific Publications, Oxford.

Cummins, K.W. (1992). Catchment characteristics and river ecosystems, in *River Conservation and Management* (Eds P.J. Boon, P. Calow and G.E. Petts), pp. 125–135, Wiley, Chichester.

Cummins, K.W. and Klug, M.J. (1979) Feeding ecology of stream invertebrates, *Annual Review of Ecology and Systematics*, **10**, 147–172.

Cummins, K.W. and Lauff, G.H. (1969) The influence of substrate particle size on the microdistribution of stream macrobenthos, *Hydrobiologia*, **34**, 145–181.

Cummins, K.W., Petersen, R.C., Howard, F.O., Wuychek, J.C. and Holt, V.I. (1973). The utilization of leaf litter by stream detritivores, *Ecology*, **54**, 336–345.

Cuppen, J.G.M. (1983) On the habitats of three species of *Hygrotus* Stephens (Coleoptera: Dytiscidae), *Freshwater Biology*, **13**, 579–588.

Cyr, H. and Downing, J.A. (1988) The abundance of phytophilous invertebrates on different species of submerged macrophytes, *Freshwater Biology*, **20**, 365–374.

Dand, I.W. and White, W.R. (1977). Design of navigation canals, in *Symposium on Aspects of Navigability of Constraint Waterways including Harbour Entrances*, pp. 1–9. Hydraulics Research Station, Wallingford, UK.

Davidson, W.S., Birt, T.P. and Green, J.M. (1989). A review of genetic variation in Atlantic salmon, *Salmo salar* L., and its importance for stock identification, enhancement programmes and aquaculture, *Journal of Fish Biology*, **34**, 547–560.

Davies, B.R., Bonthuys, B., Fourie, W., Marcus, A,. Rossouw, S., Sellick, C., Theron, L., Uys, W., Van Rooyen, L., Van Zyl, M. and Viljoen, A. (1991). The Sabie sand system, in *Flow Requirements of Kruger National Park Rivers* (Ed C. Bruwer, Technical Report TR 149, Department of Water Affairs, Pretoria.

Davies, W.D. (1983) Sampling with toxicants, in *Fisheries Techniques* (Ed L.A. Neilsen and D.L. Johnson), pp. 199–214, American Fisheries Society, Bethesda.

Davies-Colley, R.J., Hickey, C.W., Quinn, J.M. and Ryan, P.A. (1992). Effects of clay discharges on streams, I. Optical properties and epilithon, *Hydrobiology*, **48**, 215–234.

Davis, J.A. (1986). Boundary layers, flow microenvironments and stream benthos, in *Limnology in Australia* (Eds P. De Deckker and W.D. Williams), pp. 293–312, CSIRO, Dr W. Junk Publishers, Dortrecht.

Davis, J.A. and Barmuta, L.A. (1989). An ecologically useful classification of mean and near-bed flows in streams and rivers, *Freshwater Biology*, **21**, 271–282.

Davison, W. (1990). Treatment of acid waters by inorganic bases, fertilizers and organic material, *Transactions of the Institution of Mining and Metallurgy (A)*, **99**, 153–157.

Davison, W. and Woof, C. (1990). The dynamics of alkalinity generation by an anoxic sediment exposed to acid water, *Water Research*, **24**, 1537–1543.

Davison, W., Reynolds, C.S., Tipping, E. and Needham, R.F. (1989). Reclamation of acid waters using sewage sludge, *Environmental Pollution*, **57**, 251–274.

Dawson, F.H. (1976). The annual production of the aquatic macrophyte *Ranunculus penicillatus var. calcareous* (R.W. Butcher) C.D.K. Cook, *Aquatic Botany*, **2**, 51–73.

Dawson, F.H. (1989). Ecology and management of water plants in lowland streams, *Annual Report of the Freshwater Biological Association*, **57**, 43–60.

Dawson, F.H. and Charlton, F.G. (1988). *Bibliography on the Hydraulic Resistance of Roughness of Vegetated Watercourses*, Freshwater Biological Association, Occasional Publication No. 25, Ambleside, Cumbria.

Dawson, F.H. and Haslam, S.M. (1983) The management of river vegetation with particular reference to shading effects of marginal vegetation, *Landscape Planning*, **10**, 147–169.

Dawson, F.H. and Kern-Hansen, V. (1979). The effect of natural and artificial shade on the macrophytes of lowland streams and the use of shade as a management technique, *Internationale Revue der Gesamten Hydrobiologie*, **64**, 437–55.

Dawson, F.H. and Robinson, W.N. (1984). Submerged macrophtyes and the hydraulic roughness of a lowland chalkstream, *Verhandlungen der Internationalen Vereinigung für Theoretische und Angewandte Limnologie*, **22**, 1944–1948.

Day, T.J. and Wood, I.R. (1976). Similarity of the mean motion of fluid particles dispersing in a natural channel, *Water Resources Research*, **12**, 655–666.

Dedonder, A. and van Sumere, C.F. (1971). The effects of phenolics and related compounds on the growth and respiration of *Chlorella vulgaris*, *Z. Pflanzenphysiol*, **65**, 70–80.

DeGraaf, D.A. and Bain, L.H. (1986). Habitat use by and preferences of juvenile Atlantic salmon in two Newfoundland rivers, *Transactions of the American Fisheries Society*, **115**, 671–681.

Delibes, M., Macdonald, S.M. and Mason, C.F. (1991). Seasonal marking, habitat and organochlorine contamination in otters (*Lutra lutra*): a comparison between catchments in Andalucia and Wales, *Mammalia*, **55**, 567–578.

DeLury, D.B. (1947). On the estimation of biological populations, *Biometrics*, **3**, 45–167.

Demars, J.J. (1985). Repercussion of small hydroelectric power stations on populations of brown trout (*Salmo trutta*) in rivers in the French Massif-Central, in *Habitat Modification and Freshwater Fisheries* (Ed J.S. Alabaster), pp. 52–61, FAO and Butterworths, London.

Department of the Environment (1992). *Guidance for Control of Weeds on Non-agricultural Land*, HMSO, London.

Department of the Environment and the Forestry Commission (1990). *Forests and Surface Water Acidification*, Report of the Darlington Workshop, 22–25 June 1980, Department of the Environment, London.

Department of the Environment and Welsh Office (1992). *River Quality. The Government's Proposals: A Consultation Paper*. Department of the Environment and Welsh Office, London.

Department of the Environment and Welsh Office (1994). *Statutory Instruments 1994 No 1057 Water Resources, England and Wales*: The Surface Waters (River Ecosystem) (Classification) Regulations 1994. HMSO, London.

Department Of Water Affairs (1986). *Management of the Water Resources of the Republic of South Africa*, Department of Water Affairs, Pretoria.

Department Of Water Affairs (1990). Hydrology, Sabie River Catchment, Report of a hydrological survey of the Sabie/Sand catchment, undertaken by Chunnett, Fourie and Partners, Pretoria.

De Paw, N. and Vanhooren, G. (1983). Method for biological water quality assessment of water courses in Belgium, *Hydrobiologia*, **100**, 153–158.

Descy, J.-P. (1987). Phytoplankton composition and dynamics in the River Meuse (Belgium). *Archiv für Hydrobiologie (Suppl.)*, **78**, 225–245.

Devon County Council (1993). *Grand Western Canal Country Park Management Plan*, Devon County Council, Exeter, UK.

De Vries, M.B. (1987). *UPTAQE — a mathetical model for simulation of the accumulation of micro-pollutants in aquatic food-chains*, TOW-IW T250, Waterloopkundig laboratorium, Delft.

De Wit, J.A.W., van der Gaag, M.A., van der Guchte, C., van Leeuwen C.J. and Koeman, J. (1991). *The Effect of Micropollutants on Components of the Rhine Ecosystem*, Publication 35, Publications and reports of the project Ecological Rehabilitation Rhine, R.I.Z.A., R.I.V.M., R.I.V.O., Institute of Inland Water Management and Waste Water Treatment, Lelystad, and Netherlands Institute for Fisheries Research, The Netherlands.

Diamond, M. (1985). Some observations of spawning by roach, *Rutilus rutilus* L., and bream, *Abramis brama* L., and their implications for management, *Aquaculture and Fisheries Management*, **16**, 359–367.

Diamond, M. and Brown, A.F. (1984). Predation by the eel, *Anguilla anguilla*, on the eggs and spawning population of the roach, *Rutilus rutilus*, *Fisheries Management*, **15**, 71–73.

Dietrich K. (1974). Investigation into the pollution of water by two-stroke outboard motors, *Gesundheits-Ingenieur*, **85**, 342–347.

Dietrich, W.E., Kirchner, J.W., Ikeda, H. and Iseya, F. (1989). Sediment supply and the development of the coarse surface layer in gravel-bedded rivers, *Nature*, **340**, 215–217.

Dill, W.A. (1978). Patterns of change in recreational fisheries: their determinants, in *Proceedings of the Conference on Recreational Freshwater Fisheries* (Ed. J.S. Alabaster), pp. 1–22, Water Research Centre, Stevenage.

Dionne, M. and Folt, C.L. (1991). An experimental analysis of macrophyte growth forms as fish foraging habitat, *Canadian Journal of Fisheries and Aquatic Sciences*, **48**, 123–131.

Diplas, P. and Parker, G. (1992). Deposition and removal of fines in gravel-bed streams, in *Dynamics of Gravel-bed Rivers* (Eds P. Billi, R.D. Hey, C.R. Thorne, and P. Tacconi), pp. 313–329, John Wiley, Chichester.

Dirksen S., Boudewijn, T.J., Slager L.K., Mes, R.G. , van Schaick M.J.M. and De Voogt, P. (in press). Breeding succes of Cormorants (*Phalacrocorax carbosinensis*) in relation to organochlorine pollution of aquatic habitats in the Netherlands, *Environmental Pollution*, submitted.

Ditlhogo, M.K.M., James, R., Laurence, B.R. and Sutherland, W.J. (1992). The effects of conservation management of reed beds. I — The invertebrates, *Journal of Applied Ecology*, **29**, 265–276.

Doadrio, I., Elvira, B. and Bernat, Y. (1991). *Peces Continentales Españoles*, Inventario y clasificación de zonas fluviales, ICONA, Madrid.

Dobson, M. and Hildrew, A.G. (1992). A test of resource limitation among shredding detritivores in low order streams in southern England, *Journal of Animal Ecology*, **61**, 69–78.

Dodd, F.S., Waal, L.C. de Wade, P.M. and Tiley, G.E.D. (1994) Control and management of *Heracleum mantegazzianum* (Giant Hogweed), in *Ecology and Management of Invasive Riverside Plants*, (Eds L.C. de Waal, L.E. Child, P.M. Wade and J.H. Brock), pp. 111–126, Wiley, Chichester.

Dodds, W.K. (1991a). Micro-environmental characteristics of filamentous algal communities in flowing freshwaters, *Freshwater Biology*, **25**, 199–209.

Dodds, W.K. (1991b). Community interactions between the filamentous alga *Cladophora glomerata* (L.) Kuetzing, its epiphytes, and epiphyte grazers, *Oecologia*, **85**, 572–580.

Doeg, T.J., Marchant, R., Douglas, M. and Lake, P.S. (1989) Experimental colonization of sand, gravel and stones by macroinvertebrates in the Acheron River, southeastern Australia, *Freshwater Biology*, **22**, 57–64.

Dogger J.W., Balk, F., Bijlmakers, L.L. and Hendriks, A.J. (1992). *Schatting van de risicos van microverontreinigingen in de Rijn voor de soorten van de rivier — AMOEBE (Estimation of the risks of microcontaminants in the Rhine of species of the river — AMOEBA)*, Publications and Reports of the Project Ecological Rehabilitation Rhine, 38.

Dole-Olivier, M.J. and Marmonier, P. (1992). Patch distribution of interstitial communities: prevailing factors, *Freshwater Biology*, **27**, 177–191.

Dolmen, D. (1987). *Gyrodactylus salaris* (Monogenea) in Norway; infestations and management, in *Parasites and Diseases in Natural Waters and Aquaculture in Nordic Countries* (Eds A. Stenmark and O. Malmberg), pp. 63–69, University of Stockholm, Stockholm.

Donald, D.B. and Alger, D.J. (1989). Evaluation of exploitation as a means of improving growth in a stunted population of brook trout. *North American Journal of Fisheries Management*, **9**, 177–183.

Draulans, D. (1987). The effectiveness of the attempts to reduce predation by fish-eating birds: a review, *Biological Conservation*, **41**, 219–232.

Du Buys, P. (1879). Études due régime du Rhone et l'achan exercée par les eaux sur un lit à fond de graviers indéfiniment affouillable, *Annales des Ponts et Chaissees ser, 5*, **18**, 141–195.

Dudgeon, D. (1982). An investigation of physical and biological processing of two species of leaf litter in Tai Po Kau Forest Stream, New Territories, Hong Kong, *Archiv für Hydrobiologie*, **96** (1), 1–32.

Dudley, T.L. and Anderson, N.H. (1982) A survey of invertebrates associated with wood debris in aquatic habitats, *Melanderia*, **39**, 1–21.

Dudley, T.L. and Anderson, N.H. (1987) The biology and life cycles of Lipsothrix spp. (Diptera: Tipulidae) inhabiting wood in Western Oregon streams, *Freshwater Biology*, **17**, 437–451.

Dudley, T.L., Cooper, S.D. and Hemphill, W. (1986). Effects of macroalgae on a stream invertebrate community, *Journal of the North American Benthological Society*, **5**, 93–106.

Duff, D.A. and Cooper, J.I. (1976). *Techniques for conducting stream habitat survey on national resource land*, US Department of the Interior, Bureau of Land Management, Technical Note 283, pp. 1–72.

Dunn, T.C. (1970). The Himalayan Balsam (*I. glandulifera* Royle). *Vasculum*, **55**, 23.

Dunstone, N. and Ireland, M. (1989). The mink menace? A reappraisal, in *Mammals as Pests* (Ed. R.J. Putman), pp. 225–250, Chapman and Hall, London.

Dvorak, J. (1970). Horizontal zonation of macrovegetation, water properties and macrofauna in a littoral stand of *Glyceria aquatica* (L.) Wahlb. in a pond in South Bohemia, *Hydrobiologia*, **35**,

17–30.

Dvorák, J. (1978). Macrofauna of invertebrates in helophyte communities, in *Pond Littoral Ecosystems Structure and Functioning. Methods and Results of Quantitative Ecosystem Research in the Czechoslovakian IBP Wetland Project*, (Ed. D. Dykyjová and J. Kvet), pp. 389–392, Springer-Verlag, Berlin.

Dvorak, J. and Best, E.P.H. (1982) Macro-invertebrate communities associated with the macrophytes of Lake Vechten: structural and functional relationships, *Hydrobiologia*, **95**, 115–126.

Eaton, J.W. (1986). Waterplant ecology in landscape design, in *Ecology and Design in Landscape* (Eds. A.D. Bradshaw, D.A. Goode and E. Thorpe), pp. 285–306, Blackwell, Oxford.

Eaton, J.W., Murphy, K.J. and Hyde, T.M. (1981). Comparative trials of herbicidal and mechanical controil of aquatic weeds in canals, in *Proceedings of the Symposium Aquatic Weeds and Their Control*, pp. 105–116. Association of Applied Biologists, Oxford.

Ebdon, L., Evans, K. and Hill, S. (1989). The accumulation of organotins in adult and seed oysters from selected estuaries prior to the introduction of UK regulations governing the use of tributyltin-based antifouling paints. *Science of the Total Environment*, **88**, 63

Economic Commission for Europe (1958). *Annual Bulletin of Transport Statistics for Europe, 1957*, Geneva.

Economic Commission for Europe (1988). *Two Decades of Co-operation on Water, Declarations and Recommendations by the Economic Commission for Europe*, United Nations, New York.

Edington, J.M. (1968). Habitat preferences in net-spinning caddis larvae with special reference to the influence of water velocity, *Journal of Animal Ecology*, **37**, 675–692.

Edwards, A.M.C. (1973a). The variation of dissolved constituents which discharge in some Norfolk rivers, *Journal of Hydrology*, **18**, 219–242.

Edwards, A.M.C. (1973b). Dissolved load and tentative solute budgets of some Norfolk catchments, *Journal of Hydrology*, **18**, 201–217.

Edwards, A.M.C. (1974). Silicon depletions in some Norfolk rivers, *Freshwater Biology*, **4**, 267–274.

Edwards, C.J., Hanbury, R.G. and Burt, A.J. (1987). The ecology of the Oxford Canal (South), Report of the Oxford Canal (South) Survey: May 1983–August 1984, British Waterways Board and Environmental Advisory Unit, Gloucester.

Edwards, D. (1969). Some effects of siltation upon aquatic macrophyte vegetation in rivers, *Hydrobiologia*, **34**, 29–37.

Edwards, R.W. (1964). Some effects of plants and animals on the conditions in fresh-water streams with particular reference to their oxygen balance. *International Journal of Air and Water Pollution*, **6**, 505–20.

Edwards, R.W. (1968). Plants as oxygenators in rivers, *Water Research*, **2**, 234–48.

Edwards, R.W. (1984). Predicting the environmental impact of a major reservoir development, in *Planning and Ecology* (Eds R.D. Roberts and T.M. Roberts), pp. 55–81, Chapman and Hall, London.

Edwards, R.W. (1987). Ecological assessment of the degradation and recovery of rivers from pollution, in *Ecological Assessment of Enviornmental Degradation, Pollution and Recovery* (Ed. O. Ravera), pp. 159–94, Elsevier Science, London.

Edwards, R.W. (1990). The impact of angling on conservation and water quality, in *Proceedings of the Institute of Fisheries Management 21st Anniversary Conference, 1990*, pp. 41–50.

Edwards, R.W. and Crisp, D.T. (1982). Ecological implications of river regulation in the United Kingdom, in *Gravel Bed Rivers: Fluvial Processes, Engineering and Management* (Eds R.D. Hey, J.C. Bathurst and C.R. Thorne), pp. 843–865, Wiley, Chichester.

Edwards, R.W. and Cryer, M. (1987). Angler litter, *Institute of Terrestrial Ecology Symposium*, **19**, 7–14.

Edwards, R.W. and Fouracre, V.A. (1983). Is the banning of ground baiting in reservoirs justified?, *Proceedings of the British Freshwater Fisheries Conference*, **3**, 89–94.

Edwards, R.W. and Garrod, D.J. (1972). *Conservation and Productivity of Natural Waters*, The Zoological Society of London, Academic Press, London.

Edwards, R.W. and Owens, M. (1965). The oxygen balance of streams, in *Ecology and the Industrial Society* (Eds G.T. Goodman, R.W. Edwards and Lambert), pp. 49–72, Blackwell Scientific, Oxford.

Edwards, R.W. and Rolley H.L.J. (1965). The oxygen consumption of river muds, *Journal of Ecology*, **53**, 1–19.

Edwards, R.W., Densem, J.W. and Russell, P.A. (1979). An assessment of the importance of temperature as a factor controlling the growth rate of brown trout in streams, *Journal of Animal Ecology*, **48**, 501–507.

Edwards, R.W., Gee, A.S. and Stoner, J.H. (Eds) (1990). *Acid Waters in Wales*, Kluwer Academic, Holland.

Edwards, R.W., Hughes, B. and Read, M.W. (1975). Biological survey in the detection and assessment of pollution, in *Degraded Environments and Resource Renewal* (Eds M.J. Chadwick and G.T. Goodman), pp. 139–56, Blackwell Scientific, Oxford.

Egglishaw, H.J. (1964). The distributional relationship between the bottom fauna and plant detritus in streams, *Journal of Animal Ecology*, **33**, 463–476.

Egglishaw, H.J. (1969). The distribution of benthic invertebrates on substrata in fast-flowing streams, *Journal of Animal Ecology*, **38**, 19–33.

Egglishaw, H.J., Gardiner, W.R., Shackley, P.E. and Struthers, G. (1984). *Principles and practices of stocking streams with salmon eggs and fry*. DAFS Scottish Fisheries Information Pamphlet 10, 1–22.

Einarsson, S.M., Mills, D.H. and Johannsson, V. (1990). Utilisation of fluvial and lacustrine habitat by anadromous Atlantic salmon, *Salmo salar* L., in an Icelandic watershed, *Fisheries Research*, **10**, 53–71.

El-Fiky, N. and Wieser, W. (1988). Life styles and patterns of development of gills and muscles in larval cyprinids (Cyprinidae; Teleostei), *Journal of Fish Biology*, **33**, 135–145.

Ellenberg, H. (1988). *Vegetation Ecology of Central Europe*, Cambridge University Press, Cambridge.

Elliott, J.M. (1969). Life history and biology of *Sericostoma personatum* Spence, *Oikos*, **20**, 110–118.

Elliott, J.M. (1971). The distances travelled by drifting invertebrates in a Lake District stream, *Oecologia*, **6**, 350–379.

Elliott, J.M. (1975a). The growth rate of brown trout (*Salmo trutta* L.) fed on maximum rations. *Journal of Animal Ecology*,

Elliott, J.M. (1975b). The growth rate of brown trout (*Salmo trutta* L.) fed on reduced rations. *Journal of Animal Ecology*, **44**, 823–842.

Elliott, J.M. (1979). Energetics of freshwater teleosts, in *Fish Phenology: Anabolic Adaptiveness in Teleosts* (Ed. P.S. Miller), pp. 29–61, Academic Press, London.

Elliott, J. M. (1981). Some aspects of thermal stress on freshwater teleosts, in *Stress and Fish* (Ed. A.D. Pickering), Academic Press, London.

Elliott, J.M. (1982). The effects of temperature and ration size on growth and energetics of salmonids in captivity, *Comparative Biochemistry and Physiology*, **73**, 81–91.

Elliott, J.M. (1984a). Growth, size, biomass and production of young migratory trout *Salmo trutta* in a Lake District stream, 1966–83, *Journal of Animal Ecology*, **53**, 979–994.

Elliott, J.M. (1984b). Numerical changes and population regulation in young migratory trout *Salmo trutta* in a Lake District stream, 1966–83, *Journal of Animal Ecology*, **53**, 327–350.

Elliott, J.M. (1985a). Population regulation for different life stages of migratory trout *Salmo trutta* in a Lake District stream, 1966–83, *Journal of Animal Ecology*, **54**, 617–638.

Elliott, J.M. (1985b). Growth, size, biomass and production for different life-stages of migratory trout *Salmo trutta* in a Lake District stream, 1966–83, *Journal of Animal Ecology*, **54**, 985–1001.

Elliott, J.M. (1987). Population regulation in contrasting populations of trout *Salmo trutta* in two Lake District streams. *Journal of Animal Ecology*, **58**, 83–98.

Elliott, J.M. (1988). Growth, size, biomass and production in contrasting populations of trout *Salmo trutta* in two Lake District streams, *Journal of Animal Ecology*, **57**, 49–60.

Elliott, J.M. (1989a). The critical-period concept for juvenile survival and its relevance for population regulation in young sea trout, *Salmo trutta*. *Journal of Fish Biology*, (Suppl. A), **35**, 91–98.

Elliott, J.M. (1989b). Mechanisms responsible for population regulation in young migratory trout, *Salmo trutta*. I. The critical time for survival, *Journal of Animal Ecology*, **58**, 987–1001.

Elliott, J.M. (1989c). Growth and size variation in contrasting populations of trout *Salmo trutta*: an experimental study on the role of natural selection, *Journal of Animal Ecology*, **58**, 45–58.

Elliott, J.M. (1989d). The natural regulation of numbers and growth in contrasting populations of brown trout, *Salmo trutta*, in two Lake District streams, *Freshwater Biology*, **21**, 7−19.

Elliott, J.M. (1989e). Mechanisms responsible for population regulation in young migratory trout, *Salmo trutta*. I. The critical time for survival, *Journal of Animal Ecology*, **58**, 987−1001.

Elliott, J.M. (1990a). Mechanisms responsible for population regulation in young migratory trout, *Salmo trutta*. II. Fish growth and size variation, *Journal of Animal Ecology*, **59**, 171−185.

Elliott, J.M. (1990b). Mechanisms responsible for population regulation in young migratory trout, *Salmo trutta*. III. The role of territorial behaviour, *Journal of Animal Ecology*, **59**, 803−818.

Elliott, J.M. (1992). Variation in the population density of adult sea-trout, *Salmo trutta*, in 67 rivers in England and Wales, *Ecology of Freshwater Fish*, **1**, 5−11.

Elliott, J.M. and Persson, L. (1978). The estimation of daily rates of food consumption for fish, *Journal of Animal Ecology*, **47**, 977−991.

Elson, P.F. (1962). *Predator−prey relationships between fish eating birds and Atlantic salmon (with a supplement on the fundamentals of merganser control)*, Fisheries Research Board of Canada, Bulletin No. 133, 1−87.

Elson, P. F. (1975a). *The Foyle Fisheries — New Basis for Rational Management*. Foyle Fisheries Commission Special Publication, 1−224.

Elson, P. F. (1975b). Atlantic salmon rivers, smolt production and optimal spawning: an overview of natural production, *International Atlantic Salmon Foundation Special Publication Series*, **6**, 96−119.

Emery, M.J. (1983). *The ecology of Japanese Knotweed* (Reynoutria japonica *Houtt.), its herbivores and pathogens and their potential as biological control agents*. MSc Dissertation, UCNW Bangor, University of Wales.

Engel, S. (1985). Aquatic commmunity interactions of submerged macrophytes. *Wisconsin Department of Natural Resources Technical Bulletin*, **156**.

Engel, S. (1988) The role and interactions of submersed macrophytes in a shallow Wisconsin lake, *Journal of Freshwater Ecology*, **4**, 329−341.

English, J., McDermott, G. and Surber, E. (1963). Pollutional effects of outboard motor exhausts — field studies, *Journal of the Water Pollution Control Federation*, **35**, 1121−1135.

EPA. (1990). *Water Quality Program Highlights: Ohio EPA's use of biological survey information*, Assessment and Watershed Protection Division, US EPA.

Erman, N.A. (1983) The use of riparian systems by aquatic insects, in *Ecology, Conservation and Productive Management* (Ed. by R.E. Warner and C.M. Hendrix), pp. 177−182, University of California Press.

Erman, D.C. and Lignon, F.K. (1988). Effect of discharge fluctuation and the addition of fine sediment on stream fish and macroinvertebrates below a water-filtration facility, *Environmental Management*, **12**, 85−97.

Ersèus, C. and Paoletti, A. (1986). An Italian record of the aquatic oligochaete *Monopilephorus limosus* (Tubificidae), previously known only from Japan and China. *Boll. Zool.*, **53**, 115−118.

European Communities (1985). Council Directive of 27 June 1985 on the assessment of the effects of certain public and private projects on the environment (85/337/EEC), *Official Journal of the European Communities*, **L175**, 40−48.

European Communities (1992). Council Directive of 21 May 1992 on the conservation of natural habitats and of wild fauna and flora (92/43/EEC), *Official Journal of the European Communities*, **L206**, 7−50.

Evans, R., Brown, C. and Kellett, J. (1990). Geology and groundwater, in *The Murray* (Eds N. Mackay and D. Eastburn), Murray Darling Basin Commission, Canberra, Australia.

Ewing, M.S. (1991). Turbidity control and fisheries enhancement in a bottomland backwater system in Louisiana (USA), *Regulated Rivers Research and Management*, **6**, 87−99.

Extence, C.A. and Ferguson, A.J.D. (1989). Aquatic invertebrate surveys as a water quality management tool in the Anglian Water Region, *Regulated Rivers Research and Management*, **4**, 139−146.

Fahy, E. (1978). Variation in some biological characteristics of British sea trout, *Salmo trutta* L., *Journal of Fish Biology*, **13**, 123−138.

Fairchild, G.W. (1981) Movement and microdistribution of *Sida crystallina* and other littoral microcrustacea, *Ecology*, **62**, 1341−1352.

Farmer, A.M. (1990). The effects of lake acidification on aquatic macrophytes — a review, *Environmental Pollution*, **65**, 219–240.

Fausch, K.D., Hawkes, C.L. and Parsons, M.G. (1988). *Models that predict standing crop of stream fish from habitat variables: 1950–85*. US Forest Service, Pacific Northwest Field Station, General Technical Report, 213.

Fausch, K.D., Karr, J.R. and Yant, P.R. (1984). Regional application of an index of biotic integrity based on stream fish communities, *Transactions of the American Fisheries Society*, **113**, 39–55.

Fausch, K.D., Lyons, J., Karr, J.R. and Angermeier, P.L. (1990). Fish communities as indicators of environmental degradation, *American Fisheries Society Symposium*, **8**, 123–144.

Faust, M.A. (1982). Contribution of pleasure boats to faecal bacteria concentration in the Rhode River Estuary, *Science of the Total Environment*, **25**, 255–262.

Feminella, J.W. and Resh, V.H. (1991) Herbivorous caddisflies, macroalgae, and epilithic microalgae: dynamic interactions in a stream grazing system, *Oecologia*, **87**, 247–256.

Ferguson, A. (1980). *Biochemical Systematics and Evolution*, Blackie, Glasgow.

Ferguson, A. (1986). Lough Melvin, a unique fish community, *Royal Dublin Society Occasional Papers in Irish Science and Technology*, **1**, 1–17.

Ferguson, A. (1989). Genetic differences among brown trout, *Salmo trutta*, stocks and their importance for the conservation and management of the species, *Freshwater Biology*, **21**, 35–46.

Ferguson, R.I. (1981). Channel forms and channel changes, in *British Rivers* (Ed J. Lewis), pp. 90–125, Ceorge Allen and Unwin, London.

Feth, J.H. (1966). Nitrogen compounds in natural waters — a review, *Water Resources Research*, **2**, 41–58.

Fisher, S.G. and Carpenter, S.R. (1976). Ecosystem and macrophyte primary production of the Fort River, Massachusetts, *Hydrobiologia*, **47**, 175–187.

Fisher, S.G., Gray, L.J., Grimm, N.B. and Busch, D.E. (1982). Temporal succession in a desert stream ecosystem following flash flooding, *Ecological Monographs*, **52**, 93–110.

Fitch, I.C. (1976) Dispersal of Himalayan Balsam, *Impatiens glandulifera*, *North Western Naturalist*, **13**.

Forestry Commission (1991). *Forests and Water Guidelines*, Forestry Commission, HMSO, London.

Forman, R.T.T. and Godron, M. (1986). *Landscape Ecology*, John Wiley, New York.

Foster-Turley, P., Macdonald, S.M. and Mason, C.F. (Eds) (1990). *Otters: An Action Plan for Their Conservation*, I.U.C.N., Gland.

Fowles, A.P. (1988). *An ecological study of the distribution of cursorial invertebrates on polluted riparian shingle*, National Rivers Authority (Welsh Region).

Francis, R.I. (1990). Back-calculation of fish length: a critical review, *Journal of Fish Biology*, **36**, 883–902.

Fraser, J.C. (1972). Regulated discharge amd the stream environment, in *River Ecology and Man* (Eds R.T. Oglesby, C.A. Carlson and J.A. McCann), pp. 263–286, Academic Press, New York.

Fraser, P. (1984). Epidemiology and water quality, in *Environmental Protection, Standards, Compliance and Costs* (Ed. T.J. Lack), pp. 85–93, Ellis Horwood, Chichester.

Freeman, M.C. and Wallace, J.B. (1984). Production of net-spinning caddisflies (Hydropsychidae) and blackflies (Simuliidae) on rock outcrop substrate in a small southeastern Piedmont stream, *Hydrobiologia*, **112**, 3–15.

French, R.H. (1986). *Open Channel Hydraulics*, McGraw-Hill.

Friday, L.E. (1987) The diversity of macroinvertebrate and macrophyte communities in ponds, *Freshwater Biology*, **18**, 87–104.

Frissell, C.A. and Nawa, R.K. (1992). Incidence and causes of physical failure of artificial habitat structures in streams of Western Oregon and Washington, *North American Journal of Fisheries Management*, **12**, 182–197.

Frissell, C.A., Liss, W.J., Warren, C.E. and Hurley, M.D. (1986). A hierarchical framework for stream habitat classification: viewing streams in a watershed context, *Environmental Management*, **10**, 199–214.

Frost, W.E. (1942). R. Liffey survey IV. The fauna of submerged "mosses" in an acid and an alkaline water, *Proceedings of the Royal Academy of Ireland (B)*, **13**, 293–369.

Frostick, L.E., Lucas, P.M. and Reid, I. (1984). The infiltration of fine matrices into coarse-grained alluvial sediments and its implications for stratigraphical interpretation, *Journal of the Geological*

Society of London, **141**, 955–965.

Fruget, J.F. (1992). Ecology of the lower Rhône after 200 years of human influence: a review, *Regulated Rivers: Research and Management*, **7**, 233–246.

Gaevskaya, N.S. (1966). *The role of higher aquatic plants in the nutrition of animals of freshwater basins*. Translated 1969 by D.G. Maitland Muller, Boston Spa: National Lending Library for Science and Technology.

Gambi, M.C., Nowell, A.R.M. and Jumars, P.A. (1990). Flume observations on flow dynamics in *Zostera marina* (eelgrass) beds, Marine Ecology Progress Series, **61**, 159–169.

Gangmark, H.A. and Bakaala, R.G. (1960). A comparative study of unstable and stable (artificial channel) spawning streams for incubating King salmon at Mill Creek, *California Fish and Game*, **46**, 151–164.

García, A. and Braña, F. (1988). Reproductive biology of brown rout (*Salmo trutta* L.) in the Aller river (Asturias; northern Spain), *Polskie Archiwum Hydrobiologii*, **35**, 361–373.

García de Jalón, D. (1990). Tecnicas hidrobiologicas para la fijacion de caudales ecologicos minimos, in *Libro Homenaje al Profesor D.M. Garcia de Viedma* (Eds A. Ramos, A. Notario and R. Baragaio), pp. 83–196, Fucovasa UPM, Madrid.

García de Jalón, D. (1992). Din mica de las poblaciones piscicolas en los rios de montana ibiricos, *Ecologia*, **6**, 281–296.

García Marín, J.L. (1992). Diferenciacion genetica de la trucha comun (*Salmo trutta*) en Espana, Doctoral Thesis, Univ. Auton, Barcelona.

Gardiner, J.L. (Ed.) (1991). *River Projects and Conservation. A Manual for Holistic Appraisal*, John Wiley, New York.

Gardiner, R.W. (1991). Scottish Grayling: history and biology of the populations, *Proceedings of the Institute of Fisheries Management Annual Study Course*, **22**, 171–178.

Garland, J.H.N. and Hart, I.C. (1972). *Effects of Pollution on River Quality, The Trent Research Programme*, Vol. 4, Water Resources Board, Reading.

Garrad, P.N. and Hey, R.D. (1988a). The effect of boat traffic on river regime, in *International Conference on River Regime* (Ed. W.R. White), pp. 395–409, Wiley, Chichester.

Garrad, P.N. and Hey, R.D. (1988b). River management to reduce turbidity in navigable broadland rivers. *Journal of Environmental Management*, **27**, 273–288.

Gaschignard, O., Persat, H. and Chessel, D. (1983). Répartition transversale des macroinvertébrés benthiques dans un bras du Rhone, *Hydrobiologia*, **106**, 209–215.

Gatz, A.J., Sale, M.J. and Loar, J.M. (1987). Habitat shifts in rainbow trout: competitive influences of brown trout, *Oecologia*, **74**, 7–19.

Gauldie, R.W. (1991). Taking stock of genetic concepts in fisheries management, *Canadian Journal of Fisheries and Aquatic Sciences*, **48**, 722–731.

Gee, A.S. and Milner, N.J. (1980). Analysis of 70 year catch statistics for Atlantic salmon (*Salmo salar*) in the River Wye and implications for management of stocks. *Journal of Applied Ecology*, **17**, 41–57.

Gee, J.H.R. (1988). Population dynamics and morphometrics of *Gammarus pulex* L.: evidence of seasonal food limitation in a freshwater detritivore, *Freshwater Biology*, **19**, 333–343.

Gerisch, R.M. and Brusven, M.A. (1982). Volcanic ash accumulation and ash-voiding mechanisms of aquatic insects. *Journal of the Kansas Entomological Society*, **55**, 290–296.

Gerking, S.D. (1957) A method of sampling the littoral macrofauna and its application, *Ecology*, **38**, 219–226.

Ghetti, P.F. (1986). I macroinvertebrati nellanalisi dei corsi dacqua, Provincia Autonoma di Trento.

Gibbons, D.W. and Pain, D. (1992) The influence of river flow rate on the breeding behaviour of *Calopteryx* damselflies, *Journal of Animal Ecology*, **61**, 283–289.

Gibson, M.T., Welch, I.M., Barrett, P.R.F. and Ridge, I. (1990). Barley straw as an inhibitor of algal growth II: laboratory studies, *Journal of Applied Phycology*, **2**, 241–248.

Gibson, R.J. and Cunjak, R.A. (1986). An investigation of competitive interactions between brown trout (*Salmo trutta* L.) and juvenile Atlantic salmon (*Salmo salar* L.) in rivers of the Avalon Peninsula, Newfoundland, *Canadian Technical Reports in Fisheries and Aquatic Sciences*, **121**.

Gilbert, O.L. (1989). *The Ecology of Urban Habitats*, Chapman and Hall, London.

Giles, N., Phillips, V. and Barnard, S. (1991). *Ecological Effects of Low Flows in Chalk Streams*, Wiltshire Trust for Nature Conservation, Devizes, Wiltshire.

Giles, N., Street, M., Wright, R., Phillips, V. and Traill-Stevenson, A. (1989). Food for wildfowl increases after fish removal, *Game Conservancy Review*, **1988**, 137–140.

Gilvear, D.J. (1987). Suspended solids transport within regulated rivers experiencing periodic reservoir releases, in *Regulated Streams: Advances in Ecology* (Eds J.F. Craig and J.B. Kemper), pp. 245–255, Plenum, New York.

Glime, J.M. and Clemons, R.M. (1972). Species diversity of stream insects on *Fontinalis* spp. compared to diversity on artificial substrates, *Ecology*, **53**, 458–464.

Goldspink, G. (1977). The return of marked roach *Rutilus rutilus* to spawning grounds in Tjeukemeer, the Netherlands, *Journal of Fish Biology*, **11**, 599–603.

Gordon, N.D., McMahon, T.A. and Finlayson, B.L. (1992). *Stream Hydrology: An Introduction for Ecologists*, Wiley, Chichester.

Gore, J.A. (Ed) (1985). *The Restoration of Rivers and Streams: Theories and Experience*, Butterworth, Boston.

Gore, J.A. and Judy, R.D. Jr (1981). Predictive models of benthic macroinvertebrate density for use in in-stream flow studies and regulated flow management, *Canadian Journal of Fisheries and Aquatic Sciences*, **38**, 1363–1370.

Gore, J.A. and Petts, G.E. (Eds) (1989). *Alternatives in Regulated River Management*, CRC Press, Boca Raton, FL.

Gore, J.A., Layzer, J.B. and Russell, I.A. (1992). Non-traditional applications of instream flow techniques for conserving habitat of biota in the Sabie River of Southern Africa, in *River Conservation and Management* (Eds P.J. Boon, P. Calow and G.E. Petts), pp. 161–178, John Wiley, Chichester.

Gorman, O.T. and Karr, J.R. (1978). Habitat structure and stream fish communities, *Ecology*, **59**, 507–515.

Gosling, L.M. (1974). The coypu in East Anglia, *Transactions of the Norfolk and Norwich Naturalists' Society*, **23**, 49–59.

Gosling, L.M. and Baker, S.J. (1991). Coypu *Myocaster coypus*, in *The Handbook of British Mammals*, 3rd edn (Eds G.B. Corbet and S. Harris), pp. 267–274, Blackwell, Oxford.

Graham, A.A. (1990). Siltation of stone-surface periphyton by clay-sized particles from low concentrations in suspension, *Hydrobiologia*, **36**, 177–184.

Gray, L.J. and Ward, J.V. (1979). Food habits of stream benthos at sites of differing food availability, *American Midland Naturalist*, **102**, 157–167.

Gray L.J. and Ward J.V. (1982). Effects of sediment releases from a reservoir on stream macroinvertebrates, *Hydrobiologia*, **36**, 177–184.

Grayson, R.B., Bloschl, G., Barling, R.D. and Moore, I.D. (1993). Process, scale and constraints to hydrological modelling in GIS, in *Applications of Geographic Information Systems in Hydrology and Water Resources Management* (Eds K. Kovar and H.P. Nachtnebel), International Association of Hydrological Sciences Publication No. 211, 83–92.

Greger, P.D. and Deacon, J.E. (1988) Food partitioning among fishes of the Virgin River, *Copeia*, **1988** (2), 314–323.

Gregory, K.J. (1982) River power, in *Papers in Earth Studies* (Eds B.H. Adlam, C.R. Fenn and L. Morris), pp. 1–20, Geobooks, Norwich.

Gregory, K.J. (1992) Vegetation and river channel process interactions, in *River Conservation and Management* (Ed. P.J. Boon, P. Calow and G.E. Petts), pp. 255–269, John Wiley, Chichester.

Gregory, K.J. and Davis, R.J. (1992). Coarse woody debris in stream channels in relation to river channel management in woodland areas, *Regulated Rivers: Research and Management*, **7**, 117–136.

Gregory, K.J. and Gurnell, A.M. (1988). Vegetation and river channel form and process, in *Biogeomorphology* (Ed. H.A. Viles), pp. 11–42, Basil Blackwell, Oxford.

Gregory, K.J. and Park, C.C. (1976). Stream channel morphology in north-west Yorkshire, *Revue de Geomorphologie Dynamique*, **25**, 63–72.

Gregory, K.J., Gurnell, A.M. and Hill, C.T. (1985) The permanence of debris dams related to river channel processes, *Hydrological Sciences Journal*, **30**, 371–381.

Green, J., Green, R. and Jefferies, D.J. (1984). A radiotracking survey of otters *Lutra lutra* on a Perthshire river system, *Lutra*, **27**, 85–145.

Grier, C.C. and Logan, R.S. (1977) Old-growth *Pseudotsuga menziesii* communities of a western

Oregon watershed: Biomass distribution and production budgets, *Ecological Monographs*, **47**, 373–400.

Grime, J.P., Hodgson, J.G. and Hunt, R. (1988) *Comparative Plant Ecology*, pp. 488–489, Unwin Hyman, London.

Grimmet, R.F.A. and Jones, T.A. (1989). *Important Bird Areas in Europe*, I.C.B.P., Technical Publication no. 9.

Groom, A.P. and Hildrew, A.G. (1989). Food quality for detritivores in streams of contrasting pH, *Journal of Animal Ecology*, **58**, 863–881.

Gucinski, H. (1981). *Sediment suspension and resuspension from craft-induced turbulence*, Final Report EPA Project No. EPA-78-D-XO426, Chesapeake Bay Program, Office of Research and Development, Middle Atlantic III, USA.

Guiver, K. (1973). River authorities and fishery byelaws, in *Proceedings of the Sixth British Coarse Fish Conference*, University of Liverpool, pp. 35–42, Janssen Services, London.

Gulland, J.A. (1971). Appraisal of a fishery, in *Methods for Assessment of Fish Production in Freshwaters*, (Ed. W.E. Ricker), Blackwell Scientific Publications, Oxford.

Guma'a, S.A. (1978). The effects of temperature on the development and mortality of eggs of perch, *Perca fluviatilis*, *Freshwater Biology*, **8**, 221–227.

Gurnell, A.M. and Gregory, K.J. (1987). Vegetation characteristics and the prediction of runoff: analysis of an experiment in the New Forest, Hampshire, *Hydrological Processes*, **1**, 125–142.

Gurnell, A.M. and Midgley, P. (1993). Aquatic weed growth and flow resistance: influence on the relationship between discharge and stage over a 25 year river gauging station record. *Hydrological Processes*, **7**.

Gurnell, A.M., Gregory, K.J., Hollis, S. and Hill, C.T. (1985). Detrended correspondence analysis of heathland vegetation: the identification of runoff contributing areas, *Earth Surface Processes and Landforms*, **10**, 343–351.

Gurnell, A.M. Simmons, P.J., Edwards, J., Ball, J., Feaver, A., McLellan and Ogle, C. (1993). GIS and multivariate ecological analysis: incorporating plant ecological data into integrated river management, in *Applications of Geographic Information Systems in Hydrology and Water Resources Management* (Eds K. Kovar and H.P. Nachtnebel), pp. 363–374, International Association of Hydrological Sciences Publication 211.

Gurney, R. (1932). *British Fresh-water Copepoda*, Vol. II. Ray Society, London.

Gurney, R. (1933). *British Fresh-water Copepoda*, Vol. III. Ray Society, London.

Gurtz, M.E. and Wallace, J.B. (1984). Substrate-mediated response of stream invertebrates to disturbance, *Ecology*, **65**, 1556–1569.

Gustard, A. (1992). Analysis of river regimes, in *The Rivers Handbook*, Vol. I (Eds P. Calow and G.E. Petts), pp. 29–47, Blackwell Scientific, Oxford.

Gustard, A., Cole, G., Marshall, D. and Bayliss, A. (1987). *A study of compensation flows in the UK*, Institute of Hydrology, Report 99. (A summary is given in Gustard, A., (1989) Compensation flows in the UK: a hydrological review, *Regulated Rivers*, **3**, 4).

Gutleb, A.C. (1992). The otter in Austria: a review of the current state of research, *I.U.C.N. Otter Specialist Group Bulletin*, **7**, 4–9.

Gyllensten, U. and Wilson, A.C. (1987). Mitochondrial DNA of salmonids: inter and intraspecific variability detected with restriction enzymes, in *Population Genetics and Fishery Management* (Ed. N. Ryman and F. Utter), pp. 301–318, University of Washington Press, Seattle.

Hack, J.T. (1957). *Studies of Longitudinal Stream Profiles in Virginia and Maryland*, US Geological Survey, Professional Paper, 294B.

Haldane, J.B.S. (1953). Animal populations and their regulation, *New Biology*, **15**, 9–24.

Haldane, J.B.S. (1956). The relation between density regulation and natural selection, *Proceedings of The Royal Society, Series B*, **145**, 306–308.

Hale, J.G. (1965). An evaluation of trout stream habitat improvement in a North Shore tributary of Lake Superior, *Minnesota Fisheries Investations*, **5**, 37–50.

Hall, G.H., Jones, J.G., Pickup R.W. and Simon B.M. (1990). Methods to study the bacterial ecology of freshwater environments, *Methods in Microbiology*, **22**, 182–209.

Hall, J.D. and Lantz, R.L. (1973). Effects of logging on the habitat of the coho salmon and cutthroat trout in coastal streams, in *Proceedings of the Symposium on Salmon and Trout in Streams* (Ed. G. Northcote), pp. xxx–xxx, University of British Columbia, Vancouver.

Hall, R.J., Likens, G.E., Fiance, S.B. and Hendrey, G.R. (1990). Experimental acidification of a stream in the Hubbard Brook Experimental Forest, New Hampshire, *Ecology*, **61**, 976–989.

Hallerman, E.M. and Beckmann, J.S. (1988). DNA-level polymorphism as a tool in fisheries science, *Canadian Journal of Fisheries and Aquatic Sciences*, **45**, 1075–1087.

Ham, S.F., Wright, J.F. and Berrie, A.D. (1981). The effect of cutting on the growth and recession of the freshwater macrophyte *Ranunculus penicillatus* (Dumort.) Bab. var. calcareous (R.W. Butcher) C.D.K. Cook, *Journal of Environmental Management*, **15**, 263–271.

Hamilton, K.E., Ferguson, A., Taggart, J.B., Tomasson, T., Walker, A. and Fahy, E. (1989). Post-glacial colonization of brown trout, *Salmo trutta* L.: Ldh-5 as a phytogeographic marker locus, *Journal of Fish Biology*, **35**, 651–664.

Hammerton, A. M. (1990). *Sedimentation and vegetation encroachment of the channel-bed downstream of a tributary confluence on an impounded river: the case of the Upper Clyde*, unpublished BSc dissertation, Geography Department, Loughborough University.

Hammond, C.O. (1983). *The Dragonflies of Great Britain and Ireland*, Harley Books, Colchester.

Hanbury, R.G. (1982). *Reinforced vegetative bank protection: 1982 trials report*. British Waterways Board, Gloucester, UK.

Hanbury, R.G. (1986). Conservation on canals. A review of the present status and management of British navigable canals with particular reference to aquatic plants, in *Proceedings of the European Weed Research Society/Association of Applied Biologists 7th Symposium on Aquatic Weeds*, pp. 143–149.

Hanley, N. (1991). The economics of nitrate pollution control in the UK, in *Farming and the Countryside* (Ed. N. Hanley), CAB International, London.

Hansard (House of Lords) (1993). Debate on Environmental Legislation: ECC Report, 21 January 1993, Vol. 541, Column 1010.

Hansen, E.A., Alexander, G.R. and Dunn, W.H. (1983). Sand sediment in a Michigan trout stream. I. A technique for removing sand bed lead from streams, *North American Journal of Fisheries Management*, **3**, 355–364.

Harding, J.P. and Smith W.A. (1960). A key to the British freshwater cyclopoid and calanoid copepods, with ecological notes, *Freshwater Biological Association Scientific Publications*, **18**, 1–54.

Harmon, M.E., Franklin, J.F., Swanson, F.J., Sollins, P., Gregory, S.V., Lattin, J.D., Anderson, N.H., Cline, S.P., Aumen, N.G., Sedell, J.R., Lienkaemper, G.W., Cromack, K. and Cummins, K.W. (1986). Ecology of coarse woody debris in temperate ecosystems, *Advances in Ecological Research*, **15**, 133–302.

Harper, D.M., Smith, C.D. and Barham, P.J. (1992). Habitats as the building blocks for river conservation assessment, in *River Conservation and Management* (Eds P.J. Boon, P. Calow and G.E. Petts), pp. 311–319, John Wiley, Chichester.

Harrel, R.C. and Hall, M.A. (1991). Macrobenthic community structure before and after pollution abatement in the Nechez river estuary (Texas), *Hydrobiologia*, **211**, 241–252.

Harris, G. S. (Ed.) (1978). *Salmon propagation in England Wales*. Report of the Association of River Authorities/National Water Council, London, 1–62.

Harris, J.H. (1984). Impoundment of coastal drainages of south-eastern Australia and a review of its relevance to fish migration, *Australian Zoologist*, **21**, 235–250.

Harrison, A.D. and Elsworth, J.F. (1958). Hydrobiological Studies on the Great Berg River, Western Cape Province. Part I. General description, chemical studies and main features of the flora and fauna, *Transactions of the Royal Society of South Africa*, **35**, 125–226.

Harrod, J.J. (1964). The distribution of invertebrates on submerged aquatic plants in a chalk stream, *Journal of Animal Ecology*, **33**, 335–348.

Hart, B.T. (1992). Ecological condition of Australia's rivers, *Search*, **23**, 33–37.

Hart, D.D. (1978). Diversity in stream insects: regulation by rock size and microspatial complexity, *Verhandlungen Internationale Vereinigung für Theoretische und Angewandte Limnologie*, **20**, 1376–1381.

Hart, R.C. (1986). Aspects of the feeding ecology of turbid water zooplankton: *in situ* studies of community filtration rates in silt laden Lake Le Roux, Orange River, S. Africa, *Journal of Plankton Research*, **8**, 401–426.

Hart, R.C. (1988). Zooplankton feeding rates in relation to suspended sediment content: potential influences on community structure in a turbid reservoir, *Freshwater Biology*, **19**, 123–139.

Harvey, A.M. (1991). The influence of sediment supply on the channel morphology of upland streams: Howgill Fells, Northwest England, *Earth Surface Processes and Landforms*, **16**, 675–684.

Harvey, B.C. (1987). Susceptibility of young-of-the-year fishes to downstream displacement by flooding, *Transactions of the American Fisheries Society*, **116**, 851–855.

Harvey, B.C. (1991a). Interaction of abiotic and biotic influences larval fish survival in an Oklahoma stream, *Canadian Journal of Fisheries and Aquatic Sciences*, **48**, 1476–1480.

Harvey, B.C. (1991b). Larval fish survival in streams: importance of biotic interactions, *Oecologia (Berlin)*, **87**, 29–36.

Harvey, B.C. and Stewart, A.J. (1991). Fish size and habitat depth relationships in headwater streams, *Oecologia*, **87**, 336–342.

Hasenfuss, I. (1960) Die Larvalsystematik der Zauunsler (Pyralidae), *Abh. Larvalsyst. Insekt.*, **5**, 139–149.

Hasfurther, V.R. (1985). The use of meander parameters in restoring hydrologic balance to reclaimed stream beds, in *The Restoration of Rivers and Streams: Theories and Experience* (Ed. J.A. Gore), pp. 21–40. Butterworth, Boston.

Haslam, S.M. (1978). *River Plants*, Cambridge University Press, Cambridge.

Hausle, D.A. and Coble, D.W. (1976). Influence of sand in redds on survival and emergence of brook trout (*Salvelinus fontinalis*). *Transactions of the American Fisheries Society*, **105**, 57–63.

Hawkes, H.A. (1975). River zonation and classification, in *River Ecology* (Ed. B.A. Whitton), pp. 312–374. Blackwell, Oxford.

Hawkins, C.P., Murphy, M.L., Anderson, N.H. and Wilzbach, M.A. (1983). Density of fish and salamanders in relation to riparian canopy and physical habitat in streams of the Northwestern United States, *Canadian Journal of Fisheries and Aquatic Sciences*, **40**, 1173–1185.

Hay, D.W. (1989). Effect of adult stock penetration on juvenile production of *Salmo salar* L. in a Scottish stream, in *Proceedings of the 2nd International Symposium on Salmonid Migration and Distribution* (Eds E. Brannon and B. Jonsson), pp. 93–100, University of Washington.

Hay, D.W. (1991). *Stock and recruitment relationships*. Lessons from the Girnock Project, Altantic Salmon Trust/Royal Irish Academy Workshop, Dublin, 1–3.

Haycock, N.E. and Pinay, G. (in press). Nitrate retention in grass and poplar vegetated riparian buffer strips during the winter, *Journal of Environmental Quality*.

Haycock, N.E., Pinay, G. and Walker, C. (1993). Nitrogen retention in river corridors: European perspective, *Ambio*.

Healey, M.C. (1978). Fecundity changes in exploited populations of lake whitefish (*Coregonus clupeaformis*) and lake trout (*Salvelinus namaycush*). *Journal of the Fisheries Research Board of Canada*, **35**, 945–950.

Healey, M.C. (1980). Growth and recruitment in experimentally exploited lake whitefish (*Coregonus clupeaformis*) populations. *Canadian Journal of Fisheries and Aquatic Sciences*, **37**, 255–267.

Hearnden, M.N. and Pearson, R.G. (1991). Habitat partitioning among the mayfly species (Ephemeroptera) of Yuccabine Creek, a tropical Australian stream, *Oecologia*, **87**, 91–101.

Hearne, J.W. and Armitage, P.D. (1993). Implications of the annual macrophyte growth cycle on habitat in rivers, *Regulated Rivers: Research and Management*, **8**, 313–322.

Heede, B.H. and Rinne, J.N. (1990). Hydrodinamic and fluvial morphologic processes: Implications for fisheries management and research, *North American Journal of Fisheries Management*, **10**, 249–268.

Heggenes, J. (1988). Effect of short-term flow fluctuations on displacement of, and habitat use by, Brown Trout in a small stream, *Transactions of the American Fisheries Society*, **117**, 336–344.

Heggenes, J. (1988). Physical habitat selection by brown trout (*Salmo trutta*) in riverine systems, *Nordic Journal of Freshwater Research*, **64**, 74–90.

Heggenes, J. and Borgstrøm, R. (1991). Effect of habitat types on survival, spatial distribution and reproduction of an allopatric cohort of Atlantic salmon, *Salmo salar* L., under conditions of low competition, *Journal of Fish Biology*, **38**, 2767–280.

Heggenes, J. and Saltveit, S.J. (1990). Seasonal and spatial microhabitat selection and segregation in young Atlantic salmon, *Salmo salar* L., and brown trout, *Salmo trutta* L., in a Norwegian river, *Journal of Fish Biology*, **36**, 707–720.

Hellawell, J.M., (1978). *The Biological Surveillance of Rivers; A Biological Monitoring Handbook*, Water Research Centre, Stevenage.

Hellawell, J.M. (1986). *Biological Indicators of Freshwater Pollution and Environmental Management*, Elsevier, London.

Hemphill, R.W. and Bramley, M.E. (1989). *Protection of River and Canal Banks, A Guide to Selection and Design*, CIRIA Water Engineering Report, Butterworths.

Hendrey, G.R. (1976). *Effects of pH on the growth of periphytic algae in artificial stream channels*, Research Report 25/76, SNSF Project, Oslo, Norway.

Hendriks, A.J. (1993). Monitoring concentrations of microcontaminants in sediment and water in the Rhine delta: a comparison to reference values, *European Water Pollution Control*, 3, 33–38.

Hendriks, A.J. (in press). Modelling concentrations of conservative microcontaminants in aquatic organisms in the Rhine delta: steady state specification, laboratory calibration and field validation.

Hendriks, A.J., (in press). Monitoring and estimating concentrations and effects of microcontaminants in the Rhine-delta: chemical analysis, biological laboratory assays and field observations, Symposium on the Rehabilitation of the river Rhine, March 1993, Arnhem, *Water Science and Technology*, pp. 15–19.

Hendriks, A.J. and Pieters, H. (1993). Monitoring microcontaminants in aquatic organisms in the Rhine delta: a comparison to reference values, *Chemosphere*, in press.

Herbert, D.W., Alabaster, J.S., Dart, M.C. and Lloyd, R. (1961). The effect of china-clay wastes on trout streams, *International Journal of Air and Water Pollution*, 5, 56–74.

Herbert, D.W.M. and Shurben, D.S. (1964). The toxicity to fish of mixtures of poisons. I — Salts of ammonia and zinc. *Annals of Applied Biology*, 53, 33–41.

Herbst, G.N. (1980). Effects of burial on food value and consumption of leaf detritus by aquatic invertebrates in a lowland forest stream, *Oikos*, 35, 411–424.

Herrington, R.B. and Dunham, D.K. (1967). A technique for sampling general fish habitat characteristics of streams, *Research Paper of the US Forest Service*, 41, 1–12.

Hershberger, W.K. (1992). Genetic variability in rainbow trout populations, *Aquaculture*, **100**, 51–71.

Hester, H.F. and Dendy, J.S. (1962). A multiple plate sampler for aquatic invertebrates, *Transactions of the American Fisheries Society*, 91, 420–431.

Hey, R.D. (1982). Gravel-bed rivers: form and process, in *Gravel-bed Rivers*, (Eds R.D. Hey, J.C. Bathurst and C.R. Thorne), pp. 5–13, Wiley, Chichester.

Hey, R.D. (1990). Environmental River Engineering, *Journal of the Institute of Water and Environmental Management*, 4, 335–340.

Hey, R.D. (1994). Environmentally sensitive river engineering, in *The Rivers Handbook*, Vol. 2 (Eds P. Calow and G.E. Petts), pp. 337–362, Blackwell Scientific, Oxford.

Hickley, P. (1980). *An ecological investigation of benthic invertebrates and fish in a small lowland river*, PhD thesis, University of London.

Hickley, P. (1986). Invasion by zander and the management of fish stocks, *Philosophical Transactions of the Royal Society of London, Series B*, 314, 571–582.

Hickman, T. and Raleigh, R.F. (1982). *Habitat suitability index models: cutthroat trout*. United States Fish and Wildlife Service Biological Services Program, 82/10.5.

Hildrew, A.G. (1992) Food webs and species interactions, in *The Rivers Handbook*, Vol. 1 (Eds P. Calow and G.E. Petts), pp. 309–330, Blackwell Scientific, Oxford.

Hildrew, A.G. and Giller, P.S. (1994). Patchiness, species interactions and disturbance in the stream benthos, in *Aquatic Ecology: Scale Pattern and Process* (Eds P.S. Giller, A.G. Hildrew and D.G. Raffaelli), Symposia of the British Ecological Society, 21–62.

Hildrew, A.G., Townsend, C.R., Francis, J.E. and Finch, K. (1984). Cellololytic decomposition in streams of contrsting pH and its relationship with invertebrate community structure, *Freshwater Biology*, 14, 323–328.

Hill, D.J. (1994). A practical strategy for the control of Japanese Knotweed (*Reynoutria japonica*) in Swansea and surrounding area, in *Ecology and Management of Invasive Riverside Plants* (Ed L.C. de Waal, L.E. Child, P.M. Wade and J.H. Brock), pp. 195–198, John Wiley, Chichester.

Hill, M.O. (1979), TWINSPAN — A Fortran program for arranging multivariate data in an ordered two-way table by classification of the individuals and attributes, Cornell University, Section of Ecology and Systematics, Ithaca, New York.

Hill, M.O. and Gauch, H.G. (1980). Detrended correspondence analysis: an improved ordination technique, *Vegetatio*, **42**, 47–58.

Hill, M.T., Platts, W.S. and Beschta, R.L. (1991). Ecological and geomorphological concepts for instream and out-of-channel flow requirements, *Rivers*, **2**, 198–210.

Hill, T.K. and Shell, E.W. (1975). Some effects of a sanctuary on an exploited fish population, *Transactions of the American Fisheries Society*, **104**, 441–445.

Hills, D. (1991). Ephemeral introductions and escapes, in *The Handbook of British Mammals*, 3rd edn (Eds G.B. Corbett and S. Harris), pp. 576–580, Blackwells, Oxford.

Hills, J., Clement, B., Murphy, K., Obrdlik, P., Castella, E., Speight, M. and Schneider, E. (1992). Biotic indicators of wetland ecosytem functioning. Paper presented at Global Wetlands Old World and New; IVth International Wetlands Conference, Columbus, Ohio.

Hilton, J. and Phillips, G.L. (1982). The effect of boat activity on turbidity in a shallow Broadland river, *Journal of Applied Ecology*, **19**, 143–150.

Hilton, J., Irish, A.E. and Reynolds, C.S. (1992). Active reservoir management: a model solution, in *Eutrophication: Research and Application to Water Supply* (Eds D.W. Sutcliffe, and J.G. Jones), pp. 185–196, Freshwater Biological Association, Ambleside.

Hindar, K., Ryman, N. and Utter, F. (1991). Genetic effects of cultured fish on natural fish populations, *Canadian Journal of Fisheries and Aquatic Sciences*, **48**, 45–57.

Hindar, K., Jonsson, B., Ryman, N. and Stahl, G. (1991). Genetic relationships among landlocked, resident, and anadromous brown trout, *Salmo trutta* L., *Heredity*, **66**, 83–91.

Hindley, D.R. (1973). The definition of dry weather flow in river flow measurements, *Journal of the Institute of Water Engineering and Scientists*, **27**, 438–440.

Hino, M., Fujita, K. and Shutto, H. (1987). A laboratory experiment on the role of grass for infiltration and runoff processes, *Journal of Hydrology*, **90**, 303–325.

Hjulstrom, F. (1935). Studies of the morphological activity of rivers as illustrated by the River Fyris, *Bulletin of the Geological Institute University of Uppsala*, **25**, 221–527.

Hlavac, V. (1991). Finding of dead otters (*Lutra lutra*) and preliminary results of analyses of dead animals, *Vydra*, **2**, 7–13.

Hoey, T.B. and Sutherland, A.J. (1991). Channel morphology and bedload pulses in braided rivers: a laboratory study, *Earth Surface Processes and Landforms*, **16**, 447–462.

Holcik, J. and Hruska, V. (1966). On the spawning substrate of the roach *Rutilus rutilus* (Linnaeus, 1758) and bream *Abramis brama* (Linnaeus, 1758) and notes on the ecological characteristics of some European fishes, *Vestnik Ceskoslovenske Spolecnosti Zoologicke*, **30**, 22–29.

Holden, A.N.G., Fowler, S.V. and Schroeder, D. (1992). Invasive weeds of amenity land in the UK: Biological control — the neglected alternative, *Aspects of Applied Biology*, **29**, 325–332.

Holdgate, M.W. (1979). *A Perspective of Environmental Pollution*, Cambridge University Press, Cambridge.

Holland, D.G. (1976). The distribution of the freshwater Malacostraca in the area of the Mersey and Weaver River Authority. *Freshwater Biology*, **6**, 265–276.

Holland, S.M. and Ditton, R.B. (1992). Fishing trip satisfaction: a typology of anglers, *North American Journal of Fisheries Management*, **12**, 28–33.

Holmes, N.T.H. (1983). *Typing British Rivers According to their Flora*, Focus on Nature Conservation No. 4, Nature Conservancy Council, Peterborough.

Holmes, N.T.H. (1989). British rivers — a working classification, *British Wildlife*, **1**, 20–36.

Holmes, N.T.H. and Rowell, T.A. (1993). Typing British rivers according to their flora: update, unpublished report to Scottish Natural Heritage, Edinburgh.

Holmes, N.T.H. and Whitton, B.A. (1981). Phytobenthos of the River Tees and its tributaries, *Freshwater Biology*, **11**, 139–168.

Holomuzki, J.R. and Short, T.M. (1988) Habitat use and fish avoidance behaviors by the stream-dwelling isopod *Liricus fontinalis*, *Oikos*, **52**, 79–86.

Home Office (1990). *Guidance on the Operation of the Animals (Scientific Procedures) Act 1986*, HMSO, London.

Hooke, J.M. (1980). Magnitude and distribution of rates of river bank erosion, *Earth Surface Processes and Landforms*, **5**, 143–157.

Hooke, J.M. (1989). River channel change in England and Wales, *Journal of the Institute of Water and Environmental Management*, **3**, 328–335.

Hornung, M., Brown, S.J. and Ranson, A.(1990). Amelioration of surface water acidity by catchment management, in *Acid Waters in Wales* (Eds R.W. Edwards, A.S. Gee and J.H. Stoner), pp. 311–328, Kluwer Academic, Holland.

Houlihan, D.F. (1969). The structure and behaviour of *Notiphila riparia* and *Erioptera squalida*, two root-piercing insects, *Journal of Zoology, London*, **159**, 249–267.

House, M.A., Ellis, J.B. and Shutes, R.B.E. (1991). *Urban Rivers: Ecological Impacts and Management. Urban Waterside Regeneration*, University of Manchester.

Howard, A.D. (1992). Modelling channel migration and floodplain sedimentation in meandering streams, in *Lowland Floodplain Rivers: Geomorphological Perspectives* (Eds P.A. Carling and G.E. Petts), pp. 1–41, Wiley, Chichester.

Howarth,W. (1987). *Freshwater Fishery Law*, Financial Training Publications, London.

Howells, G. and Dalziel, T.R.K. (1991). *Restoring Acid Waters: Loch Fleet 1984–1990*, Elsevier Applied Science, London and New York.

Hubbs, C.L., Greeley, J.R. and Tarzwell, C.M. (1932). *Methods for the improvement of Michigan trout streams*, Michigan Dept. of Conservation Institute of Fishery Research Bulletin No.1.

Hudgins, M.D. (1984). Structure of the angling experience, *Transactions of the American Fisheries Society*, **113**, 350–359.

Hughes, B.D. (1984). The influence of factors other than pollution on the value of Shannons diversity index for benthic macroinvertebrates in streams, *Water Research*, **12**, 359–364.

Hume, R. (1976). Reactions of goldeneye to boating, *British Birds*, **69**, 178–179.

Humpesch, U.H. (1985). Inter- and intra-specific variation in hatching success and embryonic development of five species of salmonids and *Thymallus thymallus*, *Archiv für Hydrobiologie*, **104**, 129–144.

Hunt, P.C. and Jones, J.W. (1974). A population study of *Barbus barbus* (L.) in the River Severn, England. I. Densities, *Journal of Fish Biology*, **6**, 255–267.

Hunt, R.L. (1988). *A compendium of 45 Trout Stream Habitat Development Evaluations in Wisconsin during 1953–1985*. Technical Bulletin No. 162, Department of Natural Resources, Madison, Wisconsin, USA.

Hupp, C.R. (1986). The headward extent of fluvial landforms and associated vegetation on Massanutten Mountain, Virginia, *Earth Surface Processes and Landforms*, **11**, 545–556.

Hupp, C.R. and Bazemore, D.E. (1993). Temporal and spatial patterns of wetland sedimentation, West Tennessee, *Journal of Hydrology*, **141**, 179–196.

Hupp, C.R. and Osterkamp, W.R. (1985). Bottomland vegetation distribution along Passage Creek, Virginia in relation to fluvial landforms, *Ecology*, **66**, 670–681.

Huryn, A.D. and Wallace, J.B. (1988). Community structure of Trichoptera in a mountain stream: spatial patterns of production and functional organization, *Freshwater Biology*, **20**, 141–155.

Husmann, S. and Teschner, D. (1970). Ecology, morphology and history of distribution of subterranean water-mites from Sweden, *Archiv für Hydrobiologie*, **67** (2), 242–267.

Hussein Ayoub, S.M. and Yankov, L.K. (1985). Algicidal properties of tannins, *Fitoterapia*, **LVI**, 227–228.

Hutchinson, G.E. (1957). Concluding remarks, *Cold Spring Harbor Symposia in Quantitative Biology*, **22**, 415–427.

Hutchinson, G.E. (1978). *An Introduction to Population Biology*, Yale University Press.

Hynes, H.B.N. (1960). *The Biology of Polluted Waters*, Liverpool University Press, Liverpool.

Hynes, H.B.N. (1970a). The ecology of flowing waters in relation to management, *Journal of Water Pollution*, **42**, 418–424.

Hynes, H.B.N. (1970b). *The Ecology of Running Waters*, Liverpool University Press, Liverpool.

Hynes, H.B.N. (1975). Edgardo Baldi Memorial Lecture. The stream and its valley, *Verhandlungen der Internationalen Vereinigung für theoretische und angewandte Limnologie*, **19**, 1–15.

Ibañez, C. (1993). *Dinàmica hidrològica i funcionament ecològie del tram estuarí del rio Ebre*, PhD thesis, University of Barcelona.

Ibañez, C. and Prat, N. (1993). *Consequences of salt wedge dynamics on nutrient transport in the Ebro river estuary (Catalonia, Spain)*, Actes du Colloque Scientifique International des 4èmes Rencontres de l'Agence Regionales Pour l'Environment, Provence-Alps-Côte d'Azur, France.

Ibañez, C., Escosa, R., Muñoz, I. et al. (1991). Life cycle and production of *Ephoron virgo* (Ephemeroptera: Polymitarcidae) in the lower river Ebro, in *Overview and Strategies in Plecoptera and Ephemeroptera*, pp. 83—92, The Sandhill Crane Press, Florida.

Illies, J. and Botosaneanu, L. (1963). Problèmes et méthodes de la classification et de la zonation écologique des eaux courantes, considerées surtout du point de vue faunistique, *Mitteilungen der Internationalen Vereinigung für theoretische und angewandte Limnologie*, **12**, 1—57.

Inland Waterways Amenity Advisory Council (1983). *Waterways Ecology and the Design of Recreational Craft*, IWAAC, London.

Inland Waterways Association (1993). *Restoration Committee Table of Restorations completed, in progress and proposed*, 21 May 1993, IWA, London.

International Commission for the Protection of the Rhine against Pollution (1987). *Action Programme Rhine*, Strasbourg.

International Commission for the Protection of the Rhine against Pollution (1991). *Ecological Master Plan for the Rhine*, Salmon 2000, Koblenz.

International Union for Conservation of Nature and Natural Resources (1980). *World Conservation Strategy*, IUCN, Gland, Switzerland.

Internationale Kommission zum Schutze des Rheins (1989). *Aktionsprogramm Rhein (ICPR), Bericht des Prä sidenten der Internationalen Kommission zum Schutze des Rheins gegen Verunreinigung an die 10*. Ministerkonferenz, Brussel.

Internationale Kommission zum Schutze des Rheins (1989). *Aktionsprogramm Rhein, Synthesebericht über die z.Z. laufenden und bereits geplanten Manahmen zur Verbesserung des ekosystems Rhein inkl. seiner Nebengewässer*, Brüssel.

International Kommission zum Schutze des Rheins (1993). *Statusbericht Rhein. Chemisch-physikalische und biologische Untersuchungen bis 1991, Vergleich Istzustand 1990, Zielvorgaben*, Koblenz.

IRSA (1990). *Po Basin Water Resources Plan: Criteria And Guidelines*, Quad. Ist. Ric. Acque, 86.

Isaksson, A. and Bergman, P.K. (1978). An evaluation of two tagging methods and survival rates of different age and treatment groups of hatchery reared Atlantic salmon smolts. *Journal of Agricultural Research in Iceland*, **10**, 74—99.

Iseya, F. and Ikeda, H. (1987). Pulsations in bedload transport rates induced by a longitudinal sediment sorting: a flume study using sand and gravel mixtures, *Geografiska Annaler*, **69A**, 15—27.

IUCN/UNEP/WWF (1991). *Caring for the Earth. A Strategy for Sustainable Living*, IUCN, Gland, Switzerland.

Iversen, T.B., Thorup, J., Kjeldsen, K. and Thyssen, N. (1991). Spring bloom development of microbenthic algae and associated invertebrates in two reaches of a small lowland stream with contrasting sediment stability, *Freshwater Biology*, **26**, 189—198.

Iversen, T.M., Thorup, J., Hansen, T., Lodal, J. and Olsen, J. (1985). Quantitative estimates and community structure of invertebrates in a macrophyte rich stream, *Archiv fauur Hydrobiologie*, **102**, 291—301.

Iwamoto, R.N., Salo, E.O., Madej, M.A. and McComas, R.L. (1978). *Sediment and water quality: a review of the literature, including a suggested approach for water quality criteria*, US Environment Protection Agency, Report 910/9-78-048.

Jackivicz, T.P. and Kuzminski, L.N. (1973). The effects of the interaction of outboard motors with the aquatic environment — a review, *Environmental Research*, **6**, 436—454.

Jackson, D.J. (1958). Egg-laying and egg-hatching in *Agabus bipustulatus* L., with notes on oviposition in other species of *Agabus* (Coleoptera: Dytiscidae), *Transactions of the Royal Entomological Society of London*, **110** (3), 53—80.

Jackson, D.J. (1960). Observations on egg-laying in *Ilybius fuliginosus* Fabricius and *I. ater* DeGeer (Coleoptera: Dytiscidae), with an account of the female genitalia, *Transactions of the Royal Entomological Society of London*, **112** (3), 37—52.

Jackson, W., Hillaby, J., Pendlebury, D., Reeves, R. and Shelton, P. (1979). *A survey of aquatic habitats in Warwickshire 1977—78*. Warwickshire Nature Conservation Trust, Warwick, UK.

Jackson, W.L., Shelby, B., Martinez, A. and Van Haveren, B.P. (1989). An interdisciplinary value-based process for protecting instream flows, *Journal of Soil and Water Conservation*, **44**, 121—126.

Jain, S.C. (1990). Armour or pavement, *Journal of Hydraulic Engineering*, **119**, 436—440.

Jefferies, D.J. (1987). The effects of angling interests on otters, with particular reference to disturbance, *Institute of Terrestrial Ecology Symposium*, **19**, 23–30.

Jefferies, D.J., Morris, P.A. and Mulleneux, J.E. (1989). An enquiry into the changing status of the water vole *Arvicola terrestris* in Britain, *Mammal Review*, **19**, 111–131.

Jefferies, D.J., Wayre, P., Jessop, R.M. and Mitchell-Jones, A.M. (1986). Reinforcing the native otter *Lutra lutra* population in East Anglia: an analysis of behaviour and range development of the first group, *Mammal Review*, **16**, 65–79.

Jeffray, D.J. (Ed.) (1980). *The River Avon in mid-Warwickshire — an ecological survey*, Warwickshire Nature Conservation Trust, Warwick, UK.

Jenkins, D. and Burrows, G.O. (1980). Ecology of otters in northern Scotland. III. The use of faeces as indicators of otter (*Lutra lutra*) density and distribution, *Journal of Animal Ecology*, **49**, 755–774.

Jenkins, R.A. (1975). Occurrence of *Baëtis atrebatinus* (Etn.) (Ephemeroptera) in a river in south west Wales, *Entomologist's Monthly Magazine*, **110**, 21.

Jenkins, R.A. (1977). Notes on the distribution of psychomyiid larvae (Trichoptera) in South-West Wales, *Entomologist's Record and Journal of Variation*, **89**, 57–61.

Jenkins, R.A. and Cooke, S. (1978). Further notes on *Oecetis notata* (Rambur) (Trichoptera: Leptoceridae) in south west Wales, *Entomologist's Record and Journal of Variation*, **90**, 65–66.

Jenkins, R.A., Wade, K.R. and Pugh, E. (1984). Macroinvertebrate-habitat relationships in the River Teifi catchment and the significance to conservation, *Freshwater Biology*, **14**, 23–42.

Jennings, V.M. (1980). *Japanese Polygonum weed control*, Iowa State University, October 1980, Pm-762.

Jensen, A.J. (1990). Growth of young migratory brown trout *Salmo trutta* correlated with water temperature in Norwegian rivers, *Journal of Animal Ecology*, **59**, 603–614

Jensen, A.J. Johnsen, B.O. and Heggberget, T.G. (1991). Initial feeding time of Atlantic salmon, *Salmo salar*, alevins compared to river flow and water temperature in Norwegian streams, *Environmental Biology of Fishes*, **30**, 379–385.

Jensen, A.J., Johnsen, B.O. and Saksgard, L. (1989). Temperature requirements in Atlantic salmon (*Salmo salar*) and brown trout (*Salmo trutta*) and Arctic char (*Salvelinus alpinus*) from hatching to initial feeding compared with geographic distribution. *Canadian Journal of Fisheries and Aquatic Sciences*, **46**, 786–789.

Jensen, S, Kihlström, J.E., Olsson, M., Lunberg, C. and Orberg, J. (1977). Effects of PCB and DDT on mink (*Mustela vison*) during the reproductive season, *Ambio*, **6**, 239.

Jeppesen, E., Iversen, T.M., Sand-Jensen, K. and Jørgensen, C.P. (1984). økologiske konsekvenser af reduceret vandføring i Susåen. Bind 2: Biologiske processer og vandkvalitets-forhold. Miljøstyrelsen, Copenhagen.

Jimenez, J. and Lacomba, J. (1991). The influence of water demands on otter (*Lutra lutra*) distribution in Mediterranean Spain, in *Proceedings of the Vth International Otter Colloquium, Habitat*, 6, (Eds C. Reuther and R. Röchert), pp. 249–259, Otter Zentrum, Hankensbüttel.

Jobling, M. (1983). A short review and critique of methodology used in fish growth and nutrition studies, *Journal of Fish Biology*, **23**, 685–703.

Johns, A.D. (1992). Conservation in managed tropical forests, in *Tropical Deforestation and Species Extinction* (Eds T.C. Whitmore and J.A. Sayer), pp. 15–63, Chapman and Hall, London.

Johnston, C.A. and Naiman, R.J. (1987). Boundary dynamics at the aquatic–terrestrial interface: the influence of beaver and geomorphology, *Landscape Ecology*, **1**, 47–57.

Jones, J.G., Horne, J.E.M., Moorhouse, P. and Powell, D.L. (1980). *Petroleum Hydrocarbons in Fresh Waters — A Preliminary Desk Study and Bibliography*. Occasional Publications of the Freshwater Biological Assocation, No. 9, FBA, Ambleside, Cumbria, UK.

Jones, J.R.E. (1943). The fauna of the River Teifi, West Wales, *Journal of Animal Ecology*, **12**, 115–123.

Jones, J.R.E. (1950). A further ecological study of the River Rheidol: the food of the common insects of the mainstream, *Journal of Animal Ecology*, **19**, 159–174.

Jones, J.R.E. (1951). An ecological study of the River Towy, *Journal of Animal Ecology*, **20**, 68–86.

Jones, J.W. and King, G.M. (1950). Further observations on the spawning behaviour of the Atlantic salmon (*Salmo salar*), *Proceedings of the Zoological Society of London*, **120**, 317–323.

Jonsson, B. (1977). Demographic strategy in a brown trout population in Western Norway, *Zool. Scr.*, **6**, 255–263.

Jonsson, B. and Sandlund, O.T. (1979). Environmental factors and life histories of isolated river stocks of brown trout (*Salmo trutta m. fario*) in Søre Osa river system, *Environmental Biology of Fishes*, **4**, 43–54.

Jorga, W. and Weise, G. (1981). Aquatic plants and their importance for embankment stabilization and improving water quality, *Acta Hydrochimica et Hydrobiologia*, **9**, 37–56.

Juggins, S., Watson, D., Waters, D., Patrick, S.T. and Jenkins, A. (Eds) (1989). The United Kingdom Acid Waters Monitoring Network; introduction and data report for 1988–1989 (year 1), ENSIS, London.

Jungwirth, M. and Winkler, H., (1984). The temperature dependence of embryonic development of grayling (*Thymallus thymallus*), Danube salmon (*Hucho hucho*), arctic char (*Salvelinus alpinus*) and brown trout (*Salmo trutta fario*), *Aquaculture*, **38**, 315–327.

Jutila, E. (1992). Restoration of salmonid rivers in Finland, in: *River Conservation and Management* (Eds P.J. Boon, P. Calow and G.E. Petts), pp. 353–363, John Wiley, Chichester.

Kairesalo, T. and Koskimies, I. (1987). Grazing by oligochaetes and snails on epiphytes, *Freshwater Biology*, **17**, 317–324.

Kajak, Z. (1992). The River Vistula and its floodplain valley (Poland): its ecology and importance for conservation, in *River Conservation and Management* (Eds P.J. Boon, P. Calow and G.E. Petts), pp. 35–49, John Wiley, Chichester.

Kallenerg, H. (1950). Observations in a stream tank of territoriality and competition in juvenile salmon and trout (*Salmo salar* L. and *Salmo trutta* L.), *Reports of the Institute of Freshwater Research, Drottningholm*, **33**, 55–98.

Kamler, E. (1992). *Early Life History of Fish. An Energetics Approach*, Chapman and Hall. London.

Karaki, S. and van Hofen, J. (1974). *Resuspension of bed material and wave effects on the Illinois and Upper Mississippi Rivers caused by boat traffic*, Contract Report No. LMSSD 75-881, for US Army Engineers District, St. Louis. Engineering Research Center, Colorado State University, Fort Collins, CO, USA.

Karr, J.R. (1981). Assessment of biotic integrity using fish communities, *Fisheries*, **6**, 21–26.

Kaushik, N.K. and Hynes, H.B.N. (1968). Experimental study on the role of autumn-shed leaves in aquatic environments, *Journal of Ecology*, **56**, 229–242.

Kaushik, N.K. and Hynes, H.B.N. (1971) The fate of the dead leaves that fall into streams, *Archiv für Hydrobiologie*, **68**, 465–515.

Kees, H. and Krumrey, G. (1983). *Heracleum mantegazzianum* — Zier — Staude, Unkraut und "Giftpflanze", Gesunde Pflanzen, **4**.

Keilin, D. (1944). Respiratory systems and respiratory adaptations in larvae and pupae of Diptera, *Parasitology*, **36**, 1–66.

Keller, E.A. (1971). Areal sorting of bed load material: the hypothesis of velocity reversal, *Bulletin of the Geological Society of America*, **83**, 1531–1536.

Keller, E.A. and Swanson, F.J. (1979). Effects of large organic material on channel form and fluvial processes, *Earth Surface Processes and Landforms*, **4**, 361–380.

Keller, E.A. and Tally, T. (1979). Effects of large organic debris on channel form and fluvial processes in the coastal Redwood environment, in *Adjustments of the Fluvial System* (Ed. D.D. Rhodes and G.P. Williams), pp. 169–197, Kendall Hunt, Debuque, IA.

Kellert, S.R. (1984). Assessing wildlife and environmental values in cost-benefit analysis, *Journal of Environmental Management*, **18**, 355–363.

Kelly, M.G. and Whitton, B.A. (1987). Growth rate of the aquatic moss *Rhyncostegium riparoides* in northern England, *Freshwater Biology*, **18**, 461–468.

Kelly, M.G. and Whitton, B.A. (1989). Interspecific differences in Zn, Cd and Pb accumulation by freshwater algae and bryophytes, *Hydrobiologia*, **175**, 1–11.

Kemp, K.K. (1993). Environmental modelling and GIS: dealing with spatial continuity, in *Applications of Geographic Information Systems in Hydrology and Water Resources Management* (Eds K. Kovar and H.P. Nachtnebel), pp. 107–115, International Association of Hydrological Sciences Publication 211.

Kempf, T., Ludemann, D. and Pflaum, W. (1967). Verschmutzung der Gewasser durch motorischen Betrieb, insebesondere durch Aussenbordmotoren, *Schr. Reiche Ver. Wasser.Boden. u. Lufthyg.*, **26**, 3–47.

Kennedy, C.R. (1965). The distribution and habitat of *Limnodrilus* Clarapede (Oligochaeta:

Tubificidae), *Oikos*, **16**, 26—38.

Kennedy, G.J.A. (1982). Factors affecting the survival and distribution of salmon (*Salmo salar* L.) stocked in upland trout (*Salmo trutta* L.) streams in Northern Ireland, Symposium on Stock Enhancement in the Management of Freshwater Fish, Budapest, 31 May—2 June 1982, EIFAC Technical Paper 42 (Suppl.), Vol. 1, pp. 227—242.

Kennedy, G.J.A. (1984). Evaluation of techniques for classifying habitats for juvenile Atlantic salmon (*Salmo salar* L.), Atlantic Salmon Trust Workshop on Stock Enhancement, pp. 1—23.

Kennedy, G.J.A. (1988). Stock enhancement of Atlantic salmon, in *Atlantic Salmon: Planning for the Future* (Eds D. Mills and D. Piggins), pp. 345—372, Proceedings of the 3rd International Atlantic Salmon Symposium, Biarritz, 1986, Croom Helm, London.

Kennedy, G.J.A. and Crozier, W.W. (1991). Strategies for the rehabilitation of salmon rivers. Post-project appraisal, in *Proceedings of a Joint Conference Atlantic Salmon Trust/Institute of Fisheries Management/Linnean Society*, pp. 46—62, The Linnean Society, London,

Kennedy, G.J.A. and Crozier, W.W. (1993). Juvenile atlantic salmon (*Salmo salar*) — production and prediction, in *Proceedings of the International Symposium on Production of Juvenile Atlantic Salmon in Natural Waters* (Eds R.J. Gibson and R.E. Cutting), Canadian Special Publications of Fisheries and Aquatic Sciences, **118**, 179—187.

Kennedy, G.J.A. and Greer, J.E. (1988). Predation by cormorants, *Phalacrocorax carbo* (L.), on the salmonid populations of an Irish river, *Aquaculture and Fisheries Management*, **19**, 159—170.

Kennedy, G.J.A. and Johnston, P.M. (1986). A review of salmon (*Salmo salar* L.) research on the River Bush, in *Proceedings of Institute of Fisheries Management* (NI Branch) 17th Annual Study Course, University of Ulster, 9—11th September 1986, pp. 49—69.

Kennedy, G.J.A. and Strange, C.D. (1980). Population changes after two years of salmon (*Salmo salar* L.) stocking in upland trout (*Salmo trutta* L.) streams, *Journal of Fish Biology*, **17**, 577—586.

Kennedy, G.J.A. and Strange, C.D. (1982). The distribution of salmonids in upland streams in relation to depth and gradient, *Journal of Fish Biology*, **20**, 579—591.

Kennedy, G.J.A. and Strange, C.D. (1986a). The effects of intra and inter-specific competition on the survival and growth of stocked juvenile Atlantic salmon (*Salmo salar* L.) and resident trout (*Salmo trutta* L.) in an upland stream, *Journal of Fish Biology*, **28**, 479—489.

Kennedy, G.J.A., and Strange, C.D. (1986b). The effects of intra- and inter-specific competition on the distribution of stocked juvenile Atlantic Salmon, *Salmo salar* L., in relation to depth and gradient in an upland trout, (*Salmo trutta* L.) stream, *Journal of Fish Biology*, **29**, 199—214.

Kennedy, M. and Fitzmaurice, P. (1968). The biology of the bream, *Abramis brama* L. in Irish waters, *Proceedings of the Royal Irish Academy, B*, **67**, 95—157.

Kern, K. (1992). Rehabilitation of streams in South-west Germany, in *River Conservation and Management* (Eds Boon, P.J., Calow, P. and Petts, G.E.), pp. 321—335, John Wiley, Chichester.

Kern-Hansen, U. and Dawson, F.H. (1981). The standing crop of aquatic plants of lowland streams in Denmark and the inter-relationships of nutrients in plant, sediments and water, in *Proceedings of the European Weed Research Society 5th Symposium on Aquatic Weeds 1978*, pp. 143 150.

Keymer, I.F., Wells, G.A.H., Mason, C.F. and Macdonald, S.M. (1988). Pathological changes and organochlorine residues of wild otters (*Lutra lutra*), *Veterinary Record*, **122**, 153—155.

Kidson, C. (1953). The Exmoor storm and Lynmouth floods, *Geography*, **38**, 1—9.

Kinhill Engineers (1988). *Techniques for determining environmental water requirements — a review*, Technical Report Series, Report No. 40, report by Kinhill Engineers Pty. Ltd. to the Department of Water Resources, Victoria.

Kirk, K.L. (1985). Effects of suspensoids (turbidity) on penetration of solar radiation in aquatic ecosystems, *Hydrobiologia*, **125**, 195—208.

Kirk, K.L. (1991). Suspended clay reduces *Daphnia* feeding rate: Behavioural mechanisms, *Freshwater Biology*, **25**, 357—365.

Kivaisi, A.K., Op den Camp, H.J.M., Lubberding, H.J., Boon, J.J. and Vogels, G.D. (1990). Generation of soluble lignin-derived compounds during degradation of barley straw in an artificial rumen reactor, *Applied Microbiology and Biotechnology*, **33**, 93—98.

Knighton, D. (1984). *Fluvial Forms and Processes*, Edward Arnold, London.

Knöpp, H. (1957). Die heutige biologische Gliederung des Rheinstroms, *Deutsche Gewässerkundl. Mitt.*, **1** (3), 56—63.

Koeman J.H., Van Velzen-Blad, H.C.W., de Vries, R. and Vos, J.G. (1973). Effects of PCB and

DDE in cormorants and evaluation of PCB residues from an experimental study, *Journal of Reproduction and Fertility Supplement*, **19**, 353–364.

Kolkwitz, R. and Marsson, M (1908). Okologie der pflanzlichen Saprobien, *Ber. Dt. Botan. Ges*, **261**, 505–519.

Kornas, J. (1990). Plant invasions in Central Europe: historical and ecological aspects, in *Biological Invasions in Europe and the Mediterranean Basin*, (Eds F. di Castri, A.J. Hansen and M. Debussche), Monographiae Biologicae **65**, pp. 19–36. Kluwer, Dordrecht.

Kostadinov, S., Stanojevic, G. and Topalovic, M. (1992). Gubici organske materije i hranljivih elementata usled vodne erozije, Glasnik Sumarskog Faculteta, 74, Beograd, 645–654.

Kouki, J. (1991). The effect of the water-lily beetle, *Galerucella nymphaeae*, on leaf production and leaf longevity of the yellow water-lily, *Nuphar lutea, Freshwater Biology*, **26**, 347–353.

Kozel, S.J. and Hubert, W.A. (1989a). Factors influencing the abundance of brook trout (*Salvelinus fontinalis*) in forested mountains streams, *Journal of Freshwater Ecology*, **5**, 113–122.

Kozel, S.J. and Hubert, W.A. (1989b). Testing of habitat assessment models for small trout streams in the Medicine Bow National Forest, Wyoming. *North American Journal of Fisheries Management*, **9**, 458–464.

Krantzberg, G. (1985). The influence of bioturbation on physical, chemical and biological parameters in aquatic environments: a review, *Environmental Pollution (Series A)*, **39**, 99–122.

Krebs, C.J. (1972). *Ecology*, Harper and Row, New York.

Krecker, F.H. (1939). A comparative study of the animal population of certain submerged aquatic plants, *Ecology*, **20**, 553–562.

Kreshner, J.L., Forsgren, H.L. and Meehan, W.R. (1991) Managing salmonid habitats, in *Influences of Forest and Rangeland Management on Salmonid Fishes and their Habitats* (Ed. W.R. Meehan), pp. 599–606, American Fisheries Society Special Publication No. 19, Bethesda.

Kruuk, H. and Conroy, J.W.H. (1991). Mortality of otters (*Lutra lutra*) in Shetland, *Journal of Applied Ecology*, **28**, 83–94.

Kubecka, J., Duncan, A. and Butterworth A. (1992). Echo counting or echo integration for fish biomass assessment in shallow waters, in *Underwater Acoustics* (Ed. M. Weydert), pp. 129–132, Elsevier Applied Science, London.

Kuhnle, R.A. (1992). Bedload transport during rising and falling stages on two small streams, *Earth Surface Processes and Landforms*, **17**, 191–197.

Kumm, K.I. (1991). The effects of Swedish price support, fertiliser and pesticide policies on the environment, in *Towards Sustainable Agricultural Development*, (Ed. Michael D. Young), Belhaven Press for OECD.

Kuzminski, L.N. and Mulcahy, C.T. (1974). *A study on the fate of lead emitted from two-cycle outboard motor subsurface exhausts*. Report No. ENV-E-38-74-1, Environmental Engineering, Department of Civil Engineering, University of MA.

L'Abée-Lund, J.H. (1990). Variation within and between rivers in adult size and sea age at maturity of anadromous brown trout, *Salmo trutta, Canadian Journal of Fisheries and Aquatic Sciences*, **48**, 1015–1021.

L'Abée-Lund, J.H., Jonsson, B., Jensen, A.J., Sættem, L.M., Heggberget, T.G., Johnsen, B.O. and Næsje, T.F. (1989). Latitudinal variation in life-history characteristics of sea-run migrant brown trout *Salmo trutta, Journal of Animal Ecology*, **58**, 525–542.

Laane, R.W.P.M. and Ten Brink, B.J.E. (1990) Data-rich, information-poor. The modern monitoring syndrome?, *Land Water Int.*, **68**, 12–16.

Ladle, M. and Casey, H. (1971). Growth and nutrient relationships of *Ranunculus penicillatus* var *calcareus* in a small chalk stream, in *Proceedings of the European Weed Research Council 3rd International Symposium on Aquatic Weeds*, 1971, 53–64.

Lagler, K.F., Hazzard, A.S., Hazen, W.E. and Tompkins W.A. (1950). Outboard motors in relation to fish behaviour, fish production and angling success, *Transactions of the 15th North American Wildlife Conference*, 280–303.

Lancaster, J. and Hildrew, A.G. (1993). Characterizing in-stream flow refugia. *Canadian Journal of Fisheries and Aquatic Sciences*, **50**, 1663–1675.

Lane, E.W. (1955). The importance of fluvial morphology in hydraulic engineering, *Proceedings of American Society of Civil Engineers*, **81**, 1–17.

Langston, W.J., Burt, G.R. and Zhon, M.J. (1987). Tin and organotin in water, sediments and benthic organisms of Poole Harbour, *Marine Pollution Bulletin*, **18**, 634−659.

Lanka, R.P., Hubert, W.A. and Wesche, T.A. (1987). Relations of geomorphology to stream habitat and trout standing stock in small Rocky Mountains streams, *Transactions of the American Fisheries Society*, **116**, 21−28.

Lapointe, M. (1992). Burst-like sediment suspension events in a sand bed river, *Earth Surface Processes and Landforms*, **17**, 253−270.

Large, A.R. and Petts, G.E. (1993). *Restoration of Floodplains in Semi-arid Situations*.

Large, A.R.G. and Petts, G.E. (1994). Rehabilitation of river margins, in *The Rivers Handbook*, Vol. II (Eds P. Calow and G.E. Petts) pp. 401−418, Blackwell Scientific, Oxford.

Larkin, P.A. (1981). A perspective on population genetics and salmon management, *Canadian Journal of Fisheries and Aquatic Sciences*, **38**, 1469−1475.

Larsen, P. (1994). Restoration of river corridors: German experiences, in *The Rivers Handbook*, Vol. II (Eds P. Calow and G.E. Petts) pp. 419−438, Blackwell Scientific, Oxford. `

Last, F.T. and Watling, R. (Eds.) (1991). *Acid Deposition: Its Nature and Impacts*, The Royal Society of Edinburgh.

Laughlin, R.B. and Linden, O. (1987). Tributyltin−contemporary environmental issues, *Ambio*, **26**, 252−256.

Lauterborn, R. (1917). Die geographische und biologische Gliederung des Rheinstroms II, *S. Ber. Heidelb. Akad. Wiss. Math.-natw.-Kl.*, **B. 7** (5), 1−70, Heidelberg.

Lauterborn, R. (1918). Die geographische und biologische Gliederung des Rheinstroms III, *S. Ber. Heidelb. Akad. Wiss. Math.-natw.-Kl.* **B 9** (1), 187, Heidelberg.

Layher, W.G. and Maughan, O.E. (1984). Comparison of efficiencies of three sampling techniques for estimating fish populations in small streams, *Progressive Fisheries Culturist*, **46**, 180−184.

Lazauski, H.G. (1984). *An evaluation of pulsed D.C. electrofishing as a method of estimating fish population structure*, PhD Thesis, University of Auburn, AL.

Lear, D.W., Marks, J.W. and Schminke, C.S. (1966). Evaluation of coliform contribution by pleasure boats, CB-SRBP Technical Paper No. 10. Middle Atlantic Region, Federal Water Pollution Control Association, USA.

Learner, M.A., Densem, J.W. and Iles, T.C. (1983). A comparison of some classification methods used to determine benthic macroinvertebrate species associations in river survey work based on data obtained from the River Ely, South Wales, *freshwater Biology*, **13**, 12−36.

Learner, M.A., Lochhead, G. and Hughes, B.D. (1978). A review of the biology of the British Naididae (Oligochaeta) with emphasis on the lotic enviornment, *Freshwater Biology*, **8**, 357−375.

Le Cren, E.D. (1969). Estimates of fish populations and production in small streams in England, in *Symposium on Salmon and Trout in Streams* (Ed. R. Northcote), pp. 269−280, University of British Columbia, Vancouver.

Le Cren, E.D. (1972). A commentary on uses of a river: past and present, in *River Ecology and Man* (Eds R.T. Oglesby, C.A. Carlson and J.A. McCann), pp. 251−260, Academic Press, New York.

Le Cren, E.D. (1973). The population dynamics of young trout (*Salmo trutta*) in relation to density and territorial behaviour, *J. Conseil Int. pour L' explor. de la Mer*, **164**, 241−246.

Le Cren, E.D., Kipling, C. and McCormack, J. (1967). A study of the numbers, biomass and year class strengths of Perch (*Perca fluviatilis* L.) in Windermere from 1941−1966, *Journal of Animal Ecology*, **46**, 281−307.

Leisure Consultants (1989). Boating and Water Sports in Britain, Market Profiles and Prospects, Leisure Consultants, Sudbury, UK.

Leland, H.V., Carter, J.L. and Fend, S.V. (1986). Use of detrended correspondence analysis to evaluate factors controlling spatial distribution of benthic insects, *Hydrobiologia*, **132**, 113−123.

Leland, H.V., Fend, S.V., Dudley, T.L. and Carter, J.L. (1989). Effects of copper on species composition of benthic insects in a Sierra Nevada stream, *Freshwater Biology*, **21**, 163−79.

Lelek, A. (1966). The field experiment on the receptivity of chub *Leuciscus cephalus* (L.) to the repeated effects of pulsating direct current, *Verhandlungen der Internationalen Veringung für Theoretische und Argewandte Limnologie* **16**, 1217−1222.

Lelek, A. and Buhse, G. (1992). *Fische des Rheins — früher und heute*, Springer Verlag, Berlin, Heidelberg.

Leonard, P.M. and Orth, D.J. (1986). Application and testing of an index of biotic integrity in small, coolwater streams, *Transactions of the American Fisheries Society*, **115**, 401−414.

Leonard, P.M. and Orth, D.J. (1988). Use of habitat guilds of fishes to determine instream flow requirements, *North American Journal of Fisheries Management*, **8**, 399–409.

Leopold, L.B., Wolman, M.G. and Miller, J.P. (1964). *Fluvial Processes in Geomorphology*, Freeman, San Francisco.

Leslie, P.H. and Davis, D.H.S. (1939). An attempt to determine the absolute number of rats in a given area, *Journal of Animal Ecology*, **8**, 94–113.

Levin, S. (1992). The problem of pattern and scale in ecology, *Ecology*, **73**, 1943–1967.

Lewin, J. (1982). British floodplains, in *Papers in Earth Studies* (Eds B.H. Adlam, B.H. Fenn and L. Morris), pp. 21–37, Geobooks, Norwich.

Lewin, J. (1987). Historical river channel changes, in *Palaeohydrology in Practice*, (Eds K.J. Gregory and J.B. Thornes), pp. 161–175, Wiley, Chichester.

Lewin, J., Macklin, M.G. and Newson, M.D (1988). Regime theory and environmental change — irreconcilable concepts?, in *International Conference on River Regime* (Ed. W.R. White), pp. 431–445, Wiley, Chichester.

Lewis, G. and Williams, G. (1984). *Rivers and Wildlife Handbook: A Guide to Practices which Further the Conservation of Wildlife on Rivers*, Royal Society for the Protection of Birds/Royal Society for Nature Conservation, Bedford.

Lewis, V. (1990). Enhancement of brown trout stocks in the Cotswold rivers, *Fish (Journal of the Institute of Fisheries Management)*, **20**, 42–44.

Liddle, M.J. and Scorgie, H.R.A. (1980). The effects of recreation on freshwater plants and animals: a review, *Biological Conservation*, **17**, 183–206.

Lightfoot, G.W. and Jones, N.V. (1979). The relationship between the size of 0 group roach (*Rutilus rutilus* (L)), their swimming capabilities, and their distribution in a river, *Proceedings of the First British Freshwater Fisheries Conference*, 230–236.

Likens, G.E. (1984). Beyond the shoreline: A watershed-ecosystem approach. *Verhandlungen der Internationalen Vereinigung fur Theoretische und Angewandte Limnologie*, **22**, 1–22.

Likens, G.E., Bormann, F.H., Johnson, N.M., Fisher, D.W. and Pierce R.S. (1970). Effects of forest cutting and herbicide reatment on nutrient budgets in the Hubbard Brook watershed-ecosystem, *Ecological Monographs*, **40**, 23–47.

Lillehammer, A. and Saltveit, S.J. (Eds) (1984). *Regulated Rivers*, Universitetsforlaget As, Oslo.

Linfield R.S.J. (1985). The effect of habitat modification on freshwater fisheries in lowland areas of Eastern England, in *Habitat Modification and Freshwater Fisheries* (Ed. J.S. Alabaster), pp. 147–156, Butterworths, London.

Liou, Y.C. and Herbich, J.B. (1976). Sediment movement induced by ships in restricted waterways, SEA Grant Publication No. TAMU-SG-76-209. Report No. COE 188, College Station Engineering Program, Texas A and M University, TX.

Lisle, T.E. (1982). Effect of aggradation and degradation on riffle–pool morphology in natural gravel channels, northwestern California, *Water Resources Research*, **18**, 1643–1651.

Lisle, T.E. and Hilton, S. (1992). The volume of fine sediment in pools: an index of sediment supply in gravel-bed streams, *Water Resources Bulletin*, **28**, 371–383.

Lister, D.B. and Walker, C.E. (1966). The efect of flow control on freshwater survival of chum, coho and chinook salmon in Big Qualicum river, *Canadian Fish Culture*, **40**, 41–49.

Litvak, M.K. and Hansell, R.I.C. (1990). A community perspective on the multidimensional niche, *Journal of Animal Ecology*, **59**, 931–940.

Lloyd, R. and Swift D.J. (1976). Some physiological responses by freshwater fish to low dissolved oxygen, high carbon dioxide, ammonia and phenol with particular reference to water balance, in *Effects of Pollutants on Aquatic Organisms'* (Ed. A.P. Lockwood), pp. 47–69.

Lobón-Cerviá, L., Montañés, C. and Sostoa, A. (1986). Reproductive ecology and growth of a population of brown trout (*Salmo trutta* L.) in an aquifer-fed stream of Old Castile (Spain), *Hydrobiologia*, **135**, 81–94.

Locandro, R.R. (1973). *Reproduction ecology of Polygonum cuspidatum*, PhD Thesis Rutgers University, The State University of New Jersey.

Lodge, D.M. (1985). Macrophyte-gastropod associations: observations and experiments on macrophyte choice by gastropods, *Freshwater Biology*, **15**, 695–708.

Lodge, D.M. (1986). Selective grazing on periphyton: a determinant of freshwater gastropod microdistributions, *Freshwater Biology*, **16**, 831–841.

Logan, P. and Brooker, M.P. (1983). The macroinvertebrate faunas of riffles and pools, *Water Research*, **17** (3), 263–270.

Lott, D. (1992). *A survey report on the terrestrial beetles of riparian habitats along the river Soar near Loughborough, Leicestershire, March–October 1991*, Unpublished report to Severn-Trent region, National Rivers Authority, Solihull, Birmingham.

Luiten, J.P.A. and Groot, S. (1992). Modelling quantity and quality of surface waters in the Netherlands: Policy Analysis of Water Management for the Netherlands, *European Water Pollution Control*, (6), 23–33.

Lundström, H. (1984). Giant hogweed, *Heracleum mantegazzianum*, A threat to the Swedish Countryside, 25th Swedish Weed Conference, Uppsala, 1, 191–200.

Lundström, H. (1989). New experiences of the fight against the giant hogweed, *Heracleum mantegazzianum*, 30th Swedish Crop Protection Conference, Weeds and Weed Control, 2, 51–58.

Macan, T.T. (1961). A review of running water studies, *Verhandlungen Internationale Vereinigung für Theoretische und Angewandte Limnologie*, **14**, 587–602.

Macan, T.T. (1962). Ecology of aquatic insects, *Annual Review of Entomology*, 7, 261–268.

MacCrimmon, H.R. and Gots, B.L. (1979). World distribution of Atlantic salmon, *Salmo salar*, *Journal of the Fisheries Research Board of Canada*, 36, 422–457.

MacCrimmon, H.R. and Gots, B.L. (1986) Laboratory observations on emergence patterns of juvenile Atlantic salmon, *Salmo salar*, relative to sediment loadings of test substrate, *Canadian Journal of Zoology*, **64**, 1331–1336.

MacCrimmon, H.R. and Marshall, T.L. (1968). World distribution of brown trout, *Salmo trutta*, *Journal of the Fisheries Research Board of Canada*, **25**, 2527–2548.

Macdonald, S.M. (1991). The status of the otter in Europe, in *Proceedings of the Vth International Otter Colloquium, Habitat*, Vol. 6 (Ed. C. Reuther and R. Röchert), pp. 1–3, Otter Zentrum, Hankensbüttel.

Macdonald, S.M. and Mason, C.F. (1982a). Otters in Greece, *Oryx*, **16**, 240–244.

Macdonald, S.M. and Mason, C.F. (1982b). The otter in central Portugal, *Biological Conservation*, **22**, 207–215.

Macdonald, S.M. and Mason, C.F. (1983). Some factors influencing the distribution of otters (*Lutra lutra*), *Mammal Review*, **13**, 1–10.

Macdonald, S.M. and Mason, C.F. (1984). Otters in Morocco, *Oryx*, **18**, 157–159.

Macdonald, S.M. and Mason, C.F. (1985). Otters, their habitat and conservation in north-east Greece. *Biological Conservation*, **31**, 191–210.

Macdonald, S.M. and Mason, C.F. (1987). Seasonal marking in an otter population, *Acta Theriologica*, **32**, 449–462.

Macdonald, S.M. and Mason, C.F. (1988). Observations on an otter population in decline, *Acta Theriologica*, **33**, 415–434.

Macdonald, S.M. and Mason, C.F. (1994). *Status and Conservation Needs of the Otter (Lutra lutra) in the Western Palearctic*, Council of Europe, Nature and Environment Series, Strasbourg.

Macdonald, S.M., Mason, C.F. and de Smet, K. (1985). The otter in north-central Algeria, *Mammalia*, **49**, 215–219.

Mackay, D.W. (1987). Angling and wildlife conservation, *Institute of Terrestrial Ecology Symposium*, **19**, 72–75.

Mackay, R.J. and Kalff, J. (1973). Ecology of two related species of caddisfly larvae in the organic substrates of a woodland stream, *Ecology*, **54**, 499–511.

Mackay, R.J. and Kersey, K.E. (1985). A preliminary study of aquatic insect communities and leaf decomposition in acid streams near Dorset, Ontario, *Hydrobiologia*, **122**, 3–11.

Mackay, R.J. and Wiggins, G.B. (1979). Ecological diversity in Trichoptera, *Annual Review of Entomology*, **24**, 185–208.

MacLennan, D.N. and Simmonds, E.J. (1992). *Fisheries Acoustics*, Chapman and Hall. London.

Macumber, P. (1990). The salinity problem, in *The Murray*, (Eds N. Mackay and D. Eastburn), Murray Darling Basin Commission, Canberra, Australia.

Maddock, I. (1992). Instream habitat assessment, in *The River Glen: a Catchment Assessment*, (Eds G.E. Petts), National Rivers Authority, Anglian Region, Peterborough, Annex E.

Madsen, A.B. (1991). Otter (*Lutra lutra*) mortalities in fish traps and experiences with using stop-grids in Denmark, in *Proceedings of the Vth International Otter Colloquium, Habitat*, Vol. 6 (Ed. C.

Reuther and R. Röchert), pp. 237–241, Otter Zentrum, Hankensbüttel.

Mahon R. (1980). Accuracy of catch-effort methods of estimating fish density and biomass in streams, *Environmental Biology of Fishes*, **5**, 343–360.

Maisse, G. and Baglinière, J.L. (1990). The biology of brown trout, *Salmo trutta* L. in the River Scorff, Brittany: a synthesis of studies from 1973 to 1984, *Aquaculture and Fisheries Management*, **21**, 95–106.

Maisse, G. and Baglinière, J.L. (1991). Biologie de la truite commune (*Salmo trutta* L.) dans les rivières françaises, in *La Truite, Biologie et Écologie* (Eds. J.L. Baglinière and G. Maise), pp. 25–45, INRA, Paris.

Maitland, J.R.G. (1892). *On stocking rivers, streams, lakes, ponds and reservoirs with Salmonidae*, Howietoun Fishery, Stirling.

Maitland, P.S. (1971). A population of coloured Goldfish, *Carassius auratus*, in the Forth and Clyde Canal, *Glasgow Naturalist*, **18**, 565–568.

Maitland, P.S. (1974). The conservation of freshwater fishes in the British Isles, *Biological Conservation*, **6**, 7–14.

Maitland, P.S. (1989) *The Genetic Impact of Farmed Atlantic Salmon on Wild Populations*, Nature Conservancy Council, Edinburgh.

Maitland, P.S. and East, K. (1989). A increase in numbers of Ruffe, *Gymnocephalus cernua* (L.), in a Scottish loch from 1982 to 1987. *Aquaculture and Fisheries Management*, **20**, 227–228.

Maitland, P.S. and Lyle, A.A. (1992). Conservation of freshwater fish in the British Isles: proposals for management, *Aquatic Conservation: Marine and Freshwater Ecosystems*, **2**, 165–183.

Maitland, P.S. and Morris, K.H. (1978). A multi-purpose modular limnological sampler, *Hydrobiologia*, **59**, 187–195.

Maitland, P.S. and Price, C.E. (1969). *Urocleidus principalis* (Mizelle, 1936), a North American monogenetic trematode new to the British Isles, probably introduced with the Largemouth Bass *Micropterus salmoides* (Lacepede, 1802), *Journal of Fish Biology*, **1**, 17–18.

Maitland, P.S. and Turner, A.K. (1987). Angling and Wildlife conservation — are they incompatible?, *Institute of Terrestrial Ecology Symposium*, **19**, 76–81.

Maitland, P.S., East, K. and Morris, K.H. (1983). Ruffe *Gymnocephalus cernua* (L.), new to Scotland, in Loch Lomond, *Scottish Naturalist*, **1983**, 7–9.

Majerus, M.E.N. and Fowles, A.P. (1989). The rediscovery of the 5-spot ladybird (*Coccinella 5-punctata* L.) (Col., Coccinellidae) in Britain, *Entomologists Monthly Magazine*, **125**, 177–181.

Malanson, G.P. (1993). *Riparian Landscapes*, Cambridge University Press, Cambridge.

Malgorzata, N. (1990). Structure and dynamics of fish communities in temperate rivers in relation to the abiotic-biotic regulatory continuum concept, *Polskie Archiwum Hydrobiologii*, **37**, 151–176.

Malicky, H. (1990). Feeding tests with caddis larvae (Insecta: Trichoptera) and amphipods (Crustacea: Amphipoda) on *Platanus orientalis* (Platanaceae) and other leaf litter, *Hydrobiologia*, **206**, 163–173.

Malmqvist, B. and Otto, C. (1987). The influence of substrate stability on the composition of stream benthos: an experimental study, *Oikos*, **48**, 33–38.

Malmqvist, M. and Sjauostrauom, P. (1984) The microdistribution of some lotic insect predators in relation to their prey and abiotic factors, *Freshwater Biology*, **14**, 649–656.

Maltby, L. (1992). Heterotrophic microbes, in *The Rivers Handbook*, Vol. 1, (Eds P. Calow and G.E. Petts) pp. 165–194, Blackwell Scientific, Oxford.

Maltby, L. and Calow, P. (1989). The application of bioassays in the resolution of environmental problems: past, present and future, *Hydrobiologia*, **188–189**, 65–76.

Maltby, L., Naylor, C. and Calow, P. (1990). Field deployment of a Scope for Growth assay involving *Gammarus pulex*, a freshwater benthic invertebrate, *Ecotoxicology and Environmental Safety*, **19**, 292–300.

Malvestuto, S.P. (1983). Sampling the recreational fishery, in *Fisheries Techniques* (Ed. L.A. Neilsen and D.L. Johnson), pp. 397–420, American Fisheries Society, Bethesda.

Malvestuto, S.P. (1991). The customization of recreational fishery surveys for management purposes in the United States, in *Catch Effort Sampling Strategies* (Ed. I.G. Cowx), pp. 201–213, Fishing News Books, Blackwell Scientific, Oxford.

Mann, R.H.K. (1976). Observations on the age, growth, reproduction and food of the chub *Squalius cephalus* (L.) in the River Stour, Dorset, *Journal of Fish Biology*, **8**, 265–288.

Mann, R.H.K. (1979). Natural fluctuations in fish populations, in *Proceedings of the First British Freshwater Fisheries Conference*, pp. 146–150, The University of Liverpool, Liverpool.

Mann, R.H.K. (1980). The numbers and production of pike (*Esox lucius*) in two Dorset rivers, *Journal of Animal Ecology*, **49**, 899–915.

Mann, R.H.K. (1991). Growth and production, in *Cyprinid Fishes: Systematics, Biology and Exploitation* (Eds I.J. Winfield and J.S. Nelson), pp. 456–482, Chapman and Hall, London and New York.

Mann, R.H.K. and Blackburn, J.H. (1991). The biology of the eel (*Anguilla anguilla* L.) in an English chalk stream and interactions with juvenile trout (*Salmo trutta* L.) and salmon (*Salmo salar* L.), *Hydrobiologia*, **218**, 65–76.

Mann, R.H.K. and Mills, C.A. (1986). Biological and climatic influences on the dace *Leuciscus leuciscus* in a southern chalk-stream, *Freshwater Biological Association Annual Report*, **54**, 123–136.

Mann, R.H.K., Blackburn, J.H. and Beaumont, W.R.C. (1989). The ecology of brown trout *Salmo trutta* in English chalk streams, *Freshwater Biology*.

Mantle, G. and Mantle, A. (1992). Impacts of low flows on chalk streams and water meadows, *British Wildlife*, **4**, 4–14.

Marchant, J.H. and Hyde, P.A. (1980). Aspects of the distribution of riparian birds on the waterways in Britain and Ireland, *Bird Study* **27**, 183–202.

Marchetti, R. (1988). Relazione di sintesi, Atti del Convegno Fiume Lambro, situazione e prospettive.

Margalef, R. (1951). Diversidad de especies en las comunidades naturales. *Publicaciones de Instituto de Biologia Aplicada*, **6**, 59–72.

Marshall, E.J.P. and Westlake, D.F. (1978). Recent studies on the role of aquatic macrophytes in their ecosystem, in *Proceedings of the European Weed Research Symposium 5th Symposium on Aquatic Weeds*, pp. 183–188.

Marshall, E.J.P. and Westlake, D.F. (1990). Water velocities around water plants in chalk streams, *Folia Geobotanica et Phytotaxonomica, Praha*, **25**, 279–289.

Marshall, G.T.H., Beaumont, A.R. and Wyatt, R. (1992). Genetics of brown trout (*Salmo trutta* L.) stocks above and below impassable falls in the Conwy river system, North Wales, *Aquatic Living Resources*, **5**, 9–13.

Martin, D. (1991). *Macroinfauna de una bahía mediterránea*, PhD Thesis, University of Barcelona.

Marty, C., Beall, E. and Parot, G. (1986) Influence de quelque paramaètres du milieu du incubation sur la survie dalevins de saumon atlantique, *Salmo salar* L., en ruisseau experimental, *Internationale Revue der Gesamten Hydrobiologie*, **71**, 349–361.

Mason, C.F. (1988). Concentrations of organochlorine residues and metals in tissues of otters *Lutra lutra* from the British Isles, *Lutra*, **31**, 62–67.

Mason, C.F. (1989). Water pollution and otter distribution: a review, *Lutra*, **32**, 97–131.

Mason, C.F. (1991) *Biology of Freshwater Pollution*, Longman Scientific and Technical, London.

Mason, C.F. (1992). Do otter releases make sense? The experience in Great Britain, in *Otterschutz in Deutschland* (Ed. C. Reuther), pp. 157–161, Otter Zentrum, Hankensbüttel.

Mason, C.F. (1993). Regional trends in PCB and pesticide contamination in northern Britain as determined in otter (*Lutra lutra*) scats, *Chemosphere*, **26**, 941–944.

Mason, C.F. and Barak, N.A.-E. (1990). A catchment survey for heavy metals using the eel (*Anguilla anguilla*), *Chemosphere*, **21**, 695–699.

Mason, C.F. and Macdonald, S.M. (1986). *Otters: Ecology and Conservation*, Cambridge University Press, Cambridge.

Mason, C.F. and Macdonald, S.M. (1987). Acidification and otter (*Lutra lutra*) distribution on a British river, *Mammalia*, **51**, 81–87.

Mason, C.F. and Macdonald, S.M. (1989). Acidification and otter (*Lutra lutra*) distribution in Scotland, *Water, Air and Soil Pollution*, **43**, 365–374.

Mason, C.F. and Macdonald, S.M. (1993a). Impact of organochlorine pesticide residues and PCBs on otters (*Lutra lutra*): a study from western Britain, *Science of the Total Environment*, **138**, 127–145.

Mason, C.F. and Macdonald, S.M. (1993b). Impact of organochlorine pesticide residues and PCBs on otters (*Lutra lutra*) in eastern England, *Science of the Total Environment*, **138**, 147–160.

Mason, C.F. and Macdonald, S.M. (1993c). PCBs and organochlorine pesticide residues in otter (*Lutra lutra*) spraints from Welsh catchments and their significance to otter conservation strategies,

Aquatic Conservation, **3**, 43−51.

Mason, C.F. and Macdonald, S.M. (1994). PCB and organochlorine pesticide residues in otters (*Lutra lutra*) and in otter spraints from southwest England and their likely impact on populations, *Science of the Total Environment*, **144**, 305−312.

Mason, C.F. and Madsen, A.B. (1992). Mercury in Danish otters (*Lutra lutra*), *Chemosphere*, **25**, 865−867.

Mason, C.F. and Madsen, A.B. (1993). Organochlorine pesticide residues and PCBs in Danish otters (*Lutra lutra*), *Science of the Total Environment*, **133**, 73−81.

Mason, C. F., Ratford, J. R. (in press). PCB congeners in tissues of European otter, *Bulletin of Environmental Toxicology and Contamination*.

Mason, C.F. and O'Sullivan, W.M. (1992). Organochlorine pesticide residues and PCBs in otters (*Lutra lutra*) from Ireland, *Bulletin of Environmental Contamination and Toxicology*, **48**, 387−393.

Mason, C.F. and O'Sullivan, W.M. (1993a). Heavy metals in the livers of otters, *Lutra lutra*, from Ireland, *Journal of Zoology*, **231**, 675−678.

Mason, C.F. and O'Sullivan, W.M. (1993b). Further observations on PCB and organochlorine pesticide residues in Irish otters (*Lutra lutra*), *Biology and Environment, Proceedings of the Royal Irish Academy*, **93B**, 187−188.

Mason, C.F., Ford, T.C. and Last, N.I. (1986). Organochlorine residues in British otters, *Bulletin of Environmental Contamination and Toxicology*, **36**, 656−661.

Mason, C.F., Last, N.I. and Macdonald, S.M. (1986). Mercury, cadmium and lead in British otters, *Bulletin of Environmental Contamination and Toxicology*, **37**, 844−849.

Mason, C.F., Macdonald, S.M., Bland, H.C. and Ratford, J. (1992). Organochlorine pesticide and PCB contents in otter (*Lutra lutra*) scats from western Scotland, *Water, Air and Soil Pollution*, **64**, 617−626.

Mason, J.C. (1976). Response of underyearling coho salmon to supplemental feeding in a natural stream, *Journal of Wildlife Management*, **40**, 775−788.

Mathur, D., Bason, W.H., Purdy, Jr., E.J., and Silver, C.A. (1985). A critique of the instream flow incremental methodology, *Canadian Journal of Fisheries and Aquatic Sciences*, **42**, 825−831.

Matthews, C.P. and Kowalczewski, A. (1969). The disappearance of leaf litter and its contribution to production in the River Thames, *Journal of Ecology*, **57**, 543−552.

Matthews, R.A, Buikema, A.L., Cairns, J. and Rodgers, J.H. (1982). Review Papers Biological Monitoring Part II A. Receiving system functional methods, relationships and indices.

Maurer, M.A. and Brusven, M.A. (1983) Insect abundance and colonization rate in *Fontinalis neomexicana* (Bryophyta) in an Idaho Batholith stream, USA, *Hydrobiologia*, **98**, 9−15.

Mawle, G.E., Winstone, A. and Brooker, M.P. (1985). Salmon and sea-trout in the Taff − past, present, future, *Nature in Wales*, **4**, 36−45.

McAuliffe, J.R. (1984). Competition for space, disturbance, and the structure of a benthic stream community, *Ecology*, **65**, 894−908.

McCabe G.D. and O'Brien, W.J. (1983). The effects of suspended silt on feeding and reproduction of *Daphnia pulex*. *American Midland Naturalist*, **110**, 324−337.

McCahon, C.P. and Pascoe, D. (1988a). Cadmium toxicity to the freshwater amphipod *Gammarus pulex* (L) during the moult-cycle, *Freshwater Biology*, **19**, 197−203.

McCahon, C.P. and Pascoe, D. (1988b). Increased sensitivity to cadmium of the freshwater amphipod *Gammarus pulex* (L) during the reproductive period, *Aquatic Toxicology*, **13**, 183−194.

McCahon, C.P. and Pascoe, D. (1989). Short-term experimental acidification of a Welsh stream: toxicity of different forms of aluminium at low pH to fish and invertebrates, *Archives of Environmental Contamination and Toxicology*, **18**, 233−242.

McCulloch, D.L. (1986). Benthic macroinvertebrate distributions in the riffle−pool communities of two east Texas streams, *Hydrobiologia*, **135**, 61−70.

McElhorne, M.J. and Davies, R.W. (1983). The influence of rock surface area on the microdistribution and sampling of attached riffle dwelling Trichoptera in Hartley Creek, Alberta, *Canadian Journal of Zoology*, **61**, 2300−2304.

McEwen, L.J. and Werritty, A. (1988). The hydrology and long-term geomorphic significance of a flash flood in the Cairngorm Mountains, Scotland, *Catena*, **15**, 361−377.

McGaha, Y.J. (1952) The limnological relations of insects to certain aquatic flowering plants,

Transactions of the American Microscopical Society, **71**, 355–381.

McLachlan, A.J., Pearce, L.J. and Smith, A.J. (1979). Feeding interactions and cycling of peat in a bog lake, *Journal of Animal Ecology*, **48**, 851–861.

McMahon, T.A. and Finlayson, B.L. (1992). Australian surface and groundwater hydrology — regional characteristics and implications, in *Water Allocation for the Environment* (Eds J.J. Pigram and B.P. Hooper), Centre for Water Policy Research, University of New England, Armidale.

McMillan N. (1990). The history of alien freshwater Mollusca in North West England, *Naturalist*, **115**, 123–131.

Meffe, G.K. (1987). Conserving fish genomes: philosophies and practices, *Environmental Biology of Fishes*, **18**, 3–9.

Meffe, G.K. and Vrijenhoek, R.C. (1988). Conservation genetics in the management of desert fishes, *Conservation Biology*, **2**, 157–169.

Melquist, W.E. and Hornocker, M.E. (1983). Ecology of river otters in west central Idaho, *Wildlife Monographs*, **83**, 1–60.

Meredith, E.K. and Malvestuto, S.P. (1991) An evaluation of survey designs for the assessment of effort, catch rate and catch in two contrasting rivers, in *Catch Effort Sampling Strategies* (Ed. I.G. Cowx), pp. 223–232, Fishing News Books, Blackwell Scientific, Oxford.

Metcalfe-Smith, J.L. (1994). Biological water-quality assessment of rivers: use of macroinvertebrate communities, in *The Rivers Handbook*, Vol. 2, (Eds P. Calow and G.E. Petts), pp. 144–170. Blackwell Scientific, Oxford.

Milhous, R.T., Updike, M.A. and Schneider, D.M. (1989). *Physical habitat simulation system reference manual — Version II*, Instream Flow Information Paper No. 26, US Fish and Wildlife Service, Biological Report 89 (16), Washington DC.

Miller, C. (1985). Correlates of habitat favourability for benthic macroinvertebrates at five stream sites in an Appalachian Mountain drainage basin, USA, *Freshwater Biology*, **15**, 709–733.

Mills, C.A. (1981a). The spawning of roach, *Rutilus rutilus* (L.) in a chalk stream, *Fisheries Management*, **10**, 49–54.

Mills, C.A. (1981b). The attachment of dace, *Leuciscus leuciscus* L., eggs to the spawning substratum and the influence of changes in water current on their survival, *Journal of Fish Biology*, **19**, 129–134.

Mills, C.A. (1981c). Egg population dynamics of naturally spawning dace, *Leuciscus leuciscus* (L.), *Environmental Biology of Fishes*, **6**, 151–158.

Mills, C.A. (1982). Factors affecting the survival of dace, *Leuciscus leuciscus* (L.), in the early post-hatching period, *Journal of Fish Biology*, **20**, 645–655.

Mills, C.A. (1991). Reproduction and life history, in *Cyprinid Fishes: Systematics, Biology and Exploitation* (Eds I.J. Winfield and J.S. Nelson), pp. 483–508, Chapman and Hall, London and New York.

Mills, C.A. and Mann, R.H.K. (1985). Environmentally induced fluctuations in year class strength and their implications for management, *Journal of Fish Biology*, **27**, (Suppl. A), 209–226.

Mills, C.P.R. (1991). Estimates of exploitation rates by rod and line of Atlantic salmon (*Salmo salar* L.) and sea trout (*Salmo trutta* L.) in the British Isles. Atlantic Salmon Trust/Royal Irish Academy Workshop, Dublin, 1–6.

Mills, D. (1979). The impact of angling on the environment. *Proceedings of the Institute of Fisheries Management Annual Study Course*, Nottingham University, pp. 5–12, Janssen Services, London.

Mills, D.H. (1987). Predator control, *Institute of Terrestrial Ecology Symposium*, **19**, 53–56.

Mills, D. H. (1989). *Ecology and Management of Atlantic Salmon*, Chapman and Hall, London, 1–351.

Milner, A. (1994). System recovery, in *The Rivers Handbook*, Vol. 2 (Eds P. Calow and G.E. Petts), pp. 76–98, Blackwell Scientific, Oxford.

Milner, N.J. and Varallo, P.V. (1990). Effects of acidification on fish and fisheries in Wales, in *Acid Waters in Wales* (Eds R.W. Edwards, A.S. Gee and J.H. Stoner), pp. 121–143, Kluwer Academic, Dordrecht.

Milner, N.J., Hemsworth, R.J. and Jones, B.E. (1985). Habitat evaluation as a fisheries management tool, *Journal of Fish Biology*, **27**, (Suppl. A), 85–108.

Milner, N.J., Scullion, J., Carling, P.A. and Crisp, D.T. (1981). A review of the effects of discharge on sediment dynamics and consequent effects on invertebrates and salmonids in upland rivers,

Advances in Applied Biology, **6**, 153—220.

Ministry of Agriculture Fisheries and Food (1985). Guidelines for the use of herbicides on weeds in or near water courses and lakes, MAFF, B2078, London.

Ministry of Transport, Public Works and Water Management (1989). *Summary: Third National policy document on Water Management, A Time for Action*, Ministry of Transport, Public Works, The Hague.

Minnesota Department Of Natural Resources (1992). A field guide to aquatic exotic plants and animals, Minnesota Department of Natural Resources, USA.

Minshall, G.W. (1967) Role of allochthonous detritus in the trophic structure of a woodland springbrook community, *Ecology*, **48**, 139—149.

Minshall, G.W. (1988). Stream ecosystem theory: a global perspective, *Journal of the North American Benthological Society*, **7**, 263—288.

Mittelbach G.G. (1981). Patterns of invertebrate size and abundance in aquatic habitats, *Canadian Journal of Fisheries and Aquatic Science*, **38**, 896—904.

Mooij, W.M. and van Tongeren, O.F.R. (1990). Growth of 0+ roach (*Rutilus rutilus*) in relation to temperature and size in a shallow eutrophic lake: comparison of field and laboratory observations, *Canadian Journal of Fisheries and Aquatic Sciences*, **47**, 960—967.

Moore, C.A. and Chutter, F.M. (1988). *A survey of the conservation status and benthic biota of the major rivers of the Kruger National Park*, Confidential Contract report of the National Institute for Water Research.

Moore, K.M.S. and Gregory, S.V. (1988). Response of young-of-the-year cutthroat trout manipulation of habitat structure in small stream, *Transactions of the American Fisheries Society*, **117**, 162—170.

Morgan, W.S.G. and Kuhn, P.C. (1984). Aspects of utilizing continuous automatic fish biomonitoring systems for industrial effluent control, in *Freshwater Biological Monitoring* (Eds D. Pascoe and R.W. Edwards), Pergamon Press, Oxford.

Moring, J.R. (1982). Decrease in stream gravel permeability after clear-cut logging: an indication of intragravel conditions for developing salmonid eggs and alevins, *Hydrobiologia*, **88**, 295—298.

Morris, J.T. and Lajtha, K. (1986). Decomposition and nutrient dynamics of litter from four species of freshwater emergent macrophytes, *Hydrobiologia*, **131**, 215—223.

Morris, R., Taylor, E.W., Brown, D.J.A. and Brown, J.A. (Eds) (1989). *Acid Toxicity and Aquatic Animals*, Society for Experimental Biology, Seminar Series 34, pp. 1—282, Cambridge University Press, Cambridge.

Morrison, B.R.S. (1977). The effects of rotenone on the invertebrate fauna of three hill streams in Scotland, *Fisheries Management*, **18**, 128—138.

Morrison, B.R.S. (1979). An investigation into the effects of the piscicide antimycin A on the fish and invertebrates of a Scottish stream, *Fisheries Management*, **10**, 111—122.

Morrison, B.R.S. (1987). Use and effects of piscicides, *Institute of Terrestrial Ecology Symposium*, **19**, 47—52.

Morrison, B.R.S. and Struthers, G. (1975). The effects of rotenone on the invertebrate fauna of three Scottish freshwater lochs, *Fisheries Management*, **6**, 81—91.

Mortensen, E. (1977). Population, survival, growth and production of trout in a small Danish stream, *Oikos*, **28**, 9—15.

Mortensen, E., Geertz-Hansen, P. and Marcus, E. (1988). The significance of temperature and food as factors affecting the growth of brown trout, *Salmo trutta* L., in four Danish streams, *Polskie Archiwum Hydrobiologii*, **35**, 533—544.

Morton, J.K. (1978). Distribution of giant Cow Parsnip (*Heracleum mantegazzianum*) in Canada. *Canadian Field Naturalist*, **93**(1), 82—83.

Moss B. (1977). Conservation problems in the Norfolk Broads and rivers of East Anglia, England — phytoplankton, boats and the causes of turbidity, *Biological Conservation*, **12**, 95—114.

Moss B. (1986). Restoration of lakes and lowland rivers, in *Ecology and Design in Landscape* (Ed. A.D. Bradshaw, D.A. Goode and E. Thorpe), pp. 399—415, Blackwell, Oxford.

Moss, B. (1988). *Ecology of Freshwaters: Man and Medium*, 2nd edn, Blackwell Scientific, Oxford.

Moss, B. (1990). Engineering and biological approaches to the restoration from eutrophication of shallow lakes in which aquatic plant communities are important components. *Hydrobiologia*, **200/201**, 367—377

Moss, D., Furze, M.T., Wright, J.F. and Armitage, P.D. (1987). The prediction of the macroinvertebrate fauna of unpolluted running water sites in Great Britain using environmental data, *Freshwater Biology*, **17**, 41−52.

Moyle, P.B. and Baltz, D.M. (1985). Micro-habitat use by an assemblage of California stream fishes: developing criteria for instream flow determinations, *Transactions of the American Fisheries Society*, **114**, 695−704.

Mueller, G. (1980). Effects of recreational river traffic on nest defence by longear sunfish, *Transactions of the American Fisheries Society*, **109**, 248−252.

Mullholland, P.J., Elwood, J.W., Palumbo, A.V. and Stevenson, R.J. (1986). Effects of stream acidification on periphyton composition, chlorophyll, and productivity, *Canadian Journal of Fisheries and Aquatic Sciences*, **43**, 1846−1858.

Mumford, P.M. (1990) Dormancy break in seeds of *Impatiens glandulifera* Royle, *New Phytologist*, **115**, 171−176.

Munawar, M., Norwood, W.P. and McCarthy, L.M. (1991). A method for evaluating the impact of navigationally-induced suspended sediments from the Upper Great Lakes connecting channels on primary productivity, *Hydrobiologia*, **219**, 325−332.

Mundie, J.H. (1974). Optimisation of the salmonid nursery stream, *Journal of the Fisheries Research Board of Canada*, **31**, 1827−1837.

Muñoz, I. (1990). *Limnologia de la part baixa del riu Ebre i els canals de reg: els factors fisico-químics, el fitoplancton i els macroinvertebrats bentònics*, PhD thesis, University of Barcelona.

Muñoz, I. and Prat, N. (1989). Effects of river regulation on the lower Ebro river (NE Spain), *Regulated Rivers: Research and Management*, **3**, 345−354.

Munro, W.R. (1957). The Pike of Loch Choin, *Freshwater Salmon Fisheries Research, Scotland*, **16**, 1−16.

Murdoch, W. (1963). *The population ecology of certain carabid beetles living in marshes and near fresh water*, DPhil Thesis, Oxford University.

Murphy, K.J. (1992). *Reopening of the Forty Foot Navigation: environmental impacts of boat traffic during the first season of boat use 1991*, report to National Rivers Authority, Anglian Region, University of Glasgow, UK.

Murphy, K.J. and Eaton, J.W. (1981). Water plants, boat traffic and angling in navigable canals, in *Proceedings of the 2nd British Freshwater Fisheries Conference*, pp. 173−187, Liverpool, UK.

Murphy, K.J. and Eaton, J.W. (1983). Effects of pleasure-boat traffic on macrophyte growth in canals, *Journal of Applied Ecology*, **20**, 713−729.

Murphy, K.J. and Pearce, H.G. (1987). Habitat modification associated with freshwater angling, *Institute of Terrestrial Ecology Symposium*, **19**, 31−46.

Murphy, K.J., Bradshaw, A.D. and Eaton, J.W. (1980) *Plants for the protection of canal banks*, Report of the Department of Botany, University of Liverpool.

Murphy, K.J., Eaton, J.W. and Hyde, T.M. (1982). The management of aquatic plants in a navigable canal system used for amenity and recreation, in *Proceedings of the European Weed Research Society 6th Symposium on Aquatic Weeds*, pp. 141−151, Novi Sad, Yugoslavia.

Murphy, M.L. and Meehan, W.R. (1991). Stream ecosystems, in *Influences of Forest and Rangeland Management on Salmonid Fishes and their Habitats* (Ed W.R. Meehan), pp. 17−46, American Fisheries Society Special Publication No. 19, Bethesda.

Naiman, R.J. (1992). *Watershed Management: Balancing Sustainability and Environmental Change*, Springer-Verlag, New York.

Naiman, R.J., Johnston, C.A. and Kelly, J.C. (1988). Alteration of North American streams by beaver, *BioScience*, **38**, 753−762.

Naiman, R.J., Melillo, J.M. and Hobbie, J.E. (1986). Ecosystem alteration of boreal forest streams by beaver (*Castor canadensis*), *Ecology*, **67**, 1254−1269.

Naiman, R.J., Lonzarich, D.G., Beechie, T.J. and Ralph, S.C. (1992). General principles of classification and the assessment of conservation potential in rivers, in *River Conservation and Management* (Eds P.J. Boon, P. Calow and G.E. Petts), pp. 93−123, Wiley, Chichester.

National Federation of Anglers (1989). *Close Season for Coarse Fish*, National Federation of Anglers, Derby.

National Rivers Authority (1990). *Toxic Blue-Green Algae*, Water Quality Series No. 2, NRA, London.

National Rivers Authority (1991a). *Environmental Statement: Lower River Colne Improvement Scheme: Wraysbury River at Poyle*, National Rivers Authority, Thames Region, Kings Meadow House, Reading RG1 8DQ.

National Rivers Authority (1991b). NRA acts to stop rivers running dry, *News Release*, 8 May 1991.

National Rivers Authority (1991c). *Proposals for Statutory Water Quality Objectives*, Water Quality Series No. 5, National Rivers Authority, Bristol.

National Rivers Authority (1991d). *A biological assessment of the impact of the Rhymney Valley Trunk Sewer Overflows on the River Rhymney*, unpublished report, National Rivers Authority, Welsh Region.

National Rivers Authority (1991e). *Monitoring of rivers and effluents*, unpublished report, National Rivers Authority, Thames Region.

National Rivers Authority (1992a). *Water Resources Development Strategy — a discussion document*, National Rivers Authority, Bristol.

National Rivers Authority (1992b). *Recommendations for a Scheme of Water Quality Classification for setting Statutory Water Quality Objectives*, National Rivers Authority, Bristol.

National Rivers Authority (1992c). *Grading schemes for River Water Quality*, National Rivers Authority, Bristol.

National Rivers Authority (1992d). *Sewage treatment works remedial strategy*, unpublished report, National Rivers Authority, Welsh Region.

National Rivers Authority (1992e). *Further investigation into the frequency of operation of CSOs in the Maesteg area*, unpublished report, National Rivers Authority, Welsh Region.

National Rivers Authority (1992f). *A biological and water quality investigation into a possible surface water acidification problem on the Nant Sychbant, a tributary of the River Llynfi*, unpublished report, National Rivers Authority, Welsh Region.

National Rivers Authority (1992g). *The influence of Agriculture on the Quality of Natural Waters in England and Wales*, Water Quality Series No. 6, National Rivers Authority.

National Rivers Authority (1992h). *A Fair Assessment: A New Charging Scheme for Beneficiaries of Fisheries Work*, National Rivers Authority, London.

National Rivers Authority (1992i). *River Channel Typology for Catchment and River Management*, undertaken by GeoData Institute, University of Southampton for NRA Thames Region.

National Rivers Authority (1993a). *River Corridor Manual for Surveyors*, National Rivers Authority, Rivers House, Aztec Park, Bristol.

National Rivers Authority (1993b). *NRA Navigation Strategy*, National Rivers Authority, Aztec Park, Bristol, UK.

Natural Environment Research Council (1975). *Flood Studies Report*, NERC, Wallingford, England.

Nature Conservancy Council (1984). *Surveys of Wildlife in River Corridors, draft methodology*, Nature Conservancy Council, Peterborough.

Nature Conservancy Council (1989). *Guidelines for Selection of Biological SSSIs*, Nature Conservancy Council, Peterborough.

Nature Conservancy Council (1990). *Handbook for Phase 1 Habitat Survey: A Technique for Environmental Audit*, England Field Unit, Nature Conservancy Council, Peterborough.

Neiland, R., Proctor, J. and Sexton, R. (1987). Giant hogweed (*H. mantegazzianum* Somm and Levier) by the River Allen and Part of the River Forth, *Forth Naturalist and Historian*, **9**, 51−56.

Neill, C.R and Hey, R.D. (1982). Gravel-bed rivers: Engineering problems, in *Gravel-bed Rivers* (Eds R.D. Hey, J.C. Bathurst and C.R. Thorne), Wiley, Chichester.

Nelson, D.A. (Ed.) (1982). *Proceedings of the Workshop on the Environmental Effects of Navigation Traffic*, Final report, US Army Engineers Experimental Station, Environmental Laboratory, Vicksburg, MS.

Nestler, J.M., Milhous, R.T. and Layzer, J.B. (1989). Instream habitat modelling techniques, in *Alternatives in Regulated River Management* (Eds J.A. Gore and G.E. Petts), pp. 295−315, CRC Press Inc., Boca Raton, FL.

Newbold, C., Purseglove, J. and Holmes, N. (1983). *Nature Conservation and River Engineering*, Nature Conservancy Council, Shrewsbury.

Newbold, J.D. (1992). Cycles and spirals of nutrients, in *The Rivers Handbook*, Vol. 1 (Eds P. Calow and G.E. Petts), pp. 379−408, Blackwell Scientific, Oxford.

Newbold, J.D., Elwood, J.W., O'Neill and Van Winkle, W. (1981). Measuring nutrient spiraling in streams, *Canadian Journal of Fisheries and Aquatic Sciences*, **38**, 860−863.

Newcombe, C.P. and MacDonald, D.D. (1991). Effects of suspended sediments on aquatic ecosystems. *N. Am. J. Fish Mgmt*, **11**, 72–82.

Newman, J. and Barrett, P.R.F. (1993). Control of *Microcystis aeruginosa* by decomposing barley straw, *Journal of Aquatic Plant Management*, **31**, 203–206.

Newman, P.J., Piavaux, M.A. and Sweeting, R.A. (Eds.) (1992). *River Water Quality*, Office for Official Publications of the European Communities, Brussels.

Newson, M.D. (1975). *Flooding and Flood Hazard in the United Kingdom*, Oxford University Press, Oxford.

Newson, M.D. (1992). River conservation and catchment management: a UK perspective, in *River Conservation and Management* (Eds P.J. Boon, P. Calow and G.E. Petts), pp. 385–396, John Wiley, Chichester.

Nickelson, T.E., Beidler, W.M. and Mitchell, M.J. (1979). *Streamflow requirements of salmonids*, Department of Fish and Wildlife Report AFS-62, Portland, Oregon, US.

Nielsen, G. (1986). Dispersion of brown trout (*Salmo trutta* L.) in relation to stream cover and water depth, *Polskii Archiwum Hydrobiologia*, **33**, 475–488.

Nilsson, C. (1987). Distribution of stream-edge vegetation along a gradient of current velocity, *Journal of Ecology*, **75**, 513–522.

Nixon, M. (1959). A study of bankful discharges of rivers in England and Wales, *Proceedings of the Institute of Civil Engineers*, **12**, 157–175.

Noppert F., Dogger, J.W., Balk, F. and Smits, A.J.M. (1993). Milieurisico's voor de natuur in rivieruiterwaarden: een schattingsmethode (Environmental risks for nature in river floodplains), *H₂O*, **26**, 120–125 (in Dutch).

Norrgren, L., Wickland Glynn, A. and Malmborg, O. (1991). Accumulation andeffects of aluminium in the minnow (*Phoxinus phoxinus* L.) at different pH levels. *Journal of Fish Biology*, **39**, 833–847.

North, E. (1983). Mortality of salmon parr caught by different angling techniques, *Fisheries Management*, **14**, 93–95

North, E. and Hickley, P. (1989). An appraisal of anglers' catch composition in the River Severn (England), *Journal of Fish Biology*, **34**, 299–306.

Northcote, T.G. and Hartman, G.F. (1988). The biology and significance of stream trout populations (*Salmo* spp.) living above and below waterfalls, *Polskie Archiwum Hydrobiologii*, **35**, 409–442.

Nuttall, P.M. (1972). The effects of sand deposition upon the macro-invertebrate fauna of the River Camel, Cornwall, *Freshwater Biology*, **2**, 181–186.

Nyman, L. (1991). Conservation of freshwater fish, Institute of Freshwater Research, Drottningholm, *Fisheries Development Series*, **56**, 1–38.

Oates, D. (1978). The effects of boating upon lead concentrations in fish, *Transactions of the Kansas Academy of Sciences*, **79**, 149–153.

Oborne, A.C. (1981). The application of a water-quality model to the River Wye, Wales, *Hydrology*, **52**, 59–70.

Oborne, A.C., Brooker M.P., and Edwards, R.W. (1980). The chemistry of the River Wye, *Journal of Hydrology*, **45**, 233–52.

O'Connell, M.F. and Bourgeois, C.E. (1987). Atlantic salmon enhancement in the Exploits River, Newfoundland, 1957–1984. *North American Journal of Fisheries Management*, **7**, 207–214.

O'Connell, M.F., Davis, J.P. and Scott, D.C. (1983). An assessment of the stocking of Atlantic salmon (*Salmo salar* L.) fry in the tributaries of the middle Exploits River, Newfoundland, Canadian technical reports of Fisheries and Aquatic Sciences, No. 1225, 1–142.

O'Connor, N.A. (1991). The effects of habitat complexity on the macroinvertebrates colonising wood substrates in a lowland stream, *Oecologia*, **85**, 504–512.

O'Connor, N.A. (1992). Quantification of submerged wood in a lowland Australian stream system, *Freshwater Biology*, **27**, 387–95.

Odum, H.T. (1956). Primary production in flowing waters, *Limnology and Oceanography*, **1** (2), 102–117.

Oglesby, R.T., Carlson, C.A. and McCann, J.A., (1972). *River Ecology and Man*, Academic Press, New York.

O'Hara, K. (1981). Some observations of the estimation of fish abundance in different water bodies, in *Seminar on the Assessment of Freshwater Fish Stocks*, pp. 57–66, Water Space Amenity Commission, London.

O'Keeffe, J.H. (1989). Conserving rivers in Southern Africa, *Biological Conservation*, **49**, 255−274.

O'Keeffe, J.H. and Davies, B.R. (1991). The conservation and management of the rivers of the Kruger National Park: suggested methods for calculating instream flow needs, *Aquatic Conservation, Marine and Freshwater Ecosystems*, **1**, 55−71.

O'Keeffe, J.H., Danilewitz, D.B. and Bradshaw, J.A. (1987). An "expert system" approach to the assessment of the conservation status of rivers, *Biological Conservation*, **40**, 69−84.

O'Keeffe, J.H., Byren, B.A., Davies, B.R. and Palmer, R.W. (1990). The effects of impoundments on the physico-chemistry of two contrasting southern african river systems, *Regulated Rivers: Research and Management*, **5**, 97−110.

O'Keeffe, J.H., van Ginkel, C.E., Hughes, D.A,. Hill, T. and Ashton, P.A. (in press). *A situation analysis of water quality in the catchment of the Buffalo River, eastern Cape, with special emphasis on the impacts of low-cost, high-density urban development on water quality*, Report to the Water Research Commission, Pretoria.

Oksiyuk, O.P., Merezhko, A.I. and Volkova, T.F. (1978). Use of aquatic vegetation for improving water quality and stabilizing canal banks, *Water Res.*, **5**, 556−562.

Oliff, W.D. (1960). Hydrological studies on the Tugela River system, Part I. The Main Tugela River, *Hydrobiologia*, **14**, 281−385.

Olson, C.G. and Hupp, C.R., (1986). Coincidence and spatial variability of geology, soils and vegetation, Mill Run watershed, Virginia, *Earth Surface Processes and Landforms*, **11**, 619−629.

Olsson, T.I. and Persson, B.G. (1986). Effects of gravel size and peat material concentration on embryo survival and alevin emergence of brown trout, *Salmo trutta* L., *Hydrobiologia*, **135**, 9−14.

Olsson, T.I. and Persson, B.G. (1988). Effects of deposited sand on ova survival and alevin emergence in brown trout (*Salmo trutta* L.), *Archiv für Hydrobiologie*, **113**, 621−627.

Olsson, M., Reutergardh, L. and Sandegren, F. (1981). Var är uttern?, *Sveriges Natur*, **6**, 234−240.

Olsson, M., Anderson, O., Bergman, A., Blomkvist, G., Frank, A. and Rappe, C. (1992). Contaminants and diseases in seals from Swedish waters, *Ambio*, **21**, 561−562.

Omodeo, P. (1984). On aquatic Oligochaeta Lumbricomorpha in Europe, *Hydrobiologia*, **115**, 187−190.

O'Riordan, T. (Ed.) (1994). *Environmental Science for Environmental Management*, Longman, Harlow.

Ormerod, S.J. (1985). *The distribution of macroinvertebrates in the upper catchment of the River Wye in relation to ionic composition*, PhD Thesis, University of Wales.

Ormerod, S.J. (in press). Assessing critical levels of ANC for stream organisms using empirical data, in *Critical Loads and Levels* (Ed. M. Hornung and R. Skeffington), Institute of Terrestrial Ecology.

Ormerod, S.J. and Edwards, R.W. (1987). The ordination and classification of macroinvertebrate assemblages in the catchment of the River Wye in relation to environmental factors, *Freshwater Biology*, **17**, 533−546.

Ormerod, S.J. and Jenkins, A. (1994). The biological effects of acid episodes, in *Acidification: Past, Present and Future* (Eds R. Wright and C. Steinberg), Dahlem Workshop, John Wiley, Chichester, 259−272.

Ormerod, S.J. and Wade, K.R. (1990). The role of acidity in the ecology of Welsh lakes and streams, in *Acid Waters in Wales* (Eds R.W. Edwards, A.S. Gee and J.H. Stoner), pp. 93−119, Kluwer Academic, Dortrecht, Holland.

Ormerod, S.J., Donald, A.P., and Brown, S.J. (1989). The influence of plantation forestry on the pH and aluminium concentration of upland Welsh streams: a re-examination, *Environmental Pollution*, **62**, 47−62.

Ormerod, S.J., Mawle, G.E. and Edwards, R.W. (1987). The influence of forest on aquatic fauna, in *Environmental Aspects of Plant Forestry in Wales* (Ed. J.E. Good), pp. 37−49, Symposium No. 22, Institute of Terrestrial Ecology, Grange-over-Sands, UK.

Ormerod, S.J., Wade, K.R. and Gee, A.S. (1987). Macro-floral assemblages in upland Welsh streams in relation to acidity and their importance to invertebrates, *Freshwater Biology*, **18**, 545−558.

Ormerod, S.J., Weatherley, N.S., Varallo, P.V. and Whitehead, P. (1988). Preliminary empirical models of the historical and future impact of acidification on the ecology of Welsh streams, *Freshwater Biology*, **20**, 127−140.

Ormerod, S.J., Weatherley, N.S., Merrett, W.J., Gee, A.S. and Whitehead, P.G. (1990). Restoring acidified streams in upland Wales: a modelling comparison of the chemical and biological effects of liming and reduced sulphate deposition. *Environmental Pollution*, **64**, 1−91.

Ormerod, S.J., O'Halloran, J., Gribbin, S.D. and Tyler, S.J. (1991). The ecology of dippers Cinclus cinclus in relation to stream acidity in upland Wales: breeding performance, calcium physiology and nestling growth, *Journal of Applied Ecology*, **28**, 419−433.

Ormerod, S.J., Rundle, S.D., Clare-Lloyd, E. and Douglas, A.A. (1993). The influence of riparian management on the habitat structure and macroinvertebrate communities of acid streams draining plantation forests, *Journal of Applied Ecology*, **30**, 13−24.

Orsborn, J.F. and Anderson, J.W. (1986). Stream improvements and fish response. A Bio-Engineering assesment, *Water Resources Bulletin*, **22**, 381−388.

Orth, D.J. (1989). Ecological considerations in the development and application of instream-flow habitat models, *Regulated Rivers: Research and Management*, **5**, 171−181.

Orth, D.J. and Leonard, P.M. (1990). Comparison of discharge methods and habitat optimisation for recommending instream flows to protect fish habitat, *Regulated Rivers: Research and Management*, **5**, 129−138.

Orth, D.J. and Maughan, O.E. (1982). Evaluation of the incremental methodology for recommending instream flows for fishes, *Transactions of the American Fisheries Society*, **111**, 413−445.

Osborn, J.F. and Allman, C.H. (1976). *Instream Flow Needs*, Vol. II, Proceedings of Boise Symposium, Idaho, May 1976, American Fisheries Society, Bethesda, MD.

Osborne, L.L. and Kovacic, D.A. (1993). Riparian vegetated buffer strips in water quality restoration and stream management, in *Lowland Stream Restoration: Theory and Practice* (Eds L.L. Osborne, P.B. Bayley and L.W. Higler), *Freshwater Biology*, **29**, 243−258.

O'Sullivan, W.M., Macdonald, S.M. and Mason, C.F. (1993). Organochlorine pesticide residues and PCBs in otter spraints from southern Ireland, *Biology and Environment, Proceedings of the Royal Irish Academy*, **93B**, 55−57.

Otis, D.L., Burnham, K.P., White, G.C. and Anderson, D.R. (1978). Statistical influence from capture data on closed animal populations, *Journal of Wildlife Management*, **15**, 88−98.

Owens, M. (1970). Nutrient balances in rivers, *Water Treatment and Examination*, **19**, 239−252.

Owens, M. and Edwards, R.W. (1961). The effects of plants on river conditions II. Further crop studies and estimates of net productivity of macrophytes in a chalk stream, *Journal of Ecology*, **49**, 119−126.

Owens, M., Garland, J.H.N., Hart, I.C. and Wood, G. (1972). Nutrient budgets in rivers, *Symposium of the Zoological Society of London*, **29**, 21−40.

Palmer, C.G. and O'Keeffe, J.H. (1992). Feeding patterns of four macroinvertebrate taxa in the headwaters of the Buffalo River, eastern Cape, *Hydrobiologia*, **228**, 157−173.

Palmer, C.G., O'Keeffe, J.H. and Palmer, A.R. (1991). Are macroinvertebrate assemblages in the Buffalo River, southern Africa, associated with particular biotopes?, *Journal of the North American Benthological Society*, **10**, 349−357.

Palmer, L. (1976). River management criteria for Oregon and Washington, in *Geomorphology and River Engineering* (Ed. D.R. Coates), pp. 329−346, George Allen and Unwin, London.

Palmer, M. (1981). Relationship between species richness of macrophytes and insects in some water bodies in the Norfolk Breckland, *Entomologist's Monthly Magazine*, **117**, 35−46.

Palmer, R.W. and O'Keeffe, J.H. (1989). Temperature effects of impoundments on the downstream reaches of a river in south eastern Africa, *Archiv für Hydrobiologie*, **116**, 471−485.

Palmer, R.W. and O'Keeffe, J.H. (1990a). Downstream effects of impounds on the water chemistry of the Buffalo River (eastern Cape), South Africa, *Hydrobiologia*, **202**, 71−83.

Palmer, R.W. and O'Keeffe, J.H. (1990b). Transported material in a small river with multiple impoundments, *Freshwater Biology*, **24**, 563−575.

Parker, G. and Andres, D. (1976). Detrimental effect of river channelization, *Rivers, American Society of Engineers*, **76**, 1248−1266.

Parkyn, L., Harper, D.M. and Smith, C.D. (1992). *Instream and Riparian Species−Habitat Relationships*, National River Authority, Rivers House, Aztec West, Bristol, England.

Parry, M.L. (1978). Stock conservation by regulatory measures, *Proceedings of the Conference on Recreational Freshwater Fisheries* (Ed. J.S. Alabaster), pp. 153−166, Water Research Centre, Stevenage.

Parry, M.L. (1979). Why have Fisheries Acts anyway?, in *Proceedings of the Institute of Fisheries*

Management Annual Study Course, Nottingham University, pp. 152–158, Janssen Services, London.

Parry, M.L. (1987). Multi-purpose use of waters, *Institute of Terrestrial Ecology Symposium*, **19**, 66–71.

Pascoe, D. and Edwards, R.W. (1990). Single species toxicity tests, in *Aquatic Ecotoxicology: Fundamental Concepts and Methodologies*, Vol. 2 (Eds A. Bondou and Ribeyre), pp. 93–126, CRC Press.

Patrick, R., Rhyne, C.F., Richardson, R.W. III, Larson, R.A., Bott, T.T. and Rogenmuser, K. (1983). The potential for biological controls of *Cladophora glomerata*. Report No. 600/3-83-065, US Environmental Protection Agency.

Pearce, D., Markandya, A. and Barbier, E. (1989). *Blueprint for a Green Economy*, Earthscan, London.

Pearce, H.G. and Eaton, J.W. (1983). Effects of recreational boating on freshwater ecosystems — an annotated bibliography, Appendix B in: *Waterway Ecology and the Design of Recreational Craft*, Inland Waterways Amenity Advisory Council, London.

Pearson, R.G. and Jones, N.V. (1975). The effects of dredging operations on the benthic community of a chalk stream, *Biological Conservation*, **8**, 273–278.

Pearsons, T.N., Hiram, W.L., and Lamberti, G.A. (1992). Influences of habitat complexity on resistance to flooding and resilience of stream fish assemblages, *Transactions of the American Fisheries Society*, **121**, 427–436.

Peeters, E.T.H.M. and Tachet, H. (1989). Comparison of macrobenthos in braided and channelized sectors of the Drome River, France, *Regulated Rivers: Research and management*, **4**, 317–325.

Pennak, R.W. (1971). Towards a classification of lotic habitats, *Hydrobiologia*, **38**(2), 321–334.

Pennak, R.W. and Van Gerpen, E.D. (1947). Bottom fauna production and physical nature of the substrate in a northern Colorado trout stream, *Ecology*, **28**, 42–48.

Penczak, T., Forbes, I., Coles, T.F., Atkin, T. and Hill, T. (1991). Fish community structure in the rivers of Lincolnshire and South Humberside, England, *Hydrobiologia*, **211**, 1–9.

Pepper, V.A., Oliver, N.P. and Blundon, R. (1985). Evaluation of an experiment in lacustrine rearing of juvenile anadromous Atlantic salmon, *North American Journal of Fisheries Management*, **5**, 507–525.

Percival, E. and Whitehead, H. (1929). A quantitative study of the fauna of some types of stream-bed, *Journal of Ecology*, **17**, 282–314.

Perrins, J., Fitter, A. and Williamson, M. (1990). What makes *Impatiens glandulifera* invasive?, in *Biology and Control of Invasive Plants*, (Ed. J. Palmer), pp. 8–33, British Ecological Society, University of Wales, Cardiff.

Peterjohn, W.T. and Correll, D.L. (1984). Nutrient dynamics in an agricultural watershed: observations on the role of a riparian forest, *Ecology*, **65**, 1466–1475.

Peters, J.C. (1962). The effects of stream sedimentation on trout embryo survival, in *Transactions of the Third Seminar on Biological Problems in Water Pollution* (Ed. C.M. Tarzewell).

Petersen, R.C. (1981). *The multiple-plate, artificial substrate sampler: a users guide with bibliography on artificial substrates*, Institute of Limnology, University of Lund.

Petersen, R.C., Petersen, L.B.-M. and Lacoursière, J. (1992). A building-block model for stream restoration, in *River Conservation and Management* (Eds P.J. Boon, P. Calow and G.E. Petts), pp. 293–309, John Wiley, Chichester.

Peterson, R.H. (1978). *Physical characteristics of Atlantic salmon spawning gravel in some New Brunswick streams*, Technical Report, Fisheries and Marine Service, Canada, 785, 1–28.

Peterson, R.H. and Metcalfe, J.L. (1981). *Emergence of Atlantic salmon fry from gravels of varying composition: a laboratory study*, Canadian Technical Report of Fisheries and Aquatic Sciences, No. 1020.

Petts, G.E. (1979). Complex response of river channel morphology subsequent to reservoir construction, *Progress in Physical Geography*, **3**, 329–362.

Petts, G.E. (1984). Sedimentation within a regulated river, *Earth Surface Processes and Landforms*, **9**, 125–134.

Petts, G.E. (1984). *Impounded Rivers*, John Wiley, Chichester.

Petts, G.E. (1987). Times-scales for ecological changes, in *Regulated Streams: Advances in Ecology* (Eds J.F. Craig and J.B. Kemper), pp. 257–266, Plenum, New York.

Petts, G.E. (1989). Perspectives for ecological management of regulated rivers, in *Alternatives in Regulated River Management* (Eds J.A. Gore and G.E. Petts), pp. 3–24, CRC Press, Boca Raton, FL.

Petts, G.E. (1990a). Regulation of large rivers: problems and possibilities for envoironmentally-sound river development in South America, *Interciencia*, **15**, 388–395.

Petts, G.E. (1990b). *The River Glen*, Report for the National Rivers Authority, Anglian Region, Peterborough, UK.

Petts, G.E. (1990c). The role of ecotones in aquatic landscape management, in *The Ecology and Management of Aquatic–Terrestrial Ecotones* (Eds R.J. Naiman and H. Décamps), pp. 227–261, MAB Series, Vol. 4, Unesco, Paris and Parthenon Publishing Group, New Jersey.

Petts, G.E. (1992). *The River Glen: a catchment assessment*, Report for the National Rivers Authority, Anglian Region, Peterborough, UK.

Petts, G.E. (1994). Large-scale river regulation, in *The Changing Global Environment* (Ed. C.N. Roberts), pp. 262–284, Blackwell, Oxford.

Petts, G.E. and Maddock, I. (1994) Flow allocation for in-river needs, in *Rivers Handbook*, Vol. II, (Eds P. Calow and G.E. Petts), pp. 289–307, Blackwell Scientific, Oxford.

Petts, G.E. and Thoms, M.C. (1986). Channel aggradation below Chow Valley Lake, Somerset, UK, *Catena*, **13**, 305–320.

Petts, G.E. and Wood, R. (Eds) (1988). River Regulation in the United Kingdom, *Regulated Rivers Special Issue*, 2.

Petts, G.E., Armitage, P. and Gustard, A. (Eds) (1989). Proceedings of the Fourth International Symposium on Regulated Rivers, *Regulated Rivers: Research and Management*, 3.

Petts, G.E., Moller, H. and Roux, A.L. (Eds) (1989). *Historical Changes of Large Alluvial Rivers: Western Europe*, John Wiley, Chichester.

Petts, G.E., Armitage, P.D., Forrow, D., Bickerton, M., Castella, E., Gunn, R. and Blackburn, J.H. (1991). *The Effects of Abstractions from Rivers on Benthic Macroinvertebrates*, CSD Report No. 1230, Nature Conservancy Council, Peterborough.

Phillips, R.W. and Koski, K.V. (1969). A fry trap method for estimating salmonid survival from egg deposition to emergence, *Journal of the Fisheries Research Board of Canada*, **26**, 133–141.

Phillips, R.W., Lantz, R.L., Claire, E.W. and Moring, J.R. (1986). Some effects of gravel mixtures on emergence of coho salmon and steelhead trout fry, *Transactions of the American Fisheries Society*, **104**, 461–466.

Pienaar, U. de V. (1978). *The Freshwater Fishes of the Kruger National Park*, Published by the National Parks Board, South Africa.

Pieters H. and van de Guchte, C. (in press). Assessment of priority pollutants in zebra mussel and European eel from the lower Rhine area, poster Symposium on the Rehabilitation of the river Rhine, 15–19 March, 1993, Arnhem.

Pieterse, A.H. and Murphy, K.J. (Eds) (1990). *Aquatic Weeds*, Oxford University Press, Oxford.

Pillinger, J.M., Cooper, J.A. and Ridge, I. (1994) Role of phenolic compounds in the antialgal activity of barley straw, *Journal of Chemical Ecology*, (in press).

Pillinger, J.M., Gilmour, I. and Ridge, I. (1993) Control of algal growth by lignocellulosic material. Abstract of the FEMS Symposium on Lignin Biodegradation and Transformation, Lisbon, 18–23 April 1993.

Pillinger, J.M., Cooper, J.A., Ridge, I. and Barrett, P.R.F. (1992). Barley straw as an inhibitor of algal growth III: the role of fungal decomposition, *Journal of Applied Phycology*, **4**, 353–355.

Pinay, G., Décamps, H., Chauvet, E. and Fustec, E. (1990). Functions of ecotones in fluvial systems, in *The Ecology and Management of Aquatic-Terrestrial Ecotones* (Eds R.J. Naiman and H. Décamps), pp. 141–169, MAB Series, Vol. 4, Unesco, Paris and Parthenon Publishing Group, New Jersey.

Pinder, L.C. (1986). Biology of freshwater Chironomidae, *Annual Reviews of Entomology*, **31**, 1–23.

Pinto, P. (1988). Variação anual da estrutura cenótica da Ribeira do Degebe (Bacia hidrográfica do Guadiana), *Actas Col. Luso-Esp. Ecol. Bacias Hidrog. e Rec. Zoológicos*, 319–326.

Pitcher, T.J. and Hart, P.J.B. (1982). *Fisheries Ecology*. Croom Helm, London.

Pitman, W.V. (1973). A mathematical model for generating monthly river flows from meteorological data in South Africa, Report No. 2/73 of the Hydrological Research Unit, University of the Witwatersrand, Johannesburg, South Africa.

Plachter, H. (1986). Composition of the carabid fauna of natural riverbanks and man-made secondary habitats, in *Carabid Beetles: Their Adaptations and Dynamics*, (Eds P.J. Den Boer, M.L. Luff, D. Mossakowski and F. Weber), pp. 509–535, Gustav Fishcher, Stuttgart.

Platts, W.S. (1991). Livestock grazing, in *Influences of Forest and Rangeland Management on Salmonid Fishes and their Habitats* (Ed. W.R. Meehan), pp. 389–423, Amererican Fisheries Society Special Publication No. 19, Bethesda.

Platts, W.S. and Megahan, W.F. (1975). Time trends in riverbed sediment composition in salmon and steelhead spawning areas: South Fork Salmon River, Idaho, in *Transactions of the 40th North American Wildlife and Natural Resources Conference*, 1975, 229–323

Platts, W.S. and Nelson, R.L. (1989). Stream canopy and its relationship to salmonid biomass in the intermountain west, *North American Journal of Fisheries Management*, **9**, 446–457.

Platts, W.S., Megahan, W.F. and Minshall, G.W. (1983). Methods for evaluating stream, riparian and biotic conditions U.S.D.A., General Technical Report. INT-138.

Poff, N.L. and Ward, J.V. (1990). Physical habitat template of lotic systems: recovery in context of historical pattern of spatiotemporal heterogeneity, *Environmental Management*, **14**, 629–645.

Popov, I.V. (1964). Hydrogeomorphological principles of the theory of channel processes and their use in hydrotechnical planning, *Soviet Hydrology*, 158–195.

Possardt, E.E. and Dodge, W.E. (1978). Stream channelization impacts on song-birds and small mammals in Vermont, *Wildlife Society Bulletin*, **6**, 18–24.

Poulton, M. and Pascoe, D. (1990). Disruption of precopula in *Gammerus pulex* L. — development of a behavioural bioassay for evaluating pollutant and parasite induced stress. *Chemosphere*, **20**, 3–4, 403–415.

Power, M.E., Marks, J.C. and Parker, M.S. (1992). Variation in the vulnerability of prey to different predators: community-level consequences, *Ecology*, **73**, 2218–2223.

Prach, K. (1994) Seasonal dynamics of *Impatiens glandulifera* in two riparian habitats in Central England, in *Ecology and Management of Invasive Riverside Plants* (Eds L.C. de Waal, L.E. Child, P.M. Wade and J.H. Brock), pp. 127–133, Wiley, Chichester.

Pratt, M.M. (1975). *Better Angling with Simple Science*, Fishing News Books, Oxford.

Prauser, N. (1985). Vorkommen von Fischottern (*Lutra lutra* L. 1758) und ihre Abhängigkeit von der Struktur verschiedener Habitat-Zonen der Wumme-Niederung/Niedersachsen, *Zeitschrift für Angewandte Zoologie*, **72**, 83–91.

Prejs, A. and Jackowska, H. (1978). Lake macrophytes as the food of roach (*Rutilus rutilus* L.) and rudd (*Scardinius erythrophthalamus* L.). I. Species composition and dominance in the lake, and food, *Ekol. Polska*, **26**, 429–438.

Prejs, K. (1977). The nematodes of the root region of aquatic macrophytes, with special consideration of nematode groupings penetrating the tissues of roots and rhizomes, *Ekologia Polska*, **25**, 5–20.

Prejs, K. (1986). Nematodes as a possible cause of rhizome damage in three species of *Potamogeton*, *Hydrobiologia*, **131**, 281–286.

Preston-Whyte, R.A. and Tyson, P.D. (1988). *The Atmosphere and Weather of Southern Africa*. Oxford University Press, Cape Town.

Price, D.R.H. (1977). The Norfolk Broads — a changing habitat, *Water Space*, **12**, 11–14.

Probst, J-L. (1989). Hydroclimatic fluctuations of some European rivers, in *Historical Change of Large European Rivers* (Eds G.E. Petts, H. Moller and A.L. Roux) pp. 41–56, Wiley, Chichester.

Prochazka, K., Stewart, B.A. and Davies, B.R. (1991). Leaf litter retention and its implications for shredder distribution in two headwater streams, *Archiv für Hydrobiologie*, **120**(3), 315–325.

Pygott, J.R. and Douglas, S. (1989). Current distribution of *Corophium curvispinum* Sars. *var. devium* Wundsch (Crustacea: Amphipoda) in Britain with notes on its ecology in the Shropshire Union Canal, *Naturalist*, **114**, 15–17.

Pygott, J.R., O'Hara, K. and Eaton, J.W. (1990). Fish community structure and management in navigated British canals, in *Management of Freshwater Fisheries*, (Eds W.L.T. van Densen, B. Steinmetz and R.H. Hughes), Pudoc, Wagenengen.

Pygott, J.R., O'Hara K., Cragg-Hine, D. and Newton, C. (1990). A comparison of the sampling efficiency of electric fishing and seine netting in two contrasting canal systems, in *Fishing with Electricity* (Ed. I.G. Cowx), pp. 130–139, Fishing News Books, Blackwells Scientific, Oxford.

QDPI and QWRC (1979). *Barker-Barambah Irrigation Project*, Queensland Department of Primary Industries and Queensland Water Resources Commission, Brisbane.

Quinn, J.M., Davies-Colley, R.J., Hickey, C.W., Vickers, M.L. and Ryan, P.A. (1992). Effects of clay discharges. II. Benthic invertebrates, *Hydrobiology*, **248**, 235–247.

Raat, A.J.P. (1985). Analysis of angling vulnerability of common carp, *Cyprinus carpio* L., in catch and release angling in ponds, *Aquaculture and Fisheries Management*, **16**, 171–188.

Raat, A.J.P. (1990). The impact of fish on aquatic ecosystems: fish stocking in the Netherlands 1950–1990, in *Proceedings of the Institute of Fisheries Management 21st Anniversary Conference*, 1990, 299–315.

Rabe, F.W. and Savage, N.L. (1979). A methodology for the selection of aquatic natural areas, *Biological Conservation*, **15**, 291–300.

Rabeni, C.F. and Gibbs, K.E. (1980). Ordination of deep river invertebrate communities in relation to environmental variables, *Hydrobiologia*, **74**, 67–76.

Racey, G.D. and Euler, D.L. (1983). Changes in mink habitat and food selection as influenced by cottage development in central Ontario, *Journal of Applied Ecology*, **20**, 387–402.

Racey, P.A. and Swift, S.M. (1985). Feeding ecology of *Pipistrellus pipistrellus* (Chiroptera: Vespertilionidae) during pregnancy and lactation. 1. Foraging behaviour, *Journal of Animal Ecology*, **54**, 205–215.

Radojevic, M. and Harrison, R.M. (Eds) (1992). *Atmospheric Acidity: Sources, Consequences and Abatement*, Elsevier Applied Science, London.

Raleigh, R.F. (1982). *Habitat suitability index models: brook trout*. US Fish and Wildlife Service Biological Reports No. 82/10, 24, Fort Collins, CO.

Raleigh, R.F., Zuckerman, L.D. and Nelson, P.C. (1986). *Habitat suitability index models and instream flow suitability curves: Brown trout*, US Fish and Wildlife Service Biological Reports No. 82/10.124, Fort Collins, CO.

Ramos, L., Nùncio, T., Borralho, M.E., Pais, J.R. and Viachos, E. (1988). *Recursos hídricos no Sul de Portugal* — primeiro diagnóstico volume II. Ministério do Planeamento e da Administração do Território. Secretaria de Estado do Ambiente e Recursos Natu.

Ratcliffe, D.A. (Ed.) (1977). *A Nature Conservation Review*, Cambridge University Press, Cambridge.

Real, M. (1993). *Ecologia del zoobentos profund als embassaments de l'estat espanyol*, PhD Thesis, University of Barcelona.

Reeves, G.H., Hall, J.D., Roelofs, T.D., Hickman, T.L., and Baker, C.O. (1991). Rehabilitation and modifying stream habitats, in *Influences of Forest and Rangeland Management on Salmonid Fishes and their Habitats* (Ed. W.R. Meehan), pp. 519–557, American Fisheries Society Special Publication No. 19, Bethesda.

Reice, S.R. (1974). Environmental patchiness and the breakdown of leaf litter in a woodland stream, *Ecology*, **55**, 1271–1282.

Reice, S.R. (1978). Role of detritivore selectivity in species-specific litter decomposition in a woodland stream, *Verhandlungen Internationale Vereinigung für Theoretische und Angewandte Limnologie*, **20**, 1396–1400.

Reice, S.R., Wissmar, R.C. and Naiman, R.J. (1990). Disturbance regimes, resilience, and recovery of animal communities and habitats in lotic systems, *Environmental Management*, **14**, 647–659.

Reichholf, J. (1974). The influence of recreational activities on waterfowl, in *Procedings of the International Conference on Conservation of Wetlands and Waterfowl* (Ed. M. Smart), pp. 364–369, Helligenhafen, Germany.

Reid, I. and Frostick, L.E. (1985). Role of settling, entrainment and dispersive equivalence and interstice trapping in placer formation, *Journal of the Geological Society of London*, **142**, 739–746.

Reid, I., Frostick, L.E. and Layman, J.T. (1985). The incidence and nature of bedload transport during flood flows in coarse-grained alluvial channels, *Earth Surface Processes and Landforms*, **10**, 33–44.

Reiser, D.W., Wesche, T.A. and Estes, C. (1989). Status of instream flow legislation and practices in North America, *Fisheries*, **14**, 22–29.

Reiser, D.W., Ramey, M.P., Beck, S., Lambert, T.R. and Gray, E.R. (1989). Flushing flow recommendations for maintenance of salmonid spawning gravels in a steep regulated stream, *Regulated Rivers: Research and Management*, **3**, 267–275.

Reuss, J.O., Cosby, B.J. and Wright, R.F. (1987). Chemical processes governing soil and water acidification. *Nature, London*, **329**, 27−32.

Reynolds, A.J. and Eaton, J.W. (1983). The role of vegetation structure in a canal fishery, *Proceedings of the 3rd British Freshwater Fisheries Conference*, Liverpool, pp. 192−202.

Reynolds, B.R. and Ormerod, S.J. (1993). *A review of the impact of current and future acid deposition in Wales*, Institute of Terrestrial Ecology, Project Report T07072L1, ITE, Bangor Research Station.

Reynolds, B.R., Neal, C., Hornung, M., Hughes, S. and Roberts, J.D. (1988). Impact of afforestation on the soil solution chemistry of stagnopodzols in mid Wales, *Water, Air and Soil Pollution*, **38**, 55−70.

Reynolds, C.S. (1986). Experimental manipulations of the phytoplankton periodicity in large limnetic enclosures in Blelham Tarn, English Lake District, *Hydrobiologia*, **138**, 43−64.

Reynolds, C.S. (1988). Potamoplankton: paradigms, paradoxes, prognoses, in *Algae and the Aquatic Environment* (Ed. F.E. Round), pp. 285−311, Biopress, Bristol.

Reynolds, C.S. (1989). Physical determinants of phytoplankton succession, in *Plankton Ecology* (Ed. U. Sommer), pp. 9−56, Springer-Verlag, Madison.

Reynolds, C.S. (1992). Algae, in *The Rivers Handbook* Vol. 1 (Eds P. Calow and G.E. Petts), pp. 195−215, Blackwell Scientific, Oxford.

Reynolds, C.S. (1994). The role of fluid dynamics in the ecology of phytoplankton, in *Ecology of Aquatic Organisms: Scale, Pattern, Process* (Eds P.S. Giller, A.G. Hildrew and D. Raffaelli), pp. 141−187, Blackwell Scientific, Oxford.

Reynolds, C.S. and Glaister, M.S. (1989). Remote sensing of phytoplankton in the River Severn, in *NERC Remote Sensing Report 1989* (Ed. S. White), pp. 138−148, UK Natural Environment Research Council, Swindon.

Reynolds, C.S. and Glaister, M.S. (1992). *Environments of Larger UK Rivers*, Report to the UK National Rivers Authority, Institute of Freshwater Ecology, Ambleside.

Reynolds, C.S. and Glaister, M.S. (1993). Spatial and temporal changes in phytoplankton abundance in the upper and middle reaches of the River Severn, *Large Rivers*, **9**, 1−22.

Reynolds, C.S., Carling, P.A. and Beven, K.J. (1991). Flow in river channels: new insights into hydraulic retention, *Archiv für Hydrobiologie*, **121**, 171−179.

Reynolds, C.S., Carling, P.A. and Glaister, M.S. (1989). Dead zones — live markers, Report to UK Department of the Environment, Freshwater Biological Association, Ambleside, Cumbria.

Reynolds, C.S., White, M.L., Clarke, R.T. and Marker, A.F.H. (1990). Suspension and settlement of particles in flowing water: comparisons of the effects of varying water depth and velocity in circulating channels, *Freshwater Biology*, **24**, 23−34.

Reynoldson, T.B., Smith, B.D. and Maitland, P.S. (1981). A species of North American triclad (Paludicola; Turbellaria) new to Britain found in Loch Ness, Scotland, *Journal of Zoology, London*, **193**, 531−539.

Rhodes, H.A. and Hubert, W.A. (1991) Submerged undercut banks as macroinvertebrate habitat in a subalpine meadow stream, *Hydrobiologia*, **213**, 149−153.

Richards, K. (1982). *Rivers: Form and Process in Alluvial Channels*, Methuen.

Richards, K.S. (1979). Channel adjustments to sediment pollution by the china clay industry in Cornwall, England, in *Adjustments of the Fluvial System* (Eds D.D. Rhodes and G.P. Williams), pp. 309−331, Kendall-Hunt, Dubuque, Iowa.

Richardson, B.A. (1986). Evaluation of instream flow methodologies for freshwater fish in New South Wales, in *Stream Protection: The Management of Streams for Instream Uses* (Ed. I.C. Campbell), pp. 23−44, Water Studies Centre, Chisholm Institute of Technology, Melbourne.

Richardson, E.V. and Simons, D.B. (1976). River response to development, *Rivers, American Society of Civil Engineers*, **76**, 1285−1300.

Richardson, J.S. (1992). Food, microhabitat, or both? Macroinvertebrate use of leaf accumulations in a montane stream, *Freshwater Biology*, **27**, 169−176.

Richardson, J.S. and Neill, W.E. (1991). Indirect effects of detritus manipulations in a montane stream, *Canadian Journal of Fisheries and Aquatic Sciences*, **48**, 776−783.

Ricker, W.E. (1975). Handbook of computation and interpretation of biological statistics of fish populations, *Bulletin of the Fisheries Research Board*, **191**.

Ricker, W.E. (1979). Growth rates and models, in *Fish Physiology*, Vol. 8 (Eds W.S. Hoar, D.J. Randall and J.R. Brett), pp. 677−743, Academic Press, New York.

Ridge, I. and Barrett, P.R.F. (1992). Algal conntrol with barley straw, in *Vegetation Management in Forestry, Amenity and Conservation Areas, Aspects of Applied Biology*, **29**, 457−462.

Rieley, J.O. and Page, S.E. (1990). *Ecology of Plant Communities*, Longman, London.

Rincón, P.A. and Lobón-Cerviá, J. (1993). Microhabitat use by stream-resident brown trout: bioenergetic consequences, *Transactions of the American Fisheries Society*, **122**, 575−587.

Rincón, P.A., Barrachina, P., and Bernat, Y. (1992). Microhabitat use by 0+ juvenile cyprinids during summer in a Mediterranean river, *Archiv für Hydrobiologie*, **125**, 323−337.

Risser, R.J. and Harris, R.R. (1989). Mitigation for impacts to riparian vegetation om western montane streams, in *Alternatives in Regulated Rivers Management* (Eds J.A. Gore and G.E. Petts), pp. 235−250, CRC Press, Boca Raton.

Ritchie, J. (1920). *The Influence of Man on Animal Life in Scotland*, Cambridge University Press, Cambridge.

Roberts, C.P.R. (1983). Environmental constraints on water resources development, *Proceedings of the South African Institution of Civil Engineers*, **1**, 16−23.

Robison, E.G. and Beschta, R.L. (1990). Coarse woody debris and channel morphology interactions for undisturbed streams in southeast Alaska, USA, *Earth Surface Processes and Landforms*, **15**, 149−156.

Romijn, C.A.F.M., Luttik, R., van der Meent, D.D., Slooff, W. and Canton, J.H. (1991). *Presentation and analysis of a general algorithm for risk assessment on secondary poisoning*, Report 679102002, National Institute of Public Health and Environmental Protection, Bilthoven, The Netherlands.

Rooke, J.B. (1984). The invertebrate fauna of four macrophytes in a lotic system, *Freshwater Biology*, **14**, 507−513.

Rooke, J.B. (1986a). Seasonal aspects of the invertebrate fauna of three species of plants and rock surfaces in a small stream, *Hydrobiologia*, **134**, 81−87.

Rooke, J.B. (1986b). Macroinvertebrates associated with macrophytes and plastic imitations in the Eramosa River, Ontario, Canada, *Archiv für Hydrobiologie*, **106**, 307−325.

Rosen, R.A. and Hales, H.C. (1980). Occurrence of scarred puddlefish in the Missouri River, South Dakota−Nebraska, *Progressive Fish Culturist*, **42**, 82−84.

Rounick, J.S. and Winterbourn, M.J. (1983). Leaf processing in two contrasting beech forest streams: Effects of physical and biotic factors on litter breakdown. *Archiv für Hydrobiologie*, **96**, 448−474.

Royal Commission on Environmental Pollution (1992). *Freshwater Quality*, 16th Report, HMSO, London.

Royal Society (1990). *The Surface Waters Acidification Project* (Ed. B.J. Mason), Proceedings of the final SWAP conference held at the Royal Society, March 1990.

Rubin, D.M., Schmidt, J.C. and Moore, J.N. (1990). Origin, structure, and evolution of a re-attachment bar, Colorado River, Grand Canyon, Arizona, *Journal of Sedimentary Petrology*, **60**, 982−991.

Ruiz-Olmo, J. (1991). Conservation and management plan for the otter in Catalonia (NE Spain), in *Proceedings of the Vth. International Otter Colloquium, Habitat*, Vol. 6, (Eds C. Reuther and R. Röchert), pp. 259−262, Otter Zentrum, Hankensbüttel.

Rundle, S.D. and Hildrew, A.G. (1990). The micro-arthropods of some southern English streams: the influence of physicochemistry, *Freshwater Biology*, **23**, 411−431.

Rundle, S.D. and Hildrew, A.G. (1992). Small fish and small prey in the food webs of some southern English streams, *Archiv für Hydrobiologie*, **125**, 25−35.

Rundle, S.D. and Ormerod, S.J. (1991). Micro-invertebrate communities in upland Welsh streams in relation to physico-chemical factors, *Freshwater Biology*, **26**, 439−451.

Russell, I.A and Rogers, K.H. (1989). The distribution and composition of fish communities in the major rivers of the Kruger National Park, in *Proceedings of the 4th South African National Hydrological Symposium*, pp. 281−288, University of Pretoria.

Rutt, G.P., Weatherley, N.S. and Ormerod, S.J. (1989). Microhabitat availability in Welsh moorland and forest streams as a determinant of macroinvertebrate distribution, *Freshwater Biology*, **22**, 247−261.

Rutt, G.P., Weatherley, N.S. and Ormerod, S.J. (1990). Relationships between the physicochemistry and macroinvertebrates of British upland streams: the development of modelling and indicator systems for predicting fauna and detecting acidity, *Freshwater Biology*, **24**, 463−480.

Ryder, R.A. and Kerr, S.R. (1990). Harmonic communities in aquatic ecosystems: a management perspective, in *Management of Freshwater Fisheries* (Eds. W.L.T. Densen, B. Steinmetz and R.H. Hughes), pp. 594–623, Pudoc, Wageningen.

Ryder R.A. and Pesendorfer, J. (1989). Large rivers are more than flowing lakes: a comparative review, in *Proceedings of the International Large River Symposium (LARS)* (Ed. D.P. Dodge), *Canadian Special Publication of Fisheries and Aquatic Sciences*, **106**, 65–85.

Ryman, N. (1983). Patterns of distribution of biochemical genetic variation in salmonids: differences between species, *Aquaculture*, **33**, 1–21.

Sabater, S. (1990). Phytoplankton composition in a medium-sized Mediterranean river: the Ter (Spain), *Limnetica*, **6**, 47–56.

Sabater, S. and Muñoz, I. (1990). Successional dynamics of the phytoplankton in the lower part of the River Ebro, *Journal of Plankton Research*, **12**, 573–592.

Saint Girons, M.C. (1991). Wild mink (*Mustela lutreola*) in Europe, Report to the Convention on the Conservation of European Wildlife and Natural Habitats, Council of Europe, Strasbourg.

Sainz de los Terreros, M., Garcia De Jalon, D. and Mayo, M. (1990). Canalizacion y Dragado de cauces: Sus efectos y tecnicas para la restauracion del rio y sus riberas. Diput. Foral de Alava, Vitoria.

Sampson, C. (1990). *Towards biological control of* Heracleum mantegazzianum, MSc Thesis, University of London, Imperial College.

Sand-Jensen, K., Jeppesen, E., Nielsen, K., van der Bijl, L., Hjermind, L., Nielsen, L.W. and Iversen, T.M. (1989). Growth of macrophytes and ecosystem consequences in a lowland Danish stream, *Freshwater Biology*, **22**, 15–32.

Sandars, G., Jones, K.C., Hamilton-Taylor, J. and Dörr, H. (1992). Historical inputs of polychlorinated biphenyls and other organochlorines to a dated laustrine sediment core in rural England. *Enviornmental Science and Technology*, **26**, 1815–1821.

Santiago, S., Thomas, R.L., Largaibt, G., Rossel, D., Echeverria, M.A., Tarradellas, J., Loizeau, J.L., McCarthy, L., Mayfield, C.I.D. and Corvi, C. (1993). Comparative ecotoxicity of suspended sediment in the lower Rhône river using algal fractionation, Microtox® and *Daphnia magna* bioassays, *Hydrobiologia*, **252**, 231–244.

Saraiva, M.G. (cord.) (1991). *Recuperação, Protecção e Valorização das Linhas de Água na Cidade de Évora*, 2° Relatório, Câmara Municipal de Évora, Évora.

Saraiva, M.G. (cord.) (1993a). *Recuperação, Protecção e Valorização das Linhas de Água na Cidade de Évora*, 3° Relatório, Câmara Municipal de Évora, Évora.

Saraiva, M.G. (1993b). *Conservação dos Sistemas Fluviais e Impactos Ambientais das Acções de Regularização*, I Jornadas Hispano-Lusas de Impacto Ambiental, Badajoz.

Sarojini, R., Indira, B. and Nagabhushanam, R. (1990). Effect of sublethal concentrations of two antifouling organometallic compounds, $CuSO_4$ and TBTO, on the eyestalks of freshwater prawn, *Caridina weberi*, *Journal of Freshwater Biolology*, **2**, 29–35.

Sawyer, F. (1985). *Keeper of the Stream — The Life of a River and its Trout Fishery*, Allen and Unwin, London.

Scarnecchia, D.L. (1983). *Trout and Char production in Colorado's small streams*, Doctoral Thesis, Colorado State University.

Scarnecchia, D.L. and Bergersen, E.P. (1987). Trout production and standing crop in Colorado's small streams, as related to environmental features, *North American Journal of Fisheries Management*, **7**, 315–330.

Schiemer, F. (1985). Bedeutung von Augewasser als Schutzzonen für die Fischfauna, *Österreichische Wässerwirtschaft*, **37**, 239–245.

Schiemer, F. (1991). Fish fauna in the Austrian Danube: aspects of conservation in the ecotone concept, in *Fish and Land/Inland Water Ecotones* (Eds M. Zalewski, J.E. Thorpe and P. Gaudin), pp. 21–24, UNESCO, University of Lodz.

Schiemer, F. and Spindler, T. (1989). Endangered fish species of the Danube river in Austria, *Regulated Rivers: Research and Management*, **4**, 397–407.

Schiemer, F. and Waidbacher, H. (1992). Strategies for conservation of a Danubian fish fauna, in *River Conservation and Management* (Eds P.J. Boon, P. Calow and G.E. Petts), pp. 363–382, John Wiley, Chichester.

Schiemer, F. and Zalewski, M. (1992). The importance of riparian ecotones for the diversity and productivity of riverine fish communities, *Netherlands Journal of Zoology*, **42**, 323–335.

Schiemer, F., Spindler, T., Wintersperger, H., Schneider, A., and Chovanec, A. (1991). Fish fry associations: important indicators for the ecological status of large rivers, *Verhandlungen der Internationalen Vereinigung für angewandte Limnologie*, **24**, 2497–2500.

Schindler, D.W., Turner, M.A., Stainton, M.P. and Linsay, G.A. (1986). Natural sources of acid neutralising capacity in low alkalinity lakes of the Precambrian Shield, *Science*, **232**, 844–847.

Schloesser, D.W. and Manny, B.A. (1989). Potential effects of shipping on submersed macrophytes in the St. Clair and Detroit Rivers of the Great Lakes, *Michigan Academy*, **21**, 101–108.

Schlosser, I.J. (1982). Fish community structure and function along two habitat gradients in a headwater stream, *Ecological Monographs*, **52**, 395–414.

Schlosser, I.J. (1987). A conceptual framework for fish communities in small warmwater streams, in *Community and Evolutionary Ecology of North American Stream Fishes* (Eds W.J. Matthews and D.C. Heins), pp. 17–24, University of Oklahoma Press.

Schlosser, I.J. (1990). Environmental variation, life history attributes, and community structure in stream fishes: implications for environmental management and assesment, *Environmental Management*, **14**, 621–628.

Schoener, T.W. (1989). The ecological niche, in *Ecological Concepts* (Ed. M. Cherrett), pp. 79–113, British Ecological Society Symposium, Blackwell Scientific, Oxford.

Schofield, K., Townsend, C.R. and Hildrew, A.G. (1988). Predation and the prey community of a headwater stream, *Freshwater Biology*, **20**, 85–96.

Scholten, M.C.T., Foekema, E., de Kock, W.Chr. and Marquenie, J.M. (1989). *Reproduction failure in tufted ducks feeding on mussels from polluted lakes*, MT-TO, Laboratory for Applied Marine Research, Den Helder, The Netherlands.

Schuldes, H. and Kübler, R. (1990). Ökologie und Vergesellschaftung von *Solidago canadensis et gigantea*, *Reynoutria japonica et sachalinense*, *Impatiens glandulifera*, *Helianthus tuberosus* und *Heracleum mantegazzianum*. Ihre Verbreitung in Baden-Wurttemberg Sowie Notwendigkeit und Möglichkeiten ihrer Bekampfung. Ministeriums für Umwelt, Baden-Wurttemberg.

Schulze, R.E. and George, W.J., (1987). A dynamic, process-based, user-oriented model of forest effects on water yield, *Hydrological Processes*, **1**, 292–307.

Schumm, S.A. and Lichty, R.W. (1958). Time, space and causality in geomorphology, *American Journal of Science*, **263**, 110–119.

Schwabe, A. and Kratochwil, A. (1991) Gewässer-begleitende Neophyten und ihre Beurteilung aus Naturschutz Sicht unter besonderer Berücksichtigung Südwestdeutschlands, *NNA Berichte*, **4/1** (special issue).

Schwank, P. (1984) Differentiation of the coenoses of helminthes and annelids in exposed lotic microhabitats of a mountain stream, *Verhandlungen Internationale Vereinigung für Theoretische und Angewandte Limnologie*, **22**, 2048.

Scott, D. and Shirwell, C.S. (1987). A critique of the instream flow incremental methodology and observations on flow determination in New Zealand, in *Regulated Streams, Advances in Ecology* (Eds J.F. Craig and J.B. Kemper), pp. 27–43, Plenum, New York.

Scottish Natural Heritage (1994). *Scottish Natural Heritage and Sites of Special Scientific Interest*, Scottish Natural Heritage, Edinburgh.

Seager, J., Milne, I., Rutt, G. and Crane, M. (1992). Integrated biological methods for river water quality assessment, in *River Water Quality* (Eds P.J. Newman, M.A. Piavaux and R.A. Sweeting), pp. 399–416, Office for Official Publications of the European Community.

Seagle, H.H. and Zumwalt, F.H. (1981). Evaluation of the effects of tow passage on aquatic macroinvertebrate drift in Pool 26, Mississippi River, report for Environmental Works Team, Upper Mississippi Basin Committee Master Plan, Minnesota, MN, USA.

Seagrave, C. (1988). *Aquatic Weed Control*, Fishing News Books, Farnham.

Sear, D.A. (1992). Impact of hydroelectric power releases on sediment transport processes in pool–riffle sequences, in *Dynamics of Gravel-bed Rivers* (Eds P. Billi, R.D. Hey, C.R. Thorne and P. Tacconi), pp. 629–650, Wiley, Chichester.

Seber, G.A.F. (1973) *The Estimation of Animal Abundance*, Griffin, London.

Seber, G.A.F. and Le Cren, E.D. (1967). Estimating population parameters from catches large relative to the population, *Journal of Animal Ecology*, **36**, 631–43.

Secombes, C.J. (1991). Current and future developments in salmonid disease control, *Proceedings of the Institute of Fisheries Management Annual Study Course*, **22**, 81–88.

Sedell, J.R. and Froggatt, J.L. (1984). Importance of streamside forests to large rivers: The isolation of the Willamette River, Oregon, USA, from its floodplain by snagging and streamside forest removal, *Verhandlungen Internationale Vereinigung für Theoretische und Angewandte Limnologie*, **22**, 1828–1834.

Sedell, J.R., Triska, F.J. and Triska, N.S. (1975). The processing of conifer and hardwood leaves in two coniferous forest streams. I. Weight loss and associated invertebrates, *Verhandlungen Internationale Vereinigung für Theoretische und Angewandte Limnologie*, **19**, 1617–1627.

Selkirk, W.T. and Hart, R.C. (1984). *The Buffalo River Catchment: eutrophication and mineralisation of the Buffalo River reservoirs*, Institute for Freshwater Studies, Special Report 84/1, Grahamstown.

Serns, S.L. (1982). Relationship of walleye fingerling density and electrofishing catch per unit effort in northern Wisconsin lakes, *North American Journal of Fisheries Management*, **2**, 38–44.

Serns, S.L. (1983). Relationship between electrofishing catch per unit effort and density in walleye fingerlings, *North American Journal of Fisheries Management*, **3**, 451–452.

Shannon, C.E. and Weaver, W. (1949). *The Mathematical Theory of Communication*, The University of Illinois Press.

Sharrock, J.T.R. (Ed.) (1976). *The Atlas of Breeding Birds in Britain and Ireland*. BTO/IWC, Tring, UK.

Sheail, J. (1984). Constraints on water-resource development in England and Wales, concept and management of compensation flows, *Journal of Environmental Management*, **19**, 351–361.

Sheail, J. (1987). Historical development of setting compensation flows, in *A Study Of Compensation Flows in the UK* (Eds A. Gustard, G. Cole, D. Marshall and Bayliss), Report No. 99, Institute of Hydrology, Wallingford, UK, Appendix 1.

Shearer, W.M. (1986). An evaluation of the data available to assess Scottish salmon stocks, in *The Status of the Atlantic Salmon in Scotland*, (Eds D. Jenkins and W.M. Shearer), pp. 91–111, ITE Symp. No. 15, Banchory, 1985 NERC.

Shearer, W.M., Cook, R.M., Dunkley, D.A., Maclean, J.C. and Shelton, R.G.J. (1987). *A model to assess the effect of predation by sawbill ducks on the salmon stock of the River North Esk*, DAFS Scottish Fisheries Research Reports No. 37, 1–12.

Sheldon, A.L. (1988). Conservation of stream fishes: patterns of diversity, rarity and risk, *Conservation Biology*, **2**, 149–156.

Sheldon, A.L. and Haick, R.A. (1981). Habitat selection and association of stream insects: a multivariate analysis, *Freshwater Biology*, **11**, 395–403.

Shetter, D. S. (1969). The effects of certain angling regulations on stream trout populations, in *Symposium on Salmon and Trout in Streams* (Ed. T.G. Northcote), pp. 333–353, University of British Columbia, Vancouver.

Short, R.A., Canton, S.P. and Ward, J.V. (1980). Detrital processing and associated macroinvertebrates in a Colorado mountain stream, *Ecology*, **61**(4), 727–732.

Shuter, B.J. (1990). Population-level indicators of stress, in *Biological Indicators of Stress in Fish* (Ed. S.M. Adams), pp. 145–166, American Fisheries Society, Bethesda.

Simmonds, M. (1986). The case against tributyltin, *Oryx*, **20**, 217–220.

Simões, P., Machado, E., Fernandes, R. and Pinto, P. (1992). Impacto da cidade de Évora sobre o rio Xarrama e ribeira da Torregela (bacia hidrográfica do Sado), *3° Conferência Nacional sobre a Qualidade do Ambiente*, **1**, 161–170.

Simon, A. (1989). The discharge of sediment in channelized alluvial streams, *Water Resources Bulletin*, **25**, 1177–1188.

Simon, B.M. and Jones, J.G. (1992). Some observations on the absence of bacteria from acid waters in northwest England, *Freshwater Forum*, **2**, 200–211.

Simpson, D.A. (1984). A short history of the introduction and spread of *Elodea* Michx in the British Isles, *Watsonia*, **15**, 1–9

Sinclair, A.E.G. (1989). The regulation of animal populations, in *Ecological Concepts* (Ed. M. Cherrett), pp. 197–241, British Ecological Society Symposium, Blackwell Scientific, Oxford.

Singh, K.P. and Broeren, S.M. (1989). Hydraulic geometry of streams and stream habitat assessment, *Journal of Water Resources Planning and Management*, **115**(5), 583–597.

Skaren, U. (1988). Chlorinated hydrocarbons, PCBs and caesium isotopes in otters (*Lutra lutra* L.) from central Finland, *Annales Zoologici Fennici*, **25**, 271−276.

Skeffington, R.A. (1991). Soil/water acidification and the potential for reversibility, in *Restoring Acid Waters: Loch Fleet 1984−1990* (Eds G. Howells and T.R.K. Dalziel), pp. 23−37, Elsevier Applied Science, London and New York.

Skempton, A.W. (1984). Engineering on English River Navigations to 1760, in *Canals, A New Look* (Eds M. Baldwin and A. Burton), pp. 23−44, Phillimore, Chichester.

Skuhrávy, V. (1978). Invertebrates: destroyers of common reed, in *Pond Littoral Ecosystems; Structure and Functioning. Methods and Results of Quantitative Ecosystem Research in the Czechoslovakian IBP Wetland Project* (Eds D. Dykyjová and J. Kvet), pp. 376−388, Springer-Verlag, Berlin.

Slack, H.D. (1936). The food of caddisfly larvae, *Journal of Animal Ecology*, **5**, 105−115.

Smart, M.M., Rada, R.G., Nielsen, D.N. and Claflin, T.O. (1985). The effect of commercial and recreational traffic on the resuspension of sediment in Navigation Pool 9 of the Upper Mississippi River, *Hydrobiologia*, **126**, 263−274.

Smith, C.D., Barham, P.J. and Harper, D.M. (1990a). *River Welland environmental survey*, Unpublished Report, NRA Operational Investigation A13−38A.

Smith, C.D., Harper, D.M. and Barham, P.J. (1990b). Engineering operations and invertebrates: linking hydrology with ecology, *Regulated Rivers: Research and Management*, **5**, 89−96.

Smith, C.D., Harper, D.M. and Barham, P.J. (1991). *Physical environment for river invertebrate communities*, Project Report, National Rivers Authority (Anglian Region), Operational Investigation A13−38A.

Smith, C.S. and Adams, M.S. (1986). Phosphorus transfer from sediments by *Myriophyllum spicatum*, *Limnology and Oceanography*, **31**, 1312−1321.

Smith, G.R. and Stearley, R.F. (1989). The classification and scientific names of rainbow and cutthroat trouts, *Fisheries*, **14**, 4−10.

Smith, K.G.V. (1989). An introduction to the immature stages of British flies. Diptera larvae, with notes on eggs, puparia and pupae, *Royal Entomological Society of London Handbooks for the Identification of British Insects*, **10**, Part 14, pp. 1−280.

Smith-Cuffney, F.L. and Wallace, J.B. (1987). The influence of microhabitat on availability of drifting invertebrate prey to a net-spinning caddisfly, *Freshwater Biology*, **17**, 91−98.

Soil Conservation Service (1963). *Guide for Selecting Roughness Coefficient 'n' Values in Channels*, US Department of Agriculture, Soil Conservation Servive, Washington DC.

Sokal, R.R. and Rohlf, F.J. (1981). *Biometry*, 2nd edn, W.H. Freeman, New York.

Solomon, D.J. (1978). Migration of smolts of Atlantic salmon (*Salmo salar* L.) and sea trout (*Salmo trutta* L.) in a chalk stream. *Enviornmental Biology of Fishes*, 3, 226−229.

Solomon, D.J. (1985). Salmon stock and recruitment, and stock enhancement. *Journal of Fish Biology*, **27**, (Suppl. A), 45−58.

Solomon, D.J. and Templeton, R.G. (1976). Movements of brown trout *Salmo trutta* L. in a chalk stream, *Journal of Fish Biology*, **9**, 411−423.

Soluk, D.A. (1985). Macroinvertebrate abundance and production of psammophilous chironomidae in shifting sand areas of a lowland river, *Canadian Journal of Fisheries and Aquatic Sciences*, **42**, 1296−1302.

Sorenson, D.L., McCarthy, M.M., Middlebrooks, E.J. and Porcella, D.B. (1977). *Suspended and dissolved solids effects on freshwater biota: a review*, Ecological Research Series, US Environmental Protection Agency, Report 600/13-77-042.

Sosiak, A.J. (1990). *An evaluation of nutrients and biological conditions in the Bow River, 1986 to 1988*. Internal report of Environmental Quality Monitoring Branch, Environmental Assessment Division, Alberta Environment, Edmonton.

Soszka, G.J. (1975). Ecological relationships between invertebrates and submerged macrophytes in the lake littoral, *Ekologia Polska*, **23**, 393−415.

Southwood, T.R.E. (1978). *Ecological Methods*, Chapman and Hall, London.

Sparks, R.E. (1975). *Possible biological impacts of wave wash and resuspension of sediments caused by boat wash in the Illinois River*, Report, US Army Corps of Engineers, St. Louis, Missouri.

Speaker, R., Moore, K. and Gregory, S. (1984). Analysis of the process of retention of organic matter in stream ecosystems, *Verhandlungen Internationale Vereinigung für Theoretische und Angewandte*

Limnologie, **22**, 1835–1841.

Spence, D.H.N. (1964). The macrophytic vegetation of freshwater lochs, swamps and associated fens, in *The Vegetation of Scotland* (Ed. J.H. Burnett), Oliver & Boyd, Edinburgh.

Spence, J.R. (1981). Experimental analysis of microhabitat selection in water-striders (Heteroptera: Gerridae), *Ecology*, **62**, 1505–1514.

Spence, J.R. and Scudder, G.G.E. (1980). Habitats, life cycles and guild structure of water-striders (Heteroptera: Gerridae) on the Fraser Plateau of British Columbia, *Canadian Entomologist*, **112**, 779–792.

Spiridonov, G. and Spassov, N. (1989). The otter (*Lutra lutra* L., 1758) in Bulgaria, its state and conservation, *Historia naturalis bulgarica*, **1**, 57–64.

Sprague, J.B. (1970). Measurement of pollutant toxicity to fish II. Utilizing andapplying bioassay results, *Water Research*, **4**, 3–32.

Sprague, J.B. (1971). Measurement of pollutant toxicity to fish III. Sublethal effects and safe concentrations, *Water Research*, **5**, 245–266.

Stabler, M.J. and Ash, S.E. (1978). *The Amenity Demand For Inland Waterways. Private Boating — Summary Report*. Report of the Amenity Waterways Study Unit, Dept. of Economics, University of Reading, UK.

Stalnaker, C.B. and Arnette, J.L. (Eds) (1976). *Methodologies For Determination Of Stream Resource Flow Requirements: An Assessment*, FWS/OBS-76/03. US Fish and Wildlife Service, Washington DC.

Stander, G. J. (1952). The quality requirements of water for the maintenance of aquatic flora and fauna and for recreational purposes, *The South African Industrial Chemist*, June 1952.

Stanford, J.A. and Gaufin, A.R. (1974). Hyporheic communities of two Montana rivers, *Science*, **185**, 700–702.

Statzner, B. (1988). Growth and Reynolds Number of lotic macroinvertebrates: a problem for adaptation of shape to drag, *Oikos*, **51**, 84–87.

Statzner, B. and Higler, B. (1986). Stream hydraulics as a major determinant of benthic invertebrate zonation patterns, *Freshwater Biology*, **16**, 127–139.

Statzner, B., Gore, J.A. and Resh, V.H. (1988). Hydraulic stream ecology: observed patterns and potential applications, *Journal of the North American Benthological Society*, **7**, 307–360.

Stebbings, R.E. (1988). *The Conservation of European Bats*, Helm, London.

Steedman, R.J. (1988). Modification and assessment of an index of biotic integrity to quantify stream quality in southern Ontario, *Canadian Journal of Fisheries and Aquatic Sciences*, **45**, 492–501.

Stephens J.C., Blackburn R.D., Seaman D.E. and Weldon L.W. (1963). Flow retardance by channel weeds and their control, *Journal of the Irrigation and Drainage Division of the American Society of Civil Engineers*, **89**, IR2.

Stern, E.M. and Stickle, W.B. (1978). *Effects of turbidity and suspended material in aquatic environments: a literature review*, Techical Report B-78-21, US Army Corps of Engineers Waterways Experimental Station, Vicksburg, MS, USA.

Stevens, M.A., Simons, D.B. and Richardson, E.V. (1975). Non-equilibrium river form, *Journal of the Hydraulics Division of the American Society of Civil Engineers*, **101**, 557–566.

Stevens, P.A., Williams, T.G., Norris, D.A. and Rowland, A.P. (1993). Dissolved inorganic nitrogen budget for a forested catchment at Beddgelert, North Wales, *Environmental Pollution*, **80**, 1–8.

Stokes, P.M. (1981). Benthic algal communities in acid lakes, in *Effects of Acid Precipitation on Benthos* (Ed. R. Singer), pp. 119–138, North American Benthological Society, Hamilton, New York.

Stone, D. (1991). Man and Desman, *BBC Wildlife*, **9**, 538–541.

Stoner, J.H., Wade, K.R. and Gee, A.S. (1984). The effects of acidification on the ecology of streams in the upper Tywi catchment in west Wales, *Environmental Pollution*, (Series A), **36**, 125–157.

Stortelder, P.B.M., van der Gaag, M.A. and van der Kooij, L.A. (1991). *Perspectives for waterorganisms, an ecotoxicological basis for quality objectives for water and sediment*, Report 89.016a+b, Institute for Inland Water Management and Waste Water Treatment R.I.Z.A., Lelystad, The Netherlands.

Stott, B. (1977). On the question of the introduction of the Grass Carp (*Ctenopharyngodon idella* Val.) into the United Kingdom, *Fisheries Management*, **3**, 63–71.

Stout, R.J., Taft, W.H. and Merritt, R.W. (1985). Patterns of macroinvertebrate colonization on fresh and senescent alder leaves in two Michigan streams, *Freshwater Biology*, **15**, 573–580.

Stoyneva, M.P. (1991). *Algoflora of the River Duna (Bulgarian Sector) and of adjacent waters*, PhD Thesis, University of Sofiya.

Street, M. (1979). The importance of invertebrates and straw, *Game Conservancy Annual Review*, **11**, 34–38.

Street, M. and Titmus, G. (1982). A field experiment on the value of allochthonous straw as food and substratum for lake macro-invertebrates, *Freshwater Biology*, **12**, 403–410.

Streeter, H.W. and Phelps E.B. (1925). *A Study of the Pollution and Natural Purification of The Ohio River III*, Bulletin of the US Public Health Service, No. 146.

Strommer, J.L. and Smock, L.A. (1989). Vertical distribution and abundance of invertebrates within the sandy substrate of a low-gradient headwater stream, *Freshwater Biology*, **22**, 263–274.

Suárez, J.L., Reiriz, L. and Anadón, R. (1988). Feeding relationships between two salmonid species and the benthic community, *Polskie Archiwum Hydrobiologii*, **35**, 341–359.

Sukopp, H. (1971). Effects of man, especially recreational activities, on littoral macrophytes, *Hydrobiologia*, **12**, 331–340.

Sukopp, H. (1972). Florenwandel und Vegetationsveranderungen in Mitteleuropa wahrend der letzten Jahrhunderte, in *Gesellschaftsentwicklung (Syndynamik)* (Ed. R. Tuxen), pp. 469–489, Cramer, Vaduz.

Surber, E.W. (1971). The effect of outboard motor exhaust on fish and their environment, *Journal of the Washington Academy of Science*, **61**, 120–123.

Suren, A.M. (1991). Bryophytes as invertebrate habitat in two New Zealand alpine streams, *Freshwater Biology*, **26**, 399–418.

Sutcliffe, D.W. and Hildrew, A.G. (1989). Invertebrate communities in acid streams, in *Acid Toxicity and Aquatic Animals* (Eds R. Morris, E.W. Taylor, D.J.A. Brown and J.A. Brown), Society for Experimental Biology, Seminar Series 34, Cambridge University.

Swales, S., Bishop, K.A. and Harris, J.H. (1994). Assessment of environmental flows for native fish in the Murray-Darling Basin — a comparison of methods, in *Proceedings environmental flows seminar*, pp. 184–192, Canberra, August 1994. Australian Water and Wastewater Association Inc., Artarmon, NSW, Australia.

Swales, S. and O'Hara, K. (1983). A short term study of the effects of a habitat improvement programme on the distribution and abundance of fish stocks in a small lowland river in Shropshire, *Fisheries Management*, **14**, 135–144.

Swanberg, O. and Tarkpea, M. (1982). The acute toxicity of motor fuels to brackish water organisms, *Marine Pollution Bulletin*, **13**, 125–127.

Sweeting, R.A., Lowson, D., Hale, P. and Wright, J.F. (1992). Biological assessment of rivers in the UK, in *River Water Quality* (Eds P.J. Newman, M.A. Piavaux and R.A. Sweeting), pp. 319-326, Office for Official Publications of the European Communities.

Symons, P.E.K. (1979). Estimated escapement of Atlantic salmon (*Salmo salar*) for maximum smolt production in rivers of different productivity, *Journal of the Fisheries Research Board of Canada*, **36**, 132–140.

Symons, P.E.K. and Heland, M. (1978). Stream habitats and behavioural interactions of underyearling and yearling Atlantic salmon (*Salmo salar*), *Journal of the Fisheries Research Board of Canada*, **32**, 633–642.

Tachet, H., Bournaud, M. and Richoux, P. (1980). *Introduction à l'étude des Macroinvertébrés de Eaux Douces (Systématique Élémentaire et Aperçut Écologique)*. Universite Lyon I, Association Française Limnologie, Paris.

Tallis, J.H. (1973). Studies on southern Pennine peats. V. Direct observations on peat erosion and peat hydrology at Featherbed Moss, Derbyshire, *Journal of Ecology*, **61**, 1–22.

Tanner, C.C., Clayton, J.S. and Wells, R.D.S. (1993). Effects of suspended solids on the establishment and growth of *Egeria densa*, *Aquatic Botany*, **45**, 299–310.

Tautz, A.F. and Groot, C. (1975). Spawning behaviour of chum salmon (*Oncorhynchus keta*) and rainbow trout (*Salmo gairdneri*), *Journal of the Fisheries Research Board of Canada*, **32**, 633–642.

Taylor K. (1984). The influence of watercourse management on moorhen breeding biology, *British Birds*, **77**, 144–148.

Taylor, E.B. (1991). A review of local adaptation in Salmonidae, with particular reference to Pacific and Atlantic salmon, *Aquaculture*, **98**, 185−207.

Teixeira, C. (cord) (1972). *Carta Geológica de Portugal*, Serviços Geológicos de Portugal, Lisboa.

Templeton, R.G. and Churchward, A. (1990). Fisheries management practices of the Severn Trent Water Authority in England 1976−1986, in *Management of Freshwater Fisheries* (Eds W.L.T. Densen, B. Steinmetz and R.H. Hughes), pp. 558−568, Pudoc, Wageningen.

Ten Brink, B.J.H. and Woudstra, J.H. (1991). Towards an effective and rational water management: the Aquatic Outlook Project − integrating water management, monitoring and research, *EWPC*, **1**(5), 20−27.

Ten Brink, B.J.H., Hosper, S.H. and Colijn, F. (1991). A quantitative method for description and assessment of ecosystems: The AMOEBA-approach, *Marine Pollution Bulletin*, **23**, 265−270,

Ter Braak, C.J.F. (1988). *CANOCO − a FORTRAN Program for Canonical Community Ordination by (Partial) (Detrended) (Canonical) Correspondence Analysis*, TNO Institute of Applied Computer Science, Wageningen.

Tesch, F.W. (1962). Witterungsabhängigkeit der Brutenwicklung und Nachwuchsforderung bei *Lucioperca lucioperca* L., *Kurze Mitteilungen aus dem Institut für Fischereibiologie der Universität Hamburg*, **12**, 37−44.

Thames Water Authority (1980). *Report of the Working Party on River Thames leisure policy*. I. Boating. Thames Water Authority, Reading, UK.

Thibodeaux, L.J. and Boyle, J.D. (1987). Bedform-generated convective transport in bottom sediment, *Nature*, **325**, 341−343.

Thomas, L. (1974). *The Lives of a Cell*, Bantam Books, New York.

Thorne R.E. (1983). Hydroacoustics, in *Fisheries Techniques* (Eds L.A. Neilsen and D.L. Johnson), pp. 239−260, American Fisheries Society, Bethesda.

Thornton, I.W.B. (1957). Faunal succession in umbels of *Cyperus papyrus* L. on the upper White Nile, *Proceedings of the Royal Entomological Society London, Series A*, **32**, 119−131.

Thorup, J. (1966). Substrate type and its value as a basis for the delimitation of bottom fauna communities in running waters, in *Organism-Substrate Relationships in Streams* (Eds K.W. Cummins, C.A. Tyron Jr. and R.T. Hartman), pp. 59−74, Pymatuning Laboratory of Ecology, Special Publication No. 4, Pittsburgh.

Thorup, J. and Lindegaard, C. (1977). Studies on Danish springs, *Folia Limnologica Scandinavica*, **17**, 7−15.

Tiley, G.E.D. and Philp, B. (1994). *Heracleum mantegazzianum* (Giant hogweed) and its control in Scotland, in *Ecology and Management of Invasive Riverside Plants* (Eds L.C. de Waal, L.E. Child, P.M. Wade and J.H. Brock), pp. 101−109, Wiley, Chichester.

Tipping, E., Woof, C. and Clarke, K. (1993). Deposition and resuspension in a riverine dead zone, *Hydrological Processes*, **7**, 263−278.

Tittizer, T. and Schöll, F. (1988). *Faunistische Erhebungen an der Rheinsohle zur Feststellung und Bewertung der Schädigung der Benthosbiozönose durch den Brand bei der Fa. Sandoz in Basel. 2. Berichtszeitraum 1.1.1988−31.12.1988*. UBA-Forschungsbericht 10607073; Koblenz.

Tittizer, T., Schöll, F. and Dommermuth, M. (1993). *Die Entwicklung der Lebensgemeinschaften des Rheins im 20. Jahrhundert, Tagungsband: Die Biozönose des Rheins im Wandel, Lachs 2000?*, Ministerium für Umwelt Rheinland-Pfalz (Hrsg.) Petersberg.

Tittizer, T., Schöll, F., Dommermuth, M., Bäthe, J. and Zimmer, M. (1991). Zur Bestandsentwicklung des Zoobenthons des Rheins im Verlauf der letzten neun Jahrzehnte, *Wasser und Alwesser* **35**, 126−166.

Titus, R.G. and Mosegaard, H. (1992). Fluctuating recruitment and variable life history of migratory brown trout, *Salmo trutta* L., in a small, unstable stream, *Journal of Fish Biology*, **41**, 239−255.

Tokeshi, M. (1986a). Population ecology of the commensal chironomid *Epoicocladius flavens* on its mayfly host *Ephemera danica*, *Freshwater Biology*, **16**, 235−243.

Tokeshi, M. (1986b). Population dynamics, life histories and species richness in an epiphytic chironomid community, *Freshwater Biology*, **16**, 431−441.

Toledo, M.M., Lemaire, A.L., Baglinière, J.L. and Braña, F. (1993). Características biológicas de la trucha marina (*Salmo trutta* L.) en dos ríos de Asturias, norte de España, *Bull. Fr. Pêche Piscic.*, **330**, 295−306.

Townsend, C.R. (1989). The patch dynamics concept of stream community ecology, *Journal of the North American Benthological Society*, **8**, 36−50.

Trimble, S.W., Weirich, F.H. and Hoag, B.L. (1987). Reforestation and the reduction of water yield on the Southern Piedmont since circa 1940, *Water Resources Research*, **23**, 425–437.

Triska, F.J. (1984). Role of wood debris in modifying channel geomorphology and riparian areas of a large lowland river under pristine conditions: A historical case study, *Verhandlungen Internationale Vereinigung für Theoretische und Angewandte Limnologie*, **22**, 1876–1892.

Tuite, C.H. (1982). *The impact of water-based recreation on the waterfowl of enclosed waters in Britain*, Sports Council and Nature Conservancy Council Report.

Turner, C. (1989). PhD Progress Report: *Episodic pollution and recovery in streams*, unpublished report, University of Wales, College of Cardiff.

Turner, C. (1992). *Episodic pollution and recovery in streams*, PhD Thesis, University of Wales.

Turnpenny, A.W.H. and Williams, R. (1980). Effects of sedimentation on the gravels of an industrial river system, *Journal of Fish Biology*, **17**, 681–693.

Twigg, H.M. (1959). Freshwater studies in the Shropshire Union Canal, *Field Studies*, **1**, 116–142.

Tydeman, C.F. (1977). The importance of the close fishing season to breeding bird communities, *Journal of Environmental Management*, **5**, 289–296.

Tyus, H.M. (1990). Effects of altered stream flows on fishery resources, *Fisheries*, **15**, 18–21.

Uetz, H.-U. (1977). Coexistence in a guild of wandering spiders, *Journal of Animal Ecology*, **46**, 531–541.

UNCED (1992). *Protection of the Quality and Supply of Freshwater Resources: Application of Integrated Approaches to the Development, Management and Use of Water Resources*, Agenda 21, Chapter 18, United Nations Conference on Environment and Development, United Nations, Conches.

United Kingdom Acid Waters Review Group (1989). *Acidity in United Kingdom Fresh Waters*, Second Report of the UK Acid Waters Review Group, Department of the Environment, HMSO, London.

United States Environmental Protection Agency (1983). *The acidic deposition phenomenon and its effects, critical assessment and review papers*, EPA-600/8-83016B, Washington DC.

Urquhart, W.J. (1975). *Hydraulics: Engineering Field Manual*, US Department of Agriculture, Soil Conservation Service, Washington DC.

Utter, F., Aebersold, P. and Winans, G. (1987). Interpreting genetic variation detected by electrophoresis, in *Population Genetics and Fishery Management* (Eds N. Ryman and F. Utter), pp. 20–45, University of Washington Press, Seattle.

Vaillant, F. (1953). Les Trichoptères à larves hygropétriques, *Trav. Lab. Hydrobiol. Piscic. Univ. Grenoble*, **45**, 33–48.

Vaillant, F. (1954). *Tinodes algirica* McLachlan, the hygropetric larvae of the *Tinodes* (Trichoptera), *Annals and Magazine of Natural History*, **7**, 58–62.

Van de Guchte, C. (1990). The sediment quality triad: an integrated approach to assess contaminated sediments, in *Environmental Sciences and Sustainable Development*, Founding conference, SETAC-Europe, Sheffield.

Van den Berg, M., Craane, B.L.H.J. Sinnige, T., Lutke-Schipholt, I.J., Spenkelink, B. and Brouwer, A. (1992). The use of biochemical parameters in comparative toxicological studies with the cormorant (*Phalacrocorax carbo*) in the Netherlands, *Chemosphere*, **25**, 7–10.

Van der Gaag, M.A. (1991). Setting environmental quality criteria for water and sediment in the Netherlands: a pragmatic ecotoxicological approach, *European Water Pollution Control*, **1**, 13–20.

Van der Kooij, L.A., Meent, D. v.d., van Leeuwen, C.J. and Bruggeman W.A., (1991). Deriving quality criteria for water and sediment from the result of aquatic toxicity and product standards: application of the equilibrium partitioning method, *Water Research*, **26**, 697–705.

Van der Merwe, W. and Grobler, D.C. (1990). Water quality management in the Republic of South Africa, *Water SA*, **16**, 49–54.

Van der Valk, F. (1989). Bioaccumulation in yellow eel (*Anguilla*) and perch (*Perca fluviatilis*) from the Dutch branches of the Rhine — mercury, organochlorine compounds and polynuclear aromatic hydrocarbons, Publication 7, *Publications and reports of the project Ecological Rehabilitation Rhine*, DBW/RIZA, RIVM, RIVO, Netherlands Institute for Fishery Research, Ijmuiden, NL.

van Niewenhuyse, E.E. and La Perriere, J.D. (1986). Effects of placer gold mining on the primary productivity of subarctic streams in Alaska, *Water Resources Bulletin*, **22**, 91–99.

Van Urk, G. (1978). The macrobenthos of the river Ijssel, *Hydrobiological Bulletin*, **12**, 21–29.

Van Urk, G., Kerkum, F.C.M. and van Leeuwen, C.J. (1993). Insects and insecticides in the lower

Rhine, *Water Research*, **27**, 205–213.

Vannote, R.L., Minshall, G.W., Cummins, K.W., Sedell, J.R. and Cushing, C.E. (1980). The river continuum concept, *Canadian Journal of Fisheries and Aquatic Sciences*, **37**, 130–137.

Verdaguer, A., Serra, J. and Canals, M. (1985). L'intéraction fluviatile et marine dans le cours inferieur de lEbre: consequences sedimentologiques, *Rapport Commission Internationale Mer Méditerranéene*, **29**(2), 185–187.

Verdonschot, P.F.M. (1992). Macrofaunal community types in ponds and small lakes (Overijssel, The Netherlands), *Hydrobiologia*, **232**, 111–132.

Verdonschot, P.F.M. and Higler, L.W.G. (1989). Macroinvertebrates in Dutch ditches: a typological characterisation and the status of the Demmerik Ditches, *Hydrobiological Bulletin*, **23**, 135–142.

Vermaat, J.E. and de Bruyne R.J. (1993). Factors limiting the distribution of submerged waterplants in the lowland River Vecht (The Netherlands), *Freshwater Biology*, **30**, 147–158.

Verspoor, E. and Jordan, W.C. (1989). Genetic variation at the Me-2 locus in the Atlantic salmon within and between rivers: evidence for its selective maintenance, *Journal of Fish Biology*, **35**, (Suppl. A), 205–213.

Via, S. and Lande, R. (1985). Genotype-environment interactions and the evolution of phenotypic plasticity, *Evolution*, **39**, 505–522.

Vickery, J.A. and Ormerod, S.J. (1991). Dippers as indicators of stream acidity, *Acta XX Congressus Internationalis Ornitologici*, **IV**, 2494–2502.

Vivian, H. (1989). Hydrological changes of the Rhone river, in *Historical Change of Large European Rivers: Western Europe* (Eds G.E. Petts, H. Noller and A.L. Roux), pp. 57–78, Wiley, Chichester.

Vladimirov, V.I. (1975). Critical periods in the development of fishes, *Voprosy Ikhtiologii*, **15**, 955–975.

Vogt Andersen, U. (1994). Sheep grazing as a method for controlling *Heracleum mantegazzianum*, in *Ecology and Management of Invasive Riverside Plants* (Eds L.C. de Waal, L.E. Child, P.M. Wade and J.H. Brock), pp. 77–91, Wiley, Chichester.

Vollestad, L.A. and L'Abee-Lund, J.H. (1987). Reproductive biology of stream-spawning roach, *Rutilus rutilus*, *Environmental Biology of Fishes*, **18**, 219–227.

Vrijhof, (1984). The selection of priority black-list substances for the river Rhine and the waters of the European community, *Water Science and Technology*, **16**, 525–528.

Waal, L.C. de (in press). Treatment of *Fallopia japonica* near water — a case study in *Plant invasions — Theory and Applications* (Eds P. Pyšek, K. Prach, M. Rejmánen and P.M. Wade), SPB-Academic Publishing, The Hague.

Waal, L.C. de, Child, L.E., Wade, P.M. and Brock, J. (eds) (1994). *The Ecology and Management of Invasive Riverside Plants*, John Wiley, Chichester.

Wade, P.M. (1994). Management of macrophytic vegetation, in *The Rivers Handbook*, Vol. II (Eds P. Calow and G.E. Petts), pp. 363–385, Blackwell Scientific, Oxford.

Wagner, R. (1984). Effects of an artificially changed stream bottom on emerging insects, *Verhandlungen Internationale Vereinigung für Theoretische und Angewandte Limnologie*, **22**, 2042–2047.

Wagstaffe, P. (1993). Boats love batteries, *Batteries International*, **15**, 24–28.

Wallace, I.D., Wallace, B. and Philipson, G.N. (1990). A key to the case-bearing caddis larvae of Britain and Ireland, *Freshwater Biological Association Scientific Publications*, **51**, 1–237.

Wallace, J.B. (1992). Catchment disturbance and stream response: an overview of stream research at Coweeta Hydrologic Laboratory, in *River Conservation and Management* (Eds P.J. Boon, P. Calow and G.E. Petts), pp. 231–253, Wiley, Chichester.

Wallace, J.B. and Benke, A.C. (1984). Quantification of wood habitat in subtropical coastal plain streams, *Canadian Journal of Fisheries and Aquatic Sciences*, **41**, 1643–1652.

Walker, A. (1975). The use of rotenone to control recruitment of juvenile Brown trout (*Salmo trutta* L.) into an 'overpopulated' loch, *Fisheries Management*, **6**, 64–72.

Walker, D.A. (1985). Vegetation and environmental gradients of the Prudhoe Bay region, Alaska, *CRREL Report*, 85-14, 239.

Walker, K.F. (1985). A review of the ecological effects of river regulation in Australia, *Hydrobiologia*, **125**, 111–129.

Wallen, I.E. (1951). The direct effects of turbidity on fishes, *Bulletin of Oklahoma A and M College*, **48**(2).

Walling, D.E. and Webb, B.W. (1992). Water Quality I, Physical Characteristics, in *The Rivers Handbook*, Vol. I (Eds P. Calow and G.E. Petts), pp. 48−72, Blackwell Scientific, Oxford.

Walshe, B.M. (1948). The oxygen requirements and thermal resistance of chironomid larvae from flowing and from still waters, *Journal of Experimental Biology*, **25**, 35−44.

Wankowsky, J.W.J. and Thorpe, J.E. (1979). Spatial distribution and feeding in Atlantic salmon, *Salmo salar* L., juveniles, *Freshwater Fisheries*, **14**, 239−247.

Ward, D. (1990). Recreation on inland lowland waterbodies: does it affect birds? *RSPB Conservation Review*, **4**, 62−68.

Ward, D. and Andrews, J. (1993). Waterfowl and recreational disturbance on inland waters, *British Wildlife*, **4**, 62−68.

Ward, G.M. and Cummins, K.W. (1979). Effects of food quality on growth rate and life history of a stream collector (*Paratendipes albimanus* (Meigen)), *Ecology*, **60**, 57−64.

Ward, J.V. (1975). Downstream fate of zooplankton from a hypolimnial release mountain reservoir, *Verhandlunen der Internationale Vereiningen für Limnologie*, **19**, 1789−1804.

Ward, J.V. (1989). The four-dimensional nature of lotic ecosystems, *Journal of the North American Benthological Society*, **8**, 2−8.

Ward J.V. (1992). *Aquatic Insect Ecology 1: Biology and Habitat*, Wiley, New York.

Ward, J.A. and Stanford, J.V. (Eds) (1979). *The Ecology of Regulated Streams*, Plenum Press, New York.

Ward, J.V. and Stanford, J.A. (1983). The serial discontinuity concept of lotic ecosystems, in *Dynamics of Lotic Ecosystems* (Eds T.D. Fontaine and S.M. Bartell), pp. 29−42, Ann Arbor Science, Ann Arbor, Michigan.

Ware, D.M. (1975). Growth, metabolism and optimal swimming of a pelagic fish, *Journal of the Fisheries Research Board of Canada*, **32**, 33−41.

Waringer, J.A. (1987). Spatial distribution of Trichoptera larvae in the sediments of an Austrian mountain brook, *Freshwater Biology*, **18**, 469−482.

Warke, G.M.A. and K.R. Day (in press). Diet and feeding site dependency of cormorants (*Phalacrocorax carbo carbo*) in Northern Ireland, *Ardea*.

Watson, D. (1987). Hydraulic effects of aquatic weeds in UK Rivers, *Regulated Rivers: Research and Management*, **1**, 211−228.

Watts, J.F. and Watts, G.D. (1990). Seasonal change in aquatic vegetation and its effect on river channel flow, in *Vegetation and Erosion* (Ed J.B. Thornes), Wiley, Chichester.

Wawrik, F. (1962). Zur Frage: Fuhrt der Donaustrom autochtones Plankton?, *Archiv für Hydrobiologie* (Suppl.), **27**, 28−25.

Weatherley, A.H. (1972). *Growth and Ecology of Fish Populations*, Academic Press, London and New York.

Weatherley, A.H. and Gill, H.S. (1987). *The Biology of Fish Growth*, Academic Press, London and New York.

Weatherley, N.S. (1988). Liming to mitigate acidification in freshwater ecosystems: a review of the biological consequences, *Water, Air and Soil Pollution*, **39**, 421−37.

Weatherley, N.S. and Ormerod, S.J. (1987). The impact of acidification on macroinvertebrate assemblages in Welsh streams: towards an empirical model, *Environmental Pollution*, **46**, 223−40.

Weatherley, N.S. and Ormerod, S.J. (1990). The utility of biological indicators of stream acidity in Wales, in *Ecological Indicators*, pp. 1341−1354, EPA symposium, Florida 1990.

Weatherley, N.S. and Ormerod, S.J. (1990). Forests and the temperature of upland streams in Wales: a modelling exploration of the biological effects, *Freshwater Biology*, **24**, 109−122.

Weatherley, N.S. and Ormerod, S.J. (1992). The biological response of acidic streams to catchment liming compared to the changes predicted from stream chemistry, *Journal of Environmental Management*, **34**, 105−115.

Webb, B.W. and Walling, D.E. (1992). Water Quality II, Chemical characteristics, in *The Rivers Handbook* Vol. II (Eds P. Calow and G.E. Petts) pp. 73−100, Blackwell Scientific, Oxford.

Webber, N.B. (1971). *Fluid Mechanics For Civil Engineers*, Chapman and Hall, London.

Weber, D. (1988). Die aktuelle Verbreitung des Iltisses (*Mustela putorius* L.) in der Schweiz, *Revue Suisse Zoologie*, **95**, 1041−1056.

Weber, W.J., Voice, T.C., Pirbazan, M., Hunt, G.E. and Ulanoff, D.M. (1983). Sorption of hydrophobic compounds by sediments, soil and suspended solids, *Water Research*, **17**, 1443−1452.

Webster, J.R. and Benfield, E.F. (1986). Vascular plant breakdown in freshwater ecosystems, *Annual Review of Ecology and Systematics*, **17**, 567–594.

Webster, J.R., Golladay, S.W., Benfield, E.F., Meyer, J.L., Swank, W.T. and Wallace, J.B. (1992). Catchment disturbance and stream response: an overview of stream research at Coweeta Hydrologic Laboratory, in *River Conservation and Management* (Eds P.J. Boon, P. Calow and G.E. Petts), pp. 231–253, Wiley, Chichester.

Weeks, K.G. (1982). Conservation aspects of two river improvement schemes in the River Thames catchment, *Journal of the Institute of Water Engineers and Scientists*, **36**, 447–458.

Welch, I.M., Barrett, P.R.F., Gibson, M.T. and Ridge, I. (1990). Barley straw as an inhibitor of algal growth I: studies in the Chesterfield Canal, *Journal of Applied Phycology*, **2**, 231–239.

Welcomme, R.L. (1979). *Fisheries Ecology of Floodplain Rivers*, Longman, London.

Wellborn, G.A. and Robinson, J.U. (1987). Microhabitat selection as an antipredator strategy in the aquatic insect *Pachydiplax longipennis* Burmeister (Odonata: Libellulidae), *Oecologia (Berlin)*, **71**, 185–189.

Welsh Development Agency (1993). *Japanese Knotweed Control: Model Specification and Knotweed Fragment Regeneration Study*, Welsh Development Agency, Cardiff.

Welton, J.S., Cooling, D.A. and Ladle, M. (1982). A comparison of two colonization samplers with a conventional technique for quantitative sampling of benthic macroinvertebrates in the gravel substratum of an experimental recirculating stream, *Internationale Revue der Eesamten Hydrobiologie*, **67**, 901–906.

Wene, G. and Wickliff, E.L. (1940). Modification of a stream bottom and its effect on the insect fauna, *Canadian Entomologist*, **72**, 131–135.

Wesche, T.A. (1980). The WRRI cover rating method–development and application University of Wyoming, Water Resources Research Institute Series, Publication 78, Laramie.

Wesche, T.A. (1985). Stream channel modifications and reclamation structures to enhance fish habitat, in *The Restoration of Rivers and Streams* (Ed. J.A. Gore) Butterworth Publishers, London.

Wesche, T.A., Goertler, C.M. and Frye, C.B. (1987). Contribution of riparian vegetation vegetation to trout cover in small streams, North American *Journal of Fisheries Management*, **7**, 151–153.

Wesche, T.A., Goertler, C.M. and Hubert, W.A. (1987). Modified habitat suitability index model for brown trout in Southeastern Wyoming, *North American Journal of Fisheries Management*, **7**, 232–237.

Westlake, D.F. (1966). The light climate for plants in rivers, in *Light as an Ecological Factor* (Eds R. Bainbridge, C. Clifford-Evans and O. Rackham), Blackwell, Oxford.

Westlake, D.F. (1967). Some effects of low velocity currents on the metabolism of aquatic macrophytes, *Journal of Experimental Botany*, **18**, 187–205.

Westlake, D.F. (1975). Macrophytes, in *River Ecology*, (Ed. B.A. Whitton), pp. 106–128, Blackwell, Oxford.

Westlake, D.F. (1981). Temporal changes in aquatic macrophytes and their environment, in *Dynamique de Populations et Qualite de l'Eau* (Ed. H. Hoestlandt), pp. 110–138, Gauthier-Villars, Paris.

Wheeler, A. (1978). *Key to the Fishes of Northern Europe*, William Clowes, London.

Wheeler, A. and Easton, K. (1978). Hybrids of chub and roach (*Leuciscus cephalus* and *Rutilus rutilus*) in English rivers, *Journal of Fish Biology*, **12**, 167–171.

Wheeler, A. and Maitland, P.S. (1973). The scarcer freshwater fishes of the British Isles. 1. Introduced species, *Journal of Fish Biology*, **5**, 49–68.

Whitaker, G.A., McCuen, R.H. and Brush, J. (1979). Channel modification and macroinvertebrate community diversity in small streams, *Water Resources Bulletin*, **15**, 874–879.

White, D.S. and Gammon, J.R. (1977). The effect of suspended solids on macroinvertebrate drift in an Indiana creek, *Proceedings of the Indiana Academy of Science*, **86**, 182–188.

White, G.C., Anderson, D.R., Burnham, K.P. and Otis, D.L. (1982). *Capture–Recapture and Removal Methods for Sampling Closed Populations*, Los Alamos National Laboratory, New Mexico.

White, R.I. and Brynildson, O.M. (1967). *Guidelines for Management of Trout Stream Habitat in Wisconsin*, Technical Bulletin No. 39, Department of Natural Resources, Madison, Wisconsin.

Whitehead, C. and Brown, J.A. (1989). Endocrine responses of brown trout, *Salmo trutta* L. to acid, aluminium and lime dosing in a Welsh hill stream, *Journal of Fish Biology*, **35**, 59–71.

Whitfield, J.B. and Cameron, S.A. (1988). A year-long survey of the bees (Hymenoptera: Apoidea) along the Dean River at Woodford, Cheshire, *Naturalist*, **113**, 65−68.

Whitman, R.L. and Clark, W.J. (1984). Ecological studies of the sand-dwelling community of an east Texas stream, *Freshwater Invertebrate Biology*, **3**(2), 59−79.

Whitton, B.A., Kelly, M.G., Harding, J.P.C. and Say, P.J. (1991). *Use of Plants to Monitor Heavy Metals in Freshwaters*, HMSO, London.

Wickett, W.P. (1952). Production of chum and pink salmon in a controlled stream, *Progress Report of the Fisheries Research Board of Canada*, **93**, 7−9.

Wilber, C.G. (1983). *Turbidity in the aquatic environment; an environmental factor in fresh and oceanic waters*, C.C. Thomas, Springfield, Illinois, USA.

Willby, N.J. and Eaton, J.W. (1993). The distribution, ecology and conservation of *Luronium natans* (L.) Raf. in Britain, *Journal of Aquatic Plant Management*, **31**, 70−76.

Williams, D.D. (1980). Some relationships between stream benthos and substrate heterogeneity, *Limnology and Oceanography*, **25**, 166−172.

Williams, D.D. (1984). The hyporheic zone as a habitat for aquatic insects and associated arthropods, in *The Ecology of Aquatic Insects* (Eds V.H. Resh and D.M. Rosenberg), pp. 430−455, Praeger, New York.

Williams, D.D. and Mundie, J.H. (1978). Substrate size selection by stream invertebrates and the influence of sand, *Limnology and Oceanography*, **23**(5), 1030−1033.

Williams, J. and Skove, F. (1981). *The effects of recreational boating on turbidity in relation to submerged aquatic vegetation*, Report by Oceanography Department, US Naval Academy, Annapolis, MD, USA. EPA 78-D-X0426, Intra-agency Study, Chesapeake Bay Program, 43.

Williams, J.J. and Tawn, J.A. (1991). Simulation of bedload transport of marine gravel, *Coastal Sediments 91 Proceedings, Speciality Conference*, pp. 703−771, Water Resources Division, ASCE, Seattle, WA, 25−27 June 1991.

Williams, J.J., Thorne, P.D. and Heathershaw, A.D. (1990). Measurements of turbulence in the benthic boundary layer over a gravel bed, *Sedimentology*, **36**, 959−971.

Wilzbach, M.A. and Cummins, K.W. (1986). Influence of habitat manipulations on interactions between cutthroat trout and invertebrate drift, *Ecology*, **67**, 898−911.

Winner, R.W., Boesel, M.W. and Farrell, M.P. (1980). Insect community structure as an index of heavy-metal pollution in lotic ecosystems, *Canadian Journal of Fisheries and Aquatic Sciences*, **37**, 647−55.

Winterbourn, M.J., Rounick, J.S. and Cowie, B. (1981). Are New Zealand stream ecosystems really different?, *New Zealand Journal of Marine and Freshwater Research*, **15**, 321−328.

Winterbourn, M.J., Collier, K.J. and Graesser, A.K. (1988). Ecology of small streams on the west coast of the South Island, New Zealand, *Internationale Vereinigung für Theoretische und Angewandte Limnologie*, **23**, 1427−1431.

Winterbourn, M.J., Hildrew, A.G. and Box, A (1985). Structure and grazing of stone surface organic layers in some acid streams of southern England, *Freshwater Biology*, **15**, 363−374.

Winterbourn, M.J., Hildrew, A.G. and Orton, S. (1992). Nutrients, algae and grazers in some British streams of contrasting pH, *Freshwater Biology*, **28**, 173−182.

Witzel, L.D. and MacCrimmon, H.R. (1983). Embryo survival and alevin emergence of brook charr, *Salvelinus fontinalis*, and brown trout, *Salmo salar*, relative to redd gravel composition, *Canadian Journal of Zoology*, **61**, 1783−1792.

Wolf, L.L. and Waltz, V. (1988). Oviposition site selection and spatial predictability of female white-faced dragonflies (*Leucorrhinia intacta*) (Odonata: Libellulidae), *Ethology*, **78**, 306−320.

Wolman, M.G. and Gerson, R. (1978). Relative scales of time and effectiveness of climate in watershead geomorphology, *Earth Surface Processes and Landforms*, **3**, 189−208.

Wolman, M.G. and Miller, J.P. (1960). Magnitude and frequency of forces in geomorphic processes, *Journal of Geology*, **68**−74.

Woodiwiss, F.S. (1978). *Biological water assesment methods*, Second technical seminar, Summary Report, Commission of the European Communities, Environment and Consumer Protection Service, ENV/787/80-EN.

Woodroffe, G.L., Lawton, J.H. and Davidson, W.L. (1990). The impact of feral mink *Mustela vison* on water voles *Arvicola terrestris* in the North Yorkshire National Park, *Biological Conservation*, **51**, 49−62.

Wootton, R.J. (1990). *Ecology of Teleost Fishes*, Chapman and Hall, London.

WRc (1991). *Sources of farm pollution and impact on river quality*. Water Research Centre, interim report ref 001/3/W, NRA R&D contract, unpublished report.

WRc (1992). *Statistical analysis of relationships between chemical and biological river quality data*, Water Research Centre, report ref 369/3/T, NRA R&D Contract, unpublished report.

Wren, C., Hunter, D.B., Leatherland, J.F. and Stokes, P.M. (1987). The effects of polychlorinated biphenyls and methyl-mercury, singly and in combination, on mink. II. Reproduction and kit development, *Archives of Environmental Contamination and Toxicology*, **16**, 449–454.

Wright, J.F. (1992). Spatial and temporal occurrence of invertebrates in a chalk stream, Berkshire, England, *Hydrobiologia*, **248**, 11–30.

Wright, J.F., Moss, D., Armitage, P.D. and Furse, M.T. (1984). A preliminary classification of running-water sites in Great Britain based on macro-invertebrate species and the prediction of community type using environmental data, *Freshwater Biology*, **14**, 221–256.

Wright, J.F., Armitage, P.D., Furse, M.T. and Moss, D. (1989). Prediction of macroinvertebrate communities using stream measurements, *Regulated Rivers: Research and Management*, **4**, 147–155.

Wright, J.F., Blackburn, J.H., Westlake, D.F., Furse, M.T. and Armitage, P.D. (1992). Anticipating the consequences of river management for the conservation of macroinvertebrates, in *River Conservation and Management* (Eds P.J. Boon, P. Calow and G.E. Petts), pp. 137–149.

Wright, J.F., Furse, M.T., Gunn, R.J.M., Blackburn, J.H., Bass, J.A.B. and Symes, K.L. (1992). *Invertebrate Survey and Classification of Rivers for Nature Conservation*, SNH Research, Survey and Monitoring Report No. 1, Scottish Natural Heritage, Edinburgh.

Wright, T.D. (1982). *Potential biological impacts of navigation traffic*. Misc. Paper No. E-82-2. Environmental lab., US Army Waterways Exp. Stn., Vicksburg, MS, USA: AD-A116-991.

Yang, C.T. (1971). Potential energy and stream morphology, *Water Resources Research*, **7**, 311–322.

Young, P.C. and Wallis, S.G. (1987). The aggregated dead-zone model for dispersion, in *BHRA, Proceedings of the Conference on Water-Quality Modelling in the Inland Natural Environment*, pp. 421–433, BHRA, Cranfield.

Young, S.A., Kovalak, W.P. and Del Signore, K.A. (1978). Distances travelled by autumn-shed leaves introduced into a woodland stream, *American Midland Naturalist*, **100**, 217–222.

Youngs, W.D. and Robson, D.S. (1978). Estimation of population number and mortality rates, in *Methods for Assessment of Fish Production in Fresh Waters* (Ed. T. Bagenal), pp. 137–164, Blackwell Scientific, Oxford.

Yousef, Y.A., McLellon, W.M. and Zebuth, H.H. (1980). Changes in phosphorus concentrations due to mixing by motorboats in shallow lakes, *Water Research*, **14**, 841–852.

Zabawa, C. and Ostron, C. (Eds) (1980). Final Report: *The role of boat wakes in shore erosion in Anne Arundel County, Maryland*, Prepared for Coastal Resources Division by Tidewater Administration, Maryland Dept. of Natural Resources, Annapolis, MD, USA.

Zalewski, M. and Naiman, R.J. (1985). The regulation of riverine fish communities by a continuum of abiotic-biotic factors, in *Habitat Modification and Freshwater Fisheries* (Ed. J.S. Alabaster), pp. 3–9, Butterworths, London.

Zalewski, M., Frankiewicz, P. and Brewińska-Zaras, B. (1986). The production of brown trout (*Salmo trutta* L.) introduced to streams of various orders in an upland watershed, *Polskie Archiwum Hydrobiologii*, **33**, 411–422.

Zalewski, M., Frankiewicz, P., Przybylski, M., Bańbura, J. and Malgorzata, N. (1990). Structure and dynamics of fish communities in temperate rivers in relation to the abiotic-biotic regulatory continuum concept, *Pol. Arch. Hydrobiol.*, **37**, 151–176.

Zar, J.H. (1984). *Biostatistical Analysis*, 2nd edn, Prentice Hall, New York.

Zaslavskij, M.N. (1987). *Eroziovdenie, osnovi protiverozionavo zemledelia*, Vissaja skola, Moskva.

Zelinka, M. (1976). Mayflies (Ephemeroptera) in the drift of trout streams in the Beskydy mountains, *Acta entomologica bohemoslavica*, **73**, 94–101.

Zieman, J.C. (1976). The ecological effects of physical damage from motor boats on turtle grass beds in southern Florida, *Aquatic Botany*, **2**, 127–139.

Zippin, C. (1956). An evaluation of the removal method of estimating animal populations, *Biometrics*, **12**, 163–169.

Zwick, P. (1992). Stream habitat fragmentation. A threat to biodiversity, *Biodiversity and Conservation*, **1**, 80–97.

Index